Alternative Mathematical Theory of Non-equilibrium Phenomena

This is volume 196 in
MATHEMATICS IN SCIENCE AND ENGINEERING
Edited by William F. Ames, *Georgia Institute of Technology*

A list of recent titles in this series appears at the end of this volume.

Alternative Mathematical Theory of Non-equilibrium Phenomena

Dieter Straub

University of the Federal Armed Forces
Munich, Germany

ACADEMIC PRESS
San Diego London Boston
New York Sydney Tokyo Toronto

This book is printed on acid free paper. ∞

ACADEMIC PRESS, INC.
525 B Street, Suite 1900, San Diego, CA 92101-4495, USA
1300 Boylston Street, Chestnut Hill, MA 02167, USA
http://www.apnet.com

ACADEMIC PRESS LIMITED
24–28 Oval Road, London NW1 7DX, UK
http://www.hbuk.co.uk/ap/

Library of Congress Cataloging-in-Publication Data

Straub, Dieter, 1934-
Alternative mathematical theory of non-equilibrium phenomena / Dieter Straub.
p. cm — (Mathematics in science and engineering)
Includes bibliographical references and index.
ISBN 0-12-673015-6 (alk.paper)
1. Irreversible processes. I. Title. II. Series.
QC174.17.I76S77 1996
530.1'3—dc20 96-22135
CIP

Printed in the United States of America
96 97 98 99 00 BB 9 8 7 6 5 4 3 2 1

To Professor **William F. Ames**,
scientist, advisor, and friend
on the occasion of his seventieth birthday

Alternative Mathematical Theory of Non-equilibrium Phenomena:

"We conceive it as a group of abstract ideas, and our course is to have a threefold character, namely: (1) the pupil is finally to be left with a precise perception of the nature of the abstractions acquired by constant use of them, illumined by explanations and finally by precise statements; (2) the logical treatment of such ideas is to be exemplified by trains of reasoning which employ them and interconnect them; and (3) the application of these ideas to the course of nature conceived in its widest sense as including human society is to be made familiar."

A.N. Whitehead

Contents

Preface

"How much entropy is contained in a scientific paper?"—**G.N. Alekseev**

This text is intended to give rise to a new theoretical approach suitable for many studies on complex phenomena of general physical interest. Founded on the mathematical formulation of J. W. Gibbs' thermodynamics and substantially extended by G. Falk, a continuum theory will be presented here that may be applied to mechanics, thermodynamics, fluid dynamics, and electrodynamics.

Falk's contribution essentially turned out to be a new concept of dynamics eliminating so-called metaphysical elements in physics. While the former aspect concerns an unusual view of time- space-coordinates, the latter intention surprisingly affects some pillars of classical physics, such as the First and Second Laws of Thermodynamics and Maxwell's equations of electromagnetism.

In the area of physical field theories, an extensive literature has been devoted to determining those nonlinear equations amenable to irreversible processes in space and time. The most relevant discrepancies between the present leading theories are preconditioned by the history of classical mechanics; they mainly concern the so-called constitutive equations that complete the set of balance equations together with the respective initial boundary constraints.

Because of these aspects, macroscopic phenomena are emphasized. Recently developed theories of non-equilibrium statistical thermodynamics related substantially to fluctuations and stability of particle systems are not considered.

The scope of this book is intentionally different from that of any existing treatise on the macroscopic theory of matter-field systems. Attention has been focused on gaseous states, although relations to any liquid or even crystalline states have also been kept in mind.

Based on the *Gibbs–Falkian thermodynamics* mentioned above, a set of partial differential equations is derived and available for a field description of a great variety of dissipative processes in fluids. The difficulties experienced with common nonlinearities may still reside in the set of equations to be solved, as well as in the prevailing boundary conditions. Notwithstanding, the crucial point is that a new Alternative Continuum Theory of Compressible Fluids (AT) is established on a set of notions, among which the two terms *non-equilibrium state* and *motion* are prominent as compared to the other macroscopic field theories.

Although a rigorous quantitative treatment of the subject must inevitably involve a certain amount of mathematical formalism, an attempt has been made to minimize the mathematical premises, particularly with respect to tensor analysis and set theory. Nevertheless, it was the intention to provide—parallel to the mathematical development—a qualitative description in purely physical terms.

The main results of the AT may be summarized as follows:

1. Motion, quantified by the respective energy form, is dependent on all relevant variables of a given system. Thus, the common separation in a first part of pure kinetics and in a second one only determined by the laws of thermostatics is excluded.
2. The non-equilibrium state of any fluid is characterized by a special vector function—the dissipation velocity—modifying the conventional definition of the mass-point momentum.
3. The concept of standard variables, combined with all other general rules of Gibbs–Falkian thermodynamics, immediately leads to some far-reaching theorems, assumed to be generally valid for the equation of motion, the Second Law, and the so-called divergence theorem, including exclusively all dissipative transfer rates of importance.
4. The resulting expression for the pressure tensor becomes independent of the local state of motion, provided some weak condition is fulfilled concerning a simple relation between the mass density and the dissipation velocity. Consequently, the pressure of a fluid may be substituted by that equation of state which is valid for the system in its state at rest and hence equivalent to a state of equilibrium. Additionally, some new aspects are introduced for the pressure terms in electromagnetism.
5. The usual friction and heat conducting effects, along with diffusion, are immediately reduced to the corresponding local entropy production density occurring in the flow field.
6. With the balance equation of entropy density, a revised version of the constitutive equations is obtained. Furthermore, the entire theory of electrodynamics may be treated in such a way that there will be no need for the common practice of separating a uniform matter-field system into a mechanical and an electromagnetic part. Thus, engineers are now able to deal with simultaneous interactions between polarized fluids and thermofluid dynamical processes.

It should be stressed that the methodology of the AT leads systematically to the fundamentals of complex phenomena, including chemically reacting flows or the complete electromagnetic behavior of fields and materials. The new theory incorporates Maxwell's equations and refers to simultaneous interactions with mechanical or thermofluid dynamical processes.

Notwithstanding, the methodology may be applied to other scientific fields quite apart from physics. It is certainly recommendable to the general system theory, as well as to theoretical economics and, particularly, to some fields in biophysics and medicine, such as hemorheology or research on topics concerning metabolism and circulation.

This meta-theoretical aspect is symbolized by the names of the nine Valkyries, which simultaneously represent individualism and coherence within a closely-knit community of western ideology. Each name appears in the upper left-hand corner of the respective chapter opening page and associates metaphorically the inherently nonphysical elements of the mathematical representation. This peculiarity of the AT implies the realization of the inverse research program recently suggested by Philip Mirowski: physics as nature's economics $\rightarrow$ economics as social physics. The first studies in this field, published in 1992 and 1995 by Höher, Lauster, and Straub, offer promising perspectives.

The text has been divided into sections and chapters addressing each of the aforementioned topics. In addition, the new theoretical results have been simplified for the sake of lucidity and practical application. These approximations will be discussed adequately with regard to competitive theories, especially the Navier–Stokes theory, the core of which is its equation of motion. The results from comparing classical work are striking and point to some serious defects in the "viscous terms" of its paradigmatic vector equation. Their consequences give rise to a discourse on some principal problems in question, illustrated in each case by a diversity of physical applications.

In summary, the objective of this book is to provide an application-oriented text that is reasonably self-contained. It has been used as the basis for a graduate-level course at the Aerospace Department of the University of the Armed Forces in Munich, Germany. The text is aimed primarily at professionals using applied mathematics and physics and working in engineering applications. Furthermore, it is intended to be a guidebook for engineers, mathematicians, and economists with a strong interest in scientific fundamentals.

Acknowledgments

Foremost among those to whom this book owes its existence is my friend K. Höher, whose early inspiration and permanent encouragement have been a constant source of guidance. I am also indebted to members of the University Institutes of Thermodynamics and Statistics. Many thanks are due to my colleagues G. Kappler, R. Waibel, and H. Wilhelmi, but, above all, to my collaborators M. Lauster, V. Lippig, and H. Steinfeld for offering a considerable number of suggestions that I was very glad to adopt.

If this project is successful, I owe it to the support of W. F. Ames at the Georgia Institute of Technology. Thanks are also extended to Linda Ratts Engelman, Jill Lazer, Amy Mayfield, and the courteous cooperation of the publishers.

In addition, the following people deserve particular acknowledgement. Without their influence and ceaseless encouragement over the years I could not have written the book. Aside from two late colleagues, my friend A. Schaber and my mentor G. Falk, I would like to extend my gratitude to Professors E. Adams, D. Geropp, P. Kramer, W. Ruppel, W. Schönauer, and A. Walz. They, above all, have taken the initial steps to familiarize me with the problems treated in this work.

I also thank W. Finke and K. Gross, without whose selfless interest and diligent support I would long ago have lost faith in the project "Alternative Theory."

Finally, my family deserves special recognition for the ceaseless support that motivated me to keep on going.

Chapter 1 Physics Today: Perspectives

"If the Universe is the Answer, what is the Question?" —**Leon Lederman**

1.1 Motivation

That part of classical flow mechanics belonging to the basics of engineering science and education is in competition today with theories relating to some branches of thermodynamics. These theories, specified as linear or extended irreversible thermodynamics or as rational thermodynamics, are ordinarily placed on the periphery of modern physics. Aside from this constellation, the range of solvable problems remains predominantly confined to the conventional Navier–Stokes theory, particularly over the last decade and especially in worldwide industrial practice.

Due to the rapid progress in computer design and efficiency, combined with sophisticated numerical methods, the solving of actual flow problems has led to a modern discipline, well-known as *Computational Fluid Dynamics*. Therefore an increasing number of young students, scientists, and engineers are interested in computer applications. When using computers for increasingly complicated **CFD**-problems, they must frequently concentrate their work on programming languages or software packages to cultivate their talents.

At present, the scientific literature substantiates the obvious tendency that the standards of computing techniques and applied mathematics are steadily growing. Unfortunately, this trend may mean that professional competence in the fundamentals of physics, combined with thorough knowledge of the empirical background, is increasingly diminishing. This would be a hazardous development, delicate and expensive for the prospective technologies.

Therefore, this book deals with those fundamentals in an essentially new way to provide genuine qualification for any professional work in science and engineering. One who is familiar with standard mathematics will have no difficulty in following the discussion. Nevertheless, some knowledge of recent progress in applied mathematics is required. Thus, one would benefit greatly from such complementary work as the mathematical construction and analysis of the full invariance group related to a given set of differential equations and appropriate boundary conditions. Once the group is known, one can study the action of the group on the complete set.

The invariance requirement, related to the respective symmetry group with characteristic properties and consisting of a set of specific transformation rules, may be utilized to generate new solutions from known ones (Rogers and Ames, 1989). Furthermore, the admissible group offers algebraic structures that often allow relevant classifications of solutions. Such information is always helpful for any computer-aided integration procedure needed, for example for numerical solutions of partial differential equations (Ames, 1992).

The main subjects of this book are briefly stated in the preface where the essential ideas are related to key words such as *instability, non-equilibrium, motion,* and *irreversibility.* Only recently, scientists have begun to deal more comprehensively with these phenomena, which act as a basic part of the atomistic world.

In this context it is remarkable that in theoretical physics, nonlinear non-equilibrium thermodynamics has only taken shape in the last 50 years or so. But this refers mainly to quite different extensions of statistical physics in connection with fluctuation-dissipation theorems and quantum generalizations of Markov processes. Model systems are preferred to extract exact results based on the premise that random thermal fluctuations play a decisive role in governing the evolution of non-equilibrium thermodynamic processes in space-time (see, e.g., Kreuzer, 1883, Lavenda, 1985, and Stratonovich, 1994). Applied sciences, such as chemistry or process and aerospace engineering, unfortunately have little interest in these topics.

In the following sections of this chapter, the key words mentioned above will be treated in depth since they are essential to a serious understanding of the physical essence and evolution in both the micro- and the macrolevels of the universe. Furthermore, in a cursory historical comment, we will see why it is so hard for the scientific community to adequately consider and accept these basic notions that are encountered in daily life.

1. 2 Origin and Importance of Non-equilibrium Phenomena

In recent years, some radical changes have occurred in the natural sciences that are evidently inconsistent with the prevalent paradigms. For the first time, it appears possible to overcome the gap between the basic micro- and macrophysical ideas of matter and motion.

The extensive research of Ilya Prigogine and his co-workers showed in a unique way how the contemporary dualistic theory of the physical world could be superseded by a uniform description (Petrosky and Prigogine, 1988). This has, indeed, far-reaching consequences: The dogma universally suggested by physics textbooks can no longer be justified by an *irreversible phenomenological macroworld* on one hand and, separately, by an *atomistic microworld* determined by linear and reversible quantum laws on the other. The former is quite obviously controlled by phenomenological laws with a broken symmetry in time, whereas the latter can be represented by particle trajectories and wave functions, assumed to be independent of reversals in time and velocity.

Physical systems, distinguished by permanently stable and reversible behavior, are rare. They manifest special cases or refer to certain borderlines in nature that can be elegantly formulated with the help of flexible mathematics, but do not provide reliable information about intricate scenarios in space and time. Therefore, when treating real problems with quantum theory, the so-called *measuring problem* arises, because the broken symmetry in time is needed paradoxically as a precondition of experimentally detecting the quantum world, for example by means of irreversible photon scattering (Petrosky and Prigogine, 1993).

In the last few years some ideas of unstable microsystems have been proven compatible with that of atomistic interaction and correlation events, provided that a break of symmetry between past and future occurs. One conclusion follows immediately: Quantum events need irreversibility as an elementary mechanism inherent in each physical system consisting of many-particle collectives. Surely, this is a crucial point, because instability is predominant in most of the physical systems for which, in Prigogine's view, the term *dynamical chaos* is an intrinsic one (Prigogine, 1993).

As a complementary notion, Prigogine introduces the tautological term *dissipative chaos* as an intermediate concept between any pure accident and redundant order (Prigogine and Stengers, 1993, p. 123).

Following Prigogine and Stengers (1993, p. 319), all resulting unstable phenomena occurring macroscopically in the physical system as an event or process are caused primarily by inherent fluctuations of the respective state quantities. Close to equilibrium the fluctuations are attenuated. As a consequence of their wave structure, the trend toward equilibrium is distinguished by asymptotically vanishing dissipative contributions. Given enough time, the system does in the end approach its equilibrium values. Only for such a condition can each fluid always be controlled.

In contrast, *non-equilibrium states* can amplify fluctuations, locally and spontaneously. For this reason, a systematic control is hardly feasible without impacting on the system in question. Due to these non-equilibrium effects, any local disturbance can move even a whole system into an unstable state. They are also substantial preconditions of phase transition. Another preeminent case refers to the turbulence phenomenon in gaseous and liquid flows.

Naturally, such macroscopic states and processes have their own atomistic equivalents at the microscopic level. Indeed, the actual behavior of every particle collective is determined by the local *correlations,* the action range and strength of which regulate how the principal properties of the whole system are influenced by local particle events.

Correlations play an important role in statistical mechanics. In general, they relate interactive connections measured in space and time to measurable properties of two or more particles. Expressed by the so-called product-moment correlation coefficient with values ranging from –1 and +1, correlation functions are referred in a complicated way to the main tool of particle mechanics: The joint probability depends in principle on many random variables. By definition it can be reduced to a product of corresponding elementary probabilities, each one assigned to the respective random variable. This leads to a relatively primitive method of simplifying the model chosen for the statistics of complex physical events. However, if there are correlation effects, then the joint probability cannot be split into a simple product.

Equilibrium states, for instance, can be classified by correlations having an averaged range of action with typical values of about 10^{-10} m (or 1 Ångström). Correlations existing far from equilibrium can extend to macroscopic distances in an order of one centimeter or more. This immense difference is a striking indication of qualitative disparity between equilibrium and non-equilibrium states.

The great variety of state properties far from equilibrium is remarkable, especially for the zones of instability, where the qualitative behavior of the system is often drastically changed. Consequently, the traditional ideas of order and disorder become questionable: The atomistic structure, controlled by far-reaching correlations, is simultaneously coherent and unpredictable.

In other words, dynamical chaos is prevalent and responsible for most of the macroscopic phenomena of fluids. The following two examples are representative of some macroscopic processes that are significantly different, but are of great interest since the same two causes are involved simultaneously: the deterministic trend toward the state of equilibrium as a special type of attractor, and all dissipative structures characterized by their coherent behavior such as the well-known Rayleigh–Bénard instability.

In addition, there are multiple types of attractors far from equilibrium. In the case of the so-called chemical clocks, for instance, they occur as the center of certain periodic processes. Other attractors, being partly of the "strange" variety, give rise to the commonly denoted dissipative chaos, which is connected with processes unpredictable in every way (Prigogine and Stengers, 1993, p. 82). Perhaps chaos of such a kind may characterize the collective phenomenon of a many-particle system. Its mathematical representation does not succeed in the limiting cases of either pure accident, mapped by the laws of contingency, or redundant order according to Shannon's information theory. Climate fluctuations and experiences with long-range weather forecasts are current illustrations.

The state of the art just described points out two main topics to be discussed here:

1. The dogma of the microscopic reversibility is no longer tenable.
2. Non-equilibrium states condition nearly all phenomena of matter.

In other words, *reversibility* and *equilibrium* are concepts valid only for limiting processes. In practice, both are useful for references and idealizations, but very insufficient for the intensive comprehension of real physical events or for efficient engineering. Consequently, it is more easily said than done for a scientist to modify the common view to incorporate *irreversibility* and *non-equilibrium* as a central part of an advanced basic concept considered. To grasp the actual difficulties in conducting such a scientific program, a concise historical review of the respective evolution in physics should be sufficient.

1.3 Today's Mechanical Worldview: A Short Historical Outline

At the end of the nineteenth century two physical events of great influence on science and technology emerged:

1. The accomplishment of classical mechanics culminated in its Hamiltonian formalism as well as in Fourier's heat conduction laws. Moreover, the Navier–Stokes equation of motion was assumed to be a prototype of realistic continuum mechanics.
2. New notions were established and the extension of scientific knowledge laid the foundation of early thermo- and electrodynamics.

It is noteworthy that classical mechanics has retained the traditional but incomplete principle of energy expressed by the conservation of only the added potential and kinetic energies. Nevertheless, its strong influence on the scientific community has led representatives of prestigious universities and scientific societies into hard dispute with the few advocates of the new scientific evolution. The traditionalists have failed to explain the theory of electromagnetism by means of mechanical terms, yet they have succeeded in essentially altering the basic ideas of thermodynamics prevailing at that time. Thus the *Mechanical Theory of Heat* was born, motivated in particular by new demands of the industrial practice (Straub, 1990, p. 42).

However, this is not quite true due to a firm and concealed intention of some prominent theoretical physicists to eliminate, by means of the general axioms of classical mechanics, all the new thermodynamic notions, theorems, and relations, set up, above all, by engineers and amateur thermodynamicists.

Far-reaching consequences resulted, especially for the so-called Carnot–Mayer thermodynamics, which was induced mainly from engineering practice. Interestingly, this theory evolved in parallel to the continuous improvements of Watt's steam engine during the early period of industrialization at the beginning of the last century (Straub, 1990, p. 42). The basic knowledge of this technical key discipline remained immature. Its fundamentals related to some theorems based on epochal inventions and complying with principles of *thermostatics* rather than dynamics of thermophysical phenomena.

That state of the art represents a considerable part of today's technical thermodynamics and is lectured worldwide—as a rule—at engineering faculties. Dynamical problems pertaining to thermodynamics in general, such as the transport phenomena, are traditionally treated in lectures only for students in engineering independent of thermostatics and even indicated by separate notations as heat and mass transfer or gaskinetics. Since the days of the famous French geometers, this sort of problem has been intensively studied in the field of nonlinear mathematics. Up to now, however, the whole field has been strongly neglected in physics education. This fact, indeed, corresponds with the triumphal procession of modern physics dominated by linear quantum mechanics and the reversible relativity theories.

In other words, after the supremacy of the Mechanical Theory of Heat, the world of thermophysical sciences is divided into two subsections according to Helmholtz's influential lectures on "Theorie der Wärme" first published in 1903. With regard to physics today three remarks may be allowed:

1. In educational curricula of modern physics, thermodynamics is equivalent in its substance to the orthodox Mechanical Theory of Heat. This has been extended by Gibbs', Van der Waals', and Planck's well-known presentations and, more recently, by Becker's or Callen's work, supplemented by relevant contributions of a host of other physicists. Moreover, methods of statistical thermodynamics raise some attention, but almost exclusively concern equilibrium properties of matter and radiation.
2. At present, lectures, originally entitled "Reine Wärmelehre" by Helmholtz, treat non-equilibrium processes such as heat transfer, heat conduction, diffusion, combustion, or thermal radiation. Such compulsory subjects are almost exclusively occupied by and reserved for engineering sciences. With rare exceptions, they do not play any important role in the current teaching of physics. It is remarkable, however, that quite recently, a steadily rising number of young students of science is interested in mathematics of nonlinear basic phenomena (West, 1985) and hence of sophisticated theories of deterministic chaos.
3. In engineering education, thermodynamics is predominantly restricted to thermostatics, resting primarily on the First Principles established mainly by Sadi Carnot and Mayer. The basics have been applied to equilibrium states by Kirchhoff, Gibbs, Planck, and Nernst. Neglecting the substantial contributions of Reech, Poincaré, Duhem, Mach, and Boltzmann, they have finally been "extrapolated" by great scientists like Schottky, Bridgman, Keenan, and Zemansky.

The diverse causes of this schism may be reduced to the paradigm of the mechanics' primacy that is still valid at present for the natural sciences and contemporary philosophy: in particular, quantum mechanics and the relativity theories as well as their direct offsprings—particle physics and cosmology, for instance.

Both of these momentous theories had been established without any relevant reference to thermodynamic foundations. Even Mayer's principle of energy has never been taken into account officially. More than that, quantum theory adheres to the tradition of linear mathematics for a quantitative description of a reversible microworld model. For solutions, its mathematical formalism is immediately connected with Hamiltonian mechanics, that is, with mass-point systems highly distinguished by their integrability.

However, in accordance with Siegel's theorem (1941), the majority of all quantum systems belongs to nonintegrable systems. For this reason, most quantum problems can only be solved by refuting Poincaré's famous proof in 1893, according to which classical particle collectives are mechanically nonintegrable systems, except for special cases. Therefore, an analytical solution is prevented by rational ratios of frequencies, so-called *resonances*, which are responsible for small disturbances and amplify unpredictably.

As Misra proved in 1978, this holds for many-particle quantum systems as well. Thus, serious problems often arise due to hidden divergences. For this reason, some dubious inconsistencies had to be reconciled with certain mathematical descriptions in quantum mechanics and its experimental verification.

Prigogine and his co-workers have drawn far-reaching conclusions from these facts. They argue as follows:

1. It is possible to formulate an alternative quantum theory (Prigogine and Stengers, 1993, p. 254) based on the well-known *ensemble* concept of statistical physics and quantified by the *Liouville–von Neumann operator equation*, the solutions of which cannot be derived from the *Schrödinger equation*.
2. Solutions of the Liouville–von Neumann operator equation eliminate the *Poincaré divergences*. This is equivalent to the fact that the observed world cannot be reduced to a microphysical substructure consisting of particle trajectories or wave functions.

Clearly, these statements might intensify the expectation that those inconsistencies noted above, or some alleged quantum paradoxes or other "strange" quantum phenomena, may be explained by conceptual insufficiencies in basic physics.

Nonetheless, to prevent any misunderstanding, it should be stressed that orthodox quantum physics works with great success. *Tunneling, radioactivity,* or *double slit* phenomena are typical examples, insolvable by classical physics. Quantum mechanics interprets atoms and molecules. Therefore, it gives chemistry a solid scientific base, out of which new disciplines like molecular biology and genetics have emerged.

The Schrödinger equation with Born's probability interpretation of the wave function has been used successfully to understand how two or more elements combine to create a molecule. In the period of 1930–1950, this formalism was carried into the nucleus and was found to be as productive there as at the atomic level.

In material science, quantum theory helps to explain the properties of metals, insulators, super- and semiconductors. The latter led to the discovery of the transistor, which directly led to computers and microelectronics. Furthermore, the recent unfolding in communications and information technology are resulting from it.

The 1950s and 1960s were great years for science. At that time physicists were concerned with the flood of subnuclear structures, detected by the particle accelerator, and reflected the associated Nobel prizes. Theorists were busy classifying the hundreds of hadrons, seeking patterns and meanings in this new frontier of matter. The idea of a *standard model* arose, which contains all the elementary particles needed to form all the matter in the universe plus the forces that act on these particles.

This standard model rests on the presupposition that matter particles are bound to each other via four different basic forces that are transferred by yet more particles: gravity, the electromagnetic force, the strong force, and the weak force. It is easy to enumerate these terms, but intricate to grasp their physical meanings and mutual interactions. A consistent theory of forces must meet two criteria: It must be a quantum field theory that incorporates the special theory of relativity and gauge symmetry having deep aesthetic appeal. Unfortunately, it is hard, perhaps impossible, to fully explain what this concept of symmetry means. Quoting Lederman (1994, p. 349), "The problem is that this book is in English, and the language of gauge theory is math."

The insufficiencies of "strange" quantum phenomena or paradoxes refer to problems that made many founders of quantum theory unhappy, including Planck, Einstein, de Broglie, Schrödinger, and even Dirac during the latter years of his life. This issue pertains to the *wave function* ψ and what it means. In spite of the extraordinary success in scientific evolution and engineering practice, it escapes contemporary orthodox physics what ψ really represents, provided that its meaning is not constituted only by agreement. In such a case the wave function formalism is simply defined by the Schrödinger equation, raising ψ to an icon. For this reason, Lederman (1994, p. 171) is fully justified in saying that "the Born interpretation of the Schrödinger equation is the single most dramatic and major change in our world view since Newton." (No wonder that Schrödinger found Born's point of view totally unacceptable.)

These physicists represented two contrary worldviews. Schrödinger, as a physical realist, believed that a theory is committed to the existence of the objects, properties, and relations postulated by the theory. Born, as an instrumentalist, took a more skeptical view of the meaning of theories. What mattered to him were the accessible and directly observable quantities. Theories were mere instruments of an effective handling of experimental data and were intended for prediction of new facts. There needed to be no indication that the causal influences assumed by the theory were really existing. Observed microphysical phenomena, as predicted by quantum mechanics, come close to the views of antirealists, such as Born.

The problem considered touches on some other sore points of quantum mechanics:

1. *Heisenberg's uncertainty relations.* There are four interpretations (Straub, 1990, p. 103), at least, which are partly of metaphysical rather than of practical relevance, such as Heisenberg's original analysis of the X-ray microscope.
2. *The Schrödinger equation as the omnipotent source of universal and total information about all microsystems.* Modern physics presupposes that it applies to electrons, atoms, molecules, protons, neutrons, and, especially important today, clusters of quarks.

It is hard to conceive how such an assumption may account for the complete subatomistic level of matter, without a deep-reaching understanding of the wave function and its various views to physical basic questions. This holds for the uncertainty relations, too. Aside from Popper's well-founded objections against their common interpretation, Heisenberg's famous principle is the euphemism of a misunderstanding obviously stimulated by Bridgman's "operational point of view," which may be summarized by the rule "do not deal with anything that cannot be measured."

Nowadays, it is a matter of fact that such a rule is in error. For instance, since Gödel published his theorem in 1931, it has been clarified that each quantitative theory holds some hidden elements that are by definition unavoidably inaccessible to any experimental program. But there are more banal arguments concerning the reader's daily experiences. The quantity *momentum* is an essential and basic example, especially in quantum theory. In connection with the electromagnetic field, it would be questionable to suggest a concrete method of measurement—even in principle (cf. Falk, 1990, p. 114).

This is even true for measurements of length according to the simple rule: apply iteratively the length scale. But what shall we apply the scale to? Indeed, only rigid bodies come into consideration. Strictly speaking, liquids and gases must be deleted. Another important case may be mentioned as seen from a metaphysical point of view: A spatial distance is nothing but an abstract geometrical concept and, therefore, an ideal candidate to be denoted as "rigid." But it is not an object of matter. Consequently, it is inherently contradictory to determine spatial distances by way of such rigid standards. For these reasons, at least the alleged universal validity of the rule quoted above is open to doubt.

Certainly, the scientific community does not seem seriously interested in this problem. Paradoxically, Einstein would have to appear just now to offer a well-equipped alibi for making it a "habit" as Landé has criticized long ago. Thanks to a series of experiments carried out by Aspect and colleagues in 1982, there seems to be a fair chance of deciding whether Bohr's or Einstein's views of quantum phenomena are wrong. What, however, is the true origin and background of this famous controversy?

A very sophisticated theorem has become almost a cult nowadays. The theorem, developed in 1964 by Bell, claims to be a specific mathematical map of the thought experiment first published in 1935 by Einstein, Podolsky, and Rosen, and improved in 1948 by Einstein alone. This famous EPR-experiment refers to different long-distance correlations between two quantum events α and β, started under well-defined conditions. The measures taken are aimed at a final answer to some abstract questions concerning locality and inseparability of quantum events, independent of their mutual distances in space-time. The real experiments do measure the number of detector α results correlated with detector β results.

Physics nowadays has established a new dogma: Aspect's experiments seem to indicate that Bell's theorem is in error. There is no denying the consequences: Bohr's analysis is correct and Einstein's is wrong. Apparently, the assumed long-distance correlations do exist to reveal the way nature works. In my book *A History of the Glass Bead Game • Irreversibility in Physics: Irritations and Inferences* (Straub, 1990, in German), I have given a concise outline of the EPR-paradoxes and their respective tests. We will not dispute details here or even the inferences of this dogma; it suffices to summarize the results:

1. Bell's theorem is based on the Kolmogorov theory of probability, using the traditional notion of correlation constrained by linear facet inequalities, and has been investigated by experts in probability and statistics since the 1930s. In the frequentists' approach, the inequalities represent a priori constraints on proportions of properties.
2. The violation of these inequalities by quantum frequencies poses a problem in science that is ignored by nearly all experts in probability theory. Most notable is the violation of Bell-type inequalities that had been derived in connection with an analysis of certain concrete quantum experiments. But these are mathematical artifacts having nothing to do with physics.
3. Probability is regarded to be a synthetic concept. Empirical adequacy turns out to be a very weak criterion. Indeed, it does not provide any far-reaching contribution to the concept chosen. Consequently, the question, "which of alternative statistical models represents the actual world?" has no scientifically founded answer. Nothing in physics seems to indicate the true alternative.
4. According to point (3), models of quantum statistics are allowed. But for such models an equivalent logic is needed. Following Wittgenstein, the only change in classical logic that makes sense in science is giving up bivalence. This has been done for constructing quantum logic, which in its serious realistic conception is nothing but a hidden variable theory in disguise or, seen from a distance, some faint Hegelian remembrance.

Thus, for practical applications quantum logic is, as a rule, irrelevant. Regarding EPR-paradoxes, however, Bell's famous theorem, reformulated by means of such quantum statistics, offers results by which all observable statistical measurements are kept intact. Additionally, the modified theorem provides a causal, local, "realistic" explanation of them. Following Pitowsky (1989), it is noteworthy that in this case the modified frequencies predicted now for the benefit of EPR-experiments agree with those expected values to be representative for Aspect's experiments. This means that Bell's theorem in its original version is irrelevant to evaluate the EPR-paradoxes.

Davies and Brown (1986) mention some notable comments made by the two protagonists of the EPR-tests, Aspect and Bell, when confronted with the conclusion that long-distance correlations do allegedly exist. Aspect, the French experimental physicist, frankly declared that the transfer rate of information could not be realized by intelligence signals running, eventually, with superluminal velocity. Consequently, he excluded the Einstein separability of material events, but considered theories that describe nonlocal quantum events with hidden variables to be admissible.

As opposed to this plain statement, Bell's comments are contradictory. Anticipating quantum theory as a temporary substitute, he accepted the common interpretation of Aspect's experimental results as the victory of the Copenhagen orthodoxy. To conserve the traditional idea of reality, established only by phenomena occurring locally, Bell was willing to sacrifice Einstein's special theory of relativity, taking into consideration even superluminal signaling. He recommended a way out of the dilemma with the help of tradition in physics: He suggested the restoration of the state that had been in existence before Einstein's epochal work eliminated the improved ether hypotheses proposed by Lorentz and Poincaré. *Honi soit qui mal y pense.*

The quantum formalism of the Prigogine school does not include those crucial problems and their interpretations discussed above. Its notation and mathematical structure, stringently derived from an adequate concept of statistical ensembles, differ significantly from orthodox quantum mechanics: There is no Hilbert space, no superposition principle, no Hamiltonian dynamics, no unitary transformation rules, no wave function, no subjective elements by an observer, and so on (Prigogine and Stengers, 1993, pp. 251, 254, 262). This alternative quantum theory is limited to high-number particle collectives identified by special properties, in particular by peculiar interactions between accidental and sometimes also coherent events within respective phase spaces. The weakly stabilized dynamics of the many-particle system in question can be mathematically described by a non-Hermitian *motion operator* Φ. The critical question is that inherent dynamical mechanisms enforce physically feasible states by a constituent symmetry break of Φ in time. As a consequence, a new microscopic operator formally results, denoted as an *entropy operator,* which is structurally assigned to this break, responsible for dissipation at a microscopic level. Naturally, the latter manifests the Second Law of Thermodynamics, causing inherently all the irreversible effects typical for many-particle collectives exemplified by particle scattering or emission processes.

To better understand quantum phenomena, there are no serious attempts to incorporate additional mathematical elements into Einstein's general theory of relativity, with the outspoken wish to consider dissipative events occurring in the space-time of a nearly homogeneous, isotropic universe. "But which is the *real* Universe?" (Harrison, 1985). The respective *cosmological standard model,* for instance, leaves things unchanged. This means that in science all thermodynamic features of the universe remain completely indeterminate, both in the past and now, although highly dissipative evolutions undoubtedly are indicated by Penzias' and Wilson's three-degree cosmic radiation and by the just as likely existence of black holes. Add to this that an unlimited but open universe is to be expected, if modern field theories should prove to be correct, provided that, further, all existing particles are irreversibly generated by a quantum vacuum. Tasaki, Nardone, and Prigogine (1993) recently published some results as to resonance and instability in such a cosmological model.

Not accepting the leading scientists' dogmatic pretense toward "universality" and the "key for the deepest mysteries," it is proper to point out the merits of the two greatest theoretical conceptions of this century. There are many fields for the orthodox versions of these theories to gain insight and valuable information on reality. Nevertheless, quantum mechanics and cosmology do concern idealizations of immediate evidence, comparable with statistical theories of the ideal gas or of thermal radiation. They refer to limiting cases of real properties that are of great importance to many engineering applications.

For this reason, it is questionable whether quantum theory can truly be explained to be only a frame of principles. In this context, the prominent physicist Gell-Mann has recently proclaimed a new fundamental theory that is enclosed in that frame together with Einstein's relativity theory of gravitation. He identified this promised *universal quantum field theory* with the actual *superstring theory* concerning models of matter that manifest particles as very small strings embedded in a highly dimensionalized space-time (Davies and Brown, 1988).

Barely known in public, similar theoretical efforts along these lines have been done, and are called such things as "Super Gravity" or "Supersymmetry" (SUSY) or, in all modesty, the "Theory of Everything" (TOE). Common to all is that theorists are searching for a theory with no parameters. Everything must emerge from the basic equations; all the parameters must come out theoretically. Unfortunately, this type of theory has no connection with any real or virtual event yet to be observed in principle.

In the meantime, disillusion began to set in, and the more traditional roads of research having stereotyped road signs like "Grand Unification," "Constituent Models," and "Technicolor" were tried again. They all have one grave problem: There are no data.

Perhaps a quote of Planck's will be helpful to estimate Gell-Mann's message that although the superstring theory may prove to be false, it is the first candidate of this universal theory: "The inscription above the entrance to the temple of science reads: 'You must believe'" (author's translation).

1.4 Continuum Theories of Mass-Point Fluids

It is strange that both great branches of physics enjoy a unanimous and high standing with the public, in spite of the fact that people normally do not grasp even the most elementary notions and relations of today's science and mathematics. In sharp contrast to this, modern technologies are often condemned by social classes with some influence on public opinion, even though all standards and prospects of modern life in Western societies are essentially dependent on the achievements of scientific engineering and industrial innovation. It is hard to discern the impact of these groups' paradoxical attitudes despite the continuous growth of scientific knowledge in such phenomena that are conventionally investigated in mechanics and flow dynamics, as well as in electromagnetism and thermodynamics.

Since the golden age of modern physics, all these traditional and classical fields have as a matter of fact lost their usual superiority through decades of physical research. Nowadays, they are cultivated and have advanced, almost exclusively, to basic subjects in engineering sciences. From this point of view, it may be plausible that, ten years or so ago, many young physicists and mathematicians were suddenly interested in *nonlinear physics*. Now, running the risk of reinventing the wheel, they often busy themselves with phenomena such as *deterministic chaos*. It is a pity that most of the respective efforts are focused on rather irrelevant or even unreal physical situations.

The extraordinary shift of emphasis in modern physics, in particular toward quantum events and cosmology, has led to many substantial results derived from a paradigmatic change in fundamental views on nature and science. Consequently, the four classical disciplines are essentially affected, regarding a lot of applications to calculate any mechanical and flow dynamical processes, as well as electrodynamic phenomena. In this context, we may ask in what way the sets of traditional equations would have to be revised, provided that "modernized" notions of old badges such as entropy, heat, work, diffusion, dissipation, non-equilibrium, and motion would be systematically incorporated into the respective set of axioms.

It is noteworthy that the problem in question has never been seriously discussed. Quite the reverse is true: One has to take into account, for instance, that the conventional *Navier–Stokes equations*, last revised in the year 1838, are still the primary tools used in current engineering practice and in some centers of mathematical research. For example, Tokaty's fine book entitled *A History and Philosophy of Fluidmechanics*, makes no mention of terms such as entropy or dissipation, and the author's ideas of pressure and temperature seem rather naive (1971, p. 190). It looks as though thermodynamics and its history did not really happen. The compilation in Table 1.1 shows the turning point in 1838, distinguished presumably in no way, yet with undisputed reference to key terms in science (Szabó, 1987).

There is an explicit relation between the local viscosity of a flow, introduced heuristically by Newton in his *Principia Mathematica*, and the entropy production density, defined by the local balance of entropy. Certainly, the kinetic theories provide some valuable information on the respective transport properties of simple matter models. But, for instance, the kinetic theory of gases is a pure mechanical theory governed by so-called constitutive laws, which relate characteristic rate properties of a fluid linearly to local gradients of the assigned mechanical state variables. For the case of a viscous pressure tensor, only under very simple flow conditions can the mechanical stress properties be linearly related to velocity gradients. Thus, in accordance with Newton's approach, viscosity is defined as a proportionality coefficient without any definite reference to a more general concept considering the thermodynamic source of this obviously dissipative effect. Indeed, this is Newtonian mechanics, but definitely not Gibbsian physics.

Nevertheless, it is true that many researchers—worldwide yet more intensively after WW II—contributed a variety of ideas and experimental informations for

Table 1.1 Excursion into History

1687	Viscosity	Newton
1736	Mass-point	Euler
1749	Hydrodynamic pressure	Euler
1750	Newton's basic law of motion	Euler
1755	Ideal equation of motion	Euler
1821 to 1828	Viscous equation of motion	Cauchy
1822	Equation of heat conduction	Fourier
1822 to 1838	*Navier–Stokes equations*	*Navier–Saint-Venant*
1824	Carnot cycle, reversibility	Carnot
1842 to 1847	Energy equivalence principle	Mayer–Helmholtz
1848	Absolute temperature	Kelvin
1850	First Law of Thermodynamics	Clausius
1852	Second Law of Thermodynamics	Kelvin
1864	Electromagnetic field	Maxwell
1865	Entropy	Clausius
1873	Equation of state for real fluids	Van der Waals
1901	Statistical mechanics	Gibbs
1905	Mass–momentum–energy relation	Einstein

advancing the theories of the phenomenological disciplines mentioned. Such work has been done even for some latent items of classical mechanics, generally believed to be thoroughly and definitively explored.

The so-called *KAM-theorem,* developed in the period 1968 to 1973, sets a first example. It is centered around the general statement that there exist two sorts of principally different trajectories of particles, both compatible with the Hamiltonian formalism: conditionally periodic orbits and collectives of trajectories that are either irregular or chaotic. In the event of minor disturbances of the respective particle motions, the phase space of the mechanical system is commonly occupied by a mixture of those two sets of trajectories. Altogether, the real states of our world is supposed to be an interpenetration of regularity and chaos.

This is fatal for orthodox quantum theory. The KAM-theorem states that for many mechanical systems there exists no way to diagonalize the respective Hamiltonian, which is the indispensable requirement for all conventional solutions of non-simple quantum problems. This fact is stated concisely in the words of the Russian mathematician V. I. Arnol'd, one of the three founders of the theorem: "Nonintegrable problems of dynamics appeared inaccessible to tools of modern mathematics."

A second, more recent but less-known example pertains to modern cosmology. In contrast with Poincaré's famous proof in 1898, Tipler (1980) proved his "no-return" theorem valid for a closed and relativistic universe, showing that there is no eternal return. Poincaré's result claims that any given state of interacting particles in a closed region will recur to within any accuracy wanted. Thus, every non-equilibrium state once experienced by a system will occur again after a long enough time. Tipler's theorem maintains the contrary: The space-time of the world cannot oscillate; moreover, time reversal must be excluded.

Both examples demonstrate significant contributions resulting from some new ideas in classical and relativistic mechanics. The following, however, proves to be more representative: To gain insight into the working method of science, one must study *continuum physics*. Currently at least five scientific *"schools"* are engaged in this highly attractive field. Each *school* may be designated by its dominating group of scientific members:

1. Specialized mathematicians,
2. Fluid dynamicists,
3. Advocates of rational mechanics,
4. Representatives of the linear and the extended irreversible thermodynamics, and
5. Experts of deterministic chaos.

This branching out was caused both by historical facts or incidents and by several facts inherent in the natural sciences. My aforementioned book on some special kind of glass bead games in science will inform readers interested in ideologies of such sources, which are founded in sociology and psychology of science.

As a rule, these *schools* are impassably closed; there is hardly any scientific communication among them. The textbooks, scientific journals, and meetings differ basically, and the discrepancy in nomenclature often is significant. Therefore, it is not surprising that most of the work done even by prominent members of groups (3) and (5) is scarcely mentioned, for example, in textbooks of theoretical physics. Furthermore, the engineering scientists of group (2) disregard, as a rule, the papers published by scientists of groups (3) and (4); the inverse is also true. Some scientific debate does take place with the mathematicians of group (1), who are exclusively concerned with pure mathematical problems such as the existence and uniqueness of weak and strong solutions of the incompressible Navier–Stokes equations, their time analyticity and upper bounds for the fractal dimensions of bounded invariant sets, and so on. Other typical fields refer to mathematical problems in kinetic theories or statistical physics.

This lack of scientific discourse is injurious for many applications of fluid dynamics in practice, especially with regard to some essential properties of the

Navier–Stokes equations that are of utmost importance to the engineering sciences. Research and education are likewise affected.

With respect to certain methodologically different contributions from traditional mechanics and thermodynamics to modern continuum theories, Truesdell (1968, p. 324) provided some lucid and interesting observations that partly explain this unsatisfactory situation. He pointed out that from the beginning, and for more than a hundred years, continuum physics was identical with continuum mechanics. During this period, leading mathematicians dominated that scientific branch. Quality standards were established to treat physical problems adequately and offer appropriate mathematical tools for their solutions. Thus, it is not accidental that physics is accessible with mathematical description and interpretation; it is the result of the careful selection of the topics and the phenomena to be observed. Thus, theories of classical mechanics were developed, applied, generalized, and recast by nearly every distinguished scientist with well-founded knowledge in mathematics. There is an uninterrupted path beginning with Archimedes via Galilei, Newton, Euler, Hamilton, Maxwell, Poincaré, Einstein, G. D. Birkhoff, ending in our epoch with the explorers of the KAM-theorem.

In sharp contrast with this line, thermodynamics never matured to its immense potential to re-examine and extend the fundamentals of classical mechanics and the other traditional branches of physics. In its comparatively short history, thermodynamics has contributed many new basic thought patterns and words that are familiar to us in everyday life. Without doubt they are much more common than terms such as *divergence* or *tensor,* which the student learned to use accurately, but they are words that never found the right place in the mathematical structure of the traditional laws of Newtonian mechanics and Maxwell's electromagnetism. Amazingly, this is also true for orthodox quantum mechanics and Einstein's theories of relativity.

There is, however, another branch of the engineering sciences which I have labeled "Carnot–Mayer thermodynamics," in memory of its most important founders. It is an overall notation of a scientific domain consisting of all relevant topics such as technical thermodynamics, heat and mass transfer, and chemical thermodynamics. Clearly, the Carnot–Mayer thermodynamics is essentially coining Western civilization to a great extent. Without the typical terms *heat* and *work,* defined by the First Law of Thermodynamics, modern steam engine plants would not exist. The same is true with regard to turbines, compressors, jet engines, combustion chambers, rocket motors, refrigeration plants, heat pumps, nuclear power stations, and so on.

The basic knowledge on these topics has been taught to engineers now working in research institutions and industry where their advancement has become the usual business of the engineering sciences. Today, physicists, especially theorists, working at universities or research institutions are by no means equipped for such work; in general, they do not seem interested, particularly when it involves the fundamentals of the engineering sciences.

Let us turn back to the five scientific schools. The first and the second ones are dominated by mathematicians (group 1) and engineers (group 2). The leading members of groups (3) to (5) are professionals in theoretical physics or mathematics. As previously stated, mutual disinterest and partial prejudice prevail; an agreement for some serious cooperation between the schools rarely happens. At present, it is typical in academic life at universities worldwide that one cannot find institutional and efficient contacts between working groups of the physics departments and the engineering faculties. It belongs to the common rituals of politicians and leading officials of Big Science to plead for interdisciplinary contact again and again. In reality, academic teamwork happens only as a consequence of more private and often casual circumstances.

However, in spite of the sharp contrast and irreconcilable incommensurability among nearly all relevant items of their basics, theories, and results, the schools agree on one crucial point: A general axiom is assumed to be valid for all field theories, stating that each position vector

- points to the local flow velocity vector defined kinematically, and
- relates to a volume element, in which the particle density equals that of an infinitesimal macroscopic system in local thermodynamic equilibrium and also in contact with its surroundings.

In other words, thermal and caloric quantities are separately taken into account, without considering an immediate influence of motion on a hypothetical equilibrium state existing in the homogeneous volume element . This basic precondition may be denoted as *Duhem–Hadamard hypothesis* (cf. Truesdell, 1984, p.60). Without question, this procedure has evolved from the historical and psychological reasons summarized as follows:

1. Each volume element is assumed to be a mass-point, the mass of which is determined by the universe of the enclosed particles in violent agitation. Additionally, it is implicitly presupposed that in reality only *one* type of physical system does exist: a material entity made up of mass-points.
2. Consequently, the classical laws for Eulerian mass-point motions are taken for granted. These refer to kinetic energy, the linear and rotational momentum, as well as Newton's law of resulting forces (Truesdell, 1984, p. 108). Thermodynamic properties are without immediate influence and hence are omitted.
3. Following the tradition of the Hamiltonian formalism, this mass-point model is also assumed to be valid for orthodox quantum theory. Prigogine and Stengers (1993, p. 160) have stressed that the dominating preference of the so-called integrable systems is bounded by the same tradition. Only for such specified quantum systems can the total interaction between the mass-points be completely eliminated. Then, their Hamiltonians are exactly soluble.
4. Traditionally, the conviction prevails that geometrical and physical quantities are basically different (Falk, 1990, p. 111). A deeply rooted custom originated

> in the supposition that a physical object is always embedded in space and time. This bias was supported by the idea that all "true" physical quantities—that is, the nongeometric quantities—are properties only of matter. As a consequence, each physical system is incorporated in space and characterized by a definite interior separated from other systems or parts of the otherwise "empty" space. The specified properties on its surface determine the modes and intensities of interaction between the environment and the system envisioned. The popular concept of "closed" and "open" systems arises (Falk, 1990, p. 116).

In addition, it might be useful to take notice of an orthodox thermodynamicist's statement suitable for spontaneous evidence: "... the potential and the velocity v are nonthermodynamic quantities and imply that they are to be determined by nonthermodynamic considerations. Thus they are to be regarded as 'external' fields which can affect the thermodynamic state but which cannot be affected by the state directly" (Zeleznik, 1981). It should be emphasized that this plain quotation does not summarize a less relevant part of Zeleznik's thermodynamics. In fact, it is one of the general axioms referring to his theory of thermodynamics as a whole.

Notably, the essence of the key term *system* used in conventional continuum physics of mass-point fluids is interpreted verbally rather than established mathematically. In other words, this basic notion is constituted both by physical and metaphysical part.[1] Following Falk (1990, p. 116), a system of such a type is assumed to be self-contained; that is, the word *system* globally denotes a set of related and much-used terms such as *thing, object, topic, body, realm, domain,* and so on.

It is difficult to comment on the four previous statements in view of the great variety of experiences and achievements made by generations of scientists and engineers throughout the last decades. Contemporary human reasoning is dominated so much by these dogmatic remonstrances that every subsequent event in science and technology is essentially influenced and therefore dependent. But it is evident that those parts of the statements concerned with dictums, themselves related to physical properties, are meanwhile lacking foundation: Evidently, there is more than one type of physical systems. Moreover, it is questionable to separate any volume element into a compact mass-point moving along the flow field and into a part in rest, involving all nonkinematic properties of a fluid. Such a concept, introduced as an a posteriori assumption and formalized mathematically, could be justified as an appropriate approximation for many applications if it were not physically wrong and conceptually too restricted for use in modern research in physics and engineering.

As an interim result, a quote of Kestin perhaps denotes the crucial point: "Even though some progress ... has been made, no general theory of nonequilibrium states has yet evolved with the exception of one for gases which is based on Boltzmann's equation."

[1] Of course, there are some fundamentals that are called the metaphysical basic laws of the natural sciences and mathematics (cf. Böhm, 1966).

In the following chapter, we will discuss an alternative concept, first proposed by Falk, which is outstanding for its mathematically substantiated claim to give a scientifically adequate notion of a "physical system." We will see that Falk's theory eliminates the metaphysical constituents noted before.

1.5 Gibbsian Thermostatics?

The conclusion from Poincaré's analysis of dynamical systems is essential for all theories of continuum physics, provided that they are based on the Newton–Eulerian laws of mass-point mechanics. Poincaré's line of reasoning, fully recognized only recently, still followed the well-known tradition of classical mechanics. It is focused on all the conservative and some nonconservative dynamical systems with an immense, but finite, number of degrees of freedom and especially their characteristic Lagrangians. These *basis functions* are fundamental to all mathematical *solution procedures* that are usually classified as *variational principles.*

Most of the nonconservative dynamical (or irreversible) systems are physical manifestations that can be adequately described by nonlinear partial differential equations, together with appropriate initial and boundary conditions. However, it is a fact that for almost all such systems the exact Lagrangian of the respective problem does not exist in the sense of classical mechanics. A well-known and very simple example is Fourier's problem of transient, spatially one-dimensional heat conduction within a solid body, described by a linear parabolic partial differential equation. This key problem cannot be described by any means of a classical variational principle of the Hamiltonian type. That is, it can be shown that no Hamiltonian action exists whose Lagrange function, via the Euler–Lagrange equation, will generate the heat conduction equation mentioned earlier.

In view of this example, however, some variational problems can be considered differently. These might serve as a pool position for the very popular and widespread application of *Galerkin's method* for finding approximate solutions of various problems involving nonconservative phenomena, for which an exact Lagrangian cannot be found. Incidentally, Galerkin's method (and some similar procedures) can easily be extended to more intricate cases. These nonlinear problems of heat and mass transfer, wherein the nonlinearity is due to temperature-dependent thermophysical coefficients or nonlinear boundary conditions or both, can be treated successfully by this method. Certainly, it is true that Galerkin's procedure is equally applicable to dynamical systems with either a finite or an infinite number of degrees of freedom (Vujanovic and Jones, 1989, p. 152). But it should be emphasized that this purely practical method can only offer approximations. In other words, it differs basically from variational principles assuming the existence of the Lagrangian for the problem in question. This is especially true with regard to some desirable properties of Lagrangian systems. The Lagrangian formulation possesses an inherent symmetry permitting an immediate transformation to canonical form. This is a substantial part of the theory, since the canonical variables often contain the essence of the solution. In addition, certain physically important integrals of motion are directly available if the system is in canonical form. These integrals in turn can then be utilized to find some relevant information about the behavior of the solution, such as stability, periodicity, and so on.

Lagrangian systems do occupy an impressive position in the development of mechanics and applied mathematics. This is true, as seen in countless references, since the search for variational principles and Lagrangians has been going on for more than two centuries. The so-called *inverse problem* of the calculus of variations follows naturally from these efforts. It refers to the conditions by which a given set of differential equations is equivalent to a Lagrangian system. The determination of suitable Lagrangians is also an important part of this problem.

Evidently, the study of conservation laws defines a further class of problems, all of great importance to quantitative theories. In particular, we refer to first integrals, constants of motions, invariant quantities, global invariants, and so on. Probably the best-known method for finding conservation laws is based on the study of the invariant properties of Hamilton's action integral. Some explanations of this topic are discussed in Section 4.3 of this book. The study of the conservation rules of purely nonconservative dynamical systems has been advanced recently by extensions concerning the common variational principles of the Hamilton or D'Alembert types.

Recently, Sieniutycz and Berry (1993) succeeded in combining the *least action principle* with invariance requirements concerning physical variables, such as energy, momentum, particle numbers, and so on. The authors proved that the Lagrange multiplier of the entropy generation balance is crucial for consistent treatment of irreversible processes via an action formalism. They concluded with the hypothesis that embedding the First and Second Laws of Thermodynamics in the context of the external behavior of action under irreversible conditions may imply accretion of an additional term to the classical energy. Extremum conditions of action yield *Clebsch representation* of temperature, chemical potentials, flow velocities, and generalized momenta, including *thermal* momentum introduced in 1968.

It is generally accepted that the variational technique is quite powerful. For this reason, current efforts are toward expanding the class of problems to be formulated variationally. Some basic problems in physics do not allow the use of differential equations and their initial values. However, even in these cases, variational methods are often useful (Falk, 1990, p. 47). But it is also true that this technique is too cumbersome for more complicated problems relevant to today's engineering practice in flow mechanics, aerodynamics, and thermofluid dynamics. It is very often more difficult to find the basis function appropriate for the problem in question than to resort to an entirely different procedure.

Now the following question arises: Will it prove possible to find a basic concept valid for nonconservative dynamical (dissipative) systems that is founded on a basis function like the Lagrangian, but is, in addition, formulated explicitly in reference to the kind of complex problems defined above? In my opinion such a concept may be established using Gibbs' thermodynamics along with decisive ingredients worked out recently by Falk (1990) and explored by Nicolis and Prigogine (1989).

One might wonder why such an alternative was never discussed in science at an earlier time. Undoubtedly, Gibbs' substantial papers were already widely known to Boltzmann, Mach, Duhem, Nernst, and Planck, the heroes of thermodynamics at the close of the nineteenth century. Even prominent thermodynamicists of that day, such as Born, Tisza, and Callen, have often referred to Gibbs. In my former book, I have tried to find the answer by looking at things in their historical and professional perspectives (Straub, 1990, pp. 46, 215). Common to all listed *schools* of modern continuum physics, an answer can be offered that might also partly explain the hidden opposition and irrational polemic against Falk's view of thermodynamics.

Following the traces of Gibbs' papers in the textbooks of physics, it seems that his complete work, although conveniently accessible, never found a broad readership. Presumably, the degree of abstraction of his explanations as well as their uncommonly high level of logic rigor has caused considerable resistance from the very beginning.

In his epochal masterpiece entitled "On the Equilibrium of Heterogeneous Substances," Gibbs (1961, p. 55) started with Clausius's work. He recognized Clausius's ideas of mechanical heat modeling in their shrunken and atrophied formulation as being no longer a theory of motion, heat, work, and diffusion. It is

> no longer thermodynamics, but only the beginnings of thermostatics. It is Gibbs' singular merit to have seen the essence of this thermostatics: a variational definition of equilibrium, including its stabilities and instabilities, in which infinitely many putative equilibria are compared, and variational criteria are imposed to select from this class of fields that which, for a given caloric equation of state, corresponds to actual equilibrium, namely that field of fixed total energy which renders the total entropy a maximum. (Truesdell, 1984, p. 20)

But there are some sources of misunderstandings in his work. For instance, a subsection of Gibbs' famous paper (1961, p. 184) noted above is entitled "The Conditions of Internal and External Equilibrium for Solids in contact with Fluids with regard to all possible States of Strain of the Solids," which apparently referred to viscous processes. Many scientists and engineers seem to have evaluated those conditions to be reliable for such strain events, although it is hard to see how Gibbs' criteria can be reasonably applied to stressed fluids.

Unquestionably, Gibbs prepared genuine minimal principles for thermostatic equilibria and stability constraints of fluids, fluid mixtures, and elastic solids. This is not always clear from the language he used. Therefore, many scientists, particularly the members of the school of rational mechanics, reduce Gibbs' work to a treatise on special states occurring exclusively in thermostatics. Such a conclusion is too narrow. The true results of Gibbs' great essay are immediately available only by perusing the text, as we will see shortly.

Conspicuously, conciseness and formal simplicity were Gibbs' leading idea of equilibrium conditions and its mathematical formulation. Combined with a set of independent variables consisting of volume, entropy, and any finite number of particles, for instance, the respective equilibrium state can be established. At a first glance, this type of problem seems to refer exclusively to thermostatic quantities. Thus, an explanation might be found that Gibbs avoided treating dynamics manifested by irreversible processes, which are supposedly considered in only one passage of his *Scientific Papers*.

Nevertheless, he clarified that for non-equilibrium phenomena, rules other than equilibrium rules are valid. For motions, of course, the energy includes kinetic parts; moreover, it is remarkable that, as Truesdell (1984, p. 22) stated, "he attributed an entropy to a body in any state of motion. Insofar as he referred to thermodynamics, he did not limit it to slow chances or to bodies subservient to 'equation of state'." In addition, we find some remarks on "states which are not homogeneous" in one of Gibbs' early papers (1961, p. 35).

Let us now examine those results, which are accessible only by the adequate interpretation of the text, as to some key notions in classical thermodynamics that are hardly or not at all treated by Gibbs. From this analysis, the great work of Gibbs can still gain an independent and actual significance.

First, in not mentioning the cycles and the engines, Gibbs' foremost intention was to tell us that he was not interested in engineering applications. An even stronger point is that the First and Second Laws of Thermodynamics had both been renounced in Gibbs' complete essay. For this reason, one wonders why these laws are nowhere discussed until one realizes that Gibbs wrote, undoubtedly, a mathematical paper.[2] Mathematics deals with relations, that is, quantities are not distinguished by inherent properties. This fact obviously molded his work. The composition of his epochal publication demonstrates Gibbs' reasoning: The First and Second Laws do not belong to the set of logic fundamentals tied to each quantitative assertion resulting from thermodynamics.

It is easy to see that the two laws do not relate to the involved quantities in the sense given to them by mathematics. Their assertions concern properties imagined to be inherent in the respective quantities. That is, the *First Law* expresses the quantitative equivalence of heat and work as well as the property of conservation assumed to be inherent in energy and to be a property thereof.

The *Second Law* is equivalent to a qualitative statement concerning the progress of real physical processes and has many verbal assertions. A quantitative alternative

[2] Of course, Gibbs dealt with the First and Second Laws. In his first paper on thermodynamics he made lucid explanations, particularly concerning the meaning of work and heat (Gibbs, 1961, p. 3) as well as the principal distinctions between these two notions (p. 10).

emerges by using entropy. Now, the Second Law can be formulated as a strict condition on accessible and inaccessible changes of state. Furthermore, it can be divided into two parts, involving separate statements on characteristic properties of each of the parts. Thus, for example, a nonnegative entropy production rate per volume unit can be introduced which is basic to the whole of process physics and is an immediate consequence of assuming mechanical, thermal, and material stability for the system in question.

Second, the structure of Gibbsian thermodynamics is essentially established on three notions:

1. standard variable,
2. state, and
3. equilibrium as extremum principle.

Only the latter is explicitly accentuated in Gibbs' treatise.

These three terms do not determine the events in nature, but they constitute the notion and symbols of the mathematical language by which natural events can be adequately quantified and represented. Gibbs' mathematical argumentation and proceeding clearly proved that all theories applied by physicists and engineers to describe any natural phenomena and experiences are languages using mathematical and, therefore, symbolic logic. This means that even "words" (notions), "nouns" (quantities), "syntax" (relations), and so on are employed. Remember that Gibbs strictly discarded Clausius's ideas and instead directly followed the principles first introduced by the Scottish chemist J. Black and the French engineering officer S. Carnot.

Third, one of Gibbs' important and implicit inventions is that of the *standard variable,* which means some quantities that are mutually related and, above all, independent of and within the system in question. More precisely, the respective relations are tied to the notion of a system in such a way that they are the same for all other systems described by an identical set of standard variables. It is evident that the corresponding equations cannot contain any parameters identifying the special system just discussed. For the latter, indeed, some additional relations are needed for the values of those quantities that are representative of the single system.

Consequently, Gibbs' "system" is defined by a finite number n of certain standard variables X_i that may be considered as the degrees of freedom of the respective system. Thus, each relation between the corresponding standard variables leads automatically to a new characteristic function in its implicit form: $\Gamma(X_1, X_2, \ldots, X_n) \equiv (0$. Gibbs called this type of relation a *fundamental equation* of the system. He perceived that such a mathematical entity can be a basic structural constituent of thermodynamic theories.

It is noteworthy that Gibbs' main results have obviously been gathered exclusively from the interpretation of his famous essay. To my knowledge, Falk was the first to do this analysis to reveal that Gibbs' paper cannot really be reduced to a mere research report on thermostatics without damaging consequences for his entire wealth of ideas. Then, Falk concluded that the methodology of Gibbs' concept is the most valuable part of his work. Amazingly, that methodology is not restricted to thermodynamics, but can be extended at least to other disciplines of physics and perhaps to different scientific fields. This conclusion will be of great value to the theoretical approach presented in this book.

In concluding Chapter 1, let us consider the attribute *alternative,* included in the text title and indicating circumstances that I discussed elsewhere (Straub, 1990, p. 252). By analogy with Gödel's theorem, strictly valid only for his epochal investigations on number theory (Nagel and Newman, 1987),[3] additional reasons for that attribute may be summarized as follows: It is presupposed that there is no finite set of axioms for all physical problems to be solved. For appropriate conditions, well-posed problems and some axioms (selected for their solutions) merge to start innovation.

In my opinion, this general view also concerns Muncaster's remarkable idea of a *principle of consistency* inasmuch as "often two or more theories can deliver conclusions pertaining to a common class of phenomena or to bodies of a similar type ... and requiring the separate theories to agree, in some sense, when applied in instances where their interpretations overlap. ... The continuum gas dynamics–kinetic theory pair is sufficiently well developed now to serve as a prototype" (Truesdell, 1984, p. 517). Chemical kinetics and the equilibrium theory of chemical thermodynamics form a second pair. Starting from a widely divergent vocabulary of notions, this pair may allow the solution of at least the very same problem: the computation of chemical equilibria.

A third example is of special interest to the subject of this book; it concerns the accepted non-equilibrium theory of continua. For the classical case there are two main variants. The field theory of irreversible thermodynamics was developed by Meixner, Prigogine, and de Groot and founded on the hypothesis of local equilibrium. In contrast, Gyarmati's theory as a generalization of the classical field theory of thermodynamics is such that the systems are assumed not to be in local equilibrium, that is, the kinetic term of the various currents in the specific entropy is not neglected. Márkus and Gambár (1989, p. 358) presented some competent comments on significant differences between these two theoretical approaches. They emphasized that both theories have nothing to do with the original Onsager formalism.

The ideas presented in this book follow this line, too. It is true that the theoretical approaches, axioms, and objectives differ in part significantly from those of competitive theories. But, on one hand, new perspectives may be opened, even for problems already solved for a long time. Consequences and results, on the other hand, are often new and unexpected; thus, they might induce, particularly in young people, curiosity followed by the intention to become initiative, true to Popper's well-known maxim: It is better to kill the contrahent 's ideas than to kill him!

[3] Given any adequate proof procedure, Gödel demonstrated how to construct a definite statement in the notation of elementary number theory that says of itself, via Gödel numbering, that it cannot be proved. "Either it is false and provable, God forbid, or true and provable; presumably the latter" (Quine, 1987, p. 85).

Chapter 2 | Falkian Dynamics: An Introduction

"I hold that if the introduction of a quantity promotes clearness of thought, then even if at the moment we have no means of determining it with precision, its introduction is not only legitimate but desirable." — ***Sir J. J. Thomson***

2.1 Falk's Principle

Connected with today's physics are some competitive and quite singular worldviews, all of them trying to describe nature by the way of experiences. The experiences are generally assumed to be confirmed by the very results of certain standardized experiments, which are repetitive and open to proof.

For this reason, measuring means nothing more than expressing observations by the alignment of numbers. But, be the notion of numbers ever so useful, it may never be enough for relevant theoretical considerations in science. From personal experience we know important physical statements for which numbers play only a subsidiary role or are even irrelevant. One example is the assertion that all matter is composed of elementary particles. Another is the statement that processes that violate the Laws of Thermodynamics cannot be realized.

On one hand, the observation of any physical subject and the processing of the respective information take place, as a rule, by means of a finite number of data. On the other hand, the scientist describing the observed subjects—that is, including them in a theoretical framework—has to do an act of voluntary cognition as an inevitable precondition for transcending the observation. Indeed, such a mental effort, referring to what is forever beyond the grasp of ordinary experience and naive explanation, opens the chance to gain a new insight in nature, assuming that an infinite number of relating data is hypothetically available.

Following Falk (1990, p. iv), this type of worldview can be alternately signified as transcendental or metaphysical. It is based on data sets that are declared obligatory even for some unknown terrain of knowledge never located by their original content and scope. It is crucial for such a statement that it concerns in no hypothesis defined as a provisional or tentative proposal for the explanation of phenomena that has some degree of empirical substantiation or probability. As opposed to this, a metaphysical proposition just cannot be disproved by a single and properly determined measurement.

Pauli's prediction of the neutrino particle may serve as an example (cf. Schirra, 1991, p. 252). The existence of this particle had been deduced from the principle of the conservation of energy, which is assumed to be universally true. This bold inference could for the first time be confirmed experimentally 25 years later by Fermi.

Of course, metaphysical statements are in no way provable, but only provided that the term *provable* is related to mathematics being the second constituent part of physics and the experiment being the first part. Considering its enormous success, the question is certainly justified if and to what extent not only the operational part of mathematics (mathematics as a pure tool, or, metaphorically, as a servant or butler), but also its innovative and structural part should be involved in physics; all this for the progress of the contemporary worldview.

This question demands a comprehensive canon concerning provable and, thereby, generally binding statements for the quantitative experiences in physics. Actually, this means manifestations of real phenomena to be formulated by numbers.

From this point of view, physics and the engineering sciences, as well as a variety of other scientific disciplines, allow for a classification into two domains of knowledge. The first domain contains metaphysical assertions. The second domain refers to the noted canon, the statements of which may be considered scientifically rational, that is, definitely nonmetaphysical.

Provided that the mathematical formulation and precision are the right way to guarantee the scientific liability of each theory, Falk's lifework was devoted to the problem of basing the fundamentals of physics on the following two-part principle.

1. Physics should minimize the number of metaphysical elements enclosed in its theories.
2. Physics should lay bare the descriptively and mathematically hard core of its theories, constituted by scientifically rational statements.

Evidently, these demands were aimed at the methodologies of physics with the intent to enable a logically more precise formulation of the fundamentals. Unquestionably these would not be imaginable without the far-reaching inference from Falk's analysis of Gibbs' work. The results of that analysis, as compiled in the last subsection of Chapter one, show that their meanings are not tied only to the case of thermostatics that Gibbs had worked out admirably.

Falk's principle perfectly conforms with the high requirements of the modern world and, above all, its innumerable phenomena that become more and more complex. Furthermore, it is not confined to the macroscopic level of description.

But obviously Falk's principle conflicts with the force of habit with which the laws of natural sciences are considered as prototypes of metaphysical propositions in the special sense used here. In Western industrial societies they are regarded as uncommonly profound. It is generally accepted by the public that these laws, at least on the macroscopic level of experience, are deterministic and logically true without exception. Indeed, by their sheer number of applications they have achieved great triumphs. Therefore, following Reichenbach (1931, p. 30), people often bear a mental attitude toward science that can be compared with religious veneration.

There is, however, another trouble with Falk's principle. Part (1) insists on a serious and even reinforced contribution of mathematics to the metaphysical constituents of physics. Taking into consideration the increasing tendency toward abstraction in modern mathematics, this is equivalent to a strong repression of the common way of reasoning by means of sensorial stimulus and mental associations.

Modern physics has long lost this sort of intuition believed to be peculiar to classical mechanics. Ever more frequently, long-established pictures of many static and dynamic situations familiar to mechanical reasoning have to be supplemented by some characteristic ad hoc assumptions. In this case an intuitively conceivable eigen-evidence can no longer be expected: As a rule, the mathematics in use today forces such ad hoc assumptions to be purely formalized and deprived of any visual quality. Intellectual visualization has lost each logic-binding character. It degenerates to any "model," that is, to an illusory illustration of mathematical relations devoid of any ability to generalize the researcher's view to changed circumstances. Only the mathematical formula derived from a set of axioms is mandatory.

For this reason, Falk's principle corresponds to an eminently important and actual program: In case mental images lose their original power, the empirical sciences can still conserve their authenticity just with the canon introduced above. At least in principle, its mathematically provable statements can in due course be subjected to the formal machinery of statistical test and eventually falsified by a finite number of experimental data. The last point is crucial and may be paraphrased as *Falk's finiteness axiom*.

Hence, Falk's consequences are radical: In continuation of Gibbs' efforts he identified on one hand the metaphysical elements of physics. Relevant examples are the Laws of Thermodynamics, Maxwell's equations, the ideas of matter by way of elementary particles, and even the space-time concepts as parts of the special and the general theories of relativity.

On the other hand, Falk suggested accomplishing the realization of the canon by a generalized method denoted (unfortunately) as *thermodynamical* or *dynamical*. This method corresponds to an abstract framework consisting of relations between some notions appropriate for theoretical considerations. It is noteworthy that this framework is binding for all branches of physics. Moreover, the notions involved never conform with visual pictures that one associates with reality by way of experiments. As a matter of fact, they are plain mathematical symbols.

Falk (1990, p. x) made this quite clear to the reader: Concerning reality, human experience is mirrored scientifically as well as quantitatively by mathematical relations alone. These only occur between invented notions reflecting experience; one cannot map the visualizations in a one-to-one way to the notions.[4] In other words, Falk's proposal intends to extend those parts of nature that can be understood, above all, by means of numbers. This approach refers to a structure of notions, the connotation and purpose of which are exclusively based on the logic of its construction.

After Gibbs, Einstein presumably was the first to have made such a construction concerning relations between things, but not the things by themselves or even what human imagination believes to be things. His basic ideas of symmetry principles (which equal, by the way, the principles of relativity in a structural-mathematical sense) appear to be quite lucid. That is, provided some facts given by experience should be related, an important inference of logic immediately follows: Assuming the quantities A, B, C, ... of a theory can be mapped on other quantities A', B', C', ... of the same theory in such a way that all relations existing between $A, B, C, \ldots$ also exist between $A', B', C', \ldots$ and vice versa, then the quantities $A', B', C', \ldots$ are completely equivalent to A, B, C in the scientifically rational meanings defined above. In other words, transformations of quantities $A, B, C, \ldots$ into the respective A', B', C', ... do not pertain to the physical statements of the theory to be constituted solely by the unaffected relations themselves. This type of transformation denotes an *automorphism* signifying a symmetry of the theory. Quantities, mutually joined by such an *automorphic transformation,* are called, in short, *equivalent quantities*.

Some automorphisms of a theory can always be united in a group, as defined in mathematics. This means that two automorphic transformations executed one by one lead once more to such a special transformation, and the same is true for the reverse. Consequently, each scientifically rational theory defines a definite transformation group of all automorphisms assigned to the quantities of that theory. In physics the knowledge of a subgroup of the corresponding group is normally sufficient.

It is easy to grasp the subject matter of this symmetry principle: Certain essentials of any physical theory can be unambiguously established by the specification of a group of automorphisms concerning all the relevant quantities of the theory in question. It should be stressed that the larger the power of the group and the number of the equivalent quantities involved, the more precise are the respective statements on the whole theory.

Let us consider two significant examples. The first relates to *Newtonian kinematics* regarding the representation of curves within the continuum of space and time. The spatial coordinates x, y, z of each point depend on the fourth coordinate, the time t. This means that one does not deal with a space of dimension 4, but rather with two spaces, one of dimension 3 and the other of dimension 1.

[4] Another approach presently dominates some other scientific branches, such as medicine and mechatronics. Thus, all basic constructions of prostheses or robots are still oriented according to vivid images, the origin of which is obviously human limbs.

According to Newton's view, there are certain *spatially rectilinear* and *time-related uniform* motions distinguished dogmatically by an absolute being that is fully independent of every relation. The resulting absolute skeleton of straight lines within the (x; y; z; t)-continuum is the reference frame according to which all other curves are fixed. As a consequence, automorphic transformations of Newtonian kinematics map straight lines of the $\mathfrak{R}_{3+1}$ into straight lines again, even equal to the original absolute value of the velocity gradient.

This is represented mathematically as follows: The $\mathfrak{R}_3$ is considered to be a Euclidian space. This means that the homogeneous linear transformations $(x, y, z) \Leftrightarrow (x', y', z')$ satisfying the relation

$$x^2 + y^2 + z^2 = x'^2 + y'^2 + z'^2 \tag{2.1}$$

make up the maximum group of the Euclidian geometry that deals with the three quantities x, y, z and their related functions. The one-dimensional t-space is likewise a Euclidian $\mathfrak{R}_1$ in such a way that the linear transformations complying with the relation

$$t^2 = t'^2 \tag{2.2}$$

make up the maximum group of automorphic mappings that refer to the theory of t-dependent processes.

Sometimes (2.1) is assumed to describe the $\mathfrak{R}_3$ as a kind of invisible and rigid body, for which rotations and reflections at each point are allowed by means of so-called symmetry operations. Along the $\mathfrak{R}_1$ similar operations allow transformations of identity and reflections at each point of the rigid body, representing the uniform flow of time. The main characteristics of Newtonian kinematics is, however, the purely *metaphysical* approach that there are no relations between the quantities of the spaces $\mathfrak{R}_3$ and $\mathfrak{R}_1$. Clearly, this mathematically manifests the *absolute* character of time and space in classical mechanics.

A second example concerns the *Lorentz transformation* following Minkowski's calculus, which allows a simple connection to Newton's symmetry principle presented above. Substituting axioms (2.1) and (2.2) by the relation

$$c^2t^2 - (x^2 + y^2 + z^2) = c^2t'^2 - (x'^2 + y'^2 + z'^2), \tag{2.3}$$

the set of homogeneous linear transformations $(x, y, z, t) \Leftrightarrow (x', y', z', t')$ is introduced. These transformations constitute the maximum symmetry group of automorphic mappings for the so-called *Minkowski geometry* as the theory dealing with the quantities x, y, z, and t along with their functions. The symbol c denotes the *vacuum velocity of light*. For the specification $t^2 = t'^2$, the relations (2.1) and (2.2) are enclosed in (2.3) as a limiting case.

Assuming Equation (2.3) corresponds to the axiom of light propagation, then both clocks $\mathbb{C}$ and $\mathbb{C}'$ can be introduced as reference bodies moving relative to each other with $t \neq t'$. For the special case given by

$$c^2t^2 - x^2 = c'^2t'^2 - x'^2; \qquad y = y', \quad z = z', \tag{2.4}$$

it is easy to show that the behavior of the linear transformations in two variables, defined by

$$ct' = \alpha(ct) + \alpha\beta x$$

$$x = \alpha\beta(ct) + \alpha x,$$

is determined by the additional condition

$$\alpha = \frac{1}{\sqrt{(1-\beta^2)}}, \qquad \text{with} \quad -1 < \beta < +1 \tag{2.5}$$

Letting v_x be the velocity (in the x-direction) of the clock $\mathbb{C}'$ moving relatively toward the clock $\mathbb{C}$, and further putting $\beta := v_x/c$, the transformations can be described by the relations

$$t' = \frac{t - \left(\frac{v_x}{c^2}\right)x}{\sqrt{1-\left(\frac{v_x}{c}\right)^2}}; \qquad x' = \frac{x - v_x t}{\sqrt{1-\left(\frac{v_x}{c}\right)^2}}; \qquad y' = y, \quad z' = z, \tag{2.6}$$

which is easily proved to be a particular case of the so-called generalized Lorentz transformations (Mittelstaedt, 1980, p. 71).

To demonstrate that (2.6) represents a symmetry group, two transformations have to be done, one behind the other, via the chain of implications $\mathbb{C} \Rightarrow \mathbb{C}' \Rightarrow \mathbb{C}''$. This can be performed by introducing the *transformation matrices*:

$$\frac{1}{\sqrt{1-\beta'^2}}\begin{pmatrix} 1 & \beta' \\ \beta' & 1 \end{pmatrix}\frac{1}{\sqrt{1-\beta^2}}\begin{pmatrix} 1 & \beta \\ \beta & 1 \end{pmatrix} = \frac{1}{\sqrt{(1-\beta^2)\,(1-\beta'^2)}}\begin{pmatrix} 1+\beta\beta' & \beta+\beta' \\ \beta+\beta' & 1+\beta\beta' \end{pmatrix}$$

$$= \frac{1}{\sqrt{1-\left(\frac{\beta+\beta'}{1+\beta\beta'}\right)^2}}\begin{pmatrix} 1 & \frac{\beta+\beta'}{1+\beta\beta'} \\ \frac{\beta+\beta'}{1+\beta\beta'} & 1 \end{pmatrix} = \frac{1}{\sqrt{1-\beta''^2}}\begin{pmatrix} 1 & \beta'' \\ \beta'' & 1 \end{pmatrix}. \tag{2.7}$$

Obviously, the result for $\mathbb{C}''$ is of the same form as the original results for $\mathbb{C}$ and $\mathbb{C}'$. We may immediately conclude from (2.7) that the relations

$$\beta'' = \frac{\beta+\beta'}{1+\beta\beta'} \qquad v''_x = \frac{v_x + v'_x}{1 + \frac{v_x v'_x}{c^2}} \tag{2.8}$$

do hold. The second transformation, Einstein's well-known *addition theorem,* guarantees that the condition $|\beta| < 1$ holds. In other words, in principle, the relative velocity v_x is always lower than the constant velocity of light. Additionally, even for the limiting case $v_x = c$, the equation $v_x'' = c$ immediately follows from (2.8), quite independent of the value of v_x'.

Both examples produce the evidence for two cornerstones of physical history: The first example reminds us of the beginnings of classical mechanics, which are still relevant for today's science and technology. The second represents the progress of symmetry principles in modern physics, particularly in quantum mechanics and the theories of relativity and elementary particles.

But another point, concerning the fundamentals of theoretical and applied physics, along with engineering is of great importance. The coordinates x, y, z, and t for the description of events in space and time are generally adopted in an unusual way in the common applications of classical mechanics and all of traditional physics or technology. These coordinates can help fix the mutually related positions of "things" or part of "things." But this is done independently of the essence inherent to the "things" involved. Falk (1990, p. 111) even expressed the suspicion that this obvious lack of reference to "things" led to the far-reaching public opinion that a basic difference in the meanings of geometrical and physical quantities exists. This fallacy that nature should be divided into geometry and physics is apparently based on daily experiences taken as a guideline. Therefore, "things" are widely assumed to be incorporated into space and time, both idealized as a pair of absolute entities considered independent and not relative.

For this reason, "things" were (and are) commonly imagined to be more or less extended bodies composed of mass-points and endowed with properties of matter. All sciences are conclusively influenced by this habit of grasping a physical quantity as a property of the material object in question. Following Falk (1990, p. 200), Gibbs was the first to state that complex physical events could be adequately described by quantities defined as independent of the special systems in question. That applies to most macroscopic phenomena compared with problems of classical mechanics and, especially, characterized by mass-point systems.

It was also Falk (1990, pp. 111, 256) who suggested that the attribute *universally physical* be assigned to each quantity supposedly autonomous, such as geometric coordinates. With regard to an earlier recommendation (Falk and Ruppel, 1976, p. 90), I prefer the terms *standard variable* or *generic quantity.* These play a significant role in the design of all further theoretical considerations presented in the following sections.

2.2 Gibbs' Fundamental Equation and System Modeling

According to Falk's principle to prefer physically rational elements rather than the metaphysical parts of each quantitative theory, any concept has to contain mathematical constituents as low as the level of basic notions. Falk introduced four such notions, related mutually by definite mathematical mapping rules:

quantity, value of a quantity, state, and *system.*

Note that these terms do not have their usual meaning, at least in the formal and quantitative parts of the theory. Their ordinary usage, coined by verbal and metaphorical paraphrases, is avoided. To understand, we must first understand Falk's key notion of a generic quantity, introduced above.

Apparently, a deeply rooted custom puts forward the idea that physical quantities—velocity, pressure, temperature, entropy, energy, and so on—would have to be interpreted as differing entities. Each of them is presumed to have a natural, substantial existence assigned to any material body and phenomenon, yet identified as a virtual and independent "thing." "A universally physical quantity coincides with that one which may usually be indicated as the same entity of different systems" (Falk, 1990, p. 256).

This idea is supported as a rule by the widespread impression that each physical property is defined by an appropriate measuring practice. With this view of the nature of physical quantities there exists only a weak connection between *quantity* and *system*. Indeed, one is rather inclined to think that a quantity relates to the respective measuring method in a reversible one-valued manner.

For example, take an ideal gas and a condensate to be specified by three variables each, say, temperature T, pressure p, and mole number n. Then any state is fixed by a definite value of the dependent variable, the volume V, where $\mathcal{R}$ is the universal gas constant. According to the common view and notation, the variables T, p, and n are universally physical quantities along Falk's lines. As to the volume V, the same idea apparently holds because we are to imagine identical devices for measuring V. It is true that the results of such measurements can be adequately represented only by two quite different functions, say,

$$V = \mathcal{R}\frac{nT}{p}, \qquad \text{for ideal gases} \tag{2.9}$$

$$V = V_0(n, T)\ [1 + \beta(T)\,p], \qquad \text{for condensates} \tag{2.10}$$

where V_0 and β denote a reference volume and an expansion coefficient, respectively.

Obviously, the "same" quantity V can be determined by two equations, each depending on the variables T, p, and n but being quite different in their mathematical shapes. Assuming in addition that the volume V is a universally physical quantity, this property cannot be supported by evidence if one considers the two different systems "gas" and "condensate."

You may ask how a practicable and generally valid method could be established in view of the great variety of materials whose properties can adequately be described by the variables T, p, and n, for instance. Gibbs answered this question by taking thermostatics as a typical example that concerns not only physics but also other branches of sciences. His approach is based on the supposition that neither the notion of quantity, commonly defined by appropriate measuring methods, nor the vague associations often tied with the word system, are sufficient to elaborate the fundamentals of complex phenomena.

Gibbs perceived that it was necessary to first establish the notions of the universally physical quantity and the system by means of mathematically rigorous relationships.

Second, Gibbs saw that these relations are needed to supplement relations between the values of the quantities involved. The second type of relations depends explicitly on the properties of the system in question or even defines the respective system. Following Falk (1990, p. 204), these two presuppositions are obligatory, not only for the mathematical version of thermodynamics, but also for the mathematically precise version of each scientific theory constructed of finite numbers of elements.

Gibbs' idea can be demonstrated with an example concerning pure perfect gases. The functional dependencies of the state properties are generally represented by expressions of the form $\mathcal{G}(T; p; n) = n\mathfrak{g}(T; p; n)$, where temperature T, pressure p, and mole number n are chosen as independent variables. Taking the internal energy U, the entropy S, and the volume V of each one-component substance, the expressions

$$\begin{aligned} U(T; p; n) &= nu(T; p); \\ S(T; p; n) &= ns(T; p); \\ V(T; p; n) &= nv(T; p) \end{aligned} \tag{2.11}$$

will result. They are not confined to perfect gases, but are an apt description for real one-component materials.

The equations (2.11) imply that, assuming prescribed values of T, p, and n, the values of the quantities U, S, and V are proportional to the value of the mole number. Such quantities, which Rankine denoted as extensive quantities, will sometimes be called *particle-numbered*.

For perfect gases, defined as ideal gases with constant molar heat capacities c_v each, the functions U, S, and V are well-known. In addition to (2.9) for $V(T; p; n)$, the following two formulas

$$U(T;n) = n\,[c_v T + e_o] \tag{2.12}$$

$$S(T;p;n) = n\mathcal{R}\,\ell n\left(a_\kappa \mathcal{R} \frac{T^{\frac{\kappa}{\kappa-1}}}{p} \right) \tag{2.13}$$

hold (Falk, 1990, pp. 186, 189). Using the familiar relation between the molar gas constant $\mathcal{R}$ and the molar heat capacity c_v, namely,

$$c_v = \mathcal{R}(\kappa - 1)^{-1}, \tag{2.14}$$

the three parameters e_0, κ, and a_κ are available to characterize the so-called *Poisson gas* as a member of the subclass of perfect gases defined by the constant value of the *isentropic coefficient* κ. As to the corresponding variables, the total derivatives of (2.12) and (2.13), combined with the equation of state (2.9), after some simple manipulations yield

$$dU = T\,dS - p\,dV + \left\{ e_0 - \mathcal{R}\,T\,\ell n\left(\frac{a_\kappa \mathcal{R}}{e^{\frac{\kappa}{\kappa-1}}} \frac{T^{\frac{\kappa}{\kappa-1}}}{p} \right) \right\} dn. \tag{2.15}$$

This equation is actually the key to Gibbs' understanding of a universally physical quantity. For any given value of T and p, Equation (2.15) combines the changes dU, dS, dV, and dn of all relevant quantities U, S, V, and n, for these are sufficient to fully characterize a Poisson gas.

If the mole number n remains unchanged, then (2.15) is reduced to

$$dU = T\,dS - p\,dV, \tag{2.16}$$

a formula that does not indicate any more information on ideal gases. Gibbs made some concise comments regarding this formula:

> The state of the body, in the sense in which the term is used in the thermodynamics of fluids, is capable of two independent variations, so that between the five quantities V, p, T, U and S [quoted in modern terminology; D.S.] there exist relations expressible by three finite equations, different in general for different substances, but always such as to be in harmony with the differential equation (4) [i.e., (2.16); D.S.]. (Gibbs, 1961, p. 2)

It is easy to see, furthermore, that there is no reference at all to any definite substance in either a solid, liquid, or gaseous phase. In this sense Equation (2.16) can be regarded as independent of any system. This is of course only true under the condition that such a relation is exactly the same for all systems with the corresponding three degrees of freedom S, V, and n.

This is indeed a remarkable statement, first voiced by Gibbs (1961, p. 63). In accordance to (2.15), this statement is thought to be derived from a generalized relation and specified for Poisson gases. If the relation

$$dU = T\,dS - p\,dV + \hat{\mu}\,dn \tag{2.17}$$

is this generalization, then the quantity $\hat{\mu}$ arises by definition and is also assumed to be of a universally physical nature. Gibbs, its creator, named $\hat{\mu}$ the (molar) *chemical potential*. Like U, S, V, T, p, and n, this molar potential $\hat{\mu}$ is a quantity, the meanings of which are entirely independent of any single system. This is true, though the latter system might yet be distinguished by a certain function $\hat{\mu}$ different from that of any other system. As an example, the function for Poisson gases results directly from the comparison between (2.15) and (2.17) and is given by

$$\hat{\mu} = e_0 - \left(\mathcal{R}\, T\, \ell n \left\{ \frac{a_\kappa\, \mathcal{R} T^{\frac{\kappa}{\kappa-1}}}{e^{\frac{\kappa}{\kappa-1}}\, p} \right\} \right) \tag{2.18}$$

Notably, $\hat{\mu}$ does not depend on the mole number n, but only on T and p, as well as on the sort of the material specified by the three constants e_0, a_κ, and κ.

Following Falk (1990, p. 207), it is obvious that the relations (2.11) and (2.17) cannot be proven to be independent of a system; as a matter of fact they can at most be made plausible. The inverse question is formed by the problem for the intended description of physical phenomena: How then do its construction elements work to establish the seven quantities U, S, V, n, T, p, and $\hat{\mu}$ as universally physical? An answer in reference to a general theory founded on purely deductive reasoning was developed by Falk, who followed Gibbs' ideas.

To understand Falk's dynamics, refer again to relation (2.17), which connects the four quantities U, S, V, and n in a way that is distinctive to the systems defined by just these quantities. In other words:

> Each (thermodynamic) system with three degrees of freedom is defined by a relation between the values of U, S, V, and n established mathematically by an implicit relation Γ, with $\Gamma(U; S; V; n) \equiv 0$.

Without exception these four quantities have one property in common: They are particle-numbered, or extensive, which means variables whose values in homogeneous systems are proportional to the mass or the particle number of the system in question. Thus, there exists an equivalence for the function $U(S; V; n)$ given by

$$U \leftrightarrow U(S; V; n) \leftrightarrow U(ns_m; nv_m; n) \leftrightarrow nU(s_m; v_m; 1). \tag{2.19}$$

A function of several variables that possesses this property is known in mathematics as a *homogeneous function,* and its properties are governed by an important theorem due to Euler. Here, it is sufficient to summarize the statement pertinent to this theorem.

A homogeneous function $y = f(x_1, \ldots, x_\alpha)$ of order ω in the α variables x_j satisfies the identity

$$f(\lambda\, x_1, \lambda\, x_2, \ldots, \lambda\, x_\alpha) \equiv \lambda^\omega\, y \tag{2.20}$$

for any factor $\lambda \neq 0$. In general, the determination of λ belongs to the *measure theory* based on the theory of sets, where the word *set* means either a set of events, a set of particles, or a set of points. The term *measure* refers to many of the fundamental concepts of physics: charge, mass, momentum, energy, and entropy. In addition, the geometric concepts of length, area, and volume are particular kinds of measures. A concise illustration of the measure theory focused on physical applications is given by Green and Leipnik (1970, p. 17).

Taking the derivative of both sides of (2.20) with respect to the parameter λ, a second identity is obtained that is true for any value of λ. Hence, it must also hold for $\lambda = 1$.[5] With this substitution, the derivative becomes

$$\sum_i \frac{\partial f}{\partial x_i} x_i = \omega y . \tag{2.21}$$

In physics we are particularly interested in the special case of $\omega = 1$, which relates the function $f(x_1, x_2, \ldots, x_\alpha)$ to its first partial derivatives. (By the way, the converse is also true, and, following Kestin (1979, p. 327), it is possible to assert that any function satisfying (2.21) for $\omega = 1$ must be homogeneous of degree 1.)

At first glance we see that the homogeneous function $U(ns;\, nv;\, n) = nu$ concerns extensive quantities, like the volume $V = nv$ or the internal energy $U = nu$, which are both of degree $\omega = 1$. Hence, the function U obeys the Euler relation (2.21)

$$U = \frac{\partial U\,(S;V;n)}{\partial S} S + \frac{\partial U\,(S;V;n)}{\partial V} V + \frac{\partial U\,(S;V;n)}{\partial n} n . \tag{2.22}$$

What follows from (2.22)?

The answer bears not only on the problem at hand, but also on all systems to be fully defined by a homogeneous function of first degree. We will call it the *Euler–Reech equation* in retrospective appreciation of the French engineer F. Reech (1805–1884) who had introduced the Gibbs' relation (2.16) and Gibbs' three further thermodynamic potentials. (Reech's work had been published in 1853, 20 years earlier than Gibbs' famous publication. But "by 1873 Reech, still alive, had fallen silent" (Truesdell, 1980, p. 300).)

The equation may be summarized by two statements:

1. Assuming that a system can be completely defined by a set of universally physical quantities, the total differential of the functionally dependent quantity exists.
2. The representation of the system according to the Euler theorem is presupposed to be compatible with the result due to (1).

Using the example $U = U(S;\, V;\, n)$, the total differential dU is given by the relation

$$dU = \frac{\partial U\,(S;V;n)}{\partial S}\, dS + \frac{\partial U\,(S;V;n)}{\partial V}\, dV + \frac{\partial U\,(S;V;n)}{\partial n}\, dn , \tag{2.23}$$

which exclusively follows from point (1). In the abbreviated form, we have

$$dU = T_*\, dS - p_*\, dV + \hat{\mu}_*\, dn \tag{2.24}$$

[5] Appropriate examples can be found in the mathematical structure of the following polynoms: $y = \Sigma_i \Sigma_k a_{ik}\, x_i x_k$ $(\omega \rightarrow 2)$ and $y = \Sigma_k x_k^{-1/2}$ $(\omega \rightarrow -1/2)$.

$$U = T_* S - p_* V + \hat{\mu}_* n \tag{2.25}$$

$$T_* := \frac{\partial U\,(S;V;n)}{\partial S}; \qquad p_* := -\frac{\partial U\,(S;V;n)}{\partial V}; \qquad \hat{\mu}_* := \frac{\partial U\,(S;V;n)}{\partial n}, \tag{2.26}$$

where the temperature T_*, pressure p_*, and molar chemical potential $\hat{\mu}_*$ are introduced by definition.

As a rule, for every concrete problem in physics or the engineering sciences that needs to be quantified the number and selection of the relevant variables must be determined. The corresponding theoretical approach is always limited by the researcher's or engineer's incomplete knowledge of the unsolved problem. This fact leads to a map assumed to be appropriately described by abstract mathematics as compared to the reality reflected by the verbally formulated problem in question. For this reason the map represents a (mathematical) model of a more or less bounded facet of reality, but is not an objective imitation of a concrete entity (Falk and Ruppel, 1976, p. 130). Such a model can only be justified by experiences or rejected by experiments. But the user may systematically improve the model by an exchange of some variables or an extension of the number of quantities, presupposed to be appropriate for an adequate solution of the underlying problem.

Now we come to a point that often gives rise to some misunderstandings. The outline of Falk's theory offered so far is solely based on the simple assumption of an implicit function that is homogeneous in its universally physical quantities. This function, first introduced by Gibbs as the fundamental equation of the system in question, represents a system modeling which is by definition the exact qualitative level of mapping the real world onto mathematics. The model does not refer to any real object of matter, but only to the fundamental equation itself. Other assumptions are neither involved nor needed.

First consequences concern the direct comparison between Equations (2.17) and (2.24), both characterized by the same set of variables. First, recall that any application of equations of state are generally confined to physical situations defined as *rest states*. A rest state is a specified state of thermodynamic equilibrium (discussed later in this book). Every experimental determination of such an equation of state has to realize this situation as its substantial precondition. The materials affected by this measure are all real gases, liquids and their mixtures, as well as phase equilibria. But the same also applies to ideal gases defined phenomenologically by the thermal equation of state (2.9) and the two caloric equations of state (2.12) and (2.13) specified by (2.14) for Poisson gases.

With this reservation it is evident that Equations (2.17) and (2.24) are unequal. Equation (2.17) is derived with the help of the three material functions (2.9), (2.12), and (2.13), which only hold for equilibrium conditions. As opposed to this, (2.24) is not affected by any constraints, aside from the two assumptions concerning the existence of a homogeneous fundamental equation and the kind of system modeling explained above. For this reason the derivatives (2.26) are marked by an asterisk, to make a distinction between the conjugate quantities appearing in (2.17) and (2.24).

As the quantities T, p, and $\hat{\mu}$ clearly refer to equilibrium situations, we will denote the corresponding quantities T_*, p_*, and $\hat{\mu}_*$ as non-equilibrium quantities. But the notation introduced here as yet has only a formal meaning. Indeed, the mathematical arguments and premises originating from the asterisk values to the equilibrium ones is a main subject of the theory presented. (Regarding this issue, some important results of Falk's dynamics concerning equilibria will be discussed in the next section.)

To conclude this section, let us examine an immediate inference of the two fundamentals (2.22) and (2.23) used in both the abbreviated forms (2.24) and (2.25). They are mathematically compatible with themselves only under the condition that, considering (2.24), the total derivative of (2.25) leads a fortiori to the differential relation

$$S\ dT_* - V\ dp_* + n\ d\hat{\mu}_* \equiv 0\,, \tag{2.27}$$

which is well-known in textbooks as *Gibbs–Duhem relation*. Notably, it represents a strict constraint put on Gibbs' idea of fundamental equations combined with the convention of using (2.22) as a mathematical model. Clearly, (2.27) demands a relationship between the quantities T_*, p_*, and $\hat{\mu}_*$ for pure substances to be thermodynamically described by $\Gamma(U;\ S;\ V;\ n) \equiv 0$ with the four extensive quantities noted. Opposed to Γ, (2.27) postulates a relationship only for so-called intensive variables (see, e.g., Prausnitz, 1969, pp. 16, 468).

One mol of a gas, with a different mass in each particular case, is frequently employed in the formation of specific quantities. These are known as *molar (specific) quantities* and do have many advantages owing to this choice of units (Kestin, 1978, p. 108). Introducing, by reference to (2.27), the *molar specific volume* $v_m := V/n$ and the *molar specific entropy* $s_m := S/n$, a simplified version of the Gibbs–Duhem relation results by

$$d\hat{\mu}_* \equiv -s_m\ dT_* + v_m\ dp_* \quad \Rightarrow \quad \hat{\mu}_*(T_*;\ p_*),$$

which can be transferred by partial differentiation,

$$\frac{\partial \hat{\mu}_*(T_*;\ p_*)}{\partial p_*} = v_m = v_m(T_*;\ p_*), \tag{2.28}$$

into a form to be denoted as the *thermal equation* of the system in question. The corresponding *caloric equation* can be obtained by the partial differential

$$-\frac{\partial \hat{\mu}_*(T_*;\ p_*)}{\partial T_*} = s_m = s_m(T_*;\ p_*). \tag{2.29}$$

Formulas (2.28) and (2.29) indicate a typical complication in understanding the characteristic difference between thermodynamics and thermostatics in view of the equations of state of a pure substance, which are due to both equilibrium states (e.g., the Equations (2.9) and (2.13) for ideal gases) and to non-equilibrium states. Perhaps there is a subtle distinction between these equations, but in practice that distinction is irrelevant. This statement, indeed, is only true because systems that may be described by a fundamental equation $\Gamma(U; S; V; n) \equiv 0$ are frequently representatives of thermostatics controlled by states near equilibria. Unfortunately, such systems are not particularly appropriate to deal with theoretical problems in thermodynamics.

For this reason, we should classify the calculation discussed in this section as a methodologically useful chance to apprehend, above all, Falk's quite abstract construction. Nonetheless, the difference addressed above will be expounded in detail later in this book. It will then be substantiated that thermodynamics is mainly dominated by non-equilibrium states. Perhaps this fact enables us to grasp the objections vehemently raised, for example, by Boltzmann and even by Planck to the adherents of the so-called energetics who dealt with some kinds of idealized thermostatics rather than with actually running processes (cf. Helm, 1898, p. 292).

2.3 Equilibria and Criteria of Stability

Assuming Gibbs' fundamental equation Γ of the system in question, then, again closely following Falk, the identity $\Gamma \equiv 0$ can in principle be resolved with respect to any of the quantities involved. For the typical case $\Gamma(U; S; V; n) \equiv 0$, four relations come into consideration:

$$U = U(S; V; n); \qquad V = V(U; S; n); \tag{2.30}$$

$$S = S(U; V; n); \qquad n = n(U; S; V). \tag{2.31}$$

According to Falk (1990, p. 216), each of them may be called a *Massieu–Gibbs* function, or an M–G function. The commonly used term *thermodynamic potential* is too narrow, for the M–G function is not restricted to thermodynamic problems, but is advantageous to all branches of physics in principle.

Each of the Equations (2.30) and (2.31), defining a certain thermodynamic system, can be substituted by a set of characteristic M–G functions corresponding to the same system. These additional M–G functions are also assumed dependent on three variables each, selected from the complete set of the seven quantities (U; S; V; n; T; p; μ) involved. In practice, it is often advisable to use an M–G function not belonging to the stock of (2.30) and (2.31).

Such a substitution can be elegantly executed by Legendre transformations (Callen, 1966, p. 90) which means, in a certain sense, some aspects of geometrization of thermodynamics (Tisza, 1966, p. 235).

As usual, a "stock" function, such as $U(S; V; n)$, might be given. To work in the variables S; p; n, with the M–G function, we use the following procedure:

$$U - V\frac{\partial U\,(S;\, V;\, n)}{\partial V} = U + p_* V := H_*. \tag{2.32}$$

The M–G function H_* so frequently plays a role in thermodynamics that it is labeled with the separate name: *enthalpy*. Combining (2.24) and (2.25), together with (2.32), it is evident that

$$dH_* = T_* dS + Vdp + \hat{\mu}_* dn \quad \text{and} \quad H_* = T_* S + \hat{\mu}_* n \tag{2.33}$$

do hold.

Moreover, the partial derivatives

$$\frac{\partial H_*(S;\, p_*;\, n)}{\partial S} = T_*(S;\, p_*;\, n); \qquad \frac{\partial H_*(S;\, p_*;\, n)}{\partial p_*} = V(S;\, p_*;\, n);$$

$$\frac{\partial H_*(S;\, p_*;\, n)}{\partial n} = \hat{\mu}_*(S;\, p_*;\, n) \tag{2.34}$$

show that now V, T_*, and $\hat{\mu}_*$ are functions of the set of variables S, p_*, and n; the same is also true for the internal energy U, provided that (2.32) up to (2.34) are considered.

It is remarkable, however, that in the case of liquids or solids, this function $U(S;\, p_*;\, n)$ only plays an insignificant role, because U is virtually independent of the pressure. In the material sciences, this sort of experience with regard to state properties is not exceptional, in so far as sometimes the dependence on a new variable is considerably altered by an exchange of its conjugate one. In some cases even divergences may arise.[6] Hence, we can often eliminate that variable. Therefore, let us suppose that the volume function $V(S;\, p_*;\, n)$ can be resolved quite definitely with respect to p_*. Then, the pressure variable of the function $U(S;\, p_*;\, n)$ may be substituted with the help of the resulting equation. A well-known mathematical theorem to be proved by the theory of functions in many variables meets the requirement for the resolution needed.

The formal condition $\partial V(S;\, p_*;\, n)/\partial p_* \neq 0$, assumed to be fulfilled, corresponds to the important observation that real substances in all stable configurations of state obey the criterion of material stability. This may be written as

$$\partial V(S;\, p_*;\, n) \,/\, (\partial p_*) < 0, \tag{2.35}$$

which reminds us that this empirical fact apparently arbitrates the true sign of the mathematical condition $\partial V/\partial p_* \neq 0$. But such an inference might be too precipitate for the following reason: A system in Gibbs' sense is not a material subject, but rather a mathematical relation between the values of the relevant universally physical quantities. There may be some hidden mathematical structures of the M–G functions relevant for the internal consistency of Falk's dynamics. In this general case, conditions like those in (2.35) should be a priori incorporated into the mathematical description of realistic phenomena. This issue will be dealt with in the following part of this section with the emphasis on equilibria and criteria of stability.

In addition to the need to sometimes replace the variables by Legendre transformations, M–G functions have another important property: Given two systems, say (1) and (2), represented by their respective M–G functions with three degrees of freedom each, a third system (3) should be composed under the exclusive condition that its M–G function is to be exactly constructed by means of the two M–G functions

[6] This remark refers to "critical points" of fluids and fluid mixtures.

$U_{(1)}$ and $U_{(2)}$. The result will have to reproduce the mathematical form as established by Equations (2.24) up to (2.26). We may call this procedure the *composite principle*.

In detail, we first have

$$\begin{aligned} U_1 + U_2 &= T_{*2}(S_1 + S_2) - p_{*2}(V_1 + V_2) + \hat{\mu}_{*2}(n_1 + n_2) \\ &\quad + S_1(T_{*1} - T_{*2}) - V_1(p_{*1} - p_{*2}) + n_1(\hat{\mu}_{*1} - \hat{\mu}_{*2}) \end{aligned} \tag{2.36}$$

where, as before in the respective original Euler–Reech Equations (2.22) or (2.25), the anticipated result

$$U_3 = T_{*3}S_3 - p_{*3}V_3 + \hat{\mu}_{*3}n_3 \tag{2.37}$$

is generated, subject to the condition

$$T_{*1} = T_{*2} = T_{*3}; \qquad p_{*1} = p_{*2} = p_{*3}; \qquad \hat{\mu}_{*1} = \hat{\mu}_{*2} = \hat{\mu}_{*3}. \tag{2.38}$$

The common equilibrium conditions are clear, with the exception that here they arise without explicit reference to any extremum principle. But indeed, it is hard to intuitively grasp the formal derivation of (2.38) with respect to its physical meaning: To realize system (3), for each the other systems (1) and (2) a sequence of changes of state must pass through various states from the beginning to the final state determined by the set of condition (2.38).

A more immediate procedure to establish the fundamental set (2.38) follows from adding (2.24) for both the systems (1) and (2). We obtain

$$\begin{aligned} d(U_1 + U_2) &= T_{*2}d(S_1 + S_2) - p_{*2}d(V_1 + V_2) + \hat{\mu}_{*2}d(n_1 + n_2) \\ &\quad + dS_1(T_{*1} - T_{*2}) - dV_1(p_{*1} - p_{*2}) + dn_1(\hat{\mu}_{*1} - \hat{\mu}_{*2}) \end{aligned} \tag{2.39}$$

Hence, there is an important consequence in that (2.38) is only compatible with (2.39), provided that the conditions

$$S_1 + S_2 = S_3 = \text{constant}; \quad V_1 + V_2 = V_3 = \text{constant}; \quad n_1 + n_2 = n_3 = \text{constant} \tag{2.40}$$

are fulfilled and, additionally, the required extremal conditions

$$\begin{aligned} &\left[\frac{\partial(U_1 + U_2)}{\partial S_1}\right]_{V_1;n_1} = 0 \\ &\left[\frac{\partial(U_1 + U_2)}{\partial V_1}\right]_{S_1;n_1} = 0 \\ &\left[\frac{\partial(U_1 + U_2)}{\partial n_1}\right]_{S_1;n_1} = 0 \end{aligned} \tag{2.41}$$

are introduced. The subscript letters after the differential coefficients indicate the quantity that is supposed to be constant in the differentiation assigned to an extremum, each with respect to the pertaining exchange variables S_1, V_1, or n_1.

To determine whether there is a maximum or a minimum in the extremal conditions (2.41), the common calculations concerning the stability of the system in question are needed. Referring to Falk (1990, p. 220) and using the differential Taylor series expansion up to second-order terms, the following expression emerges:

$$
\begin{aligned}
d(U_1 + U_2) &= \frac{1}{2}\left[\frac{\partial^2 U_1}{\partial S_1^2} + \frac{\partial^2 U_2}{\partial S_2^2}\right]_{T_{*1} = T_{*2};\, p_{*1} = p_{*2};\, \hat{\mu}_{*1} = \hat{\mu}_{*2}} dS_1^2 \\
&+ \cdots + \frac{1}{2}\left[\frac{\partial^2 U_1}{\partial n_1^2} + \frac{\partial^2 U_2}{\partial n_2^2}\right]_{T_{*1} = T_{*2};\, p_{*1} = p_{*2};\, \hat{\mu}_{*1} = \hat{\mu}_{*2}} dn_1^2 \\
&+ \cdots + \left[\frac{\partial^2 U_1}{\partial S_1 \partial V_1} + \frac{\partial^2 U_2}{\partial S_2 \partial V_2}\right]_{T_{*1} = T_{*2};\, p_{*1} = p_{*2};\, \hat{\mu}_{*1} = \hat{\mu}_{*2}} dS_1 dV_1 \\
&+ \cdots + \left[\frac{\partial^2 U_1}{\partial V_1 \partial n_1} + \frac{\partial^2 U_2}{\partial V_2 \partial n_2}\right]_{T_{*1} = T_{*2};\, p_{*1} = p_{*2};\, \hat{\mu}_{*1} = \hat{\mu}_{*2}} dV_1 dn_1.
\end{aligned}
\tag{2.42}
$$

As usual, the internal energy $U_3 = U_1 + U_2$ by definition assumes minimum values in equilibrium states. Then, the quadratic form of the right-hand side of (2.42) must be positive-definite with regard to the differentials dS_1, dV_1, and dn_1 as well as their products. As a consequence, the so-called *criteria of stability* result, assumed to be valid for real substances that can adequately be described by three degrees of freedom. In general, they are mathematically expressed in the proper form of inequalities. For example, the most relevant criteria are compiled as follows:

$$
\frac{\partial T_*(S;\, V;\, n)}{\partial S} = \frac{1}{\dfrac{\partial S(T_*;\, V;\, n)}{\partial T_*}} \geq 0; \qquad \frac{\partial p_*(S;\, V;\, n)}{\partial V} = \frac{1}{\dfrac{\partial V(S;\, p_*;\, n)}{\partial p_*}} \geq 0
$$

$$
\frac{\partial \hat{\mu}_*(S;\, V;\, n)}{\partial n} = \frac{1}{\dfrac{\partial n(S;\, V;\, \hat{\mu}_*)}{\partial \hat{\mu}_*}} \geq 0; \qquad \begin{vmatrix} \dfrac{\partial T_*}{\partial S} & -\dfrac{\partial p_*}{\partial S} \\ \dfrac{\partial T_*}{\partial V} & -\dfrac{\partial p_*}{\partial V} \end{vmatrix} \geq 0
\tag{2.43}
$$

Whereas the first three are represented by partial derivatives, the fourth is given by the nonnegative subdeterminant of the pertaining Hessian.

Certainly, some remarks on equilibria and criteria of stability may be useful:

1. As mentioned before, the intensive quantities normally marked by an asterisk each could now be written without this subscript.
2. The second criterion of stability as formulated in (2.43) equals the inequality (2.35). This follows from the two fundamentals (2.24) and (2.25) together with the composite principle introduced above, which define a thermody-

namic system with three degrees of freedom without further restrictions. For this reason, this criterion is generally valid for such material systems. Conversely, the empirical evidence of (2.35) supports the minimum principle of energy leading stringently to (2.43).

3. Gibbs' fundamental equation $\Gamma(U; S; V; n) = 0$ does not favor any of the four variables. Thus, it is easy to repeat the calculations performed above with respect to the M–G functions $S(U; V; n)$, $V(S; n; U)$, or $n(V; U; S)$. For instance, the respective Euler–Reech equation and the differential relation to model the system for the last M–G function are

$$dn = \frac{1}{\hat{\mu}_*}dU - \frac{T_*}{\hat{\mu}_*}dS + \frac{p_*}{\hat{\mu}_*}dV \,; \qquad n = \frac{1}{\mu_*}U - \frac{T_*}{\hat{\mu}_*}S + \frac{p_*}{\hat{\mu}_*}V \,. \tag{2.44}$$

Altogether, two possibilities arise to determine an equilibrium. The first is a consequence of the minimum principle, regarding energy and volume. Alternatively, the second works along the maximum principle with reference to entropy and mole number. Hence, the notion "equilibrium principle" should be preferred to the expressions "minimum principle" of energy or "maximum principle" of entropy, which, although used commonly, are a bit narrow.

4. It should be emphasized that the First and Second Laws of Thermodynamics are neither included in the derivation of the equilibrium nor are they needed for any stability conditions. This means that the maximum principle of entropy is not linked with the Second Law, even though it states increasing values of entropy at least in so-called closed systems. Falk (1990, p. 225) has given a notable explanation: If the Second Law were tied to the maximum principle, then the First Law could also be hypothetically assumed to be coupled with the minimum principle of energy in such a way that there should be a tendency of energy to decrease at all events. Such a case, indeed, would be inconsistent with the conservation law of energy supposed to be valid in closed systems at least.

5. The M–G functions (2.30) and (2.31) establish a so-called Gibbs space, each spanned by the extensive variables (cf. Tisza, 1966, p. 235). It is singled out from all phase spaces for special consideration. In view of the practical usefulness of Gibbs space, we need to ask what is its underlying mathematical structure compared with the important abstract spaces of mathematical physics. The latter are commonly characterized by a metric based on length or the orthogonality of the basis vectors, possibly in the complex domain. But neither of these concepts are thought to be reasonably definable in Gibbs space. Unexpectedly, by means of entropy the unstructured manifold can be furnished by an element of valuation that serves as a metric, formally based on volume rather than length and on parallelism rather than orthogonality.

   It is noteworthy that, in 1917, Pick and Blaschke (1923, p. 70) developed a theory of curvature that, within affine differential geometry, is based solely on an *affine fundamental form proved to be invariant under linear transformations of determinant unity*. It is indeed remarkable that such a form can be identified with the quadratic forms that play an important role in the thermostatic theory of stability. "Thus the parallel between the significant thermostatic and geometric concepts is very close. … The treatment of affine spaces and its physical interpretation differ considerably from the eigenvalue method that is customary in spaces in which orthogonal (or unitary) transformations are meaningful" (Tisza, 1966, p. 106).

The main properties of quadratic forms arising, for instance, in Equation (2.42) and concerning the affine fundamental form noted, refers to the pertaining restricted equiaffine group. The latter preserves volume even if the transformation is broken off after the first, second, … step. Furthermore, the peculiarity of the thermodynamic formalism—that the variables to be kept constant have to be particularly indicated—appears to be the consequence of the use of oblique coordinate systems. This narrow connection between physical quantities and geometric patterns is obviously reflected in the curvature of the Gibbs surface, which itself manifests the stability of the thermodynamic system in question.

A simple example should underline the given stability criteria. For an ideal gas, specified as a Poisson gas by the condition $\kappa = \text{constant}$ for the isentropic coefficient κ, we have the M–G equation

$$U(S;V;n) = n\left\{e_0 + \frac{\mathcal{R}}{\kappa-1}\frac{1}{a_\kappa^{\kappa-1}}\left(\frac{n}{V}\right)^{\kappa-1}\exp\left[\frac{S}{n\mathcal{R}}(\kappa-1)\right]\right\}, \tag{2.45}$$

where the notation of (2.12) and (2.13) is used again. Equilibrium temperature, pressure, and chemical potential are each obtained by their partial derivatives as stated by (2.26):

$$\begin{aligned} T(S;V;n) &= \frac{1}{a_\kappa^{\kappa-1}}\left(\frac{n}{V}\right)^{\kappa-1}\exp\left[\frac{S}{n\mathcal{R}}(\kappa-1)\right]; \\ p(S;V;n) &= \frac{\mathcal{R}}{a_\kappa^{\kappa-1}}\left(\frac{n}{V}\right)^{\kappa-1}\exp\left[\frac{S}{n\mathcal{R}}(\kappa-1)\right]; \\ \hat{\mu}(S;V;n) &= e_0 - \left(\frac{S}{n\mathcal{R}} - \frac{\kappa}{k-1}\right)\frac{\mathcal{R}}{a_\kappa^{\kappa-1}}\left(\frac{n}{V}\right)^{\kappa-1}\exp\left[\frac{S}{n\mathcal{R}}(\kappa-1)\right]. \end{aligned} \tag{2.46}$$

We may proceed in the same way with the remaining equations with (2.43) representing the stability criteria of the Poisson gas. To satisfy these conditions it is necessary and sufficient that the inequalities

$$\begin{aligned} \frac{\partial^2 U}{\partial S^2} &= \frac{\partial T(S;V;n)}{\partial S} = \frac{T}{n\mathcal{R}(\kappa-1)} > 0; \\ \frac{\partial^2 U}{\partial V^2} &= \frac{\partial p(S;V;n)}{\partial V} = \frac{\kappa p}{V} > 0; \ \ldots \end{aligned} \tag{2.47}$$

hold. For the parameter $\kappa \neq 1$, the right-side values can never become zero, indicating the empirically well-known fact that an ideal gas has no stability limits in the form of phase transitions.

To demonstrate such behavior, let us evaluate the characteristic derivative $\partial p_*/\partial V$ of a Van der Waals gas. According to Falk (1990, p. 230), the following result may be of interest:

$$\frac{\partial^2 F}{\partial V^2} = -\frac{\partial p_*(T_*;\, V;\, n)}{\partial V}$$
$$= \frac{1}{n\,(v-\beta)}\left[p_* - \frac{\alpha}{v^2}\left(1 - 2\frac{\beta}{v}\right)\right]\begin{cases} > 0 & \text{for stable states} \\ = 0 & \text{for stability limit} \\ < 0 & \text{for unstable states.} \end{cases} \tag{2.48}$$

Equation (2.48) was derived using a second important Legendre transformation of the internal energy $U(S; V; n)$ as a M–G function:

$$F = U - S(\partial U/\partial S) = U - T_*S = -p_*V + \hat{\mu}_*n, \tag{2.49}$$

$$dF = dU - T_*\, dS - S\, dT_* = -p_*\, dV + \hat{\mu}_*\, dn - S\, dT_*. \tag{2.50}$$

Here the new M–G function $F(T_*; V; n)$ is called *free energy* or the *Helmholtz function.*

To conclude this section from another point of view, recall the remarks at the end of the preceding section. In mechanics it is common to classify equilibria into stable, unstable, and indifferent equilibrium. Regarding the potential energy $\not{p}$ of a movable body, three types of extremum behavior serve as criteria of stability. The body is located at position χ, subject to *displacements*; $\not{p}(\chi)$ expresses the energy as a function of χ. Let us characterize the criteria by the inequalities

$$\frac{d^2\not{p}}{d\chi^2} > 0 \ (\textit{stability}); \qquad \frac{d^2\not{p}}{d\chi^2} < 0 \ (\textit{instability}); \qquad \frac{d\not{p}}{d\chi} = 0 \ (\textit{indifference}), \tag{2.51}$$

which is similar to the analysis in (2.42) with respect to the stability of a thermodynamic three-variable system.

The convention in mechanics is to define stability by a positive sign or a negative one for the pertaining inequality. In every case only one of the three conditions (2.51) can be interpreted as a stable equilibrium. Hence, the other two do not hold true for an equilibrium state. The same is valid for the criteria of stability (2.42) in reference to the internal energy which, at least in principle, could also tend toward an extreme value via maximization. Yet, directly opposed to this, past experience has shown that the Second Law quite obviously excludes such a free convention for the entropy, along with the corresponding substitution of the entropic maximum principle.

At first sight it might look as if thermostatics requires more rigorous demands on the theoretical foundation for an equilibrium state than mechanics. Conversely, one might arrive at the conclusion that the mathematical concept, as presented in this section, would be very restrictive compared with the potentials in mechanics to describe evident experiences with movable bodies. But this would be an erroneous inference, because in thermostatics any displacement of a body that is in direct contrast with daily observation and experience is disregarded. Consequently, in thermostatics the case of instability does not concern realistic events, because instability is physically not realizable with bodies at rest. In my opinion, this is a crucial point: this idealization may be a serious handicap to describing non-equilibrium phenomena in a systematic way.

To describe unstable events by mathematical means according to Falk's dynamics, it is necessary to incorporate the three-variable system, "body at rest," into the more extended system, "movable body." Its fundamental equation has seven quantities instead of the four in $\Gamma(U; S; V; n) \equiv 0$. The three additional quantities are due to the three coordinates of the *position vector* $\mathbf{r}$. Now, the energy E can no longer be reduced to the internal energy U. Moreover, it contains the *energy form of displacement*, which can be transformed into common potential energy provided that certain physical conditions are fulfilled.

Hence, the corresponding Gibbs' fundamental equation becomes $\Gamma(E; S; V; n; \mathbf{r}) \equiv 0$, from which first the M–G function $E = E(S; V; n; \mathbf{r})$ follows. Then the Pfaffian

$$dE = T_* dS - p_* dV + \hat{\mu}_* dn + (\partial E/\partial \mathbf{r})_{S,V,n} \bullet d\mathbf{r} \tag{2.52}$$

will hold with $(\partial E/\partial \mathbf{r})_{S,V,n} := -\mathbf{F}$ only temporarily used as a formal shortening.

By the way, following Falk and Ruppel (1976, p. 88), the question arises if the *displacement vector* $d\mathbf{r}$ is to be treated as extensive quantity at all. This is allowed partly because the property to be particle-numbered may be a sufficient criterion (but never a necessary one) to ensure the extensitivity of $d\mathbf{r}$.

Yet the more relevant argument refers to the property of the position vector $\mathbf{r}$ to be, above all, a quantity of the field that interacts with the body displaced. Interactions like these, however, will always have to be connected with energy transfer. When that energy is stored or released by the field, we cannot straightforwardly determine whether the responsible variable $\mathbf{r}$ is extensive or not.

Nevertheless, the evidence supports that the *scalar product* $-\mathbf{F} \bullet d\mathbf{r}$ as a whole may safely be assumed to be particle-numbered. This is plausible, since the field force $\mathbf{F}$, by the definition introduced into (2.52) via the energy form of displacement, may be substituted by a *product variable* $(V\mathbf{f}_V)$, where $\mathbf{f}_V$ is a *force density*. In this case the energy form is taken to be proportional to the mole number n via the volume V (see also Sections 5.1 and 9.1). It is evident that this combination stresses again the importance of the quantity V, considering its relationship to stability problems noted above with respect to the concept of a Gibbs space.

Taking a closer look at the theme of this section, we may draw some interesting conclusions from the mathematical method presented. Above all, it is obvious that there are no systems "per se" or "a priori"; there are only systems with a definite number r of degrees of freedom. The results, substantiated by the formulas for the equilibrium states and their stability criteria of a system of type $r = 3$, offer an idea about the respective effect of an increase of the variable number.

Apparently, there are no significant changes as to the structure of Equations (2.38) or (2.40) up to (2.43), provided the extension toward an increased number of variables commences with the M–G function $U(S; V; n)$. In this case, the number of inequalities will increase, now including the influences of the new variables on equilibrium and stability, respectively. Furthermore, it is easy to explain the manner in which the inherent feedback of the new variables may be incorporated into the relation derived above for the M–G function $U(S; V; n)$.

We may gain insight into the mathematical structure by looking at the Taylor series (2.42). Obviously, the different influences of the variables involved are explicitly considered not only by the diverse increments dS, dV, dn and those total differentials of the new variables, but also implicitly by all derivatives arising as system factors. Each derivative is determined by the complete set of variables of the system, either by an active influence of the pertaining quantity with respect to the differentiation procedure or by passive contributions of the other quantities held constant. Thus, due to the peculiar structure of Gibbs' fundamental equation, a single *additional* variable can significantly modify the stability of any physical states in the following twofold manner.

1. The stability conditions referring to the contributions of the original variables of the system [e.g., Equations (2.43) for the system equivalent to the M–G function $U(S; V; n)$] may thus be violated by a simple shift of the actual state values compared with the "original" values of the quantities S, V, and n.
2. In principle, many new additional stability conditions arise from the products of the increment of the new variable with each one of the "original" variables and itself. Of course, these changed circumstances may cause drastic reverberations on the system with unknown consequences.

This situation is perhaps one of the keys to understanding the physical problems dealt with in this book. Point (1) indicates an essential item: For phenomena other than thermodynamic equilibria, most of the properties of substances involved are actually not available. Due to this shortage, data experimentally determined for the admittedly artificial state at rest are commonly used for non-equilibrium events. As a rule, this usage is justified by appropriate hypotheses. An outstanding example is the application of equations of state measured and compiled for an immense number of substances and mixtures of fluids.

This practice has been widespread for a long time, in such a manner that the consciousness of its approximate character has been nearly lost or at least suppressed. In my opinion, the majority of the respective textbooks may serve to prove this thesis. Remarkably, the analysis condensed in point (1), leads to the conclusions that there are sufficient and theoretically well-founded reasons to apply data of equilibrium states to non-equilibrium phenomena. The examinations devoted to flow processes in the second part of this book will stringently prove that the practice mentioned can be supported even in complex flow configurations. Yet this perspective will be accompanied by some interesting insights and far-reading consequences regarding the description of non-equilibrium phenomena.

2.4 Mathematical Foundation of Falk's Dynamics I: Mappings

To generalize the results of the previous sections, we will first properly present the basic terms of Falk's mathematical method. It is perhaps not striking that this approach rests upon a *set-theoretic base,* as is common in today's mathematics since Cantor's need for such a new base (formulated around 1872) from his first investigations of function theory.

It is useful to recall the famous definition of a set that Cantor made in 1895. "By a *set* we understand any collection M of definite, distinct objects m of our perception or of our thought (which will be called the elements of M) into a whole" (quoted from Allenby, 1991, p. 1). Subsequently, we shall consider the words *set, collection,* and *aggregate* synonymous. Additionally, the elements of a set will sometimes be called its *members.*

As mentioned earlier, the basic terms in Falk's theory are quantity, value, state, and system. In this context, the word *quantity* is exclusively assumed to be the universally physical quantity introduced and already discussed in Section 2.2. Recall that Equations (2.9) and (2.10) refer to two very different classes of fluids. Each quantity describing the first class also occurs in the formula of the second one. The difference between the two substances only exists in such a way that the values of the same physical quantities arise in different combinations. Hence, the term of a state permits us to think formally of these combinations in value. Therefore, the following rule is equally useful:

Each physical quantity takes on one value for every state of a system.

Read this rule in the sense of mathematics: Notions in mathematics are never defined as individual objects, but always as being mutually related to other terms. Thus, for mere mathematical objects, questioning in mathematical physics what is meant by a state (of a system) is as nonsensical as questioning in geometry the meaning of a point.

In short, relations of a mathematical notion with other notions are established in such a way that any individual object is an element of a set, for which certain mathematical operations (addition, multiplication, etc.) are defined. Furthermore, some conditions concerning the elements are laid down, such as order- or group-type relations. For instance, numbers are elements of (number-)fields, mass-points are elements of topological spaces, and vectors are elements of vector spaces. Similar agreements are constitutive for Falk's dynamics. Each of the terms—(physical) quantity, value (of a quantity), and state (of a system)—is an element of some well-defined set. An outline of the foundation is the subject of this section.

In principle, the following basic postulates of Falk's dynamics (1990, p. 240) are introduced.

- A quantity A is an element of a **set** $\mathcal{G}$ denoted as the *domain of physical quantities* assigned to a class of systems.
- A value α is an element of a **set** W denoted as the *range of values* assigned to $\mathcal{G}$.
- A state Z is an element of a **set** $\mathcal{S}$ denoted as the *state ensemble of a system*, in other words, the system itself.

This whole framework may be summarized in a single rule:

Every element of S, that is, every state Z, is a mapping of $\mathcal{G}$ onto W.

Consequently, each element A of $\mathcal{G}$ is mapped onto the corresponding element α of W. This may be expressed symbolically and verbally as follows.

$$Z[A] = \alpha. \tag{2.53}$$

By way of the state Z, the quantity A yields the value α.

To substantiate this general postulate with mathematical tools, Falk axiomatically prescribed the mapping rules in such a manner that the results can be assumed to be true for classical physics as well as quantum mechanics.

His method, applied to the latter theory, has been elaborated by Diestelhorst (1993). Within the scope of this book at least some side steps toward quantum mechanics seem to be useful. Compared with traditional physical sciences, it is much more illustrative to demonstrate the essential difference between a quantum quantity and its value.

It is well-known that in quantum mechanics the typical quantity is an operator whereas its "value" is represented by a set of real numbers. For instance, the spectrum of all energy eigenvalues of a quantum system is assigned to the quantity *Hamiltonian operator,* the eigenfunctions of which determine the actual state of the system. Apparently, this sharp contrast between a quantity and its values cannot be observed in classical physics. Yet there is a general diversity between a quantity A

and its value α, outlined simply by reference to mathematics: It exactly corresponds with the difference between the terms of *variable* and *number*.

Although we are accustomed to dealing with numbers and variables (e.g., those of a function) by means of the same algebraic rules of addition, multiplication, and so on, note that numbers are subject to additional "rules" in regard to the axioms of order in the field of real numbers. Generally speaking, variables are assigned to an operational domain, for which certain mathematical operations are defined and applied in a definite manner. However, numbers are elements of a (number-)field onto which the operational domain is to be mapped regarding the respective values of the variables involved.

Normally, the common representation of a quantity A in physics or in the engineering sciences becomes

$$\{A\} := A \times [\mathcal{U}_A], \tag{2.54}$$

where the variable A as a dimensionless quantity is transferred into a dimensionalized property $\{A\}$ by multiplication with the proper unit $\mathcal{U}_A$ inherent to $\{A\}$.

Of course, mathematical physics is theoretically based on relations, using exclusively the set $\mathcal{G}$ of dimensionless quantities. This set $\mathcal{G}$ is nothing else but the collection of real functions $\mathcal{F}(A, B, C, \ldots)$ of any variables $A, B, C, \ldots$ in the sense of the algebraic meanings. They are generated by means of two types of binary operations, + and •, in other words, addition and multiplication.

It is apposite at this point to note that upon a set-theoretic base the aggregate $\mathcal{G}$ belongs to a class of sets where each member may be defined by a special aggregate $\mathcal{R}$ and two binary operations, + and •. Such a member was called a ring by Fraenkel in 1914, following Hilbert's use of the term *Zahlring* (number ring). We should use this notation as a mere abbreviation (like Wagner's "Ring"), particularly since its original connotation has considerably changed (Allenby, 1991, p. 83).

Currently, every ring $<\mathcal{R}^{\#} + \bullet>$ is defined by a list of axioms consisting of a finite number of rules to be established for the two binary operations, exclusively. There are many such triples $<\mathcal{R}^{\#} + \bullet>$; for any type the operations + and • on $\mathcal{R}^{\#}$ may satisfy none, some, or all of the axioms. Note that each concrete ring $\mathcal{R}$ so far encountered contains exactly one zero element.

In view of Falk's mathematical method, note that all axioms to be used are well-known in mathematical analysis: the commutative and associative laws of addition and multiplication; the two distributive laws, the additive and multiplicative laws of identity for neutral elements, and so on. The conclusive result derived from the axioms involved can be stated in compact form by *Falk's symbolic formula* (1990, p. 243)

$$Z[\mathcal{F}](A; B; \ldots)] = \mathcal{F}(Z[A]; Z[B]; \ldots) = \mathcal{F}(\alpha; \beta; \ldots). \tag{2.55}$$

As a consequence, this extension of (2.53) implies that relations between elements of $\mathcal{G}$ are not affected by mapping $\mathcal{G}$ on $\mathcal{W}$ via Z.

Let us now generate the real functions $\mathcal{F}$ of special variables $X_1, X_2, X_3, \ldots$ as elements of the set $\mathcal{G}$ defined to be independent and denoted as generic quantities. Regarding macroscopic phenomena, the variables $X_1, X_2, X_3, \ldots$ are supplemented by some parameters to be assigned to the set $\mathbb{R}$, that is, to the continuum of real numbers. In addition, the values of the generic quantities are presupposed to be real numbers. Consequently, this simply corresponds to a set representation: $\mathcal{W} = \mathbb{R}$.

Changing from these statements of pure mathematics to mathematical physics, the following point of view now appears to be self-evident: Due to the fact that every system holds for a certain finite number r of degrees of freedom introduced above, the set $\mathcal{G}$ is in every part bound to the collection $\mathcal{S}$ of physical states. These degrees of freedom are presupposed to be identical with the independent variables X_j and the index j running up to r. Selecting admissible but otherwise optional values for all X_j, the values of all other quantities of the set $\mathcal{G}$ are also completely determined. Thus, the corresponding state of the system in question is definitely fixed. It seems useful to comment on its physical meaning: "States are not purely mathematical entities, but rather physical objects in mechanics or the 'economic agent' is an idealization of economics, but this does not mean they are extra-empirical in the way numbers or vectors are. States are 'abstract' objects as compared with usual bodies, since the same state may be instantiated in two different bodies occupying different regions in space-time. However, that states are abstract objects in this sense does not mean they aren't physical" (Moulines, 1987, p. 65).

Now, it is easy to set apart two systems (1) and (2) in the following way:

- Their difference depends on two differing numbers of degrees of freedom.
- Assuming identical numbers of degrees of freedom as well as some variables X_k in (1) that do not occur in (2), the systems are distinct.
- When in both systems some values x_l of the corresponding quantities X_l do not coincide, then the systems will also differ, even if the systems are composed of the same independent variables X_j $(j = 1(1)r)$.

The inverse of this enumeration gives us a classification of systems: One domain $\mathcal{G}$ of physical quantities is assigned to one class of systems each, constituted by the very same independent quantities X_j $(j = 1(1)r)$.

Summing up the members of the collection $\mathcal{G}$, there are r generic quantities X_j, diverse functions $\mathcal{F}(X_1; X_2; X_3; \ldots)$, and the real numbers including the zero element. These members yield the complete information that can be obtained about a system defined in the way of the set $\mathcal{S}$. This statement is based on the following decisive suppositions.

(i) Gibbs' constructive proof that the complete information aforementioned is derived from a single fundamental relation between the values of $r + 1$ quantities of the system in question. This means that there are r degrees of freedom, but $r + 1$ elements $X_0; \ldots; X_r$ of the domain of physical quantities designated by $\mathcal{G}_{r+1}$.
(ii) The ring $\mathcal{R}[X_0; X_1; \ldots; X_r]$ of real polynomials as the "core" of the domain of physical quantities.

Before studying the consequences of these suppositions, let us recall precisely what polynomials are. By tradition, a polynomial in χ with, say, rational number coefficients is an expression of the form $a_0 + a_1\chi + a_2\chi^2 + \cdots + a_r\chi^r$ for some $r \in \mathbb{Z}$ and some $a_i \in \mathbb{Q}$, where, as usual, $\mathbb{Z}$ and $\mathbb{Q}$ denote the sets of integers and rational numbers, respectively. Notice that the symbols $\chi, \chi^2, \ldots, \chi^r$ are simply (dimensionless) place markers.

In set theory, a polynomial over a field of numbers, $\mathbb{Q}$ for example is any infinite sequence $(a_0; a_1; \ldots; a_k; \ldots; a_r; 0, 0, \ldots)$ of elements taken from this field. This sequence contains only a finite number of nonzero elements and is subject to the operational rules of addition and multiplication, so that according to this definition, the sum and product of two given polynomials are again polynomials. In this context polynomials may be thought of in the traditional way, namely, "secure in the knowledge that uncomfortable questions about χ can be circumvented" (Allenby, 1991, p. 40).

Therefore, we should regard the term polynomial of the set theory as a mere abbreviation. Only statements, similar to those concerning the algebraic rules of a ring noted above, have a clear and important meaning in physics. According to Falk (1990, pp. 250, 254), such statements can be expressed in a more compact form:

- Because polynomials do concern a special case of "rings," the basic axioms with respect to the two binary operations + (addition) and • (multiplication) are completely available. They are concisely discussed in connection with (2.55).
- As a first consequence, partial differentiation rules may be established by purely algebraic argumentation with respect to the generic quantities. Hence, any resort to the values of the quantities involved can be avoided in view of the strict conditions prevailing for numbers. Thus, improper mathematical terms such as *continuity* or *limits of the differential calculus* need not be considered.
- The homogeneous elements of the ring $\mathcal{R}[X_0; \ldots; X_r]$ are the homogeneous polynomials in the generic quantities $X_0; \ldots; X_r$. There is a number of exactly $(r + \omega)!/r!\omega!$ polynomials constituting a vector space of finite dimension of the same number. This multidimensional space is spanned over the field $\mathbb{R}$ of real numbers. Those polynomials are all of degree ω of homogeneity and linearly independent from each other.

To quantify the two latter statements, it should first be pointed out that the differentiation rules addressed above are all well-known from textbooks of conventional analysis and therefore need not be recapitulated here. However, it is notorious in the analysis that the partial derivatives can exhibit dependencies between functions. This is also true for $\mathcal{G}_{r+1}$, corresponding to the problem of how an ensemble of the generic quantities $X_0; X_1; \ldots; X_r$ could be transformed into an ensemble of the generic quantities $X^{\Diamond}{}_0; X^{\Diamond}{}_1; \ldots; X^{\Diamond}{}_r$. Once more, it was Falk (1990, p. 251) who dealt in some detail with this subject.

We should take particular interest in the *homogeneous* elements of $\mathcal{G}_{r+1}$ distinguished by the finite number $r + 1$ of generic quantities X_i. The term of this special element $\mathcal{I}_\omega = \mathcal{F}_{\text{hom}}(X_0; X_1; \ldots; X_r)$ may be introduced by way of the prominent *Euler relation*

$$\sum_{i=0}^{r} X_i \frac{\partial \mathcal{I}_\omega}{\partial X_i} = \omega\, \mathcal{I}_\omega , \tag{2.56}$$

where the integer ω stands for the degree of homogeneity concerning the element $\mathcal{I}_\omega$.

The differentiation of (2.56) with respect to any generic quantity X_i indicates that every derivative $\partial\mathcal{I}_\omega / \partial X_i$ is homogeneous of degree $\omega - 1$ as contrasted with the element $\mathcal{I}_\omega$ itself, which is of degree ω.

Continued differentiation procedures yield

$$\sum_{i=0}^{r} \sum_{k=0}^{r} X_i X_k \frac{\partial^2 \mathcal{I}_\omega}{\partial X_i \partial X_k} = \omega\,(\omega - 1), \ldots \tag{2.57}$$

in continuation of (2.56).

It is remarkable that the addition or multiplication of elements $\mathcal{I}_\omega$ will lead to new elements of $\mathcal{G}_{r+1}$, if the elements involved are of the degree $\omega = 0$. Clearly, such "zero degree" elements belong to a subring $\mathcal{N}_{r+1}$ of $\mathcal{G}_{r+1}$ in accordance with the definition that this subring $\mathcal{N}_{r+1}$ has no generic quantities; that is $\partial\mathcal{I}_{\omega=0}/\partial X_i = 0$.

To get something different from the general case manifested by (2.56) and (2.57) as well as the special one defined by $\omega = 0$, we can focus on the application $\omega = 1$, which is more relevant to physics and other sciences. Assuming the element $\mathcal{I}_\omega$ is homogeneous of degree ω with respect to the generic quantities $X_0; X_1; \ldots; X_r$ of $\mathcal{G}_{r+1}$, the same applies to the transformed generic quantities $X^{\Diamond}{}_0; X^{\Diamond}{}_1; \ldots; X^{\Diamond}{}_r$, again if the degree of homogeneity is equal to one.

This theorem, derived by Falk (1990, p. 256), leads to the notable conclusion that the linear transformations

$$X_i^{\Diamond} = \sum_{j=0}^{R} \alpha_{ij} X_j \quad \text{resp.} \quad X_i = \sum_{j=0}^{r} \alpha_{ij}^{\Diamond} X_j^{\Diamond}; \quad i = 0(1)r \tag{2.57a}$$

are the only homogeneous transformations of degree 1 within the frame of the polynomial ring $\mathcal{R}[X_0; X_1; \ldots; X_r]$. The α_{ij} and $\alpha^{\Diamond}{}_{ij}$ are real numbers to be calculated by the so-called contragradient derivatives $(\partial X^{\Diamond}{}_i/\partial X_j)$ and $(\partial X_i/\partial X^{\Diamond}{}_j)$, respectively.

Let us now discuss the central parts of Falk' dynamics, which are based on the statement that every set $\mathcal{S}$ of states representing the system in question is defined with respect to special elements of $\mathcal{G}_{r+1}$. According to Falk's symbolic formula (2.55), the generic quantities $X_0, \ldots, X_r$ of $\mathcal{G}_{r+1}$ in the states Z have the respective values $Z[X_0] = \xi_0, \ldots, Z[X_r] = \xi_r$. But now, these values are assumed to be mutually dependent. Between these real numbers there exists a characteristic relation, the so-called *Gibbsian fundamental equation*

$$F(Z[X_0]; Z[X_1]; \ldots; Z[X_r]) \equiv 0 \qquad \text{for all } Z \in \mathcal{S}. \tag{2.58}$$

This means that each mapping Z of $\mathcal{G}_{r+1}$ onto $\mathbb{R}$ fulfilling (2.58) will have to belong to the system $\mathcal{S}$ and, equivalently Equation (2.58) holds for each $Z \in \mathcal{S}$. Considering (2.55), the identity (2.58) then becomes

$$F(Z[X_0]; Z[X_1]; \ldots; Z[X_r]) = Z[F(X_0; X_1; \ldots; X_r)] \equiv 0 \qquad \text{for all } Z \in \mathcal{S}. \tag{2.59}$$

Evidently, the system in question—the set $\mathcal{S}$—consists of exactly those mappings Z of $\mathcal{G}_{r+1}$ onto $\mathbb{R}$ for which a definite element $F(X_0; X_1; \ldots; X_r)$ of $\mathcal{G}_{r+1}$ changes into the *zero element* of the set of polynomials. On that account, each system is plainly established by the element $F := F(X_0; X_1; \ldots; X_r)$ of $\mathcal{G}_{r+1}$, called the *M–G element* of the system by Falk (1990, p. 258).

Compared with the term M–G function concerning (2.30) and (2.31), an M–G element has a formally additional variable. If an M–G element can be resolved with respect to the generic quantity X_0, then the result $X_0 = X_0(F; X_1; \ldots; X_r)$ apparently indicates a characteristic difference in comparison with the M–G function $X_0 = X_0(X_1; \ldots; X_r)$. But, indeed, the element F does not work as a variable because it is mapped onto the zero element via (2.59).

Let us deal with an example often used for practical problems. Suppose that the M–G element in question has the form

$$F(X_0; X_1; \ldots; X_r) := X_0 - G(X_1; \ldots; X_r), \tag{2.60}$$

including the trivial resolution

$$X_0(F; X_1; \ldots; X_r) = F + G(X_1; \ldots; X_r). \tag{2.61}$$

Furthermore, it is assumed that a one-to-one transformation of the generic quantities $X_i \Rightarrow X^{\Diamond}_i$ (and vice versa) for all r can be performed in the following way:

$$X^{\Diamond}_1 = \frac{\partial F}{\partial X_1} = \frac{\partial G}{\partial X_1}, \ldots, X^{\Diamond}_m = \frac{\partial F}{\partial X_m} = -\frac{\partial G}{\partial X_m},$$

$$X^{\Diamond}_{m+1} = X_{m+1}, \ldots, X^{\Diamond}_r = X_r \tag{2.62}$$

To complete the transformation it is necessary to express the dependent quantity $X^{\Diamond}{}_0$ as a function of the generic quantities X_0; X_1; …; X_r. To do so, the only constraint is that $X^{\Diamond}{}_0$, along with the r variables $X^{\Diamond}{}_1$, …, $X^{\Diamond}{}_r$, has to make up a complete ensemble of generic quantities with respect to $\mathcal{G}_{r+1}$. In practice, the following definition is sometimes useful

$$X_0^{\Diamond} := X_0 + \sum_{j=1}^{m} X_j \frac{\partial F}{\partial X_j} \equiv X_0 - \sum_{j=1}^{m} X_j \frac{\partial G}{\partial X_j}, \tag{2.63}$$

for which the identity can obviously be verified:

$$\begin{aligned} F &= X_0 - G(X_1, X_2, \ldots, X_r) \\ &= \left[X_0 - \sum_{j=1}^{m} X_j \frac{\partial G}{\partial X_j}\right] - \left[G - \sum_{j=1}^{m} X_j \frac{\partial G}{\partial X_j}\right] \\ &= X^{\Diamond}{}_0 - G^{\Diamond}(X^{\Diamond}{}_1; X^{\Diamond}{}_2; \ldots; X^{\Diamond}{}_r). \end{aligned}$$

This means that the M–G element F is of the same simple form as (2.60), not only with respect to the original generic quantities X_0; X_1; …; X_r, but also in the transformed generic quantities $X^{\Diamond}{}_0$; $X^{\Diamond}{}_1$; …; $X^{\Diamond}{}_r$. Consequently, the M–G function (2.61) becomes

$$X_0(F; X^{\Diamond}{}_1; \ldots; X^{\Diamond}{}_r) = F + G^{\Diamond}(X^{\Diamond}{}_1; \ldots; X^{\Diamond}{}_r), \tag{2.64}$$

where the new element

$$G^{\Diamond} := G^{[m]} = G - \sum_{j=1}^{m} X_j \frac{\partial G}{\partial X_j} \tag{2.65}$$

follows from the m-fold Legendre transformation of the original element defined by (2.60).

This important result implies that by way of coordinating the M–G element F and the transformation rules (2.62) and (2.63), the first m generic quantities X_1; …; X_m are transformed uno actu into the derivatives $\partial F/\partial X_1$; …; $\partial F/\partial X_m$, whereas the remaining generics X_{m+1}; …; X_r and the respective derivatives of F continue to be invariant toward the contact transformation. This is undoubtedly an important method to operate effectively with different generic quantities in practice.

To conclude this section, let us discuss a highly relevant inference of Falk's mathematical description outlined here. With the basic ideas stated, the relations between the notions *system* and *universally physical* quantity lead us to the question of whether there exist elements that may be selected arbitrarily from the set $\mathcal{G}_{r+1}$. Notice that it was again Gibbs who gave the correct answer:

Every system $\mathcal{S}$ with the finite number r of degrees of freedom and assigned to $\mathcal{G}_{r+1}$ can be established by an M–G element $F(X_0; X_1; \ldots; X_r)$, homogeneous of degree 1 with respect to the generics $X_0; X_1; \ldots; X_r$, for which Rankine introduced the attribute "extensive."

This *Gibbsian theorem* is mathematically founded on (2.60), along with the definitions (2.62) and (2.63). Hence, we can easily see from the trivial resolution (2.61) that the element $G(X_1; \ldots; X_r)$ is a homogeneous function of degree 1. According to (2.56), G becomes

$$G = \sum_{j=1}^{r} X_j \frac{\partial G(X_1;\ldots;X_j;\ldots;X_r)}{\partial X_j} \tag{2.66}$$

and is an M–G function, the variables of which are exclusively extensive. There are three aspects to this important result:

1. With respect to the same set of generics, the M–G elements of all systems concerned are homogeneous.
2. Different systems must have different dependencies between their elements F and the generics.
3. Each generic quantity X_j represents the same meaning for all systems involved.

For the further development of the theory presented above, it should be emphasized that there are in principle other sets of generics besides $X_0; X_1; \ldots; X_r$ that are also homogeneous with respect to the M–G element F of $\mathcal{G}_{r+1}$. Due to the linear-homogeneous transformation rules (2.57), all sets of generics connected with the original generic quantities $X_0; X_1; \ldots; X_r$ (or among each other) consist of extensive elements only.

Of course, the set $\mathcal{G}_{r+1}$ is likewise equipped with additional sets of generics and also contains nonextensive quantities defining nonhomogeneous M–G elements. For convenience, let us introduce the abbreviations

$$Y_j := \frac{\partial X_0(F;X_1;\ldots;X_r)}{\partial X_j} = \frac{\partial G(X_1;\ldots;X_r)}{\partial X_j}, \qquad j = 1(1)r \tag{2.67}$$

These are to be applied to the derivatives with regard to X_0 and $G(X_1; \ldots; X_r)$, thereby referring to (2.61) and (2.66), respectively. Following the traditional usage, the notation *intensive quantity* should be applied.

Certainly, it is interesting to note that the intensive elements defined by (2.67) are not all independent. The existence of a relationship among the various intensive quantities is a consequence of the homogeneous first-order property of the related M–G element F of $\mathcal{G}_{r+1}$ given by

$$F = \sum_{j=0}^{r} X_j \frac{\partial F(X_0;X_1;\ldots;X_r)}{\partial X_j} \tag{2.68}$$

that yields, via (2.60),

$$F = X_0 - G(X_1;\ldots;X_r) = X_0 - \sum_{j=1}^{r} X_j \frac{\partial G(X_1;\ldots;X_r)}{\partial X_j} = X_0 - \sum_{j=1}^{r} X_j Y_j . \qquad (2.69)$$

We recognize from (2.57) that, according to Euler's theorem of homogeneity, each derivative $\partial F/\partial X_j$ of the first-order function F will have to be homogeneous of degree 0. Consequently, the so-called *Gibbs–Duhem–Margules relation* (Kestin, 1979, p. 232) arises:

$$\sum_{j=1}^{r} X_j \frac{\partial Y_k}{\partial X_j} \equiv 0 ; \qquad k = 1(1)r \qquad (2.70)$$

The meaning of this relation will be illustrated by means of a few concrete examples in the book.

2.5 Mathematical Foundation of Falk's Dynamics II: Systems

Let us now consider the case when M–G functions do not work in the domain of physical quantities $\mathcal{G}_{r+1}$, but in their range $\mathcal{W}$. It is not stringent that the basic set-theoretic axioms, by which polynomial rings may be introduced, are still valid if the mappings are performed from $\mathcal{G}_{r+1}$ onto $\mathcal{W}$. This is particularly true for the problem of connecting some $\mathcal{G}_{r+1}$ elements by processes of differentiation founded exclusively on the algebraic rules mentioned above. Although these rules guarantee that the mappings normally satisfy all conditions of continuity in $\mathcal{W}$, one marked exception should be observed: Making up a system $\mathcal{S}$ by all mappings Z from $\mathcal{G}_{r+1}$onto $\mathcal{W} = \mathbb{R}$, that is, onto the set of real numbers, a special mapping is thereby included according to which the M–G element $F(X_0; X_1; \ldots; X_r)$ of the system in question is transferred into the zero element. Equation (2.59), which governs this transfer in all cases, can be applied directly:

$$Z[F(X_0; X_1; \ldots; X_r) = F(\xi_0; \xi_1; \ldots; \xi_r) \equiv 0 \qquad \text{for all } Z \in \mathcal{S}. \qquad (2.71)$$

The equation $F(\xi_0; \ldots; \xi_r) \equiv 0$ is of course nothing more than Gibbs' fundamental equation of the system $\mathcal{S}$ describing the relation between the conjugate values ξ_j ($j = 1(1)r$) and their independent generics $X_0; X_1; \ldots; X_r$.

Strictly speaking, not all $r + 1$ values $\xi_0; \xi_1; \ldots; \xi_r$ are therefore allowed to be independent variables in the sense of mathematical analysis. Yet, the function F may be handled as an implicit function in view of these very variables. In other words:

> With respect to the differential calculus, the $\xi_0; \xi_1; \ldots; \xi_r$ can be used simply as a set of independent variables (as they are merely *position* markers), but with respect to the values of the underlying generics, Gibbs' fundamental equation is simply a degenerate function, in other words, the constant zero.

In a next step and as a well-founded approach, the M–G element $F = F(X_0; X_1; \ldots; X_r)$ is assumed to be resolvable with respect to X_0. Then, the system $\mathcal{S}$ need not be described directly by means of $F(X_0; X_1; \ldots; X_r)$, but can also be represented by the

M–G function $X_0 = X_0(F; X_1; \ldots; X_r)$ considering (2.61). Following the mappings (2.71), this M–G function turns into the relation

$$\xi_0 = \hat{X}_0(0; \xi_1; \ldots; \xi_r) = G(\xi_1; \ldots; \xi_r) \tag{2.72}$$

between the values $\xi_0 = Z[X_0]$ and $\xi_1 = Z[X_1], \ldots, \xi_r = Z[X_r]$.

Consequently, the intensive elements (2.59) can be mapped easily from $\mathcal{G}_{r+1}$ onto $\mathbb{R}$ by

$$Z[Y_i] := \tau_i = \frac{\partial G(\xi_1; \ldots; \xi_r)}{\partial \xi_i}; \qquad i = 1(1)r. \tag{2.73}$$

This form is in accord with common use, especially if one also regards the limits of real number series with reference to the differential calculus.

Regarding the essential inequality between the concepts of differentiation in algebra and analysis, a comment to (2.73) may be permitted because each mapping Z is a member of a family of mappings by which any generic quantity X_i of the set $\mathcal{G}_{r+1}$ is mapped onto a continuous interval of values ξ_i, whereas every other quantity X_k of the generic collection $\{F, X_1; \ldots; X_r\}$ turns into a fixed number ξ_k. For this reason, in mathematical physics the continuum of real numbers is needed, and thus, following Falk (1990, p. 275), Gibbs' well-known condition of continuity is well-founded indeed.

Consider now the first-order homogeneity of the M–G element $F(X_0; X_1; \ldots; X_r)$. Equation (2.72) immediately admits two mathematical conclusions from this property:

$$d\xi_0 = \sum_{j=1}^{r} \tau_j d\xi_j; \qquad \xi_0 = \sum_{j=1}^{r} \tau_j \xi_j \tag{2.74}$$

Clearly, both equations are basic to the general formalism originated by Gibbs and concluded by Falk. At this point we should see that the Gibbs–Falkian theory is not restricted in its applications to thermodynamics or physics but could also be used in fields of science, such as mathematical economics and the so-called general system theory.

In this context Mirowski's knowledgeable book should be mentioned. The author expounds the many and diverse efforts by economists to connect physical notions and relations with basic terms in early neoclassical economic theory. Irving Fisher's outstanding 1892 doctoral thesis yielded an amazing comparison of the relevant analogies (cf. Mirowski, 1990, p. 224). Mirowski attributes such approaches to the permanent influence of classical mechanics on leading scientists in economics. Above all, he emphasizes the influence of *energetics* mainly propagated by Helm's book (1898), which was considerably influenced by Gibbsian thermostatics. However, it is curious that energetics as an allegedly social theory (cf. Helm, 1898, p. 207) was multiform and widespread at the turn of the nineteenth century, and yet went into eclipse rapidly in the 1930s and 1940s; whereas neoclassicism has persisted down to the present day. Two reasons may be noted: "First, neoclassical economics (unintentionally?) managed to segregate itself from the larger program of energetics, to the extent of having no apparent association with it; second, events in physics undermined the pretensions of an unabashed energetics to be based on accepted and credible physical theory, thus rendering pathetic its supposed scientific advantage over other theories" (Mirowski, 1990, p.269). From this point of view the Gibbs–Falkian approach presented in this chapter offers, above all, new opportunities of investigating Fisher's concordance list for basic terms in physics and economics with the help of much more efficient tools.

By converting (2.65) with reference to (2.71), the m-fold Legendre transformation

$$\xi^{\Diamond}{}_0 = \xi_0{}^{[m]} = \xi_0 - \sum_{j=1}^{m} \xi_j \frac{\partial \xi_0}{\partial \xi_j} \tag{2.75}$$

is obtained for application to the level of states, that is to the system in question itself.

We can appreciate a further important equation with respect to both expressions in (2.74), which are mutually consistent only under the condition

$$\sum_{j=1}^{r} \xi_j d\tau_j \equiv 0 \,. \tag{2.76}$$

This identity refers to the Gibbs–Duhem–Margules relation (2.70), which states that the r intensive parameters τ_j are dependent on themselves. Consequently, the relation

$$h\,(\tau_1;\tau_2;\ldots;\tau_r) \equiv 0 \tag{2.77}$$

becomes inevitable and typical for the state ensemble $\mathcal{S}$ in the same way as the fundamental equation (2.71) or the M–G function (2.72). In other words, assuming that a system is in fact defined by its first-order homogeneous M–G function, this system is also subject to its respective equation (2.77). Therefore, Falk (1990, p. 277) suggested calling this identity the *internal fundamental equation* of the system in question.

Let us deal first with an example that is related to the thermodynamic system used as a standard case in the preceding sections. Consider the M–G function $U = U(S; V; n)$ resulting from Gibbs' fundamental equation $\Gamma(U; S; V; n) \equiv 0$. In preparation of the calculations and considering (2.54), the dimensionless variables ξ_j are multiplied by their adequate international units as follows:

$$U := \xi_0 \times [J]; \quad S := \xi_1 \times [J\,K^{-1}]; \quad V := \xi_2 \times [m^3]; \quad n := \xi_3 \times [mole];$$
$$T_* := \tau_1 \times [\mathrm{K}]; \quad p_* := \tau_2 \times [\mathrm{Pa}]. \tag{2.78}$$

The following sequence of implications comprehensively explains the complete interdependence among both the mathematical and the physical relations:

$$F\,(\xi_0;\xi_1;\ldots;\xi_r) \equiv 0 \Rightarrow \Gamma\,(U;S;V;n) \equiv 0 \Rightarrow U = \xi_0 \Rightarrow \xi_0 = G\,(\xi_1;\ldots;\xi_r)$$
$$\Rightarrow \xi_0{}^{\Diamond} = \xi_0{}^{[V]} = U - V\,(\partial U/\partial V)_{S;\,n} = U - V(-p_*) = U + p_* V := H_* \tag{2.79}$$

The result is well-known: By a Legendre transformation, according to (2.75) with respect to the volume V, the internal energy U is transformed into the enthalpy H_* now being a function of S; p_*; and n, as we may easily verify:

$$dU = (\partial U/\partial S)_{V;\,n}\, dS + (\partial U/\partial V)_{S;\,n}\, dV + (\partial U/\partial n)_{S;\,V}\, dn$$

$$= T_*\, dS - p_*\, dV + \hat{\mu}_*\, dn \equiv dH_* - p_* dV - V\, dp_*$$

$$\Rightarrow dH_* = T_*\, dS + V\, dp_* + \hat{\mu}_*\, dn \quad \Rightarrow \quad H_* = H_*(S;\, p_*;\, n). \tag{2.80}$$

The definitions used for the thermodynamic temperature T_*, the pressure p_*, and the chemical potential $\hat{\mu}_*$ are all given by (2.26). The function signified by (2.80) elucidates the various steps to be taken from the first-order homogeneous M–G function $U(S;\, V;\, n)$ to the nonhomogeneous M–G function $H_*(S;\, p_*;\, n)$.

A second example refers to the isotropic black body radiation for which photons or phonons, respectively, are in thermal and chemical equilibrium. Its internal fundamental equation is well-known (Falk, 1990, p. 281) and may be written using the common notation of this book:

$$h(\tau_1;\, \tau_2) := a \bullet \tau_1^{\,4} + 3 \bullet \tau_2 \equiv 0. \tag{2.81}$$

The number a is assumed to be a natural constant of the radiation system.

It is by Legendre transformation according to (2.75) and with respect to the extensive variable ξ_1 that the simple relations

$$\xi_0^{[1]} = \xi_0 - \xi_1 \bullet \tau_1 = \xi_2 \bullet \tau_2 = \xi_2 \bullet (-3^{-1} a \bullet \tau_1^{\,4})$$

are obtained using the homogeneity relation (2.74) and the identity (2.81). Thus, ξ_1 can easily be calculated by the derivative

$$\xi_1 = -\frac{\partial \xi_0^{[1]}(\tau_1;\xi_2)}{\partial \tau_1} = \frac{4}{3} a\, \xi_2 \tau_1^{\,3},$$

by which the dependent extensive variable ξ_0 can be expressed in the variables τ_1 and ξ_2:

$$\xi_0 = \tau_1\, \xi_1 + \tau_2\, \xi_2 = a \bullet \xi_2 \bullet \tau_1^{\,4}.$$

By elimination of τ_1, the M–G function (2.72) as well as the fundamental equation (2.71) of the radiation system can be determined in the following form

$$\xi_0 = G(\xi_1;\xi_2) = \frac{3}{4}\left(\frac{3}{4a}\right)^{\frac{1}{3}} \xi_1^{\frac{4}{3}}\, \xi_2^{-\frac{1}{3}}; \qquad F(\xi_0;\xi_1;\xi_3) = \left(\frac{\xi_0}{\xi_2}\right)^3 - \left(\frac{3}{4}\right)^3 \frac{3}{4a}\left(\frac{\xi_1}{\xi_2}\right)^4 \equiv 0.$$

Using the dimensionalized quantities according to (2.78), it is a simple procedure to get the well-known equations for the pressure p_*, entropy S, and the internal energy U of the radiating gas:

$$p_* = \frac{b}{3}T_*^4; \qquad S = \frac{4}{3}bV\,T_*^3; \qquad U = bV\,T_*^4.$$

The parameter b, common to all three terms, is connected with the dimensionless constant a by the unit relation $b = a\,[\mathrm{Pa} \times \mathrm{K}^{-4}] = a[J \times \mathrm{m}^{-3} \times \mathrm{K}^{-4}]$.

It is striking that this example may be compared immediately with a mathematically isomorphic but physically different radiation system. It is defined by the same number r of degrees of freedom and, furthermore, even by the same M–G function in $\mathcal{G}_3$. The nontrivial and decisive difference lies hidden in the fact that now the system has entirely different physical quantities, namely U, V, n, p_*, and $\hat{\mu}_*$, as compared with the dimensionalized quantities U, V, S, p_*, and T_* first introduced in (2.78). For further details, refer to Falk (1990, p. 283).

To carry on the general theory, it is appropriate to operate within the range of values assigned to the set $\mathcal{G}$. Notably, the homogeneity of an M–G element $F(X_0; \ldots; X_r)$ as well as its image function $F(\xi_0; \ldots; \xi r)$ always refers to the complete ensemble of the $r + 1$ variables. If, for instance, the value ξ_k is assumed to be fixed, then the element $F(\xi_0; \ldots; \xi_{k-1}; \xi_{k+1}; \ldots; \xi_r)$ only remains homogeneous under the special condition that $\xi_k \equiv 0$ holds. Hence, the basic relation of homogeneity is specified toward

$$F(\lambda\xi_0; \ldots; \lambda\xi_{k-1}; 0; \lambda\xi_{k+1}; \ldots; \lambda\xi_r) = \lambda^{\omega}F(\xi_0; \ldots; \xi_{k-1}; 0; \xi_{k+1}; \ldots; \xi_r), \tag{2.82}$$

which is independent of the degree of homogeneity ω and valid for each real number $\lambda \neq 0$. This simple statement leads us to an important conclusion concerning an exact procedure to establish subsets of $\mathcal{S}$ by which physical systems may be concretely represented. It results from the fact that $F(\xi_0; \ldots; \xi_{k-1}; 0; \xi_{k+1}; \ldots; \xi_r)$ for $\xi_k = 0$ represents once more a fundamental equation of a system. For brevity's sake such a system may be called a *reduced system* $\mathcal{S}'$.

Concerning each system reduction, it is noteworthy that this can be performed not only by assuming any single variable, but also by every admitted linear combination of the variables ξ_j or τ_j that is identically zero. For example the transformations

$$\begin{aligned} \xi_1' &:= 2^{-1/2}(\xi_1 + \xi_2); & \xi_2' &:= 2^{-1/2}(\xi_1 - \xi_2); & \xi_i' &:= \xi_i, & i &= 3(1)r \\ \tau_1' &:= 2^{-1/2}(\tau_1 + \tau_2); & \tau_2' &:= 2^{-1/2}(\tau_1 - \tau_2); & \tau_j' &:= \tau_j, & j &= 3(1)r \end{aligned} \tag{2.83}$$

demonstrate the equivalence of the case $\xi_2' \equiv 0$ with the relation $\xi_1 = \xi_2$ as well as for the case $\tau_2' \equiv 0$ with the relation $\tau_1 = \tau_2$. This flexibility within the system reduction is of great importance for physical applications.

The need for admissibility criteria also arises in the framework of the cardinal connections (2.74). Assuming τ_j^0 to be constant, then from the first relation

$$dG' = d(G - \tau_j^0\xi_j) = \sum_{l=1}^{r} (\tau_l - \tau_j^0\delta_{lj})\,d\xi_l = \sum_{l=1}^{r} \tau_l' d\xi_l \tag{2.84}$$

follows, to be completed by the second relation

$$G' = G - \tau_j^0\xi_j = \sum_{l=1}^{r} (\tau_l - \tau_j^0\delta_{jl})d\xi_l = \sum_{l=1}^{r} \tau_l'\xi_l \tag{2.85}$$

Correspondingly, an intensive variable τ_j presupposed to be a constant τ_j^0 immediately equals the transformed variable τ_j' assumed to be zero. This is coincidentally accompanied by the substitution of the M–G function G for the function G'. Thus, both the simplest reductions $\xi_j = 0$ or τ_j = constant mean that the pair of variables ξ_j; τ_j does not appear in the reduced system.

In general, the special case $\tau_j' = 0$ is of great interest in the definition of equilibrium events to be characterized by the basic concepts in physics and, above all, in thermodynamics. This is true for well-known examples for any force equilibria in mechanics or for thermal and chemical equilibria. Conversely, however, it cannot be readily inferred that any pertaining intensive variables observed to be zero should be related to any realizable equilibrium phenomenon.

Mathematically, it is easy to prove that the extremum property of equilibria can be guaranteed by the homogeneity of the M–G function. The intensive variable, say τ_j, presupposed to be zero, is given via (2.74) by the partial derivative of the M–G function $\xi_0 = G$ with respect to the corresponding extensive variable ξ_j. From the next equation we obtain with

$$\frac{\partial}{\partial \xi_j}\xi_0 = \tau_j(\xi_1;\ldots;\xi_r) \equiv 0$$

the condition that must be satisfied in connection with the Legendre-transformed M–G function (2.75) of the system reduced by equilibrium. Unlike $\xi_0(\xi_1; \ldots; \xi_{j-1}; \xi_j; \xi_{j+1}; \ldots; \xi_r)$, this transformed function $\xi_0^{\lozenge} = \xi_0 - \tau_j\xi_j$ depends on the variables $\xi_1; \ldots; \xi_{j-1}; \tau_j; \xi_{j+1}; \ldots; \xi_r$, wherein τ_j is additionally to be replaced by zero. Yet, to resolve this identity with respect to the ξ_j singled out, the derivative $\partial\tau_j/\partial\xi_j$ should not vanish in the neighborhood $\tau_j = 0$. Consequently, there are two classes of representative M–G functions: the maximum class and the minimum class defined by the respective sign of the derivative $\partial\tau_j/\partial\xi_j$ each.

Falk (1990, p. 295) generalized the simple case of one equilibrium to a complete system of r degrees of freedom subject to the reduction of m simultaneously running equilibria. His analysis leads to the following stability conditions to be satisfied by the M–G function of the system in question along with its m-fold Legendre transformations according to (2.75). Given

$$\Xi_{\xi}^{[m-1]} := \left(\frac{\partial^2 G}{\partial \xi_1^{\,2}}\right)_{\tau_1 = 0} \bullet \left(\frac{\partial^2 G^{[1]}}{\partial \xi_2^{\,2}}\right)_{\tau_2 = 0} \bullet \cdots \bullet \left(\frac{\partial^2 G^{[m-1]}}{\partial \xi_m^{\,2}}\right)_{\tau_m = 0}$$

we obtain

$$\Xi_{\xi}^{[m-1]} \begin{cases} >0 & \text{if } G \text{ belongs to the } \quad \textit{minimum} \text{ class (energy class) and } m \text{ is odd} \\ <0 & \qquad\qquad\qquad\qquad\qquad \textit{maximum} \text{ class (entropy class)} \\ >0 & \qquad\qquad\qquad\qquad\qquad \text{if } m \text{ is even} \end{cases} \tag{2.86}$$

To ensure that any system reduction leads once more to new systems in the sense of thermodynamics, those systems must be highly stable with regard to any possible variations in the meaning of Gibbs (1961, p. 57). Such a behavior of stability near optional equilibrium states is significant concerning common classifications of substances in a variety of non-equilibrium states. This is particularly true in view of some characteristic properties of fluids that in principle may be assumed to be insensitive to changes of state. Well-known examples are the heat capacities, the isentropic and the isothermal compressibility of gases, liquids, solids, and their mixtures as well.

Looking at some more mechanical aspects, there arises some demand for the basic properties of matter. Assuming P_i to be one component of the vectorial (linear) momentum **P**, then the stability condition

$$\frac{\partial^2 E}{\partial P_i^{\;2}} = \frac{\partial v_i}{\partial P_i} > 0 \tag{2.87}$$

directly requires that changes in momentum **P** and velocity **v** of the system in question are always equidirectional. Consequently, only positive values of the mass m are admitted in mechanics of mass-point, here defined by the energy E of one mass-point $E := \mathbf{P}^2/2m + E_0$. For the same reason, (2.87) enforces positive energy values, defined by $E^2 := c^2\mathbf{P}^2 + E_0^{\;2}$ in Einstein's mechanics. Both cases will play an important role in the following sections.

There is an option for which the reduction of a system opens some aspects of fundamental interest in mathematical physics. Bear in mind that a system is characterized by a finite number r of degrees of freedom. Therefore, any decrease in the number of variables can only be executed by the system reduction defined above, not by any continuous decrease in the sense of mathematical continuity. This is also the key to understanding the background and solution of Gibbs' famous paradox to be discussed later in this section.

The option to reduce any given system in principle is based on the pertaining property inherent in the M–G function prescribed. However, it is an exceptional feature of the $\mathcal{G}_{r+1}$ structure to enable the decomposition of a system into subsystems or to assemble some systems to a new, more complex system.

Falk (1990, p. 312) proved that both these options can be brought into a simple inverted ratio. To begin with, three systems I, II, and III are defined by their respective M–G functions

$$\begin{aligned}
\xi^{\mathrm{I}}{}_0 &= G^{\mathrm{I}}{}_\sigma(\tau^{\mathrm{I}}{}_1; \ldots; \tau^{\mathrm{I}}{}_\sigma; \xi_{\sigma+1}; 0_j\, \xi^{\mathrm{I}}{}_{\sigma+3}; \ldots; \xi^{\mathrm{I}}{}_r),\\
\xi^{\mathrm{II}}{}_0 &= G^{\mathrm{II}}{}_\sigma(\tau^{\mathrm{II}}{}_1; \ldots; \tau^{\mathrm{II}}{}_\sigma; 0_j\, \xi_{\sigma+2}; \xi^{\mathrm{II}}{}_{\sigma+3}; \ldots; \xi^{\mathrm{II}}{}_r),\\
\xi^{\mathrm{III}}{}_0 &= G^{\mathrm{III}}{}_\sigma(\tau^{\mathrm{III}}{}_1; \ldots; \tau^{\mathrm{III}}{}_\sigma; 0_j\, \xi^{\mathrm{III}}{}_{\sigma+3}; \ldots; \xi^{\mathrm{III}}{}_r),
\end{aligned} \tag{2.88}$$

where the subscript σ indicates that these functions depend on intensive variables as well as on extensive ones. Each function is distinguished by one characteristic variable that is extraneous with respect to the other two functions. Then, the total system is assumed to be established by the (unreduced) M–G system

$$G_\sigma = G_\sigma^{\mathrm{I}}(\tau_1^{\mathrm{I}}; \ldots; \tau_\sigma^{\mathrm{I}}; \xi_{\sigma+1}; \xi^{\mathrm{I}}_{\sigma+3}; \ldots; \xi^{\mathrm{I}}_r) + G^{\mathrm{II}}_\sigma(\tau^{\mathrm{II}}_1; \ldots; \tau_\sigma^{\mathrm{II}}; \xi_{\sigma+2}; \xi^{\mathrm{II}}_{\sigma+3}; \ldots; \xi^{\mathrm{II}}_r)$$
$$+ G_\sigma^{\mathrm{III}}(\tau_1^{\mathrm{III}}; \ldots; \tau_\sigma^{\mathrm{III}}; \xi^{\mathrm{III}}_{\sigma+3}; \ldots; \xi^{\mathrm{III}}_r), \tag{2.89}$$

where now a set of $3r - 4$ degrees of freedom holds.

Now equation (2.89) will be reduced in two steps: The first one takes place by means of reduction via *intensive* variables,

$$\tau_1^{\mathrm{I}} = \tau_1^{\mathrm{II}} = \tau_1^{\mathrm{III}} := \tau_1; \ldots; \tau_\sigma^{\mathrm{I}} = \tau_\sigma^{\mathrm{II}} = \tau_\sigma^{\mathrm{III}} := \tau_\sigma, \tag{2.90}$$

the second by means of *extensive* variables

$$\xi^{\mathrm{I}}_{\sigma+3} = \xi^{\mathrm{II}}_{\sigma+3} = \xi^{\mathrm{III}}_{\sigma+3} := \xi_{\sigma+3}; \ldots; \xi^{\mathrm{I}}_r = \xi^{\mathrm{II}}_r = \xi^{\mathrm{III}}_r = \xi_r. \tag{2.91}$$

Equation (2.90) thus represents equilibria in fair accordance with (2.84) and (2.85).

Combining (2.90) and (2.91) with the M–G function (2.89), a type of system synthesis arises that may be accurately regarded as the counterpart of the decomposition.

This can be easily demonstrated by using the Legendre-transformed M–G function G, this time exclusively depending on extensive variables:

$$G = \tau_1^{\mathrm{I}}\xi_1^{\mathrm{I}} + \tau_1^{\mathrm{II}}\xi_1^{\mathrm{II}} + \tau_1^{\mathrm{III}}\xi_1^{\mathrm{III}} + \cdots + \tau_\sigma^{\mathrm{I}}\xi_\sigma^{\mathrm{I}} + \tau_\sigma^{\mathrm{II}}\xi_\sigma^{\mathrm{II}} + \tau_\sigma^{\mathrm{III}}\xi_\sigma^{\mathrm{III}}$$
$$+ \tau_{\sigma+3}^{\mathrm{I}}\xi_{\sigma+3}^{\mathrm{I}} + \tau_{\sigma+3}^{\mathrm{II}}\xi_{\sigma+3}^{\mathrm{II}} + \tau_{\sigma+3}^{\mathrm{III}}\xi_{\sigma+3}^{\mathrm{III}} + \cdots + \tau_r^{\mathrm{I}}\xi_r^{\mathrm{I}} + \tau_r^{\mathrm{II}}\xi_r^{\mathrm{II}} + \tau_r^{\mathrm{III}}\xi_r^{\mathrm{III}} \tag{2.92}$$
$$+ \tau_{\sigma+1}^{\mathrm{I}}\xi_{\sigma+1}^{\mathrm{I}} + \tau_{\sigma+2}^{\mathrm{II}}\xi_{\sigma+2}^{\mathrm{II}}.$$

Using (2.90) and (2.91) the M–G function (2.92) turns into the following set of relations

$$\begin{aligned} &\xi_0 = \xi_0^{\mathrm{I}} + \xi_0^{\mathrm{II}} + \xi_0^{\mathrm{III}}; \\ &\xi_j = \xi_j^{\mathrm{I}} + \xi_j^{\mathrm{II}} + \xi_j^{\mathrm{III}}; \qquad j = 1(1)\sigma; \\ &\tau_k = \tau_k^{\mathrm{I}} + \tau_k^{\mathrm{II}} + \tau_k^{\mathrm{III}}; \qquad k = \sigma + 1(1)r, \end{aligned} \tag{2.93}$$

where the last relationship means for $k = \sigma + 1$: $\tau_{\sigma+1} = \tau_{\sigma+1}^{\mathrm{I}}$ and $\tau_{\sigma+1}^{\mathrm{II}} = \tau_{\sigma+1}^{\mathrm{III}} \equiv 0$, and so on. The conditions (2.93) stress that under the prevailing conditions, the various variables have to be combined in a unique way to verify the approach (2.89). Thus, it is evident that the decomposition of (2.89) can be executed if the restrictions exemplarily written down

$$\frac{\partial^2 G_\sigma}{\partial\xi_{\sigma+1}\partial\xi_{\sigma+2}} = \frac{\partial\tau_{\sigma+1}}{\partial\xi_{\sigma+2}} = \frac{\partial\tau_{\sigma+2}}{\partial\xi_{\sigma+1}} \equiv 0 \tag{2.94}$$

can be satisfied. In this case, the three M–G functions I, II, and III according to (2.88) will result. Extensions and variations of the items dealt with in (2.94) may be easily deduced.

The general theory of decomposition discussed above can be specialized to a case to serve as the prevalent one. For this condition, it is assumed that the M–G function G only depends on r extensive variables (i.e., σ is cancelled), which may be divided into two subgroups $\xi_1; \ldots ; \xi_q$ and $\xi_{q+1}; \ldots; \xi_r$. Hence, every mixed second derivative of G must vanish identically for any pair of variables belonging to different subgroups. This agrees with the formula

$$\frac{\partial^2 G}{\partial \xi_j \partial \xi_k} = \frac{\partial \tau_j}{\partial \xi_k} = \frac{\partial \tau_k}{\partial \xi_j} \equiv 0 ; \qquad j = 1(1)q; \quad k = q + 1(1)r, \tag{2.95}$$

which allows the simple decomposition

$$G(\xi_1; \ldots; \xi_r) = G^{\mathrm{I}}(\xi_1; \ldots; \xi_q) + G^{\mathrm{II}}(\xi_{q+1}; \ldots; \xi_r). \tag{2.96}$$

Of course, the idea is that "composition" of a system simply means the cancellation of the respective decomposition. Later on Equation (2.95) will play a relevant role for the formulation of the Alternative Theory.

To conclude this chapter, we will demonstrate how to decompose a system defined by four degrees of freedom. Selecting temperature T_*, volume V, and two molar specific quantities n_1 and n_2 measuring the pertaining partial amount of the two substances involved, the free energy F is equal to the corresponding M–G function. Two subsystems, I and II, may be described by T_*, V, and by one of the two mole numbers n_1 and n_2 each. A third system III is characterized by a substance, the amount of which is determined by the two independent variables T_* and V in such a way that the chemical potential $\hat{\mu}_*$ is chosen to be constant.

The mathematical structure of this example corresponds exactly to the theoretical approach presented by the set of equations (2.88) up to (2.96). Thus, the exclusion condition (2.94) can be formulated in terms of the two first subsystems as follows:

$$\frac{\partial^2 F}{\partial n_1 \partial n_2} = \frac{\partial \mu_{1_*}}{\partial n_2} = \frac{\partial \mu_{2_*}}{\partial n_1} \equiv 0 . \tag{2.97}$$

Consequently, the structure of the M–G function F is isomorphic to that of (2.89):

$$F(T_*; V; n_1; n_2) = F^{\mathrm{I}}(T_*; V; n_1) + F^{\mathrm{II}}(T_*; V; n_2) + F^{\mathrm{III}}(T_*; V). \tag{2.98}$$

The conditions pertinent to the reduction (2.90) and (2.91) as well as the concluding relations (2.93) lead in a straightforward way to the corresponding equilibrium conditions for temperature and volume:

$$T^{\mathrm{I}} = T^{\mathrm{II}} = T^{\mathrm{III}} := T; \qquad V^{\mathrm{I}} = V^{\mathrm{II}} = V^{\mathrm{III}} := V. \tag{2.99}$$

Finally, taking into account the restrictions for the process realization of the problem, we obtain:

$$p^{\mathrm{I}} + p^{\mathrm{II}} + p^{\mathrm{III}} := p; \qquad S^{\mathrm{I}} + S^{\mathrm{II}} + S^{\mathrm{III}} := S. \tag{2.100}$$

In this context it seems appropriate, for reason of application, to discuss the notion of the *diversity of substances*. This term is apparently comprehensible by mere intuition, yet it remits to the renowned *Gibbs' paradox*. This paradox has puzzled thermodynamicists from the very beginning. Hence, in my opinion it is one of the great advantages of Falk's dynamics to exclude this paradox in a simple yet correct and transparent way.

Consider two *ideal* gases (subscripts 1 and 2) assumed to be subject to the exclusion condition (2.97). Let us further regard two different states always with reference to the total volume. They are distinguished by the fact that in state (2) both gases are enclosed in a unified volume, whereas in state (1) they are separated by different volumes each, yet mutually brought into contact.

Obviously, state (2) corresponds to the selected example, but without considering the above subsystem. Thus, both states can be specified by the following conditions, provided that n_1 and n_2 are variables.

$$(1)\ T^{\mathrm{I}} = T^{\mathrm{II}} := T; \qquad p^{\mathrm{I}} = p^{\mathrm{II}} := p; \qquad \text{realized by} \quad V^{\mathrm{I}} + V^{\mathrm{II}} := V_{(1)}.$$

$$(2)\ T^{\mathrm{I}} = T^{\mathrm{II}} := T; \qquad p^{\mathrm{I}}{}_{(2)} + p^{\mathrm{II}}{}_{(2)} := p; \qquad \text{realized by} \quad \left.\begin{matrix} V^{\mathrm{I}} \\ V^{\mathrm{II}} \end{matrix}\right\} \Rightarrow V_{(2)} \quad .$$

The discussion of Gibbs' paradox needs the calculation of the complete difference in entropy between (1) and (2), thereby representing plainly the effect of mixing, provided that (1) and (2) are considered to be the initial state and the final one, respectively. Take the case of two gases, where the isothermal-isobaric change in entropy is given by

$$\Delta S = S_{(2)} - S_{(1)} = -\, n\, \mathcal{R}\, (\chi_1 \ln \chi_1 + \chi_2 \ln \chi_2), \tag{2.101}$$

so that $n := n_1 + n_2$ and $\chi_k := n_k/n$ (with $k = 1, 2$) denote the total mole number and, correspondingly, the mole fraction of the kth component.

Due to the type of process realization, (2.101) is valid for the entropy of mixing as a characteristic measure of irreversibility. In this case, let us first choose $n_2 << n_1$ (or vice versa), with the consequence that ΔS will nevertheless retain its finite value. Second, let us regard the physically questionable, but mathematically admissible, limits $\chi_1 \to 1$ or $\chi_2 \to 0$, respectively. Then, the entropy of mixing will abruptly tend to zero, thus indicating that diffusion cannot occur in uniform gases.

Such a formal conclusion is rather frail, particularly with regard to the interesting fact that (2.101) may also be derived and interpreted in an entirely different way, as might be the case for the states (1) and (2): State (2) is replaced by a state (3) defined by

$$(3)\ T^{\mathrm{I}} = T^{\mathrm{II}} := T; \qquad p^{\mathrm{I}}{}_{(3)} + p^{\mathrm{II}}{}_{(3)} := p; \qquad \text{realized by} \quad V^{\mathrm{I}} + V^{\mathrm{II}} := V_{(3)}.$$

Physically, (3) can be realized by the reversible expansion of V^{I} and V^{II} or even by the irreversible Gay–Lussac expansion (see Section 8.3.2) in such a manner that the pressure assigned to $V_{(3)}$ equals the value p of (1). Mixing of the two gases does not take place at all. In this case it is obviously irrelevant for ΔS if the final state (3) is reached irreversibly and, furthermore, if the ideal gases 1 and 2 differ or not.

After this preparation, some remarks should be made on the meaning of Gibbs' paradox in view of Falk's dynamics. Gibbs (1961, p. 166) first observed that the value of ΔS does not depend on the various kinds of gas under consideration "if the quantities are such as has been supposed, except that the gases which are mixed must be of different kinds. If we should bring into contact two masses of the same kind of gas, they would also mix, but there would be no increase of entropy."

In other words, (2.101) does not contain parameters typifying some kinds of matter. Nevertheless, I agree with Falk (1990, p. 319) that this fact confirms that human imagination sometimes pretends—even in physics—the relevance of a notion even though it is that very belief that should be falsified by every rational theory. Correspondingly, Falk's dynamics do not use the term material as a relevant quantity; it even rejects such an unrelated property as a source of basic information. As a matter of fact, the theory only knows degrees of freedom; at best, some variables of "amount" can be separated from the total number r of degrees of freedom. For instance, any gas with three degrees of freedom represented by T_*, V, and n could consist of an indeterminate number of materials that are all in equilibrium with each another.

It is only a common manner of speaking to apply the term *pure material* to a system with *one* independent variable of "amount" n. The same is also true for a so-called r'-component material, provided that this notation refers to r' degrees of freedom.

To use such terms within the frame of Falk's dynamics two inferences have to be observed:

1. The term *material* can only be justified if it is synonymously used in connection with the respective number r of degrees of freedom.
2. As to point (1), the theory only knows a definite number r of universally physical quantities, for which an infinitesimal change in time or by means of another control parameter is inadmissible to all intents and purposes.

The arguments leading to Gibbs' paradox rest on the idea of "a quantitative diversity of substances" intuitively originated from varying values of a finite number of quantities, for which a mathematical continuum makes up the stock of variables. For this reason the number of various substances cannot be infinitesimally varied; that is, different substances cannot be continuously transferred into identical ones, as it is strictly prohibited, for example to arbitrarily change the dimensions of space coordinates.

Of course, it is possible to diminish the number of variables, but only by system reduction, never by an infinitesimal marginal process in the sense of mathematical analysis. Ergo, *Gibbs' paradox does not exist* within a mathematically consistent theory.

An essential question is: What role does time play in Falk's dynamics? To anticipate the main part of a relevant answer, I will quote a passage from Falk's book:

> Differing from mechanics, electrodynamics, and the theory of relativity, in the method of dynamic description derived from thermodynamics, space and time do not constitute the natural pattern of order into which all events have to be placed. Consequently, space and time do not play any particular role, not even an important one. Instead, the notion

> of the universally physical quantity is elementary for the method of dynamic description. It is true, indeed, that the coordinates of space and time belong to these quantities, but they are not distinguished from all others. On the contrary, it is often useful to leave space and time aside and to focus on relations, wherein the coordinates x; y; z; t do not appear. All such connections belong to those relationships that describe what is commonly called a "system." (Falk, 1990, p. 118; author's translation)

Undoubtedly, this statement is well supported by the mathematical treatment and apparatus of Falk's dynamics. It should be stressed, however, that there exists rather a direct, but less transparent, affinity between the notion of time and the theory presented. It is exclusively due to the well-known property of time to state synchronous events. This means that operations like the decomposition of a system or the interaction between two or more systems are tacitly supposed to be executed simultaneously in time. By the way, synchronization is not tied to any fixed points of time. References to time intervals are common practice, particularly so in technical thermodynamics, but also in economics (cf. Fischer, 1996).

Of course, today scientists and engineers learn that in modern physics the time coordinate is part of an agreement about a whole family of quantities. For this reason, Falk's statement quoted above is true as long as it refers indeed to the classical Newtonian time notion, but does not include the option to establish characteristic relations between the time coordinate and other physical quantities. These will be some of the topics of the following chapters.

Chapter 3 Motion and Matter

"Paradigmata are the winners' dogmata." —***T. S. W. Salomon***

3.1 Basic Questions

These days Gibbsian theory is applied to moving systems, although Gibbs had developed it only for systems at rest, as his own comment indicates:

> When the body is not in a state of thermodynamic equilibrium, its state is not one of those which are represented by our surface. The body, however, as a whole has a certain volume, entropy, and energy which are equal to the sums of the volumes, etc., of its parts. ... As the discussion is to apply to cases in which the parts of the body are in (sensible) motion, it is necessary to define the sense in which the word energy is to be used. We will use the word as including the vis viva of sensible motions. (Gibbs, 1961, p. 39)

Even though Gibbs unquestionably knew the quantitative difference between statics and dynamics and the flagrant disparity between an equilibrium state and a non-equilibrium one, he confined himself in his fundamental investigations to thermostatics only. It is common practice, however, to apply his methods and results to flow processes taking place in many non-equilibrium problems of conventional fluid mechanics and thermofluid dynamics. Gibbs' thermostatics also forms the theoretical background of the *schools* of linear and extended irreversible thermodynamics (LIT and EIT) mentioned in Chapter 1. As a rule, this practice is justified by various questionable hypotheses (principle of local equilibrium; Duhem–Hadamard conjecture) and even some slogans (polytropic processes, quasi-static changes of state).

Only a few representatives, cognizant of rational thermodynamics (RT), allow for certain restrictions. Thus, it is a characteristic feature for this theory to introduce so-called primitive variables to settle some axiomatic elements defined to be irreducible. In such a way, for instance, the basic concept of internal energy U, used in Gibbsian thermostatics, becomes constitutive for RT.

Taking into account the numerous contradictions emerging in theoretical fluid mechanics that cannot be argued away and were addressed in detail by Birkhoff (1960, Chapters 1 and 2), we intuitively surmise that such a thermostatic basis might be used to contribute to these inconsistencies or other contradictions. Presumably, the unusually limited validity of LIT (Garcia-Colin and Uribe, 1991) must be of concern. Its oft-quoted reference to the kinetic theory of gases rather confirms this supposition. Unfortunately, neither EIT (Lebon, 1984; Ruggeri, 1989; Kremer, 1989) nor RT (see Truesdell, 1984) are able to yield well-founded information for practical applications of their present theoretical analysis. Due to the absence of a consensus between these *schools*, the basic notions and terminology encountered in contemporary writings on non-equilibrium thermodynamics are also affected (Kestin, 1990, p. 195).

To obtain an answer acceptable to both science and practice, the problem outlined above shall be restricted and made more explicit by posing the following questions:

1. Are there distinct conditions for a selected and practically important class of dynamic systems that permit an accurate characterization of certain properties with Gibbs' thermostatics?
2. What are the consequences if such conditions do not exist? And what are the consequences if constraints exist preventing a reduction of system dynamics?

To give an elementary example, a one-component single-phase is subsequently considered. It satisfies the minimum conditions of a symmetry principle, most likely first formulated by Callen in 1974 for macroscopic body-field systems, which are the main subject of this book.

The crucial relevance of *Callen's symmetry principle* results from the rational potential to establish consistently the fundamental coordinates of the system in question. They are an essential premise for defining the universally physical quantities introduced in the last part of Section 2.3. In Falk's general formalism this term is thoroughly applied. For so-called body-field systems, however, the term *standard variables* is preferred, to emphasize that the variables of the system observed are selected with Callen's principle and afterward used according to Falk's dynamics (cf. Straub, 1992).

3.2 Callen's Principle

An adequate introduction is given by Callen's statements themselves:

> The primary theorem, relating symmetry considerations to physical consequences, is Noether's theorem. According to this theorem every continuous symmetry transformation...implies the existence of a conserved function. ... The most primitive class of symmetries is the class of continuous space-time transformation. The (presumed) invariance of physical laws under time translation implies energy conservation; spatial translation symmetry implies conservation of momentum. (Callen, 1974, p. 62)

With regard to the very high particle number N characteristic for each macroscopic system, other important symmetries do exist. This important class has been termed *dynamical* by Wigner. Such symmetries give rise to conservation of *baryon numbers* and *lepton numbers*. These are the fundamental compositional coordinates used in chemical physics and combustion technology. In practice, approximate conservation theorems also apply to some particle specifications that can be well typified by so-called internal variables (Muschik, 1990).

Many additional dynamical symmetries can be identified, such as the gauge symmetry, giving rise to the conservation of the elementary electric charge. Furthermore, there are some variables which, though not conserved, are so-called broken symmetry coordinates. The *broken symmetry state* is distinguished by the appearance of a macroscopic order parameter quantifying the characteristic behavior of (infinitely) large systems to "condense" into states of lower symmetry under certain conditions. A prototype of these coordinates is the volume V conserved by definition in special cases. This means that, in general, broken symmetry coordinates are subject to very different external auxiliary conditions, in contrast to conserved coordinates that are determined by universal conservation conditions.

It is noteworthy that volume elements of identical compositions and density may be distinct because of their symmetry properties. This means that there are two phases that transform into each other with the help of inversion or mirroring, as in the case of right and left quartz. Such a symmetry relation ensures the distinctness of the two phases as well as the exact identity of their densities. If the phases are equivalent under a pure rotation, they should not be considered as distinct (Tisza, 1966, p. 106).

Assuming the existence of the particle density $\tilde{n}$ to be satisfied by the thermodynamic limit

$$\text{limes } (N/V) \to \tilde{n} \qquad \text{for} \quad \begin{cases} N \to 0 \\ V \to 0^{\cdot} \end{cases} \tag{3.1}$$

We can also select the volume V as an infinitesimal quantity. This is essential, since it provides a possibility for local process characterizations of a system in motion, together with four-dimensional space-time symmetry and its associated momentum–energy conservation. Yet there is another important aspect considering symmetries with respect to volume. The Pick–Blaschke theory, mentioned in Section 2.3, is based on affine differential geometry and leads to a restricted equiaffine group that expresses some significant properties of the Gibbs surface. Thus, this group preserves volume under linear transformations of determinant unity. This property manifests the thermodynamic stability that is exactly reflected in the topologic patterns of the Gibbs surface, particularly with respect to its curvature. The relevance of this behavior is indeed comparable to the significance of space-time transformations and dynamical symmetries. For this very reason, the volume must also be treated as an indispensable standard variable for almost all body-field systems.

Thus, each system assumed to be a mathematical approach to reality must be identified with its respective Gibbs–Euler function that possesses at least one prototype of each of the four characteristic classes of symmetry:

- Quantities such as momentum **P** and (total) energy E, conserved with references to the continuous space symmetry (**P**) and the symmetry of time transformation (E), are subject to linear affinity.
- Coordinates *generally* conserved by particle symmetry principles, like N, and *specially* conserved coordinates, like V, are introduced by broken symmetry principles and quadratic forms.

Section 4.3 of the next chapter will be devoted to the mathematical relationships between E and the time coordinate t, as well as between **P** and the vector space coordinate **r** corresponding to the famous Noether theorem.

To summarize, the presented specifications of Callen's symmetry principle defines a one-component single-phase system by the set $\{E, \mathbf{P}, V, N, S\}$ of these four coordinates. The existence of the fifth quantity, the entropy S, poses the archetypical problem of thermodynamics. By means of E—as a definite function of **P**, V, N, and S—two isolated systems that are in diathermal and semipermeable contact can be considered in equilibrium. This is due to the procedure described by Equations (2.38) and (2.40) or by the equivalent extremum principle with respect to S. In addition, by means of the Second Law, a fifth symmetry can be introduced at least for the limiting case of idealized thermodynamic systems (cf. Straub, 1992).

The set $\{E, \mathbf{P}, V, N, S\}$ is assumed to consist exclusively of extensive variables. Hence, Falk's dynamics, as described in Chapter 2, is applicable. Particularly, the complete set of the basic equations (2.72) to (2.76) is valid. To facilitate the comparison with the notation used in Chapter 2, the relevant equations should be confronted with the equivalent relations formulated by means of the five standard variables selected. Thus it should be easy to agree upon some new designations without provoking misunderstanding. Let us assume that $F(\xi_0; \xi_1; \ldots; \xi_r) \equiv 0$ means the Gibbsian fundamental equation of any system $\mathcal{S}$ expressed by the variables ξ_0; ξ_1; …; ξ_r each of which can be identified with their values in $\mathbb{R}$. If ξ_0 identifies the energy E of the simple system in question, then the standard variables **P**, S, V, N of the M–G function $E(\mathbf{P}, S, V, N)$ have to be related to the $r = 4$ dimensionless variables ξ_1; …; ξ_4 in accordance with the example given by Equation (2.78). Then the following relationships are easy to obtain

$$\begin{gathered} E - E_{\#} := \xi_0 \times [J]; \qquad \boldsymbol{P} := \xi_1 \times [J\, m^{-1}\, s]; \qquad S := \xi_2 \times [J\, K^{-1}]; \\ V := \xi_3 \times [m^3]; \qquad N := \xi_4 \times [\textit{particle number}], \end{gathered} \tag{3.2}$$

wherein a consistent set of basic units are used. With $E_{\#}$, a reference energy amount, to be specified in the next section, is marked.

The corresponding equations of Falk's dynamics can now be directly transposed. From (2.79) follows first (for convenience, $E_{\#}$ is suppressed here)

$$F(\xi_0; \xi_1; \ldots; \xi_r) \equiv 0 \rightarrow \Gamma(E; \mathbf{P}; S; V; N) \equiv 0 \rightarrow E = \xi_0 \rightarrow \xi_0 = G(\xi_1; \ldots; \xi_4). \tag{3.3}$$

Then, both the relations (2.74) yield

$$dE = \tau_1 \bullet d\mathbf{P} + \tau_2 dS + \tau_3 dV + \tau_4 dN \quad \text{and} \quad E = \tau_1 \bullet \mathbf{P} + \tau_2 S + \tau_3 V + \tau_4 N + E_{\#}. \tag{3.4}$$

where the intensive quantities—either a vector τ_1 or scalars τ_j ($j = 2, 3, 4$)—are set by Equation (2.73) $\tau_1 = \partial G(\xi_1; \ldots; \xi_j; \ldots; \xi_4)/\partial\xi_1$ or $\tau_j = \partial G(\xi_1; \ldots; \xi_j; \ldots; \xi_4)/\partial\xi_j$, respectively, and denoted by special signs:

$$\begin{aligned} &\tau_1 := \mathbf{v} = (\partial E/\partial(\mathbf{P}))_{S,V,N}; \qquad \tau_2 := T_* = (\partial E/\partial S)_{\mathbf{P},V,N}; \\ &\tau_3 := p_* = -(\partial E/\partial V)_{\mathbf{P},S,N}; \qquad \tau_4 := \mu_*{}^t = (\partial E/\partial N)_{\mathbf{P},S,V}. \end{aligned} \tag{3.5}$$

These conjugates denote velocity, temperature, pressure, and chemical potential per particle, respectively, relevant to the system.

Two relevant relationships are obtained by inserting equation (3.5) into (3.4).[7]

$$dE = \mathbf{v} \bullet d\mathbf{P} + T_* dS - p_* dV + \mu_*{}^t dN. \tag{3.6}$$

Gibbs Main Equation

$$E = v_1 P_1 + v_2 P_2 + v_3 P_3 + T_* S - p_* V + \mu_*{}^t N + E_{\#}. \tag{3.7}$$

Euler–Reech Equation

Corresponding to Equations (2.26) the asterisks used in (3.6) and (3.7) serve to denote such quantities in their non-equilibrium states that fulfill special conditions in equilibria.

For the following considerations, in supplement to the designations of (3.6) and (3.7), every M–G function specified by the general form

$$E - E_{\#} = \mathscr{E}(\textit{extensive standard variables}) \tag{3.8}$$

may preferably be called a Gibbs–Euler function (GEF). In principle it contains all information needed for the physical systems in question. For this reason the word *system* may be synonymously substituted by its GEF. In other words, an equivalence between a system and its related GEF can be established.

[7] Preferring an operational notation, the scalar product (or dot product) of the two vectors $\mathbf{v}$ and $d\mathbf{P}$ is denoted by $\mathbf{v}\bullet d\mathbf{P}$. In practice, the component description $v_i\, dP_i$ is preferred using the well-known Einstein summation convention: $v_i\, dP_i = \Sigma_i\, v_i\, dP_i$ with $i = 1(1)3$.

The Gibbs main equation and the Euler–Reech equation noted above are primary inferences of the respective GEF. In this context we should be aware that (3.6) and (3.7) do not explicitly need the individual properties of the system in question. But both relations establish the structure of the system via the special combination of the presupposed variables.

The individual properties mentioned become manifest by the distinctive parameters of the Gibbs–Euler function as well as by the characteristics of its mathematical form. They are exhibited in the direct relations (3.5) between the GEF and the conjugate variables velocity, temperature, pressure, and chemical potential of the system. This reflects six intensive quantities, since the three components of the velocity vector **v** follow directly from the three momentum variables and indirectly from the three remaining variables S, V, and N. In other words, even the simplest system typified by the five fundamental symmetries is mathematically represented by six variables and six conjugate quantities. It is remarkable that the flow velocity **v** of the system is established by a partial derivative of the two system coordinates E and **P**, that is, **v** is not kinematically defined by the time derivative of a position vector. The reference to the GEF is representative of all dynamical descriptions; it is not limited to the flow velocity, but also holds for the temperature T_*, the pressure p_*, and the chemical potential per particle $\mu_*{}'$. All these state quantities are defined by partial derivatives of the GEF itself with respect to the corresponding extensive variables. Yet those limits are calculated under the constraint of an occurring momentum **P** to be held constant only for the benefit of the derivative procedure.

Equation (3.7) gives the component representation of the velocity–momentum scalar product. Note that all state quantities of the Euler–Reech equation (ERE) are able to describe non-equilibrium events. This is evident, because motion manifested by all forms of momentum is commonly related to non-equilibrium states and processes, especially in the Gibbs space (Tisza, 1966, p. 105). Therefore, they can easily be distinguished from states in complete thermodynamic equilibrium or, synonymously, from a state of the system at rest. Such a convention definitely refers to the limiting case of *vanishing linear momentum* **P**. For $\mathbf{P} \to 0$, the corresponding velocity **v** tends toward zero as well, but the inverse is not true as a rule.

Here the reader's attention should focus on an essential feature of the theory elaborated in this book. For many centuries a long and grievous tradition in the natural sciences pertained to the idea of matter and was erroneously brought in connection with classical mechanics. In reality, however, it has been a rather singular strain of human history running from Parmenides' reflections up to Gell-Mann's quarks. For the most part, the perpetual controversies about the idea of matter have entailed ideology, fanaticism, slander, intrigue, suppression, and even murder. Indeed, the persons involved in the story were often distinguished by an extraordinary degree of intelligence, knowledge, imagination, innovation, power, and courage.

Particularly in the early times of the Western natural sciences, the notion of an atom offered an outstanding and dangerous problem. For this reason, even the great founders—for example Galilei, Gassendi, Descartes, and Leibniz—were forced to publish all philosophical and mathematical research under cover of the so-called honorable disguise. Aristotelian thoughts and ideas were also hereditary at the Universities of Cambridge and Oxford. By their direct successions even the great English mechanics as well as Newton were exposed to the respective social pressure.

In Appendix 1, a historical outline is presented to expound the strong impact of power, politics, and religious dogmatism on the idea of matter. In conformity with Schrödinger's famous lecture titled "Ist die Naturwissenschaft milieubedingt?" I suspect one of the "basic laws" of physics is the exclusive result of those historically disastrous and dismal influences. This "basic law," engraved nowadays in every pupil's mind, dominates all current physical theories assumed to be true for macroscopic and electrically neutral phenomena. Concluding this section, the suspicion mentioned will be put into the simple relation (Balian, 1992, p. 301)

$$v_i = P_i/m_N = P_i/m^t N, \qquad i = 1, 2, 3, \tag{3.9}$$

which is commonly thought to be a general principle of matter provided the mass m_N be admitted as a constant. Equation (3.9) makes $\mathbf{v}\cdot d\mathbf{P}$, the energy form of motion, integrable for all changes of state.

In classical mechanics the first part of Equation (3.9) has the status of a "law," whereas the second part takes some results from thermodynamics (N) and chemistry (m^t) into account. The proportionality $m_N = m^t N$ between the mass m_N of the system and its number of particles N introduces a (normally) constant parameter m^t (average mass per particle), characterizing the matter of the system by a property of its constituents.

To recapitulate briefly the consequences of Equation (3.9), the subsequent properties of a one-component single-phase system are obtained from the Gibbs–Euler function (Falk, 1990, pp. 321–322):

$$\frac{\partial v_i}{\partial P_k} = 0 \qquad (i \neq k); \tag{3.10}$$

$$\frac{\partial v_i}{\partial S} = \frac{\partial^2 E}{\partial S \partial P_i} = \frac{\partial T_*}{\partial P_i} = 0; \qquad \frac{\partial v_i}{\partial V} = \frac{\partial^2 E}{\partial V \partial P_i} = -\frac{\partial p_*}{\partial P_i} = 0$$

$$E(\mathbf{P};S;V;N) = \frac{\mathbf{P}^2}{2m^t N} + E_0(S;V;N). \tag{3.11}$$

$$T_* = \partial E_0/\partial S; \qquad -p_* = \partial E_0/\partial V; \qquad \begin{aligned} \mu_*^t &= \partial E_0/\partial N - m^t\mathbf{v}^2/2 \\ &= \mu_0^t - m^t\mathbf{v}^2/2. \end{aligned} \tag{3.12}$$

With the formula apparatus we may associate the following items:

1. The flow velocity $\mathbf{v}$ is independent of all "thermodynamic" variables.
2. The total system described by E can be split up into the subsystems *rest energy* E_0 and *moved body,* which is equivalent to the familiar kinetic energy $m_N\mathbf{v}^2/2$.

3. Both the temperature and pressure of the moving system only depend on the subsystem rest energy E_0. Yet, unlike these thermal state properties, the chemical potential $\mu_*{}^t$ is directly influenced by the motion.
4. The rest energy E_0 may be divided into two parts. Then, *zero-point energy* $E_\#$ and *internal energy* U, depending on the "thermodynamic" variables of the total system, arise:

$$E_0 := E_\# + U(S, V, N); \tag{3.13}$$

for elementary gases (Falk, 1990, pp. 322, 328) $U \to 0$, if $T \to 0$. Hence, $E_\#$ is the rest energy for this limiting case.

Obviously, for motionless systems, that is, $\mathbf{v} \to 0$ as $\mathbf{v} \propto \mathbf{P}$, all results agree with the conventional treatments, "which appear to grant the energy a misleading unique status." (Callen, 1974, p. 65). By the way, the results (3.11) and (3.12) are sometimes derived by use of the Galilean invariance (Balian, 1992, p. 301). But in reality they follow simply from (3.9), and the Galilean invariance becomes a triviality.

Let us now discuss these items by quoting some comments by two of the leading experts, Truesdell and Toupin (1960, pp. 25, 32): Their extensive historical and mathematical studies on the fundamentals of linear and nonlinear field theories led them to some surprising conclusions:

a. "In fact it is almost the rule that Newtonian mechanics, while not appropriate to the corpuscles making up a body, agrees with experience when applied to the body as a whole, except for certain phenomena of astronomical scale."
b. "...the theory of the flow of viscous compressible fluids should suffice to predict definite results, fit for experimental test. ... That such results have never been obtained, is only from our lack of sufficient mathematics."

Both quotations are taken from the foreword of a paradigmatical paper for rational mechanics for the Handbook of Physics. They are remarkable since each statement contains an opinionated claim, and one apparently contradicts the other. This means that Truesdell and Toupin state unmistakably that single particles in the microscale domain do not obey the rules laid down in Equation (3.9). However, according to field theory the same rules should be overtly valid only for a macroscopic body formed now by an ensemble of these very same particles, yet likewise excluded as part of a flow field. Whereas the latter statement is likely to refer only to very special cases (e.g., to billiard balls), the first part can be considered true due to the more recent conclusions in molecular and elementary particle physics.

Quotation (b), addressing real physical processes, is ambiguous and in principle cannot be validated empirically. Numerical methods and experimental feasibilities have encountered considerable improvements during the last thirty years. Yet there are still no broad and reliable data bases concerning the fundamentals of complex flows, for example, their material laws of friction. This is particularly true for compressible fluids. For this reason, Equation (3.9) at best can be regarded as a convention for most practical problems in macroscopic physics. (Poincaré professed that point of view about hundred years ago.) But this convention is dubious if motivated by the idea to represent matter by an ensemble of mass-points, as Helm already stated in 1898 contrary to Boltzmann's view (Helm, 1898, p. 215).

The difference between non-equilibrium and equilibrium is indeed paramount and plays a key role in the characterization of real systems. To represent such a difference mathematically, many alternatives are available. Although Callen's previously mentioned symmetry principle clearly controls the selection of a minimum of standard variables, the linear momentum attains profound importance, or as he explains, "it is evident that, in principle, the linear momentum does appear in the formalism in a form fully equivalent to the energy..." (Callen, 1974, p. 65). This statement is by no means restricted to relativistic motions. (By the way, the inclusion of **P** in the set of variables is explicitly recommended nowadays at the Ecole Polytechnique even though the French tradition obviously restricts itself to the evolution of quasi-equilibrium systems (Balian, 1992, pp. 244–245).)

It seems obvious from Equation (3.9) that the classical equivalence of momentum **P** and the quantity of motion $m\mathbf{v}$ always necessitates the elimination of the momentum from the set of all other variables of the system. In this case it is impossible to distinguish whether the evaluation of the textbook formulas

$$T = (\partial U/\partial S)_{V,N}; \qquad -p = (\partial U/\partial V)_{S,N}$$

for pressure and temperature of any one-component single-phase system is achieved for $\mathbf{P} \neq \mathbf{0}$ or for the state at rest ($\mathbf{P} \to 0$). Therefore, the question arises whether this indefiniteness or Callen's symmetry principle should be given preference as a generic principle.

Callen's statement quoted above has its essential origin from the symmetry-induced restrictions on the possible properties of matter: "Every continuous symmetry transformation of a system implies a conservation theorem, and vice versa." (Callen, 1974, p. 62). From this salient result of Callen's closely reasoned inquiry follows an interesting inference. It obviously contradicts the approach denoted as *extended irreversible thermodynamics* (EIT) (Lebon, 1984, p. 72), because the fundamental assumption of this theory seems to be unfounded: "namely that the dissipative fluxes are raised to the status of independent variables" (Lebon, 1984, p. 100). Indeed, such properties neither obey Callen's principle nor satisfy Falk's principle of universally physical quantities presupposed to be mutually independent. Additionally, the EIT disregards as usual the dependency on momentum although its derivatives are considered.

Since continuous time and space transformations are the common tools to describe real physical processes kinematically, "the most obvious candidates for thermodynamic coordinates are those extensive quantities which are conserved. Each such conserved coordinate bespeaks an underlying physical symmetry." (Callen, 1974, p. 64). The answer to the question posed above, however, goes beyond these conclusions:

> **Callen's symmetry principle** postulates that any system's mathematical description, valid for its continuous motion in time and space, must necessarily be established by its Gibbs–Euler function containing simultaneously the variables energy and linear momentum as constitutive basic information.

Hence, the consequence for physical macrosystems may be easily identified: Is there additional information available to find a restricted or even a universal matter law to replace the classical Equation (3.9) and to comply with Callen's symmetry principle? In other words: Does a constitutive relation exist that invalidates Equations (3.10) to (3.12)?

The answer is yes, but in this chapter the reason can only be given via example. According to the simplest class of systems, the noted relation between linear momentum $\mathbf{P}$ and the corresponding flow velocity $\mathbf{v}$ shall be introduced and the consequences for the concept of particles on the microscale level shall be demonstrated. With reference to Equations (3.10) to (3.12), a second class of systems that is of great relevance to macroscale systems in motion will be discussed. Finally, a general proof is presented for nonrelativistic systems in Chapter 6.

3.3 Energy–Momentum Transport and Matter Model

3.3.1 MATTER CONCEPTS TODAY

Strangely, the ancient ideas of atoms still occur in today's particle models, although certain characteristic restrictions are imposed. They are used in various kinetic or statistical theories to symbolize corpuscles in billiard ball arrangements or mathematical mass-points, respectively. The so-tagged particle has no spatial extension, but can be provided with a mass value and perhaps an electrical charge. The modern concept of matter is ambiguous in comparison to Democritus's atomism, who had in mind the permanence of atoms. Metaphorically speaking, one changes one's appearance and moods while retaining one's identity. Hence, permanence is in that continued identity. In other words, the basic property of things is given by their having a substance or by their being fully permanent.

Even today this permanence is denied. However, modern theories confirm the ancient idea that matter is composed of many particles (constituents), which evolve from the microscopic scale to continuously larger distances on the macroscopic level. Common matter is formed from molecules, atoms, and (free) electrons. The molecules consist of atoms, which in turn consist of (bound) electrons and a nucleus. The latter includes protons and neutrons, which in turn consist of quarks and gluons.[8] With reference to the atomic level governed by orthodox quantum theory or to the subatomic level of elementary particles, there is no evidence that the so-defined constituents are indivisible or immortal in the strict sense of the ancient philosophy of nature. Furthermore, they are assumed to take up no space, that is, they are Eulerian mass-points (see Bethge and Schröder, 1991, p. 138). Nevertheless, these two antique principles, demanding indivisibility and a timeless existence, could survive due to the high degree of abstraction of the theory. Significant now is the unchangeable "law" with respect to transformations in space, time, and many other coordinates: On the subatomic level the characteristic rules of symmetry constitute the laws of nature for the elementary particles. Their interactions are primarily dependent on the gravitational and electromagnetic forces, as well as the so-called weak and strong nuclear ones.

A crucial point of current concepts of matter concerns the conventional notion of a vacuum. The traditional notion of a general void or an empty space believed to be the platform for microscopic events has been lost. Today, on the atomic level the (Dirac-) vacuum represents a definite state of minimum energy that is consistent with the initial and boundary conditions prescribed for the system in question. This ground state has a zero-point energy giving rise to vacuum fluctuations connected with real or virtual vacuum polarizations.

On a subatomic level, the vacuum structure is entirely different. At present, there is a particular interest in the so-called *gluon vacuum*. Gluons are the gauge particles of the strong interactions between the quarks. The vacuum constituted by gluons is defined to be "invisible," indicating that these elementary particles very efficiently influence themselves mutually (they become "lumpy"). But gluons do not interact with electrons, photons, or nucleons, even though protons or neutrons consist of quarks that cannot individually exist within the electrically neutral (Dirac-) vacuum. Thus, a nucleus is considered to be an ensemble of nucleon bubbles moving unhindered inside the lumpy gluon vacuum.

Interestingly, some researchers have recently considered an internal structure of the quarks based on the idea of a further "vacuum of the unified interactions." It may be that the modern notion of a vacuum will evolve into a sort of platform on which all subatomic microphysics takes place. To a certain degree such an evolution is reminiscent of the role formerly played by the ether hypothesis for more than hundred years.[9]

[8] Very recently physicists have discovered a fresh source of many Nobel prizes for the next generation. A team of physicists at the famous Fermi lab has found by experiments under 400 billions electron volts that quarks are probably extended and might be divided. The theorists call these constituents "preons" or "haplons." If a substructure below the quark level were confirmed, all current theories (e.g., grand unified theory, superstrings, big bang, etc.) would have to be revised.

[9] It is surprising that in 1988 Bohm developed the idea of an ether characterized by a kind of holographic structure, where all information about the universe is assumed to be interfolded in each space point (cf. Bohm in Wilber, 1986, p. 50). Even in the nineties some new ether models were suggested (cf. Safe News Binder 1993, pp. 22–29; in German).

3.3.2 *BARYON–LEPTON CONSTANCY*

Equations (3.10) to (3.12) produce evidence for the fact that the paradigmatic relation (3.9) is incompatible with Callen's symmetry principle in all realistic cases. In other words, by means of the postulated identity of the flow velocity $\mathbf{v}$ with the specific momentum $\mathbf{i} = \mathbf{P}/m$, every system in motion, described by its GEF, is automatically decomposed into two additive parts. The first part refers to the kinetic energy and the second defines the rest energy. There is no exception to this mechanism for all electrically neutral systems. Undoubtedly, this necessity *sine qua non* cannot be accepted. If Callen's symmetry principle is presupposed to be mandatory, then another solution should exist. With respect to finding an adequate concept of matter, all habits in view of the "law" (3.9) must be overcome. The status quo of today's microscopic theories of matter might facilitate the following inferences: All experience and knowledge concerning matter lead to the insight that reality is not simple but notionally complicated and only capable of being prepared by mathematical structures. For this reason, we seem justified in studying the aftermath of "material laws," so different from the "mass-point law" (3.9), on the description of physical phenomena.

As a starting point, note that the basic laws on the atomic level are much simpler than the physics of elementary particles and of vacua. Here Einstein's theory of special relativity defines a particle as a special *energy–momentum transport in vacuum,* thus destroying the very distinction between matter and process. In other words, there is a well-known mathematical relation $E(\mathbf{P})$ between energy and linear momentum, which today forges the decisive basis for the experimental link to atomic reality.

The theory of special relativity is the experimentally best scrutinized and examined theory in physics (Hentschel, 1990). The corresponding Einstein mechanics can be summarized by the relation

$$\mathbf{P} = (E/c^2)\mathbf{v}, . \tag{3.14}$$

which differs markedly from Equation (3.9) (Weyl, 1977, p. 33; Jordan, 1969, p. 111).

This so-called *Einstein fundamental relation* (EFR) contains the entire transported energy E which in turn depends on the transport velocity $\mathbf{v}$. It should be stressed that the EFR is not valid for only relativistic motions. According to Falk's dynamics, (3.14) furnishes a constitutive condition that specifies a physical system as a manifestation of a few universal classes of particles. To identify such classes, a system will be selected that moves in a (Dirac-) vacuum. The latter is defined by the double constraint of zero-point pressure and temperature

$$p \to 0; \qquad T \to 0 \qquad (\textit{vacuum condition}). \tag{3.15}$$

This allows the assumption of an energy–momentum transport $\mathbf{P}(E)$ to be identical with a characteristic motion of "particles" (but also see Bethge and Schröder, 1991, p. 137).

The typical "local" status of this theory is characterized by the transport of a certain number of mass-associated "particles" without a "field effect," which means that the gravitational potential has no influence on the speed of light c. In this context, the Gibbs main equation (3.6) will shrink to the simple differential relation

$$dE = \mathbf{v} \bullet d\mathbf{P}. \tag{3.16}$$

In combination with the EFR (3.14), the well-known energy–momentum relation

$$E = \sqrt{\langle (c\mathbf{P})^2 + E_{\#}^2 \rangle} \tag{3.17}$$

results by integration starting from $\mathbf{P} = \mathbf{0}$.

Einstein's legendary equivalence relation $E_{\#} = m_{\#}c^2$ between the zero-point energy $E_{\#}$ and the (inertial or rest) mass $m_{\#}$ holds true for $\mathbf{P} = \mathbf{0}$ (Hentschel, 1990, p. 22). The derivation of a relation of the same mathematical structure as (3.17) can be easily accomplished under the condition that the constraints (3.15) are displaced by the full Euler–Reech equation (3.7) of the system. Then, the integration function E_0 is according to (3.13) and replacing $E_{\#}$. From the different features of "motion" and "state of rest," the kinetic energy of the $E(\mathbf{P})$ transport is defined as follows:

$$E_{\text{kin}} := E - E_{\#}. \tag{3.18}$$

All energy–momentum transports through a vacuum are denoted as *particles* (Falk and Ruppel, 1983, p. 53). They may be classified by means of their different values of the (inertial) mass $m_{\#}$.

Note that (3.17) can be derived directly from the required invariance for all four forces associated with the kinematics of the theory of special relativity (Hentschel, 1990, p. 24). This fact admirably confirms the efficiency of Falk's dynamics. In addition, it is clear that the starting elements (3.14) to (3.16) are valid in general and not restricted to applications related exclusively to the theory of special relativity. Both Equations (3.17) and (3.18), along with the EFR (3.14), permit the arrangement of an extraordinarily important universal classification for particles, which quantifies the discrete energy structure of atoms first demonstrated by the Frank–Hertz experiment.

- common transports: (e.g., electrons)

$$E = E_{\#}(1 - \beta)^{-1/2}; \qquad \beta := (\mathbf{v}/c)^2$$
$$E_{\text{kin}} = E_{\#}[(1 - \beta)^{-1/2} - 1] \tag{3.19}$$

- ultra-relativistic transports: (e.g., photons)

$$E_{\#}^2 << (c\mathbf{P})^2$$
$$\mathrm{E}^2 \approx E_{\text{kin}}^2 \approx (c\mathbf{P})^2 \tag{3.20}$$

- Newtonian transports: (e.g., atoms)

$$\mathbf{P} = m_{\#}\mathbf{v}; \qquad \beta << 1$$
$$E_{\text{kin}} = 1/2 m_{\#}\mathbf{v}^2 \tag{3.21}$$

Eliminating the energy E between (3.19) and (3.14), an interesting result arises with

$$\mathbf{P} = m_{\#}\mathbf{v}(1 - \beta)^{-1/2},$$

which leads to Planck's relation (1958, p. 118) concerning the notion of force $\mathbf{F}$ in its relation to the time parameter t:

$$\mathbf{F} := \frac{d}{dt}\left(\frac{m_{\#}\mathbf{v}}{\sqrt{\left(1-\left(\frac{v}{c}\right)^2\right)}}\right) = \frac{d}{dt}(m\mathbf{v}) = \sqrt{\left(1-\left(\frac{v}{c}\right)^2\right)}\mathbf{F}_{\text{Newton}}\,. \tag{3.22}$$

The most significant data of such transport processes are compiled in Table 3.1. The precise results stem from molecular relaxation processes and chemical reactions and their interactions. Conventional units are used. From the group of "bodies" for which $m_{\#} \neq 0$, only the stable particles are considered with an infinite life expectancy. Following Penrose, one might try to imagine that $m_{\#}$ would be a good measure of the "quantity of matter." Unfortunately, the rest-mass is not algebraically additive. "If a system splits into two, then the original rest-mass is not the sum of the resulting two rest-masses. The π^0 – meson has a positive rest-mass, while the rest-masses of each of the two resulting photons is zero" (Penrose, 1990, p. 220).

All four basic forces involved in elementary interactions are mentioned: electromagnetic (e), weak (w), strong (s), and gravity (g) forces; they are mediated by massless gauge particles. Proust's or Dalton's law of constant or multiple proportions could make it possible to determine particle number ratios on a molar basis for the different mass-marked baryons and leptons. These remain constant for steady

Table 3.1 Particle properties and basic forces

Family	*Particle*	*$m_{\#}$ [eV]*	*Charge*	*Spin [$\hbar$]*	*Interaction*
Photons		0	0	1	e
Leptons	Neutrino	0	0	1/2	w, g
Leptons	Electron	0.511	± e	1/2	e, w, g
Baryons	Proton	938.27	± e	1/2	all
Baryons	Neutron	939.57	0	1/2	all (≠ e)
Newtonian	Mass-points	> 0 optional	0	0	all (≠ w, s)

Note: Representative values have been compiled from Bethge and Schröder (1991, p. 12).

chemical processes. Photons are gauge particles that mediate electromagnetic interactions. They are not bodies with mass; hence a position vector cannot be assigned to the photon; it cannot be located. In contrast to the baryon–lepton constancy there is no conservation law for photons; the associated amount does not refer to the Loschmidt number.

3.3.3 ELEMENTARY SCATTERING PROCESSES: IDENTIFICATIONOF PARTICLES

In comparison to the usual method of deriving the $E(\mathbf{P})$ relationship (3.17), Falk's dynamics has the considerable advantage of achieving the same result as well as providing an opportunity to examine particle classification experimentally. This opportunity can only be mentioned here. In this book, the rigorous confirmation will be highlighted, as well as the true realism of the proposed matter model in comparison with the hypothetical billiard ball concept used in classical physics. Some general comments and three characteristic examples are proposed. The particle confirmation is methodically performed by the scattering of particle beams.

The concept of the adiabatic collision process includes the following principles:

- There are interactions (IA) between energy–momentum transports in empty space, which constitute an *elementary collision* system.
- Empty space constraints represent conditions of a *process realization* that have to be proven experimentally.
- The colliding system as a whole is not subject to the action of external forces.
- Energy and momentum are collision invariants. A time coordinate is not a relevant physical parameter.
- The number of particles involved in the scattering process need not be conserved.

The principal collision configurations are sketched by the following symbolic notation; the superscripts $\underline{a}$ and $\underline{o}$ refer to the initial position (arrival) and the outgoing position of the collision participants $\mathbf{P}_1$ and $\mathbf{P}_2$, characterized by their linear momentum, respectively.

$\mathbf{P}_1^a \searrow \nearrow \mathbf{P}_1^o$	$\mathbf{P}_1^a \searrow$	$\nearrow \mathbf{P}_1^o$
$\{\mathrm{IA}\}$	$\{\mathrm{IA}\} \Rightarrow \mathbf{P}_1^o$	$\mathbf{P}_1^a \Rightarrow \{\mathrm{IA}\}$
$\mathbf{P}_2^a \nearrow \searrow P_2^o$	$\mathbf{P}_2^a \nearrow$	$\searrow \mathbf{P}_2^o$
chemical reactions	photon absorption	radioactive decay

For an N-particle system the real inelastic and elastic collisions are determined by the following generally valid collision invariants for energy and momentum:

$$E = \underbrace{\sum_i E_{\#i} + \overbrace{\sum_i E_{i,\,\mathrm{kin}}^{c}}^{\sum_i E_{i,\mathrm{kin}}}}_{E_{\#}} + E^{c}_{\mathrm{kin}} = \text{constant}; \qquad \mathbf{P} = \sum_i \mathbf{P}_i = \text{constant} \tag{3.23}$$

$$\sum_i E_{\#i} = \text{constant}$$

→ additional constraints for ellastic collisions

$$\sum_i E_{i,\,\mathrm{kin}} = \text{constant}$$

The index *i* refers to the *i*th particle that participates in the collision process of the particle ensemble. The total energy E of an arbitrary reference system can adequately be presented as the sum of the zero-point energies $E_{\#i}$ of the particles and two more portions. The first portion $\Sigma_i E_{i,\,\mathrm{kin}}^{c}$ is the kinetic energy sum of all N particles, measured in the system's center of gravity. The second portion E^{c}_{kin}, representing the kinetic energy in the collision center of the system, is different from zero only to an outside observer.

Subsequently, three elementary collisions are regarded with respect to important partial processes taking place during every interaction of matter. They can be identified with high precision in regular experiments by particle accelerators. Experimental validations are briefly described, for instance, by Bethge and Schröder (1991, p. 86). The test conditions must conform to the postulates of Equation (3.15), which are associated with energy–momentum transports in a vacuum (1991, p. 90).

Following Falk and Ruppel (1983, pp. 88, 103, 110), in each of the following examples two different particles are considered before a collision occurs; they undergo a typical interaction after the impact, according to the scheme previously shown. In the first example the measurements of the resulting *Compton scattering angle* $\Theta(0 < \Theta \leq \pi)$ permits an exact confirmation of the fundamental theory. The second example is related to the well-known *Mössbauer effect*, which can be measured with particularly high accuracy. The third example shows that elementary particles may be created by the inelastic collision between two protons.

Example 1 Compton effect (elastic photon-electron scattering; $E_1^a \approx$ 20.000 eV).

The initial electron (particle 2), hit elastically by a photon (particle 1), is moving with the constant velocity of the center of mass, that is, $\mathbf{P}_2^a := \mathbf{0}$. From scheme (3.23) the relations

$$\mathbf{P}_1^a = \mathbf{P}_1^o + \mathbf{P}_2^o; \qquad E_1^a + E_2^a = E_1^o + E_2^o. \tag{3.24}$$

follow. The energies are expressed by means of Equations (3.17) and (3.21), in accordance with GEF and EFR:

$$\begin{aligned} E_1^a &= c|\mathbf{P}_1^a| \quad \text{and} \quad E_1^o = c|\mathbf{P}_1^o| \\ E_2^a &= E_{2\#} \quad \text{and} \quad \mathrm{E}_2^o = [E_{2\#}^2 + (c\mathbf{P}_2^o)^2]^{1/2}. \end{aligned} \tag{3.25}$$

The Compton effect (1923; see Bethge and Schröder, 1991, p. 126) refers to the outlet data of the photon resulting from the given relations

$$|\mathbf{P}_1|^{o-1} = |\mathbf{P}_1{}^a|^{-1} + E_{2\#}{}^{-1}c(1 - \cos\Theta); \quad (E_1{}^o)^{-1} = (E_1{}^a)^{-1} + E_{2\#}{}^{-1}(1 - \cos\Theta), \tag{3.26}$$

where the scattering angle Θ is defined by the directions of the momentum in front of the collision and behind it.

In the case of "soft" photons, characterized by $E_1{}^a << E_{2\#}$, the approximation $E_1{}^o \approx E_1{}^a$ is justified. Then, the kinetic energy imposed on the electron via collision becomes

$$E_{2,\mathrm{kin}} = E_1{}^a - E_1{}^o \approx E_{2\#}{}^{-1}(E_1{}^a)^2(1 - \cos\Theta).$$

Collisions with "hard" photons, defined by $E_1{}^a << E_{2\#}$, lead to the simple relation

$$E_1{}^o \approx E_{2\#}(1 - \cos\Theta)^{-1},$$

neglecting now $1/E_1{}^a$ in (3.26).

It is interesting that the photon's energy after the collision depends only on the internal energy $E_{2\#}$ of the electron; it is independent of the initial energy amount of the photon.

Example 2 Atom emitting a photon.

The atom as a Newtonian particle (1) is assumed to move according to the lab system. The balances of momentum and energy according to (3.23) change to

$$\mathbf{P}_1{}^a = \mathbf{P}_1{}^o + \mathbf{P}_2{}^o; \qquad E_1{}^a = E_1{}^o + E_2{}^o, \tag{3.27}$$

from which the energy theorem may be transformed by means of (3.20) and (3.21) into

$$E_{1\#}{}^a = E_{1\#}{}^o + (\mathbf{P}_1{}^o)^2/2m_1 + c|\mathbf{P}_2{}^o|.$$

Assuming the atom is at rest in the initial state, that is, $\mathbf{P}_1{}^a \equiv \mathbf{0}$, the energy of the emitted photon can then be addressed to the change in the internal energy of the atom, provided that $E_2{}^o << m_1c^2$ can a priori be taken for granted:

$$E_2{}^o \approx \Delta E_{1\#} - (\Delta E_{1\#})^2\,(2m_1c^2)^{-1} = \Delta E_{1\#} - E_{1,\mathrm{kin}}{}^o. \tag{3.28}$$

The amount of $E_2{}^o$ itself may be related to a characteristic emission frequency υ using the well-known expression $\Delta E_{1\#} = h\upsilon$.

If some energy $E_{1,\mathrm{kin}}{}^o$ of the repulsive forces is stored, the transition energy must be decreased below the uncertainty limit. (Note that the Mössauer effect for high energetic photons (1958) has, for instance, a kinetic energy $E_{1,\mathrm{kin}}{}^o = 0.002$ eV, as compared to the energy level's uncertainty of an emitting Fe-nucleus $|{}_{26}{}^{57}\mathrm{Fe}|$, which is about 10^{-8} eV. Absorption of this photon by a second nucleus does not take place, normally. However, it may happen if two nuclei interact as part of the complete atomic lattice. Hence, the atomic mass m_1 must be replaced by the enormous mass M of the solid body.)

Example 3 Subatomic particles: proton-proton collision.

The model consists of a proton at rest that collides with another proton having received a very high amount Q of energy by a synchrocyclotron. This experimental setup only leads to new elementary particles, but precludes the usual creation of a nucleus made up of a deuteron.

Let us assume that proton 1 is moving in a test cell attached to the lab system (and thus is thought to be fixed by the synchrocyclotron at rest) along with a proton 2 at rest. Regarding first both protons as a single particle, then, before the collision is effected, its internal energy $E_\#^c$ is simply the total internal energy of both protons in their center-of-mass system (c-system). To get new particles brought into existence by collisions, $E_\#^c$ must at least equal the sum of the internal energies of all particles generated after the collision. Consequently, the condition

$$E_\#^c = E_{1\#} + E_{2\#} + Q \geq \sum_j E_{j\#}^o \tag{3.29}$$

has to be fulfilled, where Q denotes the reaction energy needed.

Now, the energy of the "single system" in the lab system is given by (3.17)

$$E^\ell = \{c^2\mathbf{P}_1{}^2 + (E_\#{}^c)^2\}^{-1/2}, \tag{3.30}$$

whereas the same energy calculated for both protons, assumed to be two different particles, becomes

$$E^\ell = E_{2\#} + (c^2\mathbf{P}_1{}^2 + E_{1\#}{}^2)^{-1/2}. \tag{3.31}$$

By equating (3.31) with (3.30) and considering equation (3.29), the unknown reaction energy Q will result. This information, together with the kinetic energy of proton 1 in the lab system yields $E^\ell{}_{1,\mathrm{kin}} = E^\ell - (E_\#{}^c - Q)$ and, combined with inequality (3.29), the necessary condition for the experimental realization

$$E^\ell{}_{1,\mathrm{kin}} \geq \left\{ \left(\sum_j E_{j\#}{}^o\right)^2 - (E_{1\#} + E_{2\#})^2 \right\} (2E_{2\#})^{-1} \tag{3.32}$$

is given, where the rest energy of the proton is $E_{1\#} = E_{2\#} = E_{\mathrm{p}\#} = 938.27$ MeV.

For instance, the creation of an electrically neutral pion, characterized by an internal energy of $E_{\pi\#} = 134.97$ MeV and symbolically described by[10]

$$\mathrm{p} + \mathrm{p} \rightarrow \mathrm{p} + \mathrm{p} + \pi^0,$$

can be accomplished under the following condition:

$$E^\ell{}_{1,\mathrm{kin}} \geq \{(2E_{\mathrm{p}\#} + E_{\pi\#})^2 - (2E_{\mathrm{p}\#})^2\}(2E_{\mathrm{p}\#})^{-1} = E_{\pi\#}\,(2 + 0.5\,E_{\pi\#}\,\mathrm{E}_{\mathrm{p}\#}{}^{-1}) = 280 \text{ MeV}.$$

To create a proton (p)–antiproton (°p) pair, much more energy is needed. For the reaction

$$\mathrm{p} + \mathrm{p} \rightarrow \mathrm{p} + \mathrm{p} + \mathrm{p} + {}^\circ\mathrm{p}$$

the condition

[10] In particle physics a proton is usually symbolized by the letter p.

$$E^{\ell}_{1,\mathrm{kin}} \geq \{(4E_{p\#})^2 - (2E_{p\#})^2\}(2E_{p\#})^{-1} = 6E_{p\#} = 5.63 \text{ GeV}.$$

will have to be satisfied.

3.4 Realistic Concept of Real Matter

To comply with Callen's symmetry condition, the mathematical theory of a physical system needs the standard variables (total) energy and (linear) momentum when the system undergoes a continuous process in time and space. Such a requirement is inconsistent with the common hypothesis of continuum physics concerning the identity of specific momentum and flow velocity at nonrelativistic motion. Nowadays, this hypothesis is heuristically confirmed by its "success," if not simply accepted as self-evident. It is rather exceptional that at least three additional theoretical arguments have been mentioned:

1. It is tacitly assumed that the observer always moves with the center-of-mass velocity of the system.
2. The hypothesis of local equilibrium (HLE) is generally thought to also be valid for non-equilibrium processes (see: Kestin, 1990, p. 202).
3. It is asserted that the nonlinear energy–momentum relation—shown in Equation (3.17): $E^2 = (c\mathbf{P})^2 + E_{\#}{}^2$—results from principles that are exclusively valid for the special theory of relativity (STR).

Since statement (2) includes the older statement (1), but not vice versa, it seems that our own high precision experiments with flows of real gases indicate the doubtfulness of the HLE even for small local temperature gradients (cf. Neumaier, 1996).

Especially remarkable is statement (3), which simply manifests a mere prejudice. In contrast, the results discussed in Sections 3.2 and 3.3 are based on suppositions free of any immediate association with the special theory of relativity. Assuming an energy–momentum relation with reference to Falk's dynamics, the simplest model $\mathbf{v} := \mathbf{P}\, c^2/E$ is sufficient to eliminate $\mathbf{v}$ for the integration of the reduced Gibbs main equation $dE = \mathbf{v}\bullet d\mathbf{P}$. The results must of course also cover relativistic conditions to be valid in general.

Of much larger significance is the fact that physical conditions in an empty space can be represented explicitly with Gibbs–Falk dynamics. Following only these prescriptions for process realizations, even repeatable collision experiments with subatomic particles can be described analytically. In the meantime they have confirmed strikingly the results of Einstein's mechanics. But it should be noted that these experiments are executed under the condition of energy conservation; in other words, interactions between the collision system and its surroundings are assumed to be excluded.

The conclusions concerning reality can be given without great uncertainty:

- For test conditions in empty space all energy–momentum transports can be uniquely characterized.

- An initial sign of distinction is the classification into some types of transport phenomena: the common and the purely relativistic, ultra-relativistic, and Newtonian transports.
- An additional sign of distinction is provided by the typical interaction behavior of the participants involved in the elementary collision system investigated before.
- The empirical results justify the hypothesis of the universal validity of Einstein's fundamental relation (3.14) for vacuum conditions.

Further distinguishing elements come from quantum theory, particularly with reference to the dynamical behavior of particle collectives.

Do these conclusions matter? With reference to these four points, the thesis might be advocated that real matter is a conglomerate of different interacting particles. This statement is no truism if electrons, photons, atoms, and so on are unambiguously defined for a specified energy–momentum transport in an empty space.

Other idealized $E(\mathbf{P})$ relations are well-known, such as the momentum transport going through a perfect crystal. In this case, Equation (3.14) does not hold, and the respective $E(\mathbf{P})$ relation relates to quasi-particles.

The unique simplicity of such definitions is particularly convenient to discuss a matter model, consistent with Callen's and Falk's theories. However, each of these definitions refers to an artificial limiting case. In reality, energy–momentum transports continuously occur in a *pressurized* space. When this case is discussed in the following chapter, it will become apparent that the resulting relation is the correct generalization of the EFR (3.14).

3.4.1 PSEUDO-PARTICLES

Chapter 2 provides a concrete substantiation of the thesis that the universal topic of physics is "the being of the things as we encounter it under experimental conditions. Therefore, we also find in physics an 'ontological difference' that does not permit switching directly from physical statements to statements concerning the being." (Mutschler, 1990, p. 149; author's translation). The consequences of this thesis are difficult to cope with, since they render the familiar idea of natural science immediately ambiguous again, provided that the experimental conditions have to be changed. Another objective difficulty must be added, which relates to the covariance postulate of physical laws. This can be recognized at once, if the vacuum requirement (3.15) is canceled and the $E(\mathbf{P})$ transports of the one-component single-phase system proceeded with changed constraints of

$$V = \text{constant}, \qquad S = \text{constant}, \qquad N = \text{constant}.$$

Unlike the boundary conditions (3.15), these constraints allow variations of pressure and temperature over the time of state evolution. However, they refer to changes of state, where the number of particles involved in the scattering process are conserved. For this reason, the energy–momentum transport under consideration is now restricted to a substance in motion that represents a unique, definite class of particles. These are distinguished by only one of the diverse kinds of motion within the range spanned from ultra-relativistic velocities to the motion of Newtonian mass-points (see Equations (3.19)–(3.21)).

The combination of EFR (3.14) and GEF (3.6), followed by a subsequent integration, would yield an analytical solution that formally agrees with Equation (3.17). However, there is a difference in contents, requiring the zero-point energy $E_{\#}$ to be replaced by the rest energy $E_0(S, V, N)$ according to (3.13). Such a result is in conflict with the necessity to obtain a valid relationship within the framework of the special theory of relativity. Every identity $\Gamma_0(E_0, S, V, N) \equiv 0$ between E_0, S, V, and N would be destroyed by a Lorentz transformation from any reference system at rest to one in motion, since the volume V, as it will be proved below, is not a Lorentz invariant (Planck, 1910, p. 125) as are E, S, and N.

In a treatise on the photon gas in motion, Planck (1907–1908) proposed an extended formulation of the EFR (3.14):

$$\mathbf{P} := c^{-2}\,(E + p_* V)\mathbf{v} = c^{-2}\,E_*^{[V]}\mathbf{v}. \tag{3.33}$$

The enthalpy $E_*^{[V]}$ represents the Legendre-transformed energy $E(\mathbf{P}, S, V, N)$ with respect to the volume V. The associated Legendre-transformed Gibbs main equation is easily obtained:

$$dE_*^{[V]} = \mathbf{v}\bullet d\mathbf{P} + T_*\,dS + V\,dp_* + \mu_*\,dN. \tag{3.34}$$

The combination of the last two equations with respect to $\mathbf{v} = (\partial E_*^{[V]}/\partial \mathbf{P})_{S,p_*,N}$, viz.

$$c\mathbf{P} = c^{-1}E_*^{[V]}\,\mathbf{v} = E_*^{[V]}\,(\partial E_*^{[V]}/\partial c\mathbf{P}) = 1/2(\partial(E_*^{[V]})^2/\partial c\mathbf{P})$$

leads immediately to the following partial differential equation

$$\frac{\partial}{\partial\,(c\mathbf{P})}\left[\,(c\mathbf{P})^2 - (E_*^{[V]})^2\right] = 0\,. \tag{3.35}$$

The solution of this vector equation yields the M–G function of *pressurized enthalpy-momentum transports* denoted as pseudo-particles:

$$E_*^{[V]}\,(\mathbf{P}|S, p_*, N) = \sqrt{\langle\,(c\mathbf{P})^2 + H_0^2\,(S, p_*, N)\rangle}\,. \tag{3.36}$$

As will be shown below, all independent variables associated with the rest enthalpy H_0 are Lorentz invariants. In the limit for $\mathbf{P} \to \mathbf{0}$, both energetic quantities $E_*^{[V]}$ and H_0 are equal.

Compared to (3.17), together with the assigned requirements (3.15), Equation (3.36) is not confined to any constraints assumed to be true for the realization of experiments with pseudo-particles. According to Planck's relation (3.33) particles are the limiting case for pseudo-particles under the conditions (3.15), realized by high-vacuum engineering.

3.4.2 FALK'S EQUATIONS

The properties of such systems, delineated by Equations (3.34) and (3.36), can easily be determined. Some of them are shown below:

$$T_* = (\partial E_*^{[V]}/\partial S)_{\mathbf{P},p*,N} = (\partial H_0/\partial S)_{p*,N}(H_0/E_*^{[V]}) = T_0\sqrt{\langle 1-\beta\rangle}\,;$$
$$V = (\partial E_*^{[V]}/\partial p_*)_{\mathbf{P},S,N} = (\partial H_0/\partial p_*)_{S,N}(H_0/E_*^{[V]}) = V_0\sqrt{\langle 1-\beta\rangle}\,; \tag{3.37}$$
$$E_*^{[V]} = \mathbf{v}\bullet\mathbf{P} + T_*S + \mu_*{}^tN = H_0(1-\beta)^{-1/2}; \qquad \beta := (\mathbf{v}/c)^2.$$

By inserting the relation $(H_0/E_*^{[V]})^2 = 1 - ß$ into the two first equations of (3.37), we may immediately realize the property of *Lorentz invariance*:

$$E_*^{[V]}\,T_* = H_0\,T_0 \;\leftrightarrow\; \text{L invariant}; \qquad E_*^{[V]}\,V = H_0V_0 \;\leftrightarrow\; \text{L invariant};$$
$$E_*^{[V]}\,\mu_*{}^t = H_0\,\mu_0{}^t \;\leftrightarrow\; \text{L invariant}; \qquad E_*^{[V]}\,(1-\beta)^{1/2} = H_0 \;\leftrightarrow\; \text{L invariant.} \tag{3.38}$$

According to these rules the values themselves of temperature, volume, and enthalpy are not Lorentz invariants with reference to any state at rest. They experience a certain reduction, caused by the Lorentz contraction coefficient $\sqrt{\langle 1-\beta\rangle}$.

On the other hand, T_*, V, and $E_*^{[V]}$ approach the rest-values (subscript 0) for the limiting case with $\beta \ll 1$. This result indeed agrees with the expected one, but advocates a fortiori the prejudices due to the widely spread speculation that Planck's equation (3.33) should be valid only for relativistic motion anyway. However, this is not true; on the contrary, there are the mixed second derivatives of Equation (3.37) with respect to the extensive variables

$$\partial v_i/\partial P_k = \partial v_k/\partial P_i = -v_iv_k(E_*^{[V]})^{-1}$$
$$\partial v_i/\partial S = \partial T_*/\partial P_i = -v_iT_*(E_*^{[V]})^{-1} \qquad i, k = 1,2,3. \tag{3.39}$$
$$\partial v_i/\partial V = \partial\mu_*{}^t/\partial P_i = -v_i\mu_*{}^t(E_*^{[V]})^{-1}$$

These generally hold for every pressurized enthalpy–momentum transport of pseudo-particles. These equations do not vanish for nonrelativistic flows, even if $E_*^{[V]}$ tends to approximate H_0. They were first derived by Falk (1990, p. 325) and shall consequently bear his name.

Comparing Falk's Equations (3.39) with the corresponding ones in (3.10), the observed differences result exclusively from the divergent basis (3.9) and (3.33) of the momentum–velocity relation. However, these differences have a significant impact on the structure of physical macrotheories. The significance of Equations (3.39) can be directly summarized as follows:

1. The velocity **v** of a system is not only a function of **P** but also of all the other independent variables of this system consisting of pseudo-particles.
2. The associated Gibbs function $E_*^{[V]}(\mathbf{P}, S, p_*, N)$ cannot be subdivided without constraints: In all real flows it is not permitted to differentiate between internal states and separate motion effects of a physical object.

Statement (1) entirely affirms the *principle of equipresence* first used explicitly by Truesdell (1984, p. 301). Accordingly, statement (2) underlines that the simultaneous momentum–energy or enthalpy transports occur in every real process and are not restricted to relativistic motion.

3.4.3 ATOMIC STRUCTURES OF MACROSCOPIC SYSTEMS

Now the question arises whether there is the slightest indication that the obviously atomistic structure of matter becomes apparent as a foundation for the physical description of macroscopic systems. Let us not refer to emotional or intuitive reasons, but to a mathematical problem as posed by Falk's dynamics.

To get an adequate answer, we may deal with a single case that is representative for macroscopic systems. This case calls for an exact knowledge of its M–G function, and, above all, a very reliable set of experimental data. It concerns the ideal gas, which is of paramount significance for the theory of real gases. Ideal properties of atomically nonsimple substances are intensively studied by experiments and quantum theory alike.

Above all, caloric and spectroscopic methods are of special interest with respect to the dependence of heat capacities on temperature, provided that the real gas in question can be investigated under vacuum conditions. Falk (1990, p. 327) proved that the free enthalpy $G = n\hat{\mu}$ of an ideal gas when at rest may be offered as a sum of three terms:

$$\mathrm{G}(T;\, p;\, n) = E - T\,S + n\mathcal{R}\,T = G_1(n) + G_2(T;\, p;\, n) + G_3(T;\, n). \tag{3.40}$$

The G_i are the M–G functions of independent subsystems of the ideal gas corresponding to

$$G_1(n_1) = n_1 e_0 = n_1\mu_1$$

$$G_2(T_2;\, p_2;\, n_2) = -n_2\mathcal{R}\,T_2 \ell n\left(\frac{a\mathcal{R}}{e^{\frac{\kappa}{\kappa-1}}}\frac{T_2^{\frac{\kappa}{\kappa-1}}}{p_2}\right) = n_2\mu_2(T_2;\, p_2), \tag{3.41}$$

$$G_3(T_3;\, n_3) = -n_3\mathcal{R}\,T_3\int_0^{T_3}\frac{\varphi(T')}{T'}dT' = n_3\mu_3(T_3).$$

The complete system (3.40) is arrived at by the reduction of (3.41) according to the theory presented in Chapter 2. The six variables n_1, n_2, n_3, p_2, T_2, and T_3 are reduced to a set of three variables n, p, and T under the following conditions:

$$\begin{aligned} n_1 = n_2 = n_3 := n; \qquad \mu_1 = \mu_2 = \mu_3 := \mu; \qquad p_2 := p \\ T_3 = T_2 := T; \qquad S_3 = S_2 := S; \qquad V_2 := V. \end{aligned} \tag{3.42}$$

Subsystem 1 may be taken for an "amount-energy" reservoir. Subsystems 1 and 2 together define a simple ideal gas characterized by the identity $\varphi(T) \equiv 0$.

The crucial point is the integral given by G_3. The continuous function $\varphi(T)$ of temperature is defined by a characteristic relation of the molar heat capacity at constant pressure C_p

$$\mathcal{R}\,T\varphi(T) = \int_0^{T_3}[C_p(T') - C_p(0)]\,dT', \tag{3.43}$$

where the integrand is the molar heat capacity $C_{p,3}$ of subsystem 3. An expression like the integral

$$\int_0^T \frac{\varphi(T')}{T'} dT'$$

is inconsistent with the *finiteness axiom* of Falk's dynamics: The full information via the (always) finite number of experimental $C_{p,3}$ values observed are presupposed to give rise to a continuous function $\varphi(T)$. This property is the same as the fact that the above integral contributes an infinite number of φ values to the subsystem G_3. Such an integral relationship would only be allowed if the function $\varphi(T)$ was proved to contain a finite number of parameters. This option offers the key of an appropriate solution to the question posed at the beginning of this section.

Within the framework of statistical thermodynamics thermochemical equilibria of canonical ensembles are established by means of the well-known internal partition functions Z_N^{int}. For corrected Boltzmann statistics of independent particles assumed to be the basic property of ideal gases, Z_N^{int} becomes (Sonntag and Van Wylen, 1968, p. 166)

$$Z_N^{\text{int}} := \sum_{j=1}^{N} g_j e^{-\frac{\varepsilon_j/k_B}{T}} = \sum_{j=1}^{N} g_j e^{-\frac{\varpi_j}{\theta}}; \qquad \varpi_j := \frac{\varepsilon_j}{k_B T_0}; \qquad \theta := \frac{T}{T_0}, \tag{3.44}$$

where k_B and T_0 are the Boltzmann constant and any reference temperature, respectively. The parameters g_j denote the component energy level degeneracies and measure the probability to find the jth energy level ε_j within the set of all N energy levels. This may be equal to the number of particles marked "exactly" by the level ε_j of the (internal) energy mode in question (viz. vibration modes of molecules, for instance).

Taking for granted that an ideal gas may consist of N partial or "simple" gases $\mathcal{A}_j$, reflected by the two material constants a_j and e_j each, it is possible to combine this method of decomposition with statistical considerations as follows. Provided that the unreduced system consisting of all the simple gases $\mathcal{A}_j$—with a total of $3N$ degrees of freedom—can be reduced to the respective ideal gas by the equilibrium conditions

$$T_1 = \cdots = T_N := T; \qquad \mu_1 = \cdots = \mu_N := \mu; \qquad V_1 = \cdots V_N := V, \tag{3.45}$$

then it is easy to derive the connection between Z_N^{int} and the free enthalpy $G_3(T; n)$ of subsystem 3 referring to the same energy mode (Falk, 1990, p. 337):

$$G_3(T; n) = -n \mathcal{R} T \ell n Z_N^{\text{int}}. \tag{3.46}$$

Using (3.41), the relation

$$\varphi(T) = T\, d\, \ell n\, Z_N^{\text{int}}/dT \tag{3.47}$$

is generated by differentiation. We apparently obtain the wanted option by expressing the continuous function $\varphi(T)$ by means of a finite number of data combined in formula (3.44) and extending the summation over N terms, with N an integer. To carry out this option, however, we need to know whether a mathematically exact rule exists to invert (3.47) to get the data of interest,

$$\begin{gathered} g_1 = a_1/a = 1; \qquad g_j = a_j/a; \\ \varepsilon_1 = 0; \qquad \varepsilon_j = e_j - e_0; \qquad j = 2(1)N, \end{gathered} \tag{3.48}$$

from the experimentally determined $\varphi(T)$ function and the exact relations between the statistical parameters g_j, ε_j, as well as the parameters a_j, e_j of the partial gases $\mathcal{A}_j$. Indeed, an appropriate inverse function exists. According to a theorem following from the theory of Laplace transformations, the (internal) partition function can be expressed by a special case of the so-called Laplace–Stieltjes integral

$$Z_N^{\text{int}}(\theta) = \int_0^\infty e^{\frac{-\varpi}{\theta}}\, d\Gamma(\varpi) = Z_3(\theta), \tag{3.49}$$

wherein $\Gamma(\varpi)$ means a nondecreasing function in $\varpi \in \mathbb{R}$ that is called a *spectral function* assigned to Z_3. Corresponding to this reversal theorem the course of $\Gamma(\varpi)$ can be determined for differentiable functions as well as for step functions with jump points ϖ_j and heights g_j of the respective jumps.

The description of the spectral function takes place by means of a *standard step function* Λ according to

$$\Gamma(\varpi) = \sum\nolimits_{j=1} g_j \,\Lambda\,(\varpi - \varpi_j); \qquad j = 1(1)N, \tag{3.50}$$

where two restrictions hold:

$$\Gamma(\varpi) = 0 \quad \text{for } \varpi < \varpi_1 \quad \text{and} \quad \Gamma(\varpi) = \sum\nolimits_{j=1} g_j \quad \text{for } \varpi > \varpi_1.$$

The most important inference to be drawn from the analytical procedure presented here undoubtedly is the optional relation between a finite number N, the integers g_j, and the energy levels ε_j on one hand, and the empirical information $C_p(T)$ about a macroscopic system on the other. The result also confirms the statistical interpretation of thermostatics. Its notation suggests the interpretation of the integer N as identical with the number of particles involved. It should be stressed, however, that it does not concern mechanical mass-points, but rather "thermodynamic particles." According to Einstein, a single "particle" of the simple gas $\mathcal{A}_j$ forms an ensemble of states, for which the quantity n is equal to the value of the elementary amount, viz. $1.66 \cdot 10^{-24}$ mol (Falk, 1990, p. 344).

This result is, by the way, not as self-evident as it might seem. In the daily practice of quantum theory it is believed that an atom as a rule possesses an infinite number of energy levels and, consequently, infinitely many (in mathematical sense) quantum states. Some of the most renowned quantum physicists, for example Herzfeld (in 1916), Bohr (in 1923), Fermi (in 1924), and Planck (in 1924), endeavored to heuristically improve the partition function of internal excitations resulting from the Balmer series for atomic hydrogen and diverging for $N \to 0$.

3.4.4 CONCLUSIONS

It is trite to say that all problems of practical interest in the engineering sciences should deal with irreversible processes. Therefore, concepts originally developed by consideration of equilibrium states only have a very restricted range of application.

A general theory of non-equilibrium physics may be founded by means of Callen's symmetry principle and Falk's extension of Gibbs' thermostatics to form a powerful methodological tool (viz. Falk's dynamics). This chapter demonstrated that both theoretical approaches require a revision of one of the most familiar relationships in classical mechanics ($\mathbf{P} = m\mathbf{v}$). The analysis indicated how to formulate an alternative method using Falk's dynamics. The outcome completely confirms Einstein's mechanics, which primarily rests on principles of the special theory of relativity and has been validated abundantly by experiment. Such a result deserves considerable attention from a methodological point of view.

Observing artificial but well-posed test conditions, the matter model, derived from Einstein's mechanics, strictly speaking is constitutive for modern particle physics. However, with regard to macroscopic phenomena this matter model is questionable in view of realistic matter and field concepts for volume elements of a macroscopic continuum theory.

However, the theory of pseudo-particles shows the necessity of finding a connection between $\mathbf{v}$ and $\mathbf{P}$ that is also parametrically dependent on all other variables of the system, as displayed in the V-Legendre-transformed Gibbs–Euler function. In this case, additional differential equations appear to resemble Falk's equations (3.39). They prevent the balances for linear momentum and internal energy from being separately established with regard to an accurate representation of the complete (nonrelativistic) continuum theory of compressible fluids.

In addition, the basic problem of formulating a mathematical description of macroscopic phenomena by means of a finite number of empirical data leads to an amazing solution: It can be proven for a highly representative system distinguished by a well-known M–G function and a reliable data base that the atomic structure of macroscopic systems offers an option to avoid metaphysical results.

This statement is not trivial: The atomic structure refers to a finite number of particles to be defined thermodynamically with the help of an elementary amount of moles u involved. These thermodynamic particles cannot be imagined as a swarm of mechanical particles moving discriminately. Although relations exist between the properties and, particularly, the notations of the two sorts of particles (Falk, 1990, p. 339), we cannot extrapolate them to real gases or other nonidealized substances like liquids. For this reason, considerations triggered by purely mechanical imaginations seem to be misplaced in face of a consequent mathematical theory of non-equilibrium phenomena. This is especially true for mass-point physics manifested by the alleged identity (3.9) of specific momentum and flow velocity of a fluid element with mass m.

Thus, the classical idea of separating the internal state characteristics of a volume element from its center-of-mass motion cannot be seriously maintained in science, not even for low flow velocities. These facts create far-reaching consequences, for example, for the theory of the Navier–Stokes equations and turbulence phenomena.

Chapter 4 | Systems and Symmetries

"Space and time are not things, but relations of things."—**G. W. Leibniz**

4.1 An Approach to Implant Space and Time in Physics

Regarding the theory presented hitherto, it is striking that neither space nor time play any essential role in Falk's dynamics and its applications. This is true with the exception of Callen's symmetry principle, by which space- and time-coordinates are appropriate for introducing momentum and energy as standard variables of the system in question. Before this basic relationship can be concisely established, allowing only assignments from the phase spaces (from the Gibbs space, especially) to the so-called space of events, the meaning of space and time in mechanics must be discussed. For this purpose we will examine the Hamiltonian theory more closely, because its mathematical structure offers three notable advantages:

- It both contains and goes beyond essential parts of Newton–Eulerian mass-point mechanics.
- It has proven its adequacy for a diverse use of time.
- It yields a relevant example of the great efficiency and universal applicability of Falk's dynamics.

Regarding space- and time-*coordinates*, there is a relevant difference in the selections and solutions of physicists' and engineers' professional problems in daily practice. Physicists as a rule prefer to work on energetically closed systems, whereas, engineers predominantly deal with open systems. The pertaining issues have far-reaching consequences, especially due to the notion of time. To illustrate, we will discuss the First Law of Thermodynamics *in toto*. We will also offer a summary that concerns the different aspects of space- and time-coordinates with regard to the balance equations of a continuum theory of non-equilibrium phenomena.

We will not delve into the philosophical aspects of space and time. (This topic is covered in the renowned book of Whitrow (1990) and in Straub (1990; 1996).) Instead, we will discuss exclusively a few mathematical aspects of space- and time-coordinates pertaining to some physical problems. But we will also become acquainted with the attractive opportunities offered by the results of these considerations for other fields of science such as economics, medicine, and the general theory of systems.

4.2 Falk's Dynamics of Hamiltonian Systems

The number of $2s$ mutually independent variables play a dominant role in Hamiltonian theory. Primarily they are called the *canonical variables* of momentum p_i and position q_i. As a rule they are defined by a set of $2s$ coupled first-order differential equations of the form

$$\frac{dq_i}{d\lambda} = \frac{\partial H}{\partial p_i}; \qquad \frac{dp_i}{d\lambda} = -\frac{\partial H}{\partial p_i}; \qquad i = 1, \ldots, s, \tag{4.1}$$

which is incomplete as long as the meanings of H and λ are not revealed. Letting H be the Hamilton function and $\lambda \equiv t$ the time coordinate, respectively, we can readily recognize the familiar case of classical mechanics. Thus, (4.1) represents the so-called *canonical equations of motion*, which can be solved subject to the general assumption that for an arbitrary value t_0 a set of initial values $\{p_1, \ldots, q_s\}$ is given together with the special Hamiltonian $H(p_2; \ldots; p_s; q_1; \ldots; q_s)$ of the mechanical system in question.

Along with $H(p_1; \ldots; p_s; q_1; \ldots; q_s)$, the system's evolution made up by the unconstrained solutions $p_1(t); \ldots; p_s(t); \ldots; q_1(t); \ldots; q_s(t)$ determines all its other properties characterized and described by any function of the kind $\Phi_j(p_1; \ldots; p_s; q_1; \ldots; q_s)$. This follows from the total derivative of Φ_j with respect to all variables

$$\frac{d\Phi_j}{dt} = \sum_{i=1}^{s}\left(\frac{\partial\Phi_j}{\partial q_i}\frac{dq_i}{dt} + \frac{\partial\Phi_j}{\partial p_i}\frac{dp_i}{dt}\right) = \sum_{i=1}^{s}\left(\frac{\partial\Phi_j}{\partial q_i}\frac{\partial H}{\partial p_i} - \frac{\partial\Phi_j}{\partial p_i}\frac{\partial H}{\partial q_i}\right), \tag{4.2}$$

where the right-hand side denotes the so-called *Poisson bracket*. Thus, the expression

$$\frac{d\Phi_j}{dt} = [H, \Phi_j]; \qquad j = 1(1)J \tag{4.3}$$

holds for the time derivative of the function Φ_j. Identifying each of these J functions with the Hamiltonian H, then $dH/dt = 0$ arises due to the Poisson bracket $[H, H] = 0$. This means that the Hamiltonian is independent of t during the motion of the system defined by H itself. In this case the function H, as well as every other function Φ_j, is a constant of motion provided that the condition $[H, \Phi_j] = 0$ is satisfied. This conclusion allows two important statements:

The time coordinate t is a curve parameter, that is, a number indicating successively the integrals of motion involved.

There seems to be a connection between t and a conservation property of H.

By the way, there is another type of conservation property: If Hamilton's function does not depend on a particular coordinate, say q_k (with k being a fixed integer), it follows from (4.1) that the conjugated variable p_k remains unchanged. This is called a cyclic conservation rule.

Note then that general restrictions that serve to classify dynamical systems are often introduced in theoretical mechanics. It is supposed that their variables, particularly if they are assigned to mechanical particles, are not completely free to be changed, but are restricted to be in permanent contact with the properties of other material objects. Commonly, such constraints are mathematically expressed by a given set of γ functions of time and s position coordinates. These functions are, at least in principle, considered for solutions of the equations of motions. De facto they cause a decreasing number of all degrees of freedom of the system in question compared with the number $2s$ introduced in (4.1). In the literature, such positional (or holonomic) constraints are commonly subdivided into scleronomic (do not contain t explicitly) and rheonomic (contain t explicitly) constraints.

Note that these considerations do not presume that the variables p_i and q_i refer to Newton–Eulerian mass-points. For this reason the so-called generalized momentum variables p_i need not obey any conservation law. Furthermore, to dispense the original meaning of conjugated variables, consider one of the extraordinary properties of Hamiltonian systems: They are, above all, invariant toward all canonical transformations defined by changes of the variables referring to the scheme $\{p_1; \ldots; p_s; \ldots; q_1; \ldots; q_s\} \rightarrow \{P_1; \ldots; P_s; \ldots; Q_1; \ldots; Q_s\}$ and characterized by an unchanged Hamiltonian H. Therefore, the *transformation rule* becomes

$$
\begin{aligned}
H(p_1; \ldots; q_s) &= H(p_1\{P_1; \ldots; P_s; \ldots; Q_1; \ldots; Q_s\}; \ldots; q_s\{P_1; \ldots; P_s; \ldots; Q_1; \ldots; Q_s\}) \\
&= H(P_1; \ldots; Q_s). \qquad (4.4)
\end{aligned}
$$

Even if the special canonical variables $p_1; \ldots; q_s$ represent momentum and position coordinates of mass-points, the transformed variables $P_1; \ldots; Q_s$ normally do not allow their rule either as momentum or as position coordinates in the sense of Newtonian mechanics.

Surprisingly, the general formulation of the Hamiltonian theory expounded above does not indicate any constants or fixed parameters. But they undoubtedly exist and are apt to characterize the individual system. Since they are physical quantities, for example, values of masses and charges, it is clearly an unjustified restriction to suppress them. Consequently, the domain of physical quantities underlying the mathematical operations of the Hamiltonian theory consists of the pertaining functions that depend on the $2s + \ell$ independent quantities $p_1; \ldots; p_s; q_1; \ldots; q_s; \zeta_1; \ldots; \zeta_\ell$.

The main difference between the p_i and q_i and the parameters ζ_i is that only the pairs $\{p_i; q_i\}$ are conjugated in accordance to (4.1). In other words, at a first glance there does not exist a direct counterpart ζ_k of an arbitrary quantity ζ_j.

Following Falk's considerations (1990, p. 382), the $2s + \ell$ quantities $p_1; \ldots; \zeta_\ell$ may be taken to be the generics of the domain $\mathcal{G}_{2s,\ell}$, using the notation of Section 2.4. Then, by determining whether the properties of a Hamiltonian system can be adequately represented by certain elements of $\mathcal{G}_{2s,\ell}$, we can also determine whether to include the Hamiltonian formalism into Falk's dynamics.

It is a crucial point of the arguments presented above that the generalized variables p_i and q_i must be *universally physical quantities* in the sense of Falk's dynamics. Recall that in this distinguished case the single variable, say p_1 or q_1, represents the *same* quantity for *all* Hamiltonian systems described by the theory.

Yet there is another situation considering the Hamiltonian of the system in question. Different from the p_i and q_i, a special function of $p_1; \ldots; p_s; q_1; \ldots; q_s; \zeta_1; \ldots; \zeta_\ell$ is defined for each system with the parameters ζ_j being treated as latent variables. Let us discuss this basic problem by means of the original Hamiltonian system consisting of $2s$ variables.

According to the rules of Falk's dynamics, the single system is defined by its M–G element, given by a function $F(p_1^{\otimes}; \ldots; q_s^{\otimes}; \xi_0)$ of all $2s + 1$ generics of $\mathcal{G}_{2s,1}$. States Z of the system are mappings of $\mathcal{G}_{2s,1}$ onto the range of values $\mathcal{W}$, that is, the set of real numbers.

Following Equation (2.72), the element F becomes zero in conformity with the following mapping procedure:

$$Z[F(p_1^{\otimes}; \ldots; q_s^{\otimes}; \xi_0)] = F(Z[p_1^{\otimes}]; \ldots; Z[q_s^{\otimes}]; Z[\xi_0]) = F(p_1; \ldots; q_s; -E) = 0. \tag{4.5}$$

As usual the values $Z[p_1^{\otimes}], \ldots, Z[q_s^{\otimes}]$ of the generics $p_1^{\otimes}; \ldots; q_s^{\otimes}$ in the state Z are indicated by the same letters, but without the superscript $^{\otimes}$. However, the real number $Z[\xi_0]$ arising from the mapping of ξ_0 is denoted by a symbol $-E$ of its own.

Now, the Hamiltonian $H(p_1; \ldots; q_s)$, commonly given within the range of values $\mathcal{W}$, is assumed to be an element of $\mathcal{G}_{2s,1}$. However, note that such an element $H(p_1^{\otimes}; \ldots; q_s^{\otimes})$ is not an M–G element of the system because it does not consist of $2s + 1$ degrees of freedom (as any respective M–G element will). It is easy to correct this result by an explicit ansatz for the M–G element wanted:

$$F(p_1^{\otimes}; \ldots; q_s^{\otimes}; \xi_0) := \xi_0 + H(p_1^{\otimes}; \ldots; q_s^{\otimes}). \tag{4.6}$$

Then, for the states of the respective system, consisting of $2s$ degrees of freedom, the following mapping rule

$$Z[F] = Z[\xi_0] + H(Z[p_1^{\otimes}]; \ldots; Z[q_s^{\otimes}]) = -E + H(p_1; \ldots; q_s) = 0 \tag{4.7}$$

holds.

Clearly, Equation (4.7) is nothing else but Gibbs' fundamental equation of the Hamiltonian system in question. According to (2.72), the corresponding M–G function $E = H(p_1; \ldots; q_s)$ leads to the assigned total derivative

$$dE = \sum_1^s \left(\frac{\partial E}{\partial p_i} dp_i + \frac{\partial E}{\partial q_i} \right). \tag{4.8}$$

The common Hamiltonian theory postulates the identity of H with the energy E of the system in question provided that the variables p_i and q_i mean linear momentums and positions of any abstract "bodies."

In this case (and only in this case), Relation (4.8) yields two important facts about the system in complete agreement with mechanics

$$\frac{\partial H}{\partial p_i} = v_i; \qquad \frac{\partial H}{\partial q_i} = -f_i; \tag{4.9}$$

these relations concern components of the assigned vectors of velocity **v** and force **f**.

In $\mathcal{G}_{2s,1}$ of the common Hamiltonian systems, the elements representing the values of v_i and f_i can easily be expressed by a so-called *anticommutative multiplication,* defined by Falk (1990, p. 383) and applied as follows:

$$v_i^{\circledast} = [F, q_i^{\circledast}] \quad \text{and} \quad f_i^{\circledast} = [F, p_i^{\circledast}]. \tag{4.10}$$

Note that the axioms of this multiplication agree completely with the rules defining the Poisson brackets introduced above. With regard to the meaning of the M–G element F, we may infer the option to implement additional properties of the system under certain conditions. Thus, for instance, the representation of the components a_i of the acceleration in $\mathcal{G}_{2s,1}$ succeeds by writing:

$$a_i^{\circledast} = [F, v_i^{\circledast}] = [F, [F, q_i^{\circledast}]. \tag{4.11}$$

By the way, angular momentum effects are not included. In the Hamiltonian formalism they may be considered, indeed, for mass-point systems. In this case the number s of the relevant degrees of freedom must be substituted with an integer N representing the number of mass-points prescribed. Furthermore, we must fix the locations of the position variables q_i in a three-dimensional Euclidean space. Thus, additional restrictions of the pertaining M–G function arise, by which the components of the angular momentum become involved. Of course, the idea itself suggests that we successively extend the domain of physical quantities $\mathcal{G}_{2s,1}$ according to the option noted above in relation to the parameters ζ_j. In fact, we will examine a theory, specified by two quite different concepts that precisely carry the information required, to debate the role of time-coordinates in physics.

The first case has already been discussed in connection with Equation (4.3) and refers to a Hamiltonian that is independent of time. Obviously, it belongs to a domain of type $\mathcal{G}_{2s,1}$. By comparison, *time-dependent Hamiltonian systems* are represented by the extended types $\mathcal{G}_{2s,2}$ or $\mathcal{G}_{2s+2}$. This means that there exists one more quantity with respect to the generics.

The second case is distinguished by two generics, ζ_1 and ζ_2, to be identified with the representatives of energy $-E$ and time $\tau_\zeta$in $\mathcal{G}_{2s,2}$.This set thoroughly contains the Hamiltonian domain $\mathcal{G}_{2s,1}$ predominated by $2s$ canonical generics. The corresponding M–G elements $F(p_1^{\circledast}; \ldots; q_s^{\circledast}; \zeta_1; \zeta_2)$ can be used to define an extended Hamiltonian theory

$$F(p_1^{\circledast}; \ldots; q_s^{\circledast}; \zeta_1; \zeta_2) = \zeta_1 + H(p_1^{\circledast}; \ldots; q_s^{\circledast}; \zeta_2), \tag{4.12}$$

confirming (4.6) for the limiting reduction $\zeta_2 \equiv 0$.

Now, all states of the system in question obey the relations

$$Z[F] = Z[\zeta_1] + H(Z[p_1^{\circledast}]; \ldots; Z[q_s^{\circledast}]; Z[\zeta_2]) \\ = -E + H(p_1; \ldots; q_s; \tau_\zeta) = 0, \tag{4.13}$$

where the second line equals Gibbs' fundamental equation of the Hamiltonian system. The total derivative of the corresponding M–G function is given by

$$dE = \sum_1^s \left(\frac{\partial H}{\partial p_i} dp_i + \frac{\partial H}{\partial q_i} \right) + \frac{\partial H}{\partial \tau_\varsigma} d\tau_\varsigma, \tag{4.14}$$

from which the same relations as those of (4.9)–(4.11) can be derived. Assuming that the time variable τ_ζ is not conjugated to any other variables, it is obvious that time-dependent properties of the system cannot be expressed with the Poisson brackets. By use of (4.3) with respect to E, and in reference to the obligatory curve parameter t, the relation

$$\frac{dE}{dt} = [H, E] + \frac{\partial E}{\partial \tau_\varsigma} \frac{d\tau_\varsigma}{dt} \tag{4.15}$$

becomes evident.

Consequently, a conservation law for E does not hold even for the identity $H \equiv E$. Even if τ_ζ is equated with t, the time dependency of energy E does not vanish, with the exception of scleronomic dynamical systems as defined by $\partial H/\partial t \equiv 0$.

An additional case appears: The two generics ζ_1; ζ_2 are assumed to be a canonically conjugated pair of variables denoted by $p_0^{\circledast}$ and $q_0^{\circledast}$ in the domain $\mathcal{G}_{2s+2}$. This might be equivalent to a conjugated pair of energy $-E$ and time τ_q in the range of values $\mathcal{W}$. Consequently, in addition to (4.10) and (4.11), there exist two relations formulated by means of the Poisson brackets (Falk, 1990, p. 392)

$$[F, q_i^{\otimes}] = \delta_{i0} + [H, q_i^{\otimes}] = \begin{cases} 1, i = 0 \\ v_j^{\otimes}, i = 1, \ldots, s \end{cases} \qquad [F, p_i^{\otimes}] = [H, p_i^{\otimes}] = \begin{cases} f_0^{\otimes}, i = 0 \\ f_i^{\otimes}, i = 1, \ldots, s \end{cases}. \tag{4.16}$$

Thus a new universally physical quantity $f_0^{\circledast}$ is introduced. With respect to the variables taken from $\mathcal{W}$, this quantity corresponds to $f_0 = \partial H/\partial \tau_q$ and means an energy rate independent of the forces $f_1, \ldots, f_s$. Furthermore, from the first relation (4.16), the identity

$$[F, [F, q_0^{\circledast}]] \equiv 0$$

follows immediately. Hence, an acceleration term a_0 is excluded, quite in contrast to (4.11) and derived to be true for the acceleration components $a_1, \ldots, a_s$.

It is interesting to examine the equations of motion for the conjugated pair E and τ_q:

$$\frac{d(-E)}{dt} = -\frac{\partial H}{\partial \tau_q}; \qquad \frac{d\tau_q}{dt} = \frac{\partial H}{\partial E}. \tag{4.17}$$

Contrary to (4.15), Equation (4.17) only allows a linear dependency of τ_q on the curve parameter t if the identity $H \equiv E$ holds in accordance to (4.13). For this assumption the Hamiltonian H depends on t in the same way since it is a function of the internal time variable τ_q. Setting $\tau_q \equiv t$ Equation (4.17), of course, agrees with (4.15) simplified for the case τ_ζ.

Summing up the results of the three discussed cases, we have the following conclusions with respect to the time-dependent properties:

1. Hamiltonian theory can be constructed so as to become a special case of Falk's dynamics.
2. The resulting M–G function for any Hamiltonian system considers the property of its canonically conjugated variables. The energy E is the M–G function of the Hamiltonian H, provided its canonical variables represent the classical meaning of momentum and position.
3. In general, H is only conserved under the condition that it is independent of the curve parameter λ. Assuming λ portrays a time-coordinate, it must then be determined how the Hamiltonian must be modified to account for some time variables in reference to t.

These comments result from the favorable option relating Hamiltonian theory to Falk's dynamics. But it should also be noted that the connection is confined to a singular approach defined by the M–G element $F := \xi_0 - H$ according to (4.6) and (4.12). All the domains of physical quantities $\mathcal{G}_r$ evaluated above might be sufficient in classical mechanics, but they do not fully comply with the options of Falk's dynamics.

For instance, those special domains fail to consider homogeneity of the M–G elements, as far as generics are concerned. As for the more common cases of Hamiltonian theory, the generics cannot simply be assumed to be homogeneous, either in $\mathcal{G}_{r=2s,1}$, $\mathcal{G}_{r=2s,2}$, or $\mathcal{G}_{r=2s+2}$. The general method to overcome these difficulties was first demonstrated by Falk (1990, p. 382) for classical mechanics and has been more recently elaborated by Diestelhorst (1993) for quantum mechanics.

Statement (3) indicates the mechanical point of view that there is only one way to have energy conserved, that is, by means of the curve parameter λ that particularly expresses the evolutions of the generalized momentums and positions by continuously changing λ values. Naturally, λ may be equated with the time-coordinate t representing real numbers, in reference to any adequate method of measuring all λ-values required. In practice, however, the parameter t should also serve to indicate simultaneity to show the important synchronization of events without difficulty.

Evidently, this problem is part of a much more complex issue that deals with the option of systematically constructing *conservation* rules that hold for any quantities useful for efficient and reliable mathematical theories of physics and other fields of science. Of course, this general idea is beyond the scope of this book. It may suffice to solve the sketchy problem only for energy and momentum. In support of this limitation two remarks should be made:

- There are certainly some advantages to representing many non-equilibrium phenomena within the so-called space of events. Unlike the well-known phase spaces (including the Gibbs space), this coordinate system is generated by means of all the relevant physical variables over the space- and time-coordinates chosen.
- Nearly all concepts of theoretical physics to be surveyed are based on a universal restriction of events in space and time. Theoretically, this is accomplished simply by conservation of energy; practically, it is approximated by working with closed systems. In sharp contrast to this idealization, engineering work commonly has to deal with open systems characterized by certain rates of energy and matter.

4.3 Review of the Noether Theorem

In this section, we briefly discuss several general points concerning the relevant results of Emmy Noether's famous theorem first published in 1918. The considerations focus on the introduction of continuous coordinates for space and time in non-equilibrium physics on a macroscopic level. In view of this aim and considering Callen's principle, the connections between these coordinates and the conservation of momentum and energy are of particular interest.

Remarkably, neither of these quantities are conserved in general, as some mathematical research—particularly on relativistic physics—has shown in the last decades (Schmutzer, 1972, p. 50). But for the objective of this book such experiences are irrelevant as long as two assumptions are maintained:

1. The processes to be dealt with are not subject to effects ruled by laws of relativistic gravitation according to Einstein's general theory of relativity.
2. Due to the need to consider mainly open systems, the relations between the coordinates and conservation properties noted are sufficient for the limiting case of closed systems only.

Assumption (2) may especially be confined to the case for which a far-reaching and well-founded idea of matter is available, even if this case is true only for a restricted area of theory and practice. As in Section 4.2, we can use a system with properties that are exactly known and representative of systems well-proven in mechanics for a long time—for instance, Hamiltonian systems.

4.3.1 NOETHER THEOREM IN LAGRANGIAN DYNAMICAL SYSTEMS

For the problem in question certain experiences show that it is easier to work with Lagrangian systems that facilitate the presentation of Noether's theorem. In this context it should be sufficient to affirm that there is a unique relation between the characteristic functions of both these mechanical systems. Concretely expressed, the

Hamiltonian can be converted to the respective Lagrangian by Legendre transformation and vice versa (Vujanovic and Jones, 1989, p. 153).

Consider a holonomic nonconservative dynamical system whose Lagrangian function $\mathscr{L}$ is of the form

$$\mathscr{L} = \mathscr{L}(\mathbf{x}; \mathring{\mathbf{x}}; t), \tag{4.18}$$

where t is the curve parameter "time," $\mathbf{x}$ denotes the generalized independent coordinates $\mathbf{x} = \{x^1; \ldots; x^s\}$, and $\mathring{\mathbf{x}} = \{\mathring{x}^1; \ldots; \mathring{x}^s\}$ stands for the generalized velocity vector defined by $\dot{\mathbf{x}} := d\mathbf{x}/dt$.

By means of Hamilton's action integral I with respect to $\mathscr{L}$, together with the variational equation $\delta I = 0$, it is easy to show that the resulting steady-state conditions of I are equivalent to the so-called *Euler–Lagrange equations* (also known as Lagrangian equation of motion):

$$\varphi_k \equiv \frac{d}{dt}\frac{d\mathscr{L}}{d\dot{x}^k} - \frac{\partial}{\partial x^k}\mathscr{L}; \qquad k = 1. \tag{4.19}$$

According to the fundamental lemma of the calculus of variations, if all $\varphi_k(t)$ are continuous for t in the closed time interval $[t_0, t_1]$, then

$$\varphi_k(t) \equiv 0; \qquad k = 1(1)s. \tag{4.20}$$

hold for all t in $[t_0, t_1]$.

Hamilton's variational principle provides only the necessary condition for the stationary property of I along the actual trajectory of the dynamical system. The question of the type of the extremum may be answered by the sign of the second variation $\delta^2 I$. It can be demonstrated that for short intervals of time $t_1 - t_0$, this variation is positive for a scleronomic system provided the Lagrangian $\mathscr{L}$ is taken to be the classical energy difference

$$\mathscr{L} := \mathscr{T}(\mathbf{x}; \mathring{\mathbf{x}}; t) - \mathscr{P}(\mathbf{x}; t), \tag{4.21}$$

where $\mathscr{T}$ and $\mathscr{P}$ stand for the kinetic part and the potential part, respectively. Thus, Hamilton's action integral is a minimum along the actual trajectories of such systems. (By the way, the rule (4.21), although very important for Lagrangian systems, is not generally true since the Lagrangian is not unique (Vujanovic and Jones, 1989, p. 17).)

From the standpoint of Falk's dynamics, discussed above by the example of Hamiltonian systems, Equation (4.19) together with (4.20) simply defines all Lagrangian systems. The situation is comparable to Hamiltonian systems defined by their respective equations of motion (4.1). Nevertheless, there is a characteristic distinction: Whereas (4.1) describes a set of ordinary differential equations of the first order with regard to time, the Euler–Lagrange equations are of the second order. As for the rest, a change of $\mathscr{L}$ to H can easily be carried out by means of the Legendre transformation of $\mathscr{L}$ with respect to the generalized velocities $\mathring{\mathbf{x}} = \{\mathring{x}^1; \ldots; \mathring{x}^s\}$. Then, the relation

$$H(\mathbf{x};\, \mathbf{p};\, t) = \dot{x}^k \frac{\partial \mathscr{L}}{\partial \dot{x}^k} - \mathscr{L}\left(\mathbf{x};\, \dot{\mathbf{x}};\, t\right) = \dot{x}^k p_k - \mathscr{L}(\mathbf{x};\, \mathring{\mathbf{x}};\, t) \tag{4.22}$$

holds, where Einstein's summation convention with respect to repeated indices k is employed, running from $k = 1$ to $k = s$.

In accordance with the transformational behavior of H mentioned in Section 4.2, an invariance of $\mathscr{L}$ may be postulated considering all variables $\mathbf{x};\, \mathring{\mathbf{x}};\, t$, but not the mathematical form of $\mathscr{L}$ itself. This so-called *divergence-invariance* is assumed to be valid if there are ρ so-called gauge functions Ω^ν such that

$$\mathscr{L}(x'(t');\, \mathring{\mathbf{x}}'(t');\, t') - \mathscr{L}(x(t);\, \mathring{\mathbf{x}}(t);\, t)\frac{dt'}{dt} = \varepsilon^\nu \frac{d}{dt}\Omega_\nu(x(t);\, t) + O(\varepsilon) \tag{4.23}$$

is fulfilled. This means that $\mathscr{L}$ remains unchanged up to exact differentials weighted by parameters ε^ν and to first-order terms in $\varepsilon := \{\varepsilon^1;\, \ldots;\, \varepsilon^\rho\}$ on the right side of (4.23). The summation convention is to apply to index ν running from $\nu = 1$ to $\nu = \rho$.

The type of invariance transformations that will be considered are transformations of the configuration space, $\{x^1;\, \ldots;\, x^s;\, t\}$, that depend on ρ real, independent parameters $\varepsilon^1;\, \ldots;\, \varepsilon^\rho$. To be more precise, the following transformation rules

$$\begin{aligned} t' &:= \phi(t, \mathbf{x}, \varepsilon); \qquad \varepsilon := \{\varepsilon^1;\, \ldots;\, \varepsilon^\rho\} \\ x'^k &:= \psi^k(t, \mathbf{x}, \varepsilon); \qquad k := 1(1)s \end{aligned} \tag{4.24}$$

are required in conjunction with the limiting conditions:

$$\phi(t, \mathbf{x}, \varepsilon = \mathbf{0}) := t; \qquad \psi^k(t, \mathbf{x}, \varepsilon = \mathbf{0}) := x^k;\ k := 1(1)s. \tag{4.25}$$

Expanding the right-hand sides of Equations (4.24) in Taylor series around $\varepsilon = \mathbf{0}$, the transformation rules are approximated up to first-order terms in ε:

$$\begin{aligned} t' &= t + \tau_\nu(t, \mathbf{x})\,\varepsilon^\nu + O(\varepsilon); \qquad \nu := 1(1)\rho \\ x'^k &= x^k + \xi_\nu{}^k(t, \mathbf{x})\varepsilon^\nu + O(\varepsilon); \qquad k := 1(1)s. \end{aligned} \tag{4.26}$$

The principal linear parts $\tau_\nu(t, \mathbf{x})$ and $\xi_\nu{}^k(t, \mathbf{x})$ are called the infinitesimal generators of the rules (4.24) given explicitly by

$$\tau_\nu(t;\, \mathbf{x}) := \frac{\partial \Phi}{\partial \varepsilon^\nu}(t;\, \mathbf{x};\, 0); \qquad \xi_\nu{}^k(t;\, \mathbf{x}) = \frac{\partial \Psi}{\partial \varepsilon^\nu}(t;\, \mathbf{x};\, 0). \tag{4.27}$$

The definition of the divergence-invariance according to (4.23) implies that there are some relations between the derivatives $\partial\mathscr{L}/\partial\mathbf{x}^k$, $\partial\mathscr{L}/\partial\mathring{\mathbf{x}}^k$, and $\partial\mathscr{L}/\partial t$ and the generators $\tau_\nu(t, \mathbf{x})$ and $\xi_\nu{}^k(t, \mathbf{x})$. For this reason many derivatives of the transformation rules are needed with respect to the original variables of $\mathscr{L}$ and the parameter set $\varepsilon := \{\varepsilon^1;\, \ldots;\, \varepsilon^\rho\}$ around the identity $\varepsilon \equiv \mathbf{0}$. These derivatives are compiled in Appendix 2.1.1.

The total derivative of the divergence-invariance with respect to ε^ν around $\varepsilon = \mathbf{0}$ results, with some manipulations documented in Appendix 2.1.2, in the conclusive equation

$$\frac{\partial \mathscr{L}}{\partial t}\tau_\nu + \frac{\partial \mathscr{L}}{\partial x^k}\xi_\nu^{\;k} + \frac{\partial \mathscr{L}}{\partial \dot{x}^k}\left(\frac{d\xi_\nu^{\;k}}{dt} - \frac{d\tau_\nu}{dt}\dot{x}^k\right) + \mathscr{L}\frac{d\tau_\nu}{dt} = \frac{d}{dt}\Omega_\nu \tag{4.28}$$

$$\nu = 1(1)\rho \quad \textit{summation on } k.$$

Equation (4.28) is called the *fundamental invariance identity.*

Considering certain properties of the Lagrangian $\mathscr{L}$, an essential conversion of (4.28) can be easily achieved. All identities used for this formal manipulation are also compiled in Appendix 2.1.3. The alternate formulation of (4.28),

$$\begin{aligned}&\left[\frac{\partial \mathscr{L}}{\partial x^k} - \frac{d}{dt}\left(\frac{\partial \mathscr{L}}{\partial \dot{x}^k}\right)\right]\left(\xi_\nu^{\;k} - \tau_\nu \dot{x}^k\right)\\ &+ \frac{d}{dt}\left\{\left[\mathscr{L} - \left(\frac{\partial \mathscr{L}}{\partial \dot{x}^k}\right)\dot{x}^k\right]\tau_\nu + \frac{\partial \mathscr{L}}{\partial \dot{x}^k}\xi_\nu^{\;k} - \Omega_\nu\right\} = 0; \qquad \nu = 1(1)\rho,\end{aligned} \tag{4.29}$$

offers a mathematically transparent structure. Furthermore, it is quite appropriate to incorporate the Euler–Lagrangian equations (4.19–20). Inserting the latter into (4.29), the first bracketed term vanishes. Clearly, the remaining part of Equation (4.29) can be integrated immediately with respect to time. Before writing the result further simplification is applied, bearing in mind that the first part of the third bracket is identical with the negative Hamiltonian H. Let us then formulate *Noether's invariance theorem* in the light of the final formula

$$\begin{aligned}&K_\nu := -H\,\tau_\nu + \frac{\partial \mathscr{L}}{\partial x^k}\xi_\nu^{\;k} - \Omega_\nu = \textit{constant},\\ &\nu = 1(1)\rho \quad \textit{summation on } k,\end{aligned} \tag{4.30}$$

which has been derived here by means of differential calculus exclusively. In its precise meaning the theorem states (see Sato, 1981, pp. 242–251 for proof):

> If the fundamental integral of a problem in the calculus of variations is divergence-invariant under the ρ-parameter family of transformations, then ρ distinct quantities K_ν are constant along any extremal.

Equation (4.30) is true for every infinitesimal transformation with generators $\xi_\nu^{\;k}$ and τ_ν and gauge functions Ω_ν. In this case there exist some constants of motion K_ν of the holonomic dynamical system investigated, under the condition that for all indices $\nu = 1(1)\rho$ the factors $(\xi_\nu^{\;k} - \tau_\nu \mathring{\mathbf{x}}^k)$ are different from zero.

Having an appropriate reference to such properties of conservation, it is evident from the previous considerations that the possibility of finding a conservation law of a given dynamical system strictly depends on the special conditions of finding the generators of the infinitesimal transformations and the gauge functions. For the objective of this book, however, this general task is irrelevant. Here, the only important question refers to the connections between the generators $\xi_\nu{}^k$ as well as τ_ν and energy as well as momentum to be conserved.

Conservation of energy can be proved formally by reference to Noether's invariance theorem defining the following generators and gauge functions:

$$\tau_\nu \equiv 1; \qquad \xi_\nu{}^k \equiv 0; \qquad \Omega_\nu \equiv 0 \qquad \text{for all indices } \nu \text{ and } k. \tag{4.31}$$

In accordance to (4.26), the resulting transformation rule corresponds to a linear affine time

$$t' = t + \varepsilon. \tag{4.32}$$

Additionally, the suppositions (4.31) lead immediately to the conservation of the respective Hamiltonian H. With regard to the results presented in Section 4.2, the equivalence

$$E = -H \tag{4.33}$$

holds, provided that H does not explicitly depend on t. This means that t does not act as a variable, but only as a curve parameter. Conservation of linear momentum follows from a second set of generators and gauge functions with:

$$\tau_\nu \equiv 0; \qquad \xi_\nu{}^k \equiv \delta_\nu{}^k; \qquad \Omega_\nu \equiv 0 \qquad \text{for all indices } \nu \text{ and } k. \tag{4.34}$$

Hence, the transformation rule (4.26) yields linear affine position coordinates:

$$x'^k = x^k + \delta_\nu{}^k \varepsilon^\nu = x^k + \varepsilon; \qquad k := 1(1)s \quad \text{and} \quad \begin{cases} \delta_\nu{}^k = 1 & \text{for } \nu = k \\ \delta_\nu{}^k = 0 & \text{for } \nu \neq k. \end{cases} \tag{4.35}$$

Considering (4.34), Noether's theorem becomes

$$\frac{\partial \mathscr{L}}{\partial \overset{\circ}{x}{}^k} \delta_\nu{}^k = p_k \delta_\nu{}^k = \sum_{k=1}^{k=s} p_k = \text{constant}; \tag{4.36}$$

this result is equivalent to the conservation law of linear momentum.

In Appendix 2.1.4, this state of affairs is supplemented with the conservation laws derived from a mass-point model.

Note that both the combinations (4.32)–(4.33) and (4.35)–(4.36) refer to idealizations in the sense of definite approximations of real systems. This is true because Noether's theorem is exclusively formulated for a system where its Lagrangian is supposed to be available. But the crucial point is the close connection between conservation properties and coordinates that are fit to serve for the description of the system in question. For this reason both inferences from (4.32) and (4.35) with respect to the differential changes

$$dt' = dt; \qquad dx'^k = dx^k; \qquad k = 1(1)s \tag{4.37}$$

are of great importance in practice.

The decision to work within the space of events based on (4.37) is truly inevitable with regard not only to applications in practice but also to a convenient handling of some continuum concepts, along with a mathematical theory of non-equilibrium phenomena. As opposed to that, the results derived above can only be a guideline with respect to the conservation of energy and momentum. In fact, this property need not be generally true in the restricted sense expressed by (4.33) and (4.36), but rather in relation to (4.16). Consequently, adequate conditions for energy and momentum to be conserved must be established in quite a different manner. This is mainly the case if real (i.e., open) systems are regarded to be embedded in continuously running events. Nevertheless, for physics the old problem remains of how a useful concept of time should be introduced in theory and what its meaning in practice should be. The considerations expounded in Sections (4.2) and (4.3) surely instigate some stimulations. Hence, it is preferable to use time as a curve parameter in reference to (4.37) as well as to the limiting behavior for closed Hamiltonian systems. Still, as to the meaning of time in practice, we will conclude this subsection with T. S. W. Salomon's sophisticated remark at the 14th IMACS World Congress on Computation and Applied Mathematics in 1994. I can only give the gist of what he said:

> Due to the dogmatism of rational mechanics, many scientists and engineers are thoroughly convinced that the best of their theories are apt to forecast future events of reality in a mathematically suitable way. Caused by erroneous concepts of space and time, such an illusion concerns quantum mechanics, cosmology, and all the basic engineering sciences.
>
> Far more modest expectations may be fulfilled for virtual or realistic scenarios to be modeled on the one hand by restricted mathematical relations, but ruled on the other one by some more adequate ideas on motion and irreversibility.

4.3.2 PRINCIPLE OF LEAST ACTION IN MACROSCOPIC SYSTEMS

The steps that led to the differential relations (4.37) are based on Noether's theorem applied to a purely mechanical system of particles. Still, it is true that such a system of many degrees of freedom can incorporate a finite number of thermodynamic variables. Nevertheless, the Noether theorem should be extended to dissipative macroscopic systems, because there are some additional relevant conclusions that cannot be drawn solely by means of classical Hamiltonian systems. Very recently, Lauster (1995) proved that the total energy of physical systems must not be approximated by their kinetic and potential energies only, although even for them Noether's identity holds.

For a system defined by its Gibbs–Euler function $E - E_{\#} = \mathscr{E}(x^k;\, P_k;\, S;\, V;\, N)$, an extended version of this identity may be derived by starting from the Legendre-transformed energy

$$E_*^{[\mathbf{P}]} - E_\# = E - (\partial E/\partial P_k)P_k = E - \dot{x}^k P_k, \tag{4.38}$$

where the summation convention is relevant for the index k. In contrast to the classical Lagrange systems, which are distinguished by their immense number of degrees of freedom, this index k is running from $k = 1$ up to $k = 3$, corresponding to the three Cartesian space-coordinates. Clearly, $E_*^{[P]}$ is a function of the Lagrangian type, viz.

$$E_*^{[\mathbf{P}]} = \mathscr{L}_*(x^k;\ \dot{x}^k;\ S;\ V;\ N \mid E_\#); \qquad k = 1(1)3, \tag{4.39}$$

which depends on the two characteristic variables x^k; $\dot{x}^k$, and the three thermodynamic variables S, V, and N. In accordance with Helmholtz, we may call the time integral of (4.39) the *kinetic potential*, which is an extension of Hamilton's original action integral of the system concerned. (Another well-known approach to extending the variational principle to thermodynamics stems from Gyarmati (1970, p. 162). The comparison between his integral principle and the Hamiltonian principle is questionable because he deals with thermostatic events; that is, he excludes the variable **P** from the beginning. A much more realistic approach for a far-from-equilibrium nonstationary regime using a least action principle was recently published by Sieniutycz (1987).)

In accordance with Lauster's work (1996) on Noether's theorem, the following form of the extended Noether identity becomes

$$E_{EL}\left(\xi_s^{\ k} - \dot{x}^k \tau_s\right) + \frac{d}{dt}\left[\left\{E_*^{[\mathbf{P}]} - \frac{\partial E_*^{[\mathbf{P}]}}{\partial \dot{x}^k}\dot{x}^k\right\}\tau_s + \frac{\partial E_*^{[\mathbf{P}]}}{\partial \dot{x}^k}\xi_s^{\ k} - \Omega_s\right] = 0, \tag{4.40}$$

for which the formal agreement with (4.29) is evident. However, both relations are not the same in substance: Equation (4.29) is the starting point for Noether's theorem, mathematically expressed by (4.30), for a traditional Lagrange system. By use of its Euler–Lagrange set of ordinary differential equations (4.19)–(4.20), the step from (4.29) to (4.30) can immediately be executed. But what of an Euler–Lagrange set E_{EL} as defined by (4.40)?

The answer should be given by the *principle of least action* (PLA). This principle is part of a long history, paradigmatically proclaimed by Planck's famous lecture in 1915. Its use is inevitable because the notion of a process, presupposed to be basic for all physical changes in the configuration space, is yet to be defined. In this space, constructed by the chosen position- and time-coordinates, a system may pass through many events within a given interval defined by an initial space-time point and the corresponding final state. Human experiences cannot be traversed in any other way.

Above all, Planck maintained that the validity of the PLA does not apply only to mechanical phenomena, but to thermic and electrodynamic events as well. The PLA includes the energy principle, but the reverse is not true. Underlying reasons for that are that only the PLA exactly allocates the number of equations, which in itself equals the number of coordinates defining the space of events. This agreement follows from the variational principle proven to be basic for the PLA. "For it singles out from an innumerable family of virtual motions, conceivable within the framework of the constraints, a certain distinct motion by means of a simple characteristic and describes it as really occurring in Nature" (Planck, 1922, p. 105; author's translation).

Unfortunately, it is hard to find an exact mathematical formulation of the PLA. Which quantity is to vary under what conditions? Many have offered approaches to establishing this basic feature mathematically within the framework of a physical worldview: J. and D. Bernoulli (virtual displacements), D'Alembert (the passivity of all reaction forces), Gauß (least constraint), and Hertz (the straightest trajectory). All of these are properly applicable to mechanical processes only. *Leibniz's best of all worlds* formulation seems to contain the maximum of good along with the evil.

Now, the PLA may be formulated in such a way that a finite integral of motion is assumed to be extreme for the system in question. In this case it is decisive that only the energy of the system plays an important role with respect to a curve parameter—the time-coordinate. Thereby, nature is supposed to always work "economically." This means that nature should possess a dominating property, the essence of which is accurately expressed by the notion of energy. This economic idea of the least costs was perhaps best expressed in the famous qualitative version of the PLA proposed in 1744 by the Marquis de Maupertuis.

In 1788, Lagrange succeeded in formulating the first "modern" representation of a variational principle uniformly valid for most kinds of mechanical systems then known. But for theoretical investigations, the Hamiltonian version of the PLA offers several advantages to using the Lagrangian variation characterized by the fact that both variation δ and differentiation are commutative operations with respect to the time-coordinate t. This means that the property

$$\delta \mathring{x}^{\alpha} = d\,\delta x^{\alpha}/dt$$

is accepted as a postulate (Vujanovic and Jones, 1989, p. 28).

In conservation laws of dynamical systems, the variation may depend on time, on generalized coordinates, and on generalized velocities of the pertaining Lagrange function according to (4.18). Consequently, the PLA may mathematically be formulated in such a way that—among all the varied paths connecting the given states of the system for a prescribed time interval $[t_e, t_0]$—the actual motion renders the action integral I stationary. That is,

$$\delta I = \delta \int_{t_e}^{t_0} \mathscr{L}\left(t; x^1; \ldots; x^s; \mathring{x}^1; \ldots; \mathring{x}^s\right) dt = 0\,. \tag{4.41}$$

By tradition, Hamilton's principle (4.41) refers to systems that are defined by a set of discrete variables indicated by the subscripts $j = 1(1)s$. For this case the Euler–Lagrange equations (4.19)–(4.20) are the very solutions of the basic problem. The latter essentially consists of selecting the actual path assumed to be uniquely existing between two states measured at two instants of time, t_e and t_0, such that

$$\rightarrow t_e : x^j = A_j; \qquad j = 1(1)s \qquad \text{equals configuration } A$$

$$\rightarrow t_0 : x^j = B_j; \qquad j = 1(1)s \qquad \text{equals configuration } B,$$

where A_j and B_j are a given set of constants and $t_0 - t_e$ is the interval of time wherein any motion of the dynamical system is taking place. We may view these two special configurations as the "location" of the system's state experimentally determined at the initial and final instants of time. This point is crucial, as certain options of indirect experiences can now be assigned to the theory by means of the PLA.

The decisive advantage of such a general principle is often wasted, if concrete problems are to be solved following a few merely conventional rules:

a. Find a Lagrange function $\mathscr{L}$ of the pertaining system formulated explicitly and subject to the boundary conditions of the problem.
b. Vary the integral (4.41) and obtain the Euler–Lagrange equations (4.19)–(4.20).
c. Derive the corresponding partial differential equations of the problem, possibly by means of vanishing parameters introduced into $\mathscr{L}$.

Important cases (e.g., many heat conduction problems) demonstrate that it is quite impossible to put all cases into the classical framework of Hamilton's variational principle. The reason is that either no Lagrange function exists or the Lagrange function used cannot be connected with the energy of the system in question. But the latter point is substantial in view of the PLA: The question arises whether the kinetic potential of a real macroscopic system, subject to dissipative effects, can be varied in such a way that the resulting Euler–Lagrange equations E_{EL} are proven to be vanishing identically. This means that the identities

$$\varphi_j(t) \equiv 0; \qquad j = 1(1)s \tag{4.42}$$

hold for all t in $[t_e, t_o]$, in accordance to (4.20).

This general question can of course only be answered in general; that is, any answer in the sense of a principle of nature will have to be constructive within the theoretical framework. In other words, as long as there is no serious objection by experience or reasoning, the answer is yes. Naturally, this argumentation leads to a metaphysical conclusion, as it can never be proved by a finite number of data. But as with the meaning of the three laws of thermodynamics, the results justify this price physics has to pay. An alternative might be the a priori property of the principle of energy conservation, which still is questionable with respect to some theoretical aspects of space-time and gravitational relativity (Schmutzer, 1972, p. 50).

The topic under discussion additionally requires one to deal only with systems described by a set of continuous variables instead of the traditional discrete variables. In the case of a one-component single-phase system the system variables x^k, $\overset{\circ}{x}{}^k$, S, V, and N are used in connection with the Noether theorem given by Equation (4.40). They are presupposed to be continuous functions of the time- and space-coordinates t and x^k (k = 1, 2, 3). The explicit representation of the corresponding kinetic potential then becomes

$$I(\alpha) = \int_{t_e}^{t_0} L_*(\alpha)\,dt\,; \quad L_* = \iiint_V \mathscr{L}_{*V}\left(x^k;\, \dot{x}^k;\, S_V;\, N_V \,|\, x^k;\, t\right) dx^1 dx^2 dx^3\,; \quad k = 1, 2, 3. \tag{4.43}$$

The Lagrange density $\mathscr{L}_{*V}$ is determined by the set of the variables x^k (component of the displacement vector), $\dot{x}^k$ (component of the velocity vector), S_V (entropy density), and N_V (particle density). The variables themselves depend on the continuously changing time- and space-coordinates t and x^k. The difference between variables and coordinates is significant for the variational procedure δI related to the arbitrary parameter α and joined to the variational rule $\delta \to d\alpha(\partial/\partial\alpha)$.

The coordinates t and x^k are not generalized variables; they only serve as a continuous index substituting the discrete subscript j belonging to (4.41). For this reason, the variational procedure neither influences the limits of the time integral (4.43) nor the boundaries of the volume integral (4.43). Thus, any variation of the coordinates is inadmissible (but see Gyarmati 1970, p. 164: In Maupertuis' version the PLA can be attained by the convention that the time is also varied); this means that virtual displacements creating alternate paths, labeled by the parameter α, are only valid for fixed values of the coordinates t and x^k. However, it should be stressed that this is not true for the variable x^k, which determines the force field characterized by the force density

$$F_{*V,\,k} = -\left(\frac{\partial E_V}{\partial x^k}\right)_{\mathbf{P}_V,\, S_V,\, V_V}; \qquad k = 1, 2, 3, \tag{4.44}$$

according to the density E_{*V} of the Gibbs–Euler function $E - E_{\#} = \mathscr{E}(x^k;\, P_k;\, S;\, V;\, N)$ and affecting the system at position x^k.

Thus, if one applies the principle of least action, in the sense defined above, then the first term in Lauster's version (4.40) of the Noether identity vanishes as a consequence of $E_{\mathrm{EL}} = 0$. The result can be simplified by considering relation (4.38) between the system energy E and the Legendre-transformed energy $E_*^{[P]}$. Thus the *extended Noether theorem* becomes

$$E\tau_\nu - P_k \xi_\nu^{\ k} - \Omega_\nu = \text{constant}; \qquad k = 1, 2, 3 \quad \text{and} \quad \nu = 1(1)\rho, \tag{4.45}$$

where the parameters τ_ν and $\xi_\nu^{\ k}$ refer to the linear contributions of the transformation rules of the coordinates t and x^k according to (4.26)–(4.27). The physical meaning of the gauge functions Ω_ν in general is unknown and must be examined for each case. It is remarkable that, for instance, some special interpretations may be found for a few economic applications (see Sato and Ramachandran, 1990, p. 85).

Conservation properties follow immediately from (4.45) by way of specialization:

$$\begin{aligned} &\tau_\nu \equiv 1 \qquad \text{for all } \nu \text{ and } \tau_\nu \varepsilon^\nu = \varepsilon \rightarrow t' = t + \varepsilon \rightarrow E = \text{constant} \\ &\xi_\nu{}^k \equiv \delta^k{}_\nu \qquad \text{for all } \nu = k \text{ and } - P_k \xi_\nu{}^k - \Omega_\nu \equiv 0 \rightarrow x_k' = x_k + \varepsilon^k \rightarrow P_k + \Omega_k = 0. \end{aligned} \tag{4.46}$$

Equations (4.46) yield a striking result:

1. The conservation of total energy is connected with a linear affine time-coordinate.
2. The conservation of linear momentum is coupled to three linear affine position-coordinates according to the results proven to be valid for classical Lagrange–Hamiltonian systems. The gauge function, however, additionally balances the momentum of the system with regard to each component.
3. Evidently, both conservation theorems complement each other: The finite number of the space-coordinates restricts all indices to a definite number of facts according to Falk's finiteness axiom introduced in Section 2.1.

Point (2) indicates that the physical meaning of Ω_k strongly depends on the holistic approach of the mathematical theory presented. Thus, the absolute invariance condition $\Omega_\nu \equiv 0$ of Hamilton's action integral (4.41) under infinitesimal transformations (4.23) (see, e.g., (4.22)) cannot be satisfied, provided the conservation of energy and linear momentum are regarded to be true for macroscopic systems.

There is a simple explanation of $\Omega_\nu \neq 0$ for such systems with regard to (4.46). Together all the Ω_k represent the vectorial momentum $\mathbf{P}^{\text{if}}$ of the inertial field penetrating every moving body. In other words, the definition

$$\Omega_k := P_k^{\text{if}} \, \xi_\nu{}^k; \qquad k = 1, 2, 3 \tag{4.47}$$

shows how the conservation of momentum is realized in nature. Two classes of problems are considered: The first refers to motions of any single body. The second is indicated by displacements of the center of mass constituted by several mutually interacting bodies.

Note that empty space is also manifested by the inertial field, whose interaction with any other body affects nothing but the inertial events. The states of this field depend on the observer, but this is not true for the existence of the field itself. The basic property of an inertial field concerns its ability to interact not only with every material body, but also with other fields that are furnished with energy and momentum (Falk and Ruppel, 1983, p. 200). For example, an electric field only interacts with charged bodies or particles, but the inertial field represents a universal characteristic of nature closely linked with a gravitational field.

The exchange of momentum in so-called inertial systems takes place only between the bodies or fields actually involved. However, such special schemes are reference frames that work well only in uncommon situations. Generally speaking, unsteady motions or any accelerations cannot realistically be described within the framework of inertial systems.

Notwithstanding, two distinguished frames may be mentioned for systems each consisting of several mutually interacting bodies:

$$\Sigma_\ell \mathbf{P}_\ell = \mathbf{P}_{\text{syst}} = \text{constant} \Leftrightarrow \text{inertial system}; \; \Sigma_\ell \mathbf{P}_\ell = \mathbf{0} \Leftrightarrow \text{center-of-mass system}; \tag{4.48}$$

the subscript ℓ, indicating the respective linear momentum $\mathbf{P}_\ell$, runs from $\ell = 1$ up to the total number of all bodies involved.

A second application of the action integral (4.43) concerns continuum physics. The option to describe precisely a distributed variational problem in Eulerian, or a field representation of a control volume in time presupposes an exact expression for the respective kinetic potential. A formidable difficulty is that this Lagrangian of any system subject to irreversible changes of state is unknown. Moreover, systematic rules for obtaining the adequate Lagrangian are not yet available. For this reason a new approach published by Sieniutycz and Berry is remarkable. The authors construct (rather than assume) a kinetic potential describing processes of heat transfer, such that the Second Law is obeyed. With reference to some important relations to results presented in Chapter 6, it seems appropriate to summarize the answers to the authors' question: "What has been achieved in this development?" (Sieniutycz and Berry, 1993, p. 1783).

1. A unification of the First and Second Laws in connection with the extremal behavior of action. In this context, thermal and fluiddynamical changes of state may be simultaneously described by a Gibbs fundamental equation, obeying Legendre transformations with respect to both mechanical and thermodynamical variables.
2. A mathematical procedure for deriving conservation laws and equations of transport within a generalized Hamilton's variational principle, including dissipative terms and still leading to the canonical formalism.
3. A view of heat and work rate as effects of the transport of entropy in the flow region.
4. A proof that the total energy E is not the same function as the common term expressed by summing up all relevant types of energy such as kinetic, potential, and internal energy. Certain additional terms appear in the equations of motions that preserve dissipation.
5. A supposition that far from equilibrium, the energy and conservation laws may have forms different from those currently accepted, with the non-equilibrium temperature playing an essential role.

4.4 Phases, Heating, and Power as Interacting Phenomena

In Section 1.5, some remarks were made on a variety of possibilities to be activated by variational principles in favor of nonconservative dynamical systems. Sections 4.2 and 4.3 reflect some fundamental problems of analytical mechanics with regard to space- and time-coordinates, as well as to their relation to conservation laws. Recently, concise representation of this topic and its history was given by Schirra (1991, pp. 84–86).

Although the results are valuable for various aspects, bear in mind that the methods of variational calculus are very limited to a few select applications in theory and practice. Even a cursory inspection of the literature reveals that for very few of the important irreversible processes does an exact Lagrangian of the problem in question exist. For example, one cannot find a Lagrange function for the transient parabolic differential equation of heat conduction in solids, not even in the linear one-dimensional case. Of course, there are some well-founded attempts to modify the basic rules of the original variational calculus, but such attempts always concern approximate solutions whose mathematical structure reveals its exclusively pragmatic character. Refer to the fine book of Vujanovic and Jones (1989), yet be aware that the situation is similar to that of the so-called *extended irreversible thermodynamics*. That is, although we may be justified in extending the set of the Lagrangian variables by additional field functions together with their partial derivatives (Vujanovic and Jones, 1989, p. 241), these efforts are little more than an approximation at best (see Section 3.2).

To prepare the mathematical tools for the description of real processes observed along with non-equilibrium phenomena in physics, we need to address important issue. Let us assume any system to be open against its surroundings. This means that the system in question, defined by its M–G function, interacts with a set of systems beyond the surfaces bordering the volume of the system regarded.

Such interactions are traditionally described by two different methods:

1. In *theoretical physics* each system, assumed to be adjacent to the special system under consideration, is characterized by its own M–G function. This distinguished system has to transfer some energy, momentum, or mass to its environment (or vice versa), prescribed by the pertaining interaction laws. This method is preferred when dealing with the trend toward so-called phase equilibria.
2. In the *engineering sciences* the interactions take place according to the First Law of Thermodynamics.

Method (1) permits the handling of only a few cases in an exact manner. Method (2) indeed refers to a powerful tool, but also gives rise to the following discussion on a far-reaching misconception documented in many textbooks of today's physics and engineering.

4.4.1 THE DOGMA OF THE DIMINUTIVE

First let us define a simple system by its M–G function $E(\mathbf{P}; S; V; N)$ and more specifically by its Gibbs main equation

$$dE = \mathbf{v} \cdot d\mathbf{P} + T_* \, dS - p_* \, dV + \mu_* \, dN, \, . \tag{3.6}$$

Then the question arises: How is interaction with the material environment settled by the general condition $dE \neq 0$ described with the help of the First Law of Thermodynamics.

In many of today's physics or technical thermodynamics textbooks, the usual answer is an expression of the form

$$dU = đQ + đW, \tag{4.49}$$

implying the option of changing energy into two different terms called *work* and *heat*, respectively (Zemansky, 1968, p. 81). The operator đ was introduced by C. Neumann in 1875. To quote its creator, "The hooked đ represents an infinitely small quantity believed to be undescribable as an increment of any function" (1875, p. IX; author's translation). In other words, this operator đ (called diminutive by Neumann) is simply a symbolic indicator of the intention to transcend "physical truth." Unfortunately, mathematics does not enable us to do this. Note that, for instance, both the terms $đQ$ and $đW$ "need not be small themselves in order to give a small increment in the energy E" (Clarke and McChesney, 1964, p. 42).

Unfortunately, many physicists and engineers, working in thermodynamics by means of mathematics, use formulas of the following type,

$$\begin{gathered} đE_{\text{kin}} + đE_{\text{pot}} \neq đ(E_{\text{kin}} + E_{\text{pot}}); \\ W(X;Y) \Rightarrow đW \neq \left(\frac{\partial W}{\partial X}\right)_Y dX + \left(\frac{\partial W}{\partial Y}\right)_X dY\,; \\ \int_1^2 đW \neq W_2 - W_1\,; \qquad \int_1^2 \frac{đQ}{T} \neq \int_1^2 \frac{\mathring{Q}}{T}\, d\lambda, \end{gathered} \tag{4.50}$$

even though no one has mathematically proven that the left-hand side equals the right-hand one. Other "relations" of such a dubious quality are frequently used in literature. Thus, we find for infinitesimal and reversible changes of state expressions such as

$$dW := -p\, dV \quad \text{for work} \qquad \text{or} \qquad dQ := T\, dS \quad \text{for heat}, \tag{4.51}$$

even though in the majority of practical cases neither the pressure p nor the temperature T depend on a single variable alone. In other words, functions like $p = p(V)$ or $T = T(S)$ do not exist in general. For this reason, expressions like (4.51) strictly cannot be interpreted as a total derivative, but any other meaning would be mathematically incorrect. This is true in general, even in the case that the diminutive $-\,đW$ is simply replaced by $-\,p\, dV$ or another energy form (Zemansky, 1968, pp. 54, 60, 61, 64) to construct the integrating factor for the diminutive $đQ$.

It is hard to apprehend the broad use of (4.49) and (4.51), especially in formal education. Even the mathematical genius Henri Poincaré (1892, p. 66) unscrupulously employed such "equations," as did the great mechanician and philosopher Ernst Mach (1896, p. 274). Energetics was ruled by (4.51) thoroughly—see Helm (1898, p. 244). Characteristically, Max Planck at least suspected that diminutives make things difficult (1913, p. 55). But such objections were lost again in modern and renowned textbooks.

Certainly, there are some historical reasons for this broad usage and the overwhelming influence of mechanics on the development of thermodynamics may be an example (see, e.g., Straub, 1990, p. 27). But there may also be metaphysical reasons with regard to the First Law. Thus, for instance, Kestin claimed that the distinction between the perfect differential operator d and a diminutive $đ$ "constitutes an important physical truth which transcends the fact that Q and $đQ$ have been introduced to balance the equations" (Kestin, 1979, p. 164).

But we should also remember, say, Gibbs or Tisza, who never explicitly used the First Law in their salient works. In fact, the First Law is metaphysical; that is, it is not mathematical in the strict sense of Falk's dynamics: The First Law cannot be proved or disproved with a finite number of data. For every doubtful case the observer might always hold responsible some unknown or hidden constraints. Seen from a more methodological viewpoint, even Popper's famous falsification test would be obviously questionable.

Nevertheless, the First Law is of great significance for a mathematical theory of non-equilibrium phenomena. This is true because it offers some options to extend Falk's dynamics systematically to continuously running processes. To elaborate on such a concept, the following preconditions must be satisfied.

1. A space of events defined by time- and position-coordinates should offer the possibility of describing continuous physical processes.
2. The object of every process is a so-called body-field system for which all relevant conservation laws can be fulfilled.
3. Body-field systems are open systems by definition.

Point (3) refers to the trivial fact that the First Law must be treated with correct mathematics. In addition, the term *open* must be understood in a mathematically flexible way to differentiate clearly between the notions of work and heat.

It is easy to obtain a mathematically correct description of the First Law. Truesdell (1984, p. 67) explained the relationship between heat and work in the following way:

> We assume the existence of a second kind of working, Q, called the heating, which is not identified with anything from mechanics. Q is a function of time having the same dimensions as W and hence may be added to it so as to yield the rate of increase of internal energy E according to the first axiom:
>
> $$\dot{U} = W + Q. \tag{4.52}$$

Hence, it appears that he solved the problem by means of the corresponding differential equation. Although one can accept such a mathematical representation, it is obviously incomplete because it refers to systems consisting of mass-points. But this problem was fully discussed in both Chapters 1 and 3. [Truesdell (1984, p. 108) himself confirmed this supposition by using his integral expression (2.4) for linear and rotational momentum of a body approximated as a single mass-point in accordance with his equation (3.9).]

4.4.2 COMPOSITE PRINCIPLES OF PHASES

Let us now examine the possibility that the First Law only needs to include the total energy E. In principle, we may describe the concerning differential equation by the use of any adequate curve parameter λ. Yet, with regard to the results of the last sections, it is advantageous to employ the time-coordinate t. Then, the First Law in its recommended version becomes

$$\frac{dE}{dt} := W + Q \,. \tag{4.53}$$

The right-hand side of this definition contains two rates: *work rate* or power W and *heat rate* or heating Q.

Combining (4.53) with the Gibbs main equation (3.6), the relation

$$\frac{dE}{dt} := W + Q \; = \mathbf{v} \bullet \frac{d\mathbf{P}}{dt} + T_* \frac{dS}{dt} - p_* \frac{dV}{dt} + \mu_*^{\;t} \frac{dN}{dt} \tag{4.54}$$

arises, which describes the behavior of the simple system defined by its M–G function $E(\mathbf{P}; S; V; N)$ and subject to interactions with its surroundings. Let us compare the last two equations under adiabatic conditions, that is, $Q \equiv 0$. By integration of (4.53), the change of energy is found to be

$$E_{t_2} - E_{t_1} \; = \; \int_{t_1}^{t_2} \frac{dE}{dt'} dt' \; = \; \int_{t_1}^{t_2} W\,(t'; \mathbf{P}; S; V; N; \textit{external variables})\, dt' \,. \tag{4.55}$$

This expression clearly demonstrates the general problem when handling the First Law: Except for simple cases, the functional W (in the external variables) is generally unknown. For this reason it is common practice to evaluate (4.55) by the right side of (4.54), which exclusively contains system variables, combined with its boundary conditions.

To assess this option, as well as the misapprehension implied in (4.54), some comments are necessary with regard to the following possible issues.

Phase Equilibrium

Let us imagine a rigid box that is subdivided into two parts and closed completely against the whole environment. Part 1 is filled with a fluid fixed in a state at rest and defined by its M–G function $U_1(S_1; V_1; N_1)$; the same is true for part 2 with reference to $U_2(S_2; V_2; N_2)$. The fluids in 1 and 2 are separated by an unspecified wall. This constellation is fully described in Section 2.3 with regard to the resulting conditions for equilibrium and stability of the composed system starting with different states in 1 and 2.

The crucial point of the solutions offered by Equations (2.38) and (2.42) refers to their chosen presuppositions . The latter are basic as to the following items:

- Phases and phase equilibria
- Composite principle
- Partition walls

Whereas the form of the First Law is independent of the system in question, the corresponding Gibbs main equation, given in parameter form by the time-coordinate t, is directly influenced by the mathematical structure of the M–G function assigned. Thus, for instance, the M–G function $U(S; V; N)$ describes a one-component single-phase system, in other words, the simplest case of a phase. According to this, the equilibrium between parts 1 and 2 of the box may be denoted as phase equilibrium and explicitly assumed to be independent of the properties of the wall. This means that surface phenomena and shape-dependencies are not considered.

The case in question belongs to the important type of systems called *phases.* Phases are the structural elements for building up systems under certain conditions (to be presented in the next chapter). Here, it is sufficient to state that systems may be constructed from subsystems coupled via the exchange of quantities to be conserved.

The notion of a phase indicates the basic idea of establishing more complicated systems by means of the composite principle noted in Chapter 2. It concerns composite systems defined as a conjunction of spatially disjoint simple systems. A composite system is obtained by uniting separate systems or by partitioning a single system. The simple system itself is a single-phase system with one or more components. It is defined by its M–G function $\xi_0(\xi_1; \ldots; \xi_r)$ and assumed to be spatially homogeneous. The intrinsic problem of the composite principle arises from the so-called *additivity postulate*, which is defined by

$$\xi_j := \Sigma_p \xi_{j_{(p)}}, \quad j := 0,1(1)r, \quad p := 1(1)p_\#, \tag{4.56}$$

where the summation is over all $p_\#$ subsystems. The quantity on the left belongs to the composite system.

One of the main problems here is that the usual concept of an interface as a surface of discontinuity, while suitable for the description of many properties of multiphase systems, is in fact not realistic. Additionally, it leads to some paradoxes if applied to the interface itself. The naive approach of including the surface area in the set of extensive variables cannot be done if the extensivity of all the other system variables is defined by the use of the pertaining mass and the interface is assumed to be approximately a purely geometric quantity.

Actually, in the vicinity of what is observed as an interface between two phases, the mass density does not undergo a discontinuity. The mass of the surface layer is small but finite. For this reason, the density exhibits a very steep gradient over an extremely short distance compared to the radius of curvature of the interface and is directed orthogonally to it.

Clearly, the interface between two fluid phases at equilibrium can only sustain an isotropic surface tension. However, under dynamic conditions this is no longer evident: A *surface tension tensor* needs to be considered (Astarita, 1990, p. 160) that connects a vectorial surface force with a two-dimensional vector orthogonal to any line element of the surface and with a modulus proportional to its length.

A further kind of surface phenomena occurs in mixtures and concerns adsorptions at the interface. They can be incorporated in Falk's dynamics by means of the vector of the relative adsorption as an appropriate variable that is conjugated to a vectorial chemical potential of the particle numbers of a mixture determining the actual surface composition. This is relevant for solids where diffusion from and to the interface may be slow so that no equilibrium exists between the actual surface composition and the respective bulk data. Corresponding catalytic properties of solid surfaces are related to this type of phenomena (cf. Astarita, 1990, p. 164).

The thermodynamic theory is normally developed by substituting the real system with an "ideal" model consisting of, for instance, two bulk phases and a geometric surface—two-dimensional in practice, but endowed with finite mass. For this case Equation (4.56) leads to the following relations

$$\begin{aligned} U &= U_1 + U_2 = \text{constant}; \qquad S = S_1 + S_2 \neq \text{constant} \\ V &= V_1 + V_2 = \text{constant}; \qquad N = N_1 + N_2 = \text{constant}, \end{aligned} \tag{4.57}$$

where the conversion factors given by (3.2) for the M–G function $U(S; V; N)$ are used. Equations (4.57) are introduced as constraints to be predominant for the trend of the two subsystems 1 and 2 toward phase equilibrium.

Additive Invariance

Regarding additive invariance, Tisza's statement is indeed appropriate:

> The requirement of additive invariance is a characteristic feature of the present theory, in which thermodynamic systems are constructed from spatially disjoint subsystems coupled by the exchange of conserved quantities. Additive invariance is of as much importance to the present formalism as the assumption of nondissipative forces in analytical mechanics (Tisza, 1966, p. 122).

For this reason, we must consider the conditions under which these quantities can actually be taken for additive invariants. We must also consider the status of those elastic, electric, and magnetic variables that are not on the list of invariants, although they are generally regarded as extensive quantities.

It is strange that the only geometric variable listed in the set of invariants is volume. This is equivalent to the fact that the basic theory neglects surface phenomena and shape-dependent effects. Indeed, to adequately describe, for example, capillary effects or elastic strains, an extended set of quantities is needed. Of course, the volume V is an additive, and for reasons mentioned is above all a conserved variable.

Yet, we should notice that V represents a quantity with properties, the origin of which is rooted in pure intuition and convention. In particular, it may be the Newtonian idea of an absolute space, occupied by either matter or nothing, that prevents an unbiased view on the bounds of the term *volume*. As Falk has proven, there does exist a quantity $\acute{u}$, the value of which depends on the material nature and the physical state of the object assumed to be fixed within a prescribed volume V. Whereas V describes a geometric extension quite in contrast to $\acute{u}$, the quantity $\acute{u}$ describes a "physical" extension. This means that every sort of matter has its own measure of space that is, however, not automatically related and reduced to the three-dimensional space coordinates like V.

In this sense, the true importance of V is not only founded on well-established facts but also on historical arguments, mainly due to idealized imaginations about rigid bodies and mass-points (Falk, 1990, p. 367). As a whole and above all, the volume V should be considered as a scale factor, serving merely to define the size and the stability of the system in question. Furthermore, V may provide a constitutive attribute with regard to the introduction of walls (still to be discussed in connection with some remarks on the First Law).

To turn to the internal energy U as an additive invariant, remember that this term exists only for systems in a state at rest and is, moreover, subject to the qualification that the potential energy, arising from external fields, is subtracted. Experiences have disproved the validity of the additivity postulate for U to be inferred from general principles. Therefore, we must require that any special interaction energy between thermodynamic systems be negligible.

Theoretically, this conjecture can be a posteriori justified by relevant experiences with regard to the truly observed stability of phases. Thermostatics implies that this stability is an intrinsic property, virtually independent of the size and shape of the system. The crucial point is that this result follows essentially from homogeneity and just from additivity (see Section 2.3).

In practice, however, the situation differs from system to system. Within the range of common experimental data the additivity postulate is justified to such an extent that it may be taken for granted. These experiences cannot be overestimated, particularly in reference to the respective equations of state given for unit amounts of each substance and scaled to any size of interest. Yet, that postulate obviously breaks down and

> stability becomes size-dependent in small droplets because of surface effects, in stars because of long-range forces, and in heavy atomic nuclei because of the combination of both factors. In all these cases the basic theory is to be modified.
>
> We turn to the discussion of the variables describing the distribution of matter. In the absence of chemical reactions, the mole number of the chemical species are additive invariants. In the presence of chemical reactions, these numbers do not satisfy a conservation law. However, it is possible to define chemical components, the mole numbers of which are invariants of the chemical reactions, and are appropriate thermostatic variables (Tisza, 1966, p. 123).

4.4.3 *THE FIRST LAW OF THERMODYNAMICS*

The evolution of phase equilibria is restricted to such situations where the trend to the respective equilibrium is actually dominated by those variables for which the additivity postulate is valid. This is always true if the coupling between the subsystems involved is weak. Then, the characteristic time t, associated with establishing phase equilibria, takes much longer to elapse than all the times $\tau_{\mathrm{int},i}$ to bring about internal equilibria. This introduces two time scales, where one is considered to be macroscopic and the other microscopic.

There are many phenomena with interactions between the subsystems for which $t > \tau_{int,i}$ cannot be fulfilled. Hysteresis effects belong to an important class of applications, and thermal radiation, caused by a field of photons normally far from internal equilibrium, is another relevant example. The most significant cases, however, refer to latent variables for which additive invariance is neither justified theoretically nor wanted in practice.

Consequently, the guiding principle is to account for a complete energy balance and all variables that are needed to describe the work performed or received by the system. Thus, the components of the strain tensor, as well as those of the electric and magnetic polarization vectors, are joined to the set of independent variables. In general, there are spatial gradients with respect to the intensive quantities of both material subsystems, indicating the fact that phase equilibrium is not reached at all. As a rule, those cases belong to a key branch of engineering sciences, usually called *heat and mass transfer*. Its basic tools are the First and Second Laws of Thermodynamics.

It is beyond the scope of this book to give an outline of the complicated mixture of theory and experiment characterizing heat and mass transfer in practice, especially since the discussion of the First Law would demand dealing additionally with work transfer. Therefore, let us focus on a few qualitative aspects concerning the rates of the quantities "heating" and "work rate" with regard to the meanings of these notions and the significant differences between them. The differences especially can lead to considerable misconceptions due to the fact that open systems do not immediately play an essential role in the leading theories of modern physics (cf. Straub, 1990, p. 171–172). But "heating" and "work rate" are the key terms of such systems. For this reason, they are constitutive for many problems of engineering, biology, and medicine.

Starting with a preliminary definition of heating as a "transfer quantity, associated with energy, across a rigid diathermic partition" (Tisza, 1966, p. 113), let us regard three points:

- Unlike the traditional idea in physics, heating is not explicitly declared as irregular motion of particles enclosed within the boundaries of a system.
- The definition of heating is tied to the notion of a wall and refers to the transfer of energy rates between the system and its surroundings (and vice versa).
- The "wall" in reference to heating is "rigid" in the sense of volume and shape preserving, but it is also unaffected by any other variable of the system.

In Table 4.1, some characteristics of walls are compiled regarding their nomenclature and the type of energy rate to be transferred (or not). The scheme should clarify that the five types of walls represent different devices, each working by means of special mechanisms and providing the realization of the pertaining transfer process. Moreover, the concept of a wall precludes a change of shape of the system enclosed.

Table 4.1 Wall characteristics concerning heat and mass transfer

Flux Wall	*Flux of Matter*	*Flux of Energy*	*Flux of Chemical Species*
impermeable	restrictive	★	★
semipermeable	★	nonrestrictive	nonrestrictive
permeable	nonrestrictive	nonrestrictive	restrictive
adiabatic	restrictive	restrictive	★
diathermic	restrictive	nonrestrictive	★

Compared with the traditional interpretation, the generalization of this wall concept indicates that heating may be transferred not only by means of the existence of temperature gradients (e.g., by diathermic walls), but also as a consequence of gradients established by different values of chemical species in the system and its surroundings. Even "mixed" heat transfer occurs in the form of so-called thermodiffusion processes achieved by simultaneously operating gradients of mass constituents and temperature. Although the theory mainly deals with shape-independent effects, a limiting case will necessarily have to discern between "heating" and "power." For this reason, it is necessary to specify whether or not an enclosure is *rigid*. The assumption that all walls and enclosures remain unchanged while the process is underway serves as an idealization to exclude a compressional work rate (a work rate of a surface film, for example).

But there are other types of macroscopic work rates, defined by means of system variables other than volume or the shape of the enclosing surfaces. Remember that in thermodynamics a work rate always involves a boundary-crossing interaction between the system as a whole and its surroundings. Thus, for instance, variances that take place in an electric cell are not accompanied by the performance of work if the cell is on an open circuit. The situation undergoes a radical change, however, if the cell is linked to a circuit through which electricity is transferred across the boundaries of the system. Then, the current is considered to be the source of the work rate, producing, for example, the rotation of a corresponding device within the system. The same is true for a magnet passing through a change of magnetization while it is surrounded by an electric conductor. In contrast, a change of magnetization within the magnet is not associated with a produced physical work rate.

Recently, a new aspect of the complementary properties of work rate and heating is being explored (see, e.g., Stonier, 1990). Here we will only give some indications regarding these key notions of the First Law. Particularly, the notion of work rate, like all forms of energy other than heating, contains an information component: Mechanical energy activated by motion comprises distance, internal times, and direction—all exemplify forms of information. Chemical compounds are dependent on the special patterns of electronic structures of the atoms and molecules involved as reactants. The osmotic work rate is strongly conditioned by the organization of semipermeable membranes. Electrical energy is influenced by structures that build up nonrandom charges. And atomic energy relates to the organization of the atomic nucleus.

Thus, the relationship between work rate and information is twofold:

Any performance of work rate may give rise to a change in the information content of the system acted upon.

To acquire a work rate, a system must be provided with both energy and information.

Note that the concept of work rate entails in part the transformation of energy into information. Four possible changes in the system may result:

Energy is merely absorbed, thereby increasing the entropy.

Due to the input of energy, the system itself becomes organized to a higher degree.

The system accomplishes work rate by information.

The system gains information by work rate.

All four processes may occur in any combination whatever. Regarding the third change, for example, in a steam engine the energy supplied pushes the piston in a fixed direction, in accordance with the designer's plans as the relevant information input. As another example, when a photon is absorbed, an electron is shifted to an outer shell, thus causing the atom to achieve a less probable state in the proper sense of thermodynamic probability.

Let us use Stonier's approach of relating information and organization directly and even linearly. For ease in understanding, he also assumed a first quantitative approximation of information to be defined by reference to the entropy of the system (divided by the Boltzmann constant k_B). Thus, the expression

$$I := I_0 e^{-\frac{S}{k_B}} \tag{4.58}$$

"defines the fundamental relationship between information I and entropy S." (Stonier, 1990, p. 39). The quantity I_0 represents the information constant of the system at zero entropy.

To obtain the work rate, energy must either contain information itself, or act on some organized device that operates as an *energy transducer* (or both). The energy transducer possesses organization and structural information without which it could not succeed. As a rule, it is realized by contemporary standards, even for highly complex technology such as a nuclear reactor generating high-pressure steam. "Energy transducers create two conditions necessary for the production of useful work rate: (1) They create a non-equilibrium situation; and (2) they provide a mechanism of countervailing force necessary for the production of useful work" (Stonier, 1990, p. 96). Every energy transducer is characterized by its structural information. An essential property of such information is that it does not "wear out." This is best represented by a photovoltaic cell, which does not wear out by converting light into electrical energy. "Even a steam engine which does wear out after many years, for practical purposes shows zero change in organization during a single cycle in which there may be large changes in temperature T, pressure p, or volume V" (Stonier, 1990, p. 98).

In Stonier's scheme, the first step generally involves the creation of certain non-equilibrium conditions organized by adding the adequate information. As a result, a reduction in entropy of the system takes place. An efficient energy transducer in the second step tends to keep the increase in entropy to a minimum by coupling this set of non-equilibrium states to a mechanism that allows the system to do work steadily.

Let us look at three examples to clarify this scheme. The first concerns steam engines: Any mechanical device is driven by its piston interposed between the highly energetic steam and the relatively low energetic molecules of the exhaust steam. The second example pertains to a photovoltaic cell containing atoms that absorb photons. The resulting instabilities are exploited by trapping electrons into an electrical circuit that is able to produce work. The third example involves a living cell containing an enzyme that removes or adds electrons to the atoms of the reactant, thereby making it unstable. These non-equilibrium states are then utilized to couple the reactant to other ones, or to itself, to form, say, a polymer.

Such living systems provide the information needed to transform an input of energy into work under conditions not possible in technical inventions. This is especially true with regard to comparable levels of organizations. The photosynthetic mechanism of a plant cell, for instance, is able to dissociate water and strip electrons of hydrogen atoms at room temperature. In sharp contrast, in a technical system water needs to be heated to over 1000°C so that water molecules are destroyed by vehement collisions to produce ions and electrons.

Photosynthetic systems are able to accomplish this power at room temperature as a result of the ingenious pattern of molecules (chlorophyll) highly endowed to absorb light. Hence, the electrons are strongly excited to activate the entire cellular machinery based on membranes and enzyme molecules. This permits a lowering of the activation energies required for the huge number of metabolic reactions. The structure of the organization, realized by means of the properties and the mutual interactions of the molecules involved, determines the ability of the molecules to perform the work needed. The Second Law may be omnipresent, but indeed, whenever entropy is increased in living systems "as a by-product of metabolic reactions, it is more than compensated for by converting much of the 'free energy' obtained from these metabolic reactions" (Stonier, 1990, p. 100). For this reason, all living systems are extremely complex. This means that any complex system will have to be highly organized, polyphasic, and equipped with the capacity for vast stores of information. Even self-organization takes place.

A first conclusion is evident:

- The mechanical notion of work is only useful for simple applications; it represents only a primitive concept which cannot be extended to complex systems. As opposed to this, all complex systems must be considered as dissipative and interactive.
- Real systems are complex: They are able to perform some work rate provided that an inherent mechanism is available in the form of an adequate device. This device operates and controls the non-equilibrium conditions by means of structural information, which is mandatory for coherently running processes on the atomic level.

Heating is a form of energy transfer lacking information. If no work is performed, then, according to Equation (4.53), heating manifests itself by the temporal change of the total energy caused by interactions with the surroundings of the system in question. For this reason, the term *heating* is commonly believed to be equivalent to the concept of uncorrelated phonons in a solid or the random motion of molecules in a gas. But such an interpretation seems vague and meaningless considering the fact that heat transfer is ruled by the mechanisms dominating the wall properties noted above.

Nevertheless, it is true that *heat rates* cause particles (molecules, atoms, phonons, plasmons, etc.) to vibrate and to move at random in space and time. "In that sense, heat may be considered as the antithesis of organization" (Stonier, 1990, p. 74). In contrast, the application of *work rates* causes particles to be bound into fixed patterns and ordered motion. No wonder it is hard to transpose the qualitative representation of work and heat rates into a quantitative version by the use of mathematics. It is common practice to operate with the First Law, provided that the process regarded can be specialized as an adiabatic or a "workless" one. In both cases the rate quantity in question can be expressed via (4.54) by the set of relevant energy forms whose time derivatives other than heating exhibit, or are dependent on some sort of organization or pattern with respect to space or time.

A second conclusion is that at present it is impossible to realistically describe very complex systems, for instance the space-time behavior of the worldwide climatic evolution. This is true, in spite of the available capacity and actual efficiency of the computers used. There is reason to suppose that neither the complex dynamics of the multiphase multicomponent system nor the interaction laws of the boundary constraints and, above all, their temporal behavior are sufficiently known. But even single-phase multicomponent systems are complex. Therefore, it is our intent to study the relevant properties of such systems with strong emphasis on dissipative and non-equilibrium phenomena, along with compressible and even polarized fluids.

Chapter 5 | Barriers and Balances

"But how by rules shall I commence? You set the rules and follow thence."
—***Richard Wagner***

5.1 Body-Field Systems

From Equation (4.43) in the previous chapter,

$$\frac{dE}{dt} = W + Q = \mathbf{v} \bullet \frac{d\mathbf{P}}{dt} + T_* \frac{dS}{dt} - p_* \frac{dV}{dt} + \mu_*^t \frac{dN}{dt}\,, \tag{4.43}$$

the following three general instructions can be obtained:

1. A one-component single-phase system is well described for non-equilibrium processes, including interactions with its environment via heating and work rate.
2. Open systems of type (1) are mathematically determined by two adequate levels of representation: The first level always requires the constitutive Gibbs – Euler function (3.8) with the associated variables for the Gibbs space representation. The second level is then established by introducing the *configuration space* (sometimes called the *space of events*) to be spanned by time- and space-coordinates.
3. The time- and space-coordinates must not be regarded as system variables of the Gibbs space. However, they are mandatory for the explicit consideration of interactions with the surrounding environment and, furthermore, they are inevitable for the local conservation of total energy and momentum.

Undoubtedly, the definite separation between these two levels of description is mainly the result of Falk's dynamics. This method, explained extensively in Chapter 2, also permits a continuous election of state values by mapping the set of system quantities onto the set of its variables. This continuum only refers to the field of real numbers and has nothing to do with the continuum hypothesis by which field variables in the space of events are defined.

As a consequence, joining a priori these two levels of description it seems to be a basic methodological defect in mechanically founded theories. Two examples should substantiate this. The first concerns the theoretical concept propagated by the adherents of extended irreversible thermodynamics. Its variables comprise the spatial derivatives of some quantities as well as their time rates, the definition of which already presumes the explicit use of the term *time* (see, e.g., Garcia-Colin, 1984, p. 155). Our previous discussion about the correct form of the First Law of Thermodynamics confirms this assertion.

The second example, the "principle of material frame-indifference" (first introduced by Truesdell), is perhaps more affected. It postulates that the constitutive functions, commonly used in continuum physics for the representation of local dissipative phenomena, are invariant toward the Euclidian reflection-transformations. In particular, those functions have the same mathematical form in both an inertial and a noninertial frame. Consequently, it is evident "that t and $\mathbf{x}$ cannot appear as arguments of any constitutive relation" (Truesdell, 1984, p. 144).

It is well-known that this principle can lead to erroneous results. The simple kinetic theory of nonuniform gases based on the Maxwell–Boltzmann equation is the most prominent example (see, e.g., Truesdell, 1984, p. 429). Liu and Müller aimed at the very core of Truesdell's principle with their precise analysis based on the extended gas kinetics of Grad's type. "Indeed, it only reflects the material frame dependence of the basic equations of balance ..., while the theory is frame independent in the constitutive relations" (Liu and Müller, 1986, p. 114).

Unfortunately, this concise statement concerning a dogma of Rational Mechanics is incomplete, since it is restricted only to the kinetic theory of ideal gases. Moreover, the authors' comment refers to a design principle generally accepted to be convenient for the theory of any constitutive functions. However, one can obtain an alternative formulation of Truesdell's principle from Falk's equations (3.38). Due to the universally physical quantities as the key terms of Falk's dynamics (see Chapter 2), each M–G function of the system in question is frame-indifferent, provided that Lorentz invariant variables are used.

But an additional requirement must be fulfilled. The selection of the essential M–G variables must always occur in reference to the macroscopically infinitesimal time and space elements appropriate for the mathematical description of the physical processes under consideration. Of course, such a connection contains not only mathematically stringent relations, but also some intersubjective elements influenced by the special conditions of the processes and subject to the observer's decision.

Let us turn back to Equation (4.43) and remember that the conservation of energy and linear momentum reveals the existence of an inertial field that can be realized, for instance, by the local gravitational field. This is a crucial point because the system defined by its Gibbs–Euler function $\mathscr{E}(\mathbf{P}; S; V; N)$, according to Equation (3.8), must be extended in such a way that this field can be included. In order to get an impression of the operating mechanism of such a field, let us study an especially simple case of a system bearing the label "body + given field." We can describe this configuration by a special *two-body system* where the first body of the two is assumed to be at rest whereas the second one moves. This means that the first body, as a result of its relatively enormous mass, has just the single function to fix the inertial field involved. Due to this, the inertial field does not take part in the interchange events of energy and momentum between the two bodies. The field's momentum $\mathbf{P}_f$ (introduced below) is nearly zero. This special case conforms with the basic assumption that the field can be treated in the so-called *static approximation* (Falk and Ruppel, 1983, p. 203). Additionally, an inertial system is chosen as a well-defined reference of the motion of the two bodies under consideration.

The more general case concerns motion characterized exclusively by those interchange events that the two bodies settle with the field in any separate way or mutually through the intermediary of the field. The corresponding equations of motion

$$\frac{d\mathbf{P}_i}{dt} = \mathbf{F}_i; \qquad \frac{d\mathbf{r}_i}{dt} = \mathbf{v}_i, \qquad (i = 1,2) \tag{5.1}$$

complete the Gibbs–Euler function $\mathscr{E}(\mathbf{P}_1;\, \mathbf{P}_2;\, \mathbf{r}_1;\, \mathbf{r}_2)$ of the two-body system, where $\mathbf{F}_i$ is the force to be exerted on the ith body by the field. However, assuming the motions occur in an inertial system, it is easy to prove that the two position vectors $\mathbf{r}_1$ and $\mathbf{r}_2$ can be replaced by their difference $\mathbf{r}_1 - \mathbf{r}_2$. This important fact may be demonstrated by adding the two equations (5.1) and considering that the total two-body momentum is conserved in a unique inertial system:

$$\frac{d}{dt}(\mathbf{P}_1 + \mathbf{P}_2) = \mathbf{F}_1 + \mathbf{F}_2 \equiv 0. \tag{5.2}$$

The forces $\mathbf{F}_1$ and $\mathbf{F}_2$ can be represented by the Gibbs–Euler function $\mathscr{E}(\mathbf{P}_1;\, \mathbf{P}_2;\, \mathbf{r}_1;\, \mathbf{r}_2)$ of the two-body system. Thus, for all components a characteristic six-element sequence of equations results:

$$-F_{1x} = \frac{\partial E}{\partial x_1} = F_{2x} = -\frac{\partial E}{\partial x_2};\ \ldots;\ F_{2z} = -\frac{\partial E}{\partial z_2}. \tag{5.3}$$

Apart from the sign, this sequence of partial differentiations yields identical force components with respect to x_1 and x_2, respectively. The same is true for the coordinates y and z of the two bodies. Therefore, the energy E can only be dependent on all the coordinate combinations. Hence, the Gibbs–Euler function (GEF) can be simplified by the form $\mathscr{E}(\mathbf{P}_1;\, \mathbf{P}_2;\, \mathbf{r}_2 - \mathbf{r}_1)$ using the so-called *relative coordinates.*

In addition to the two special cases, assuming the field in its static approximation or as one part of an inertial system, there exists, of course, the very simple case of a body-field system. A single body moves under the conditions prescribed by the local state of the field. This motion is characterized by the momentum $\mathbf{P}$ of the body and by the location $\mathbf{r}$ where the field force acts on this very body. Clearly, this body-field system is defined by its GEF $\mathscr{E}(\mathbf{P};\, \mathbf{r})$ under the additional condition $dE \equiv 0$. But furthermore, the conservation of momentum should also hold. It is easy to fulfill this requirement by the simple identity

$$\frac{d\mathbf{P}}{dt} + \frac{d\mathbf{P}_f}{dt} \equiv 0, \tag{5.4}$$

which simultaneously introduces the momentum $\mathbf{P}_f$ of the field. Writing the Pfaffian of the body-field system

$$dE \equiv \left(\frac{\partial E}{\partial \mathbf{P}}\right)_{\mathbf{r}} \bullet d\mathbf{P} + \left(\frac{\partial E}{\partial \mathbf{r}}\right)_{\mathbf{P}} \bullet d\mathbf{r} = \mathbf{v} \bullet d\mathbf{P} - \mathbf{F} \bullet d\mathbf{r}, \tag{5.5}$$

note that the negative sign of the field force $\mathbf{F}$ is mere convention. From (5.1) we see that the field force $\mathbf{F}$ and the velocity $\mathbf{v}$ are defined as usual by their partial differentials

$$\mathbf{v} = \left(\frac{\partial E}{\partial \mathbf{P}}\right)_{\mathbf{r}}; \qquad \mathbf{F} := -\left(\frac{\partial E}{\partial \mathbf{r}}\right)_{\mathbf{P}}, \tag{5.6}$$

which do not explicitly presume any proper agreements on time and space.

Let us assume that the field vector $\mathbf{F}$ is conservative and thus only dependent on the distances $\Sigma_i(\mathbf{r}_k - \mathbf{r}_i)$ between the arbitrary, but fixed position $\mathbf{r}_k$ of the field and all its other positions $\mathbf{r}_i$. Of course, for the total differential $d\mathbf{r}_k$, all contributions to $\mathbf{F}_k$ indicated by $i \neq k$ vanish. Conversely, we can make use of this special property by identifying this distance differential $d\mathbf{r}$ with the differential $d\mathbf{x}$ with $\mathbf{x}$ bearing the coordinates x, y, and z.

The next step is straightforward if the time parameter t is introduced and applied to (5.5). The formula

$$\frac{dE}{dt} = \mathbf{v} \bullet \left(\frac{d\mathbf{P}}{dt} - \mathbf{F}\right) \tag{5.7}$$

follows when the common kinematic notion of velocity is used: $\mathbf{v} := d\mathbf{x}/dt$.

A basic assumption concerning inertial fields is that in all nontrivial cases (i.e., the cases $\mathbf{v} \equiv 0$ or $\mathbf{v} \perp (\mathbf{F} - d\mathbf{P}/dt)$ do not hold) the left side of (5.7) vanishes identically. As a consequence, Equation (5.1) results, relating the field force $\mathbf{F}$ existing at position $\mathbf{x}$ to the time derivative of the local linear momentum of the body at the same place. Considering the balance of momentum (5.4), the corresponding equation of motion of the field is given by

$$\frac{d\mathbf{P}_f}{dt} + \mathbf{F} = 0. \tag{5.8}$$

Without doubt the momentum of the field $\mathbf{P}_f$ cannot be directly related to a pertaining acceleration as it can for a mass-point.

Returning to the general theory, a one-component single-phase system defined by its GEF $\mathscr{E}(\mathbf{P}; S; V; N)$ can easily be extended to a new system by considering the two field quantities $\mathbf{P}_f$ and $\mathbf{r}$ even for the case $dE \neq 0$. According to the results presented by the equations given above for any classical two-body systems, a complex type of the body-field system (BFS) arises that is constituted by its GEF $\mathscr{E}(\mathbf{P}; \mathbf{r}; S; V; N)$, along with the balance of momentum (5.4). The corresponding Gibbs main equation

$$dE = \mathbf{v} \bullet d\mathbf{P} - \mathbf{F} \bullet d\mathbf{r} + T_* \, dS - p_* \, dV + \mu_* \, dN \tag{5.9}$$

leads to some interesting conclusions that will be discussed in the next section. But first two points should be made. First, bear in mind that in principle the variables of

Equation (5.9) refer to the whole system. Indeed, the system is set up such that the body is assumed to be infiltrated into all its parts by the field within the volume V assigned exclusively to the body. However, in the special case where the field force $\mathbf{F}$ is conservative (that is $\mathbf{F} = \mathbf{F}(\mathbf{r})$), there are no contributions of the field to the pressure and momentum of the BFS. Such effects do occur when electromagnetic variables have to be considered (see Chapter 9).

Second, note that Equation (5.9) represents a class of body-field systems that is a special case of two more extensively defined classes: the Multicomponent single-phase BFS and the Multicomponent multiphase BFS. Both classes concern processes typical for all dissipative phenomena occurring in any universe of electrically neutral particles. In addition to friction and heat conductivity, there exists diffusion along with so-called cross-effects, such as pressure or thermodiffusion. Furthermore, simultaneous chemical reactions and highly complicated surface phenomena are taking place associated with some characteristic polyphase flows. Hence, it is cogent to deal subsequently with these more complex systems, realizing that the systems defined by (5.9) are limiting cases.

5.2 Multicomponent Single-Phase and Multiphase Properties

In this section we treat mathematical description of complex phenomena in dissipative flows, including some relevant properties of the two multicomponent BFSs introduced above. The single-phase BFSs will be discussed mainly with regard to some special theoretical problems concerning mixtures. Conversely, multiphase properties will only be considered in view of their basic rules.

5.2.1 *MULTICOMPONENT SINGLE-PHASE BODY-FIELD SYSTEMS*

It is easy to extend the complex BFS defined by (5.9) and (5.4) to specify mathematically the peculiarities of a multicomponent single-phase BFS. To simplify its presentation, the representative M–G function should only be extended to different particle numbers N_1; N_2; ...; N_J of a homogeneous mixture consisting of J components in toto. Hence, the M–G function $E(\mathbf{P}; \mathbf{r}; S, V, N_j)$ leads to the corresponding Pfaffian

$$dE = \mathbf{v} \bullet d\mathbf{P} - \mathbf{F} \bullet d\mathbf{r} + T_* \, dS - p_* \, dV + \sum_{j=1}^{J} \mu^t_{*j} dN_j \tag{5.10}$$

of the BFS. The intensive state parameters are given by the following relations

$$\begin{aligned} \mathbf{v} &= \frac{\partial E(\mathbf{P}; \mathbf{r}; S; V; N_1; \ldots; N_j)}{\partial \mathbf{P}}; \qquad -\mathbf{F} = \frac{\partial E(\mathbf{P}; \mathbf{r}; S; V; N_1; \ldots; N_j)}{\partial \mathbf{r}}; \\ T_* &= \frac{\partial E(\mathbf{P}; \mathbf{r}; S; V; N_1; \ldots; N_j)}{\partial S}; \qquad -p_* = \frac{\partial E(\mathbf{P}; \mathbf{r}; S; V; N_1; \ldots; N_j)}{\partial V}; \\ \mu^t_{*j} &= \frac{\partial E(\mathbf{P}; \mathbf{r}; S; V; N_1; \ldots; N_j)}{\partial N_j}; \qquad j = 1(1)J, \end{aligned} \tag{5.11}$$

which are true for the corresponding non-equilibrium phenomenon.

According to Equation (3.7), the respective Euler – Reech equation becomes

$$E - E_{\#} = \mathbf{v} \bullet \mathbf{P} - \mathbf{F} \bullet \mathbf{r} + T_* S - p_* V + \sum_{j=1}^{J} \mu^{t}_{*j} N_j . \tag{5.12}$$

Note that the variable **r** is treated as an extensive quantity that supposedly is unable to relate in any way a mass or particle number. But such a "specific quantity" may be introduced by means of the apparently conjugated force function **F**. Hence, the scalar product **F** • **r** fulfills the precondition associated with the homogeneity of a function (see Falk, 1990, p. 269). Strictly speaking, however, the correct variable should consist of the product $V\mathbf{g}$, where V and **g** respectively denote the body volume and a field density that is completely independent of V and given as direct information on the field. Therefore, $\mathbf{g} \not\equiv \mathbf{F}/V$, the common definition of the density of a phase quantity, does not hold. Consequently, the incorporation of V guarantees the extensivity of this combined variable. Moreover, it involves a contribution to the pressure of the system via V. As for gravitational forces, this special part of the whole pressure is traditionally neglected. With electromagnetic forces, however, such a disregard can lead to serious errors. The *zero point energy* $E_{\#}$ is defined by

$$E_{\#} = m_{\#} c^2 = \sum_j m_j^{tB+L} N_j c^2 (1 + \Delta_j m_{\#} / m_j^{tB+L}) = \sum_j m_j^{B+L} c^2 (1 + \Delta_j m_{\#} / m_j^{tB+L}), \tag{5.13}$$

where the simple decomposition $m_{\#} := \Sigma_j N_j (m_j^{t\,B+L} + \Delta_j m_{\#}) = \Delta m_{\#} + \Sigma_j m_j^{B+L}$ of the total inertial mass $m_{\#}$ [see Equation (3.17)] allows us to introduce the *average mass* m_j^{B+L} of the jth component under consideration. In common practice this mass serves as a useful reference property for the definition of so-called (mass-) specific quantities. The thermodynamic notion of mass refers to the conservation of baryons and leptons. This means that the theory is confined to changes of state by which any chemical reaction might be induced at an atomic level. In other words, reactions at the level of the elementary particles are not considered. For this reason the term *mass* is commonly used in the sense of the particle numbers of the baryons and leptons involved. The average mass $m_j^{t\,B+L}$ per particle of the jth component is defined by the contributions of all sorts of baryons and leptons, weighted by their respective particle numbers $N_\lambda^{(j)}$ and assigned to the jth component of the body. In the corresponding formula,

$$m_j^{t\,B+L} \sum_\lambda N_\lambda^{(j)} := m_\lambda^{t} N_\lambda^{(j)}, \tag{5.14}$$

the subscript λ refers to all kinds marked by the mass values m^{t}_{λ} per baryon or lepton. Here, and henceforth, the symbol Σ_j will stand for a summation from $\lambda = 1$ to $\lambda = \Lambda$.

The particle number N_j of component j equals the expression $\Sigma_\lambda N_\lambda^{(j)}$ that in itself yields the total number of all baryons and leptons involved (i.e., the corresponding protons, neutrons, hyperons, electrons, neutrinos, etc.—see Table 3.1). For instance, protons and antiprotons constitute some sorts of baryons and leptons. Corresponding to Dalton, each value m'^{t}_{λ} means the integral multiple ϑ_λ of the value $m_H{}^t$ of the mass assigned to a hydrogen atom; in other words, the proportionality $m'^{t}_{\lambda} := \vartheta_\lambda \bullet m_H{}^t$ holds for all λ.

The total mass of all particles of the jth component corresponds to the left side of (5.14), viz.

$$m_j^{B+L} := m_j^{t\,B+L} \sum_\lambda N_\lambda^{(j)} := m_j^{t\,B+L} N_j; \qquad j = 1(1)J,. \tag{5.15}$$

whereas the summation over the index j yields the reference mass of the mixture

$$m^{B+L} := \sum_j m_j^{B+L} := \Sigma_j m_j^{t\,B+L} N_j. \tag{5.16}$$

Now let us introduce the mole number ñ defined by the ratio N_j/N_A using the *Avogadro number* $N_A \approx 6.023 \times 10^{23}$ particles per mole. By extending the right side of (5.15) with N_A, we can define the *mole masses* M_j and thus determine the reference mass of the mixture as follows:

$$M_j := m_j^{t\,B+L} N_A \to m^{B+L} = \sum_j M_j \tilde{n}_j. \tag{5.17}$$

Summarizing the information contained in (5.13), (5.16), and (5.17), we can see that the masses m^{B+L} and M_j are reduced to values considering changes of state in contrast to the total mass $m_\#$ of the mixture. (Incidentally, the value of the Avogadro number N_A established by Perrin in 1905 has never been altered, even though modern physics has discovered quite new particles existing partly in myriads.) It is remarkable that the use of the reference mass m^{B+L} leads to an essential inference regarding the total particle number N of the mixture forming the body of the single-phase BFS. In analogy to (5.14), an average mole mass M may be defined in connection with (5.17) as

$$m^{B+L} = \sum_j M_j \tilde{n}_j := M \sum_j \tilde{n}_j := M\tilde{n}, \tag{5.18}$$

where ñ denotes the total mole number of the mixture. Differentiation of (5.18) yields the simple relation

$$dm^{B+L} = M\, d\tilde{n} + \tilde{n}\, dM, \tag{5.19}$$

which permits the following two statements.

1. At the level of baryons or leptons the total number $\Sigma_j \Sigma_\lambda N_\lambda^{(j)}$ is conserved, but, due to chemical reactions, rearrangements of the $N_\lambda^{(j)}$ are of course permissible. Clearly the conservation law holds:

$$dm^{B+L} \equiv 0. \tag{5.20}$$

2. As a consequence of point (1), at the level of the particle numbers N_j constituting the mixture under consideration, the general rule

$$M\, d\tilde{n} + \tilde{n}\, dM \equiv 0 \tag{5.21}$$

   is valid.

Following from (5.21), the commonly used conservation theorem $dN \equiv 0$ does not dominate the exchange processes performed by means of populations of particles consisting of baryons or leptons. It is true that all types of diffusion use the condition $dN \equiv 0$, but chemical reactions (due to $dM \neq 0$) and radiation processes caused by unconserved photon rates usually obey (5.21). Dividing (5.12) and (5.13) by the constant m^{B+L} and then combining both expressions, the formula

$$e = \mathbf{v} \cdot \mathbf{i} - \mathbf{f} \cdot \mathbf{r} + T_* s - p_* \rho^{-1} + \sum_{j=1}^{J} \mu_{*j} \omega_j + c^2 \left(1 + \frac{\Delta_\# m}{m^{B+L}} \right) \tag{5.22}$$

offers a new form of the Euler–Reech equation, by which some specific quantities are introduced as follows:

$$\begin{aligned} &e := E/m^{B+L} : \text{specific (total) energy;} && \mathbf{i} := \mathbf{P}/m^{B+L} : \text{specific (linear) momentum;} \\ &\mathbf{f} := \mathbf{F}/m^{B+L} : \text{specific field force;} && s := S/m^{B+L} : \text{specific entropy;} \\ &\rho := (V/m^{B+L})^{-1} : \text{mass density;} && \mu_{*j} := \mu'_{*j} N_A/M_j : \text{specific chemical potential.} \end{aligned} \tag{5.23}$$

Additionally, the *mass fraction* ω_j of the jth component is used, the definition of which automatically leads to the so-called *closure condition*

$$\omega_j := \frac{M_j N_j}{M\ N} := \frac{\rho_j}{\rho}; \qquad j = 1(1)J \xrightarrow{\text{c.c.}} \sum_{j=1}^{J} \omega_j = 1, \tag{5.24}$$

which is equivalent to the conservation rule (5.20).

To complete the mathematical tool for the description of any multicomponent single-phase BFS, its Gibbs main equation, along with its Gibbs–Duhem relation, is derived from (5.22). The last term of this M–G function is dropped because it only contains constants. After some simple manipulations the two constitutive differential forms appear.

Gibbs main equation as a Pfaffian

$$\rho\, de = \mathbf{v} \cdot \rho\, d\mathbf{i} - \rho \mathbf{f} \cdot d\mathbf{r} + T_* \rho\, ds + (p_*/\rho) d\rho + \sum_j \mu_{*j}\, \rho\, d\omega_j; \qquad j = 1(1)J \tag{5.25}$$

Gibbs – Duhem relation

$$dp_* = \rho s\, dT_* + \sum_j \rho \omega_j\, d\mu_{*j} + \mathbf{i} \cdot \rho\, d\mathbf{v} - \mathbf{r} \cdot \rho\, d\mathbf{f}. \tag{5.26}$$

Whereas Equation (5.26) indicates that there is an equation of state constituted by the function

$$p_* = p_*(T_*; \mu_{*j}; \mathbf{v}; \mathbf{f}); \qquad j = 1(1)J \tag{5.27}$$

and assumed to be valid for non-equilibrium phenomena, the Pfaffian (5.25) shows that each term implies its own balance equation with respect to the dependence of the respective variable on an appropriate parameter. The latter case will be further examined in following sections.

5.2.2 *MULTICOMPONENT MULTIPHASE BODY-FIELD SYSTEMS*

Equation (5.25) is a special case of an M–G function $\xi_0 = G(\xi_1; \ldots; \xi_r)$, as given by Equation (2.72). Its variables $\xi_0; \xi_1; \ldots; \xi_r$ are only extensive ones. Using this set of variables, certain conditions allow us to prove that these variables themselves may result from a *system synthesis*. An example is Equation (2.92), which itself follows from the three M–G functions belonging to any three subsystems involved. Incidentally, this example is not confined to M–G functions with extensive variables only. Intensive variables are also admitted, provided that all sets of variables define the corresponding M–G functions of the system observed and of each of the respective subsystems.

A second example will explain in more detail. The last term of the right side of (5.25) refers to the sum over all J components of the fluid mixture. Suppose the jth component is subdivided into K distinct subfluids, numbered by the subscript k ($k = 1(1)K$). If these constituents are not in equilibrium with each other, then the last term must be replaced by the double sum $\Sigma_j \Sigma_k \mu_{*jk} \rho \, d\omega_{jk}$. Here μ_{*jk} is the chemical potential (or specific Gibbs function) for the kth subfluid within the jth component; for the mass fraction ω_{jk} of this component the closure rule (5.24) must be observed.

If equilibrium prevails between the constituents of the jth component, μ_{*jk} will have to be the same for all values of the integer k. Because the double sum must then shrink to the sum appearing in (5.25), it follows that the equality $\mu_{j1} = \mu_{j2} = \cdots = \mu_{jk} \equiv \mu_{*j}$ holds. As before, the asterisk (*) indicates the non-equilibrium state of the system with respect to its dissipative motion at the level of the components. This example leads to the most important case of phase equilibria, and the assumptions and postulates of which will be briefly introduced in accordance with Tisza (1966, p. 115).

Phases are the structural elements that form any body in equilibrium. They are spatially homogeneous extensions of matter. The basic idea is centered on the assumption that any system of given composition potentially exists in a certain number of phases, each of which is determined by a *phase-entropy function*

$$\eta^\alpha = \eta^\alpha(\xi_1^\alpha; \ldots; \xi_r^\alpha; \eta_1^\alpha; \ldots; \eta_K^\alpha); \qquad \alpha = 1, 2, \ldots \tag{5.28}$$

expressing the entropy of phase α as a continuous first-order homogeneous function of the extensive variables $\xi_1; \ldots; \xi_r$. Equation (5.28) represents r-dimensional hypersurfaces, called *phase surfaces* in Gibbs space. Each function is defined within a characteristic range of variables. In some cases η^α also depends on additional parameters η_k^α, running with the integers k, the number K of which is disposable. These are introduced for certain homogeneous states of matter that are discerned not by their state values but by virtue of their intrinsic symmetry properties, such as mirror-image states.

From the practical point of view, by far the most important case concerns the well-known λ-phenomenon observed in many crystals and in liquid helium. Such a phenomenon is commonly called *second order transition*. Plotting values of heat capacity or thermal compressibility of helium near its melting pressure curve against the temperature T at a constant pressure p, we find that these thermodynamic properties exhibit unexpected singularities of a shape reminiscent of the letter λ. The sharp maximum of this curve is called the λ-point. As the pressure varies, the λ-point is traced in the $(p\text{-}T)$-plane by the λ-line, which can be compared with the well-known lines of ordinary heterogeneous equilibrium.

From the theoretical point of view of Gibbs' phase theory, λ-points have contradictory properties. It is true that they are like critical points in that the specific heat and the other pertaining properties of state exhibit characteristic singularities. But instead of appearing in isolated points in the $(p\text{-}T)$-plane, the λ-points of one-component systems will form λ-lines in that plane. The failure of the classical theory to account for these lines is due to the fact that the specification of homogeneous phases in terms of extensive variables is incomplete. The variables have to be supplemented by quantities that account for their symmetry properties.

Above all, the extended definition of a phase including the set of K parameters η_k leads to a generalized version of Gibbs' famous phase rule (Tisza, 1966, p. 39). Generally speaking, an adequate incorporation of the symmetry concept into the fundamental principles of thermostatics offers concrete possibilities to explain much of the so-called cooperative phenomena as a collective term for order-disorder transitions in fluids and some kinds of magnetic bodies. Such a symmetry concept may be considered as a complement of Callen's symmetry principle (presented in Chapter 4).

There are theoretical difficulties with phenomena beyond that domain of thermostatics still governed by the *principle of determinism*. This principle claims that the intensive variables of the surroundings uniquely determine the densities of the system in question, provided that certain conditions of stability can be satisfied. The reverse is also true: the intensities of the system determine the densities of its environment. This principle, accepted as essential for basic experimental situations in thermostatics, has its reality-related limitations. For example, near critical points of matter the densities are subject to extremely large fluctuations. Another example concerns states near absolute zero, where entropy is also practically zero, and the system becomes insensitive to temperature variations. This effect is due to long-range space-time correlations in the system, which are theoretically assumed to be absent in general. "The actual occurrence of these extreme situations makes it evident that correct thermodynamics should include both the elements of randomness and of space-time correlations" (Tisza, 1966, p. 185).

At present the mathematical difficulties in accomplishing such a formalized concept cannot be reduced or eliminated. This is especially true with regard to multiphase flows of fluids, characterized by some latent symmetry properties that may be activated by the flow dynamics prevailing locally and globally. Distinguished symmetry breaks occur, represented by classes of phase flows that discern the internal distributions of very different surface patterns within the flow. Such physical situations point out the extraordinary relevance of an adequate theory of phases for continuum physics. In my opinion there is a considerable lack of insight into the leading principles regarding phase phenomena like evolution, transition, and inter-phase events.

Let us now turn back to the basic formula (5.28) for phase α. To complete its main properties, the following definitions will be useful (Tisza, 1966, p. 116).

Scale factor: One of the extensive variables, say ξ_r^{α}, is singled out to specify the size of the system. Commonly, the volume V is chosen to be this scale factor.

Densities:

$$\begin{aligned} &\rho_j^{\alpha} := \xi_j^{\alpha}/\xi_r^{\alpha} = \xi_j^{\alpha}/V; \qquad j = 1(1)r-1 \\ &\text{for example: } S_v^{\alpha} := \xi_0^{\alpha}/\xi_r^{\alpha} = \xi_0^{\alpha}/V \end{aligned} \tag{5.29}$$

The intrinsic, size-independent properties of any phase are well represented by the $(J-1)$-dimensional surface

$$S_v^{\alpha} = S_v^{\alpha}\,(\rho_1^{\alpha};\ \rho_2^{\alpha};\ \ldots;\ \rho_{J-1}^{\alpha};\ \eta_1^{\alpha};\ \ldots;\ \eta_K^{\alpha}); \qquad \alpha = 1, 2, \ldots \tag{5.30}$$

extended within the r-dimensional Gibbs space. For some options a scaling by means of a mole number or the mass of the mixture offers several advantages. This is particularly true for one-component systems. The densities (5.29) will then become molar or specific quantities.

Modification of a phase: Two modifications of a phase α are distinct if they have different densities or if they differ in the parameters η_k^{α} $(k = 1(1)K)$. There are two important examples: Vapor and liquid are two modifications of a fluid at phase equilibrium. The same is true for right- and left-hand quartz with identical densities but different values of the respective parameters η_k^{α} $(k = 1(1)K)$.

Heterogeneous equilibrium: Assuming the additive postulate (4.56) to be valid for an isolated system, it is possible to determine a heterogeneous equilibrium state that is singled out from the complete set of virtual heterogeneous states of the whole system. The system can be defined by Equation (4.56)

$$\xi_j = \sum_{\beta} \xi_j^{\beta}; \qquad j = 1(1)r, \tag{5.31}$$

where Σ_{β} stands for the summation over all phases β involved. Each term corresponds to a volume $\xi_r^{\beta} = V^{\beta}$ and to a point on the β-phase surfaces in the Gibbs space.

As usual, the equilibrium state is established by the maximum of the entropy

$$S(\xi_1;\ \ldots;\ \xi_r) = \max\{\sum_{\beta} \eta^{\beta}(\xi_1^{\beta};\ \ldots;\ \xi_r^{\beta};\ \eta_1^{\beta};\ \ldots;\ \eta_K^{\beta})\}, \tag{5.32}$$

with its value being dependent on the number m of distinct modifications in the actually realized heterogeneous state. It is noteworthy that the parameters η_k^{β} are not conserved; they are unconditionally varied in the maximization procedure.

If we let the maximum of (5.32) be attained for the heterogeneous state according to the quantities

$$\xi_1^K; \ldots; \xi_r^K; \qquad \eta_1^K; \ldots; \eta_K^K; \qquad K = 1(1)m, \tag{5.33}$$

then two special cases are of singular interest: If $m = 1$ holds, the observed state degenerates into a homogeneous equilibrium; only $m \geq 2$ corresponds to a heterogeneous state in the strict sense.

The heterogeneous equilibrium of a system is reminiscent of the composite system, where the distinct properties of the subsystems are maintained by passive forces. Some constraints, defined as the set of material or immaterial walls existing in a system, are said to exert such passive forces on the distribution of its variables and parameters.

Unfortunately, the general theory of phases is rather retarded in spite of the fact that multiphase phenomena have become increasingly relevant for many important and serious problems concerning environmental control, such as the efficient restoration of healthy soil or the regulation of waste-treatment plants. As opposed to this there are several approaches to the theoretical treatment of some classes of two-phase flows (see, e.g., Pai, 1977).

Similar to ordinary thermofluid dynamics, two-phase flows may be studied from both the microscopic point of view—the kinetic theory of two-phase flow—or from the macroscopic one—the continuum theory of two-phase flow. Yet the kinetic theory of two-phase flow has not been well established, because even the pure kinetic theory of liquid is still not in an advanced stage. Indeed, in many practical problems there are some users who do not care about the motion of individual matter particles, but are interested only in the resultant effects of the motion of a large number of particles. Thus, such users focus on the macroscopic quantities only, such as pressure, temperature, density, concentrations, flow velocities, and so on. An appropriate tool is the set of the fundamental equations of two-phase flow problems based on the conservation laws of mass, components, momentum, and energy of each phase and their interactions.

5.3 Time Parameters in Thermodynamics of Fluid Systems

It may be surprising that time does not appear explicitly in the complete Gibbs–Falkian formalism. It is only by taking into account any surroundings of a system by parameterization of the First Law, along with the Noether theorem, that leads to the use of a linear-affine time term formally introduced to convert differentials into time derivatives, as in Equation (4.43). Notwithstanding this special case, time is in fact one of the most relevant variables also in thermodynamics. Its role is not merely that of a passive parameter on which the coordinates of the Gibbs space may depend. It is rather the nature of the Gibbs space itself that is constituted by means of time t via the observer's time scale $\mathcal{T}$. Some basic concepts are indeed joined to the time coordinate: closure, adiabaticity, all kinds of equilibria, and non-equilibrium phenomena.

Unfortunately, the absence of t has led to some confusion in situations such as those experienced in gas turbines or shock waves, where rapid time changes obviously do occur. The unexpected implication cannot preclude that there are events, such as in mechanics, where time is a "natural" independent variable. But the most plausible explanation arises from the concept of "state" being absolutely rather than relatively independent of time. Certainly, this is true with respect to the kind and number of degrees of freedom involved.

At the microscopic level of description small lengths and quantities of time are characteristic features of all events. Because of the particle structure of matter and radiation, many changes may occur in the properties of the individual particles .

Under STP conditions in air, there are values of the *mean free path* ℓ and *mean free time* τ_ℓ in an order of magnitude of 10^{-5} cm and 10^{-10} s, respectively. However, the averages of particle properties, like the flow velocity $\mathbf{v}$, taken over the immense number of about 10^7 particles in a microscopic volume element of about 10^{-12} cm^3, usually change over much larger length and time scales. Such averages are called macroscopic properties and referred to correlative length and time scales $\mathcal{L}$ and $\mathcal{T}$.

It is a crucial point of continuum physics that macroscopic infinitesimals $|d\mathbf{r}|$ and dt of length and time may be introduced that are comparatively large on a microscale but small on a macroscale. Then, each macroscopic property, like $\mathbf{v}$, will normally be a smooth function of the position- and time-coordinates, changing by small increments $|d\mathbf{v}|$ over $|d\mathbf{r}|$ and dt. In other words, continuum physics is justified if the general conditions $|d\mathbf{r}| >> \ell$ and $dt >> \tau_\ell$ are satisfied. In this case, the molecular nature of matter is thoroughly below the level of description. Furthermore, these substantial changes take place locally and so can be described by differential equations. One of the main problems of defining a system in the sense of Falk's dynamics lies in how to specify the set of its variables for a given process $\wp$ and a given observer $\mathcal{O}$. This refers to the kind and the number of the relevant standard quantities as well as their selection rule, often influenced by different views of several observers on what is important in $\wp$.

The $\mathcal{O}$'s length and time scales, say $\mathcal{L}$ and $\mathcal{T}$, depend mostly on his objectives and knowledge. They must be compared with the variety of characteristic lengths and times occurring in the reality of $\wp$, due to either purely internal events or to interactions with the surroundings of the system under consideration.

Regarding non-equilibrium phenomena, the most important question is whether any partial equilibrium may be conjectured by $\mathcal{O}$ for some of the variables. To give an answer, we will discuss a few criteria below in reference to relaxation processes. Here it may suffice to explain two quite distinct kinds of partial equilibrium using *natural time scales*

$$\tau_i := (\xi_i^{\text{equ}} - \xi_i)/\overset{\circ}{\xi}_i; \qquad i = 1, 2, \ldots, m, \ldots, (m-1) + \Delta, \ldots, I; \qquad \Delta \geq 1, \qquad (5.34)$$

where ξ_i^{equ} denotes the value attained as $t/\tau_i \to \infty$ and $\overset{\circ}{\xi}_i$ is the time rate of ξ_i. An order is specified for the time scale, so that

$$\tau_1 \geq \tau_2 \geq \cdots \geq \tau_I \tag{5.35}$$

holds. In addition, we assume that $\mathcal{O}$'s time scale fulfills

$$\tau_m >> \mathcal{T} >> \tau_{m+\Delta-1}, \tag{5.36}$$

where τ_m is the smallest and $\tau_{m+\Delta-1}$ is the largest scale involved. On $\mathcal{O}$'s time scale the variables ξ_i for the subscript $i = 1, \ldots, m-1$ are said to be in *frozen equilibrium*, and $\mathcal{O}$ is assumed to observe them as mere constants in $\wp$. The alternative limit is given for the variables ξ_i, indexed by $i > (m-1) + \Delta$, which represent changes of physical events occurring so rapidly that the ξ_i hardly deviate from their equilibrium values ξ_i^{equ} for any given measurability. Such variables are in *relaxed equilibrium*. The result may be condensed in a short rule: Every system $\{\wp; \mathcal{O}\}$ (i.e., $\wp$ observed by $\mathcal{O}$) defined by means of Gibbs' dynamics is assumed to be adequately described by the set of independent variables $\xi_i(m \leq i \leq (m+\Delta-1))$ that change more or less on $\mathcal{O}$'s individual time scale.

In general it is insignificant to endow such inequalities as (5.35) or (5.36) with some precision taken out of any physical context. The recommendation to estimate roughly by factors of 10^2 is useful. In practice, however, the answer to the problem depends on whether

1. electromagnetic and surface phenomena have to be considered by separated variables in the Gibbs' fundamental equation $\Gamma(\xi_0; \xi_1; \ldots; \xi_r) \equiv 0$;
2. multiphase problems arise, for which the additivity postulate (4.56) has to be proven;
3. the dependency of Γ on the respective particle numbers N_j can be suppressed or substituted, provided that it needs no explicit consideration for physical reasons.

Well-founded decisions may be justified only for each individual case. To be sure, they benefit from professional experience with regard to both the physical items of concern and the appropriate mathematical tools to be applied. In particular, points (1) and (2) are affected by $\mathcal{O}$'s subjective evaluation. Point (3) calls for a thorough explanation, because misinterpretations arise easily. Thus, for instance, Clarke and McChesney state in two well-written books (1964 and 1976) that "there is no doubt whatsoever that $T\,ds = de + p\,d(\rho^{-1})$ is an equation which only holds for reversible processes, i.e., those taking place through a series of equilibrium states" (Clarke and McChesney, 1964, p. 169; 1976, p. 62). On the other hand, the authors claim that they "intend to say that" the same equation is also true "in non-equilibrium situations." They explain the apparent contradiction by the fact that in non-equilibrium states the pressure is a second-rank tensor $\mathbf{\Pi}$. According to their statement that "in full equilibrium situations the gas is at rest, the pressure is truly the hydrostatic pressure ... (and) in fact a scalar quantity" (1964, p. 169), it is hard to identify $\mathbf{\Pi}$. Thus, the authors do not see any difficulties with respect to the conjugate quantity of pressure, viz. the volume V: "Note that there is no ambiguity about the density ρ for any situation" (1964, p. 169). Such a remark is hardly compatible with the necessity of forming a scalar product with an energy unit, using a second-rank tensor $\mathbf{\Pi}$ and a scalar quantity V as the two factors.

We can confidently introduce other temperatures besides the temperature T, conjugated to the entropy of the system under non-equilibrium conditions. The definition of a new temperature occurs in connection with a mode of internal molecular motions (such as molecular rotations or vibrations). It is, therefore, associated with a descent from the very detailed quantum state description of the fluid properties to a less-detailed formulation that involves averaging over the quantum states applicable to a distinct mode of internal molecular energy storage. "Such a description lies halfway between the species and chemical-species methods of accounting for gaseous behavior" (Clarke and McChesney, 1976, p. 63). A well-known relevant example is from the statistical theory of equilibrium thermodynamics. According to this theory, contributions of the rotational degrees of freedom to the complete caloric state functions of an ideal diatomic gas can act as a stimulant to insert new non-equilibrium variables. For the rotational contributions to the internal molar energy $\tilde{u}_r$ and entropy $\hat{s}_r$, the following expressions are obtained (Sonntag and Van Wylen, 1968, p. 206):

$$\tilde{u}_r = \mathcal{R}T; \qquad \hat{s}_r = \mathcal{R}[\ell n(T/\sigma\theta_r) + 1], \tag{5.37}$$

where the universal gas constant $\mathcal{R}$ and two specific material parameters (symmetry number σ, characteristic rotational temperature θ_r) arise. The simple relation between these two functions,

$$d\tilde{u}_r = T\, d\,\hat{s}_r \tag{5.38}$$

evidently leads to an analogous approach

$$d\tilde{u}_r := T_r\, d\hat{s}_r. \tag{5.39}$$

Actually, Equation (5.39) represents the definition of the *rotational temperature* T_r justified for the case where $\hat{s}_r$ only depends on this variable T_r.

By a straightforward analogy with (5.39), the generalization with respect to several internal modes and to all components involved commonly takes place by means of the definition

$$T_{j\nu}\, d\, s_{j\nu} := d\, u_{j\nu}, \tag{5.40}$$

which now connects the entropy $s_{j\nu}$ per unit mass with the internal non-equilibrium energy $u_{j\nu}$ of species j contributed by the νth internal mode. It should be stressed that (5.40) can mathematically be substantiated exclusively for idealized components. This considerable restriction is reflected by the complementary definitions

$$u := \sum_j \omega_j u_j \quad \text{and} \quad s := \sum_j \omega_j s_j \tag{5.41}$$

of the specific internal energy u and specific entropy s of the mixture. This set of definitions is completed by two agreements, both concerning the pertaining contributions of the components

$$u_j = u_{j1} + \sum_{\nu=2}^{\aleph} u_{j\nu}; \qquad s_j := s_{j1} + \sum_{\nu=2}^{\aleph} s_{j\nu}, \tag{5.42}$$

where $\aleph$ is the maximum number of communicable modes of internal energy storage found in any component j in the mixture. If some components have less than $\aleph$ modes, the respective $u_{j\nu}$ and $s_{j\nu}$ values are zero. For convenience, the subscript $\nu = 1$ is adopted for both values to indicate a translational mode of energy storage. Certainly, such a picture at the molecular level seems realistic, particularly for ideal gases and the corresponding mixtures. For nonideal fluids like dense gases, however, it is hard to find physically reasonable arguments for using the last three equations. Rearrangement of Equations (5.40) to (5.42) yields an extension of the Gibbs main equation (5.25)

$$de = \mathbf{v} \bullet d\mathbf{i} - f \bullet d\mathbf{r} + T_{*1}\, ds - p_*\, d\rho^{-1} - \sum_{j=1}^{J} \omega_j \sum_{\nu=2}^{\aleph} \left(\left(\frac{T_{*1}}{T_{j\nu}} \right) - 1 \right) de_{j\nu} + \sum_{j=1}^{J} \mu_{*j}\, d\omega_j, \tag{5.43}$$

where the chemical potential of the jth component is now related to the translational temperature T_{*1} of the multicomponent single-phase body-field system:

$$\mu_{*j} := u_j + p_* \rho^{-1} - T_{*1} s_j; \qquad j = 1(1)J. \tag{5.44}$$

The analysis presented above calls for two remarks:

1. Equation (5.43) is an approximation rather than an exact expression in the sense of Falk's dynamics. To obtain agreement with Gibbs – Falkian theory, we start with the appropriate Gibbs fundamental equation $\Gamma(E; \mathbf{P}; \mathbf{r}; S; V; E_{j\nu}; N_j) \equiv 0$, provided that there is a precise concept of the *energetic variables* $E_{j\nu}$ for all components $j = 1(1)J$ and internal modes $\nu = 2(1)\aleph$. Then T_{*1} is the non-equilibrium temperature of the system—that is, a conjugate variable of its entropy S—and the subscript 1 should be suppressed. The corresponding conjugate variables of the relaxation variables $S_{j\nu}$ are the temperatures $T_{j\nu}$ using definition (5.40). Of course, the latter may be substituted by $\Delta T_{*j\nu} := T_* - T_{j\nu}$ for all subscripts j and ν.
2. In practice, the double sum term of Equation (5.43) is commonly related to the so-called relaxation phenomena, provided that the differentials are formally converted into time derivatives as in Equation (4.54). Generally, the specific energy of several internal modes $E_{j\nu}$ may be expressed by a set of balance equations of the universal form (Clarke and McChesney, 1976, p. 76)

$$\frac{\partial}{\partial t} \rho e_{j\nu} + \frac{\partial}{\partial x_k} (\rho v_k e_{j\nu} + q_{k,\, j\nu}) = \rho Q_{j\nu}; \qquad k = 1(1)3; \quad j = 1(1)J; \quad \nu = 1(1)\aleph, \tag{5.45}$$

where the terms $q_{k,j\nu}$ and $Q_{j\nu}$ define dissipative fluxes and rates of energy gain in the νth mode per unit mass, respectively. This agrees with the axiomatics of balances to be discussed below. As usual, the v_k denote the three components of the local flow velocity vector. There are considerable worldwide efforts to understand and consequently model these complicated relaxation processes in a mathematically stringent form.

For theoretical gas dynamics a considerable simplification of (5.45) is frequently offered, using the well-known *linear Landau–Teller relaxation equation,* which is deduced exactly from the so-called *Master-equation* for the case of harmonic oscillators undergoing certain physical conditions. This relaxation equation,

$$\frac{d}{dt}E_{j\nu}(t) = \tau_{j\nu}^{-1}[E_{j\nu}(T_*) - E_{j\nu}(t)] , \tag{5.46}$$

is commonly applied to vibrational relaxation, using the well-known *Landau–Teller plot approximation* $\tau_j^{\text{vib}} = A_j p_*^{-1}\exp(B_j/T_*)^{1/3}$ for different *vibrational relaxation times* of the components involved (Clarke and McChesney, 1976, p. 424). And yet, there is no a priori reason this form of relaxation equation should also hold for rotational relaxation. The exception is that the relaxation of both the translational and rotational energy modes of a rigid rotator may be approximated by Boltzmann distributions, which in turn are characterized by distinct temperatures T_* and T_j^{rot}. For this very special case—often applied in practice—the equation (cf. Clarke and McChesney, 1976, p. 452)

$$\frac{d}{dt}T_j^{\text{rot}} = \frac{T_* - T_j^{\text{rot}}(t)}{\tau_j^{\text{rot}}} \tag{5.47}$$

should be accepted as a definition of the rotational relaxation times τ_j^{rot} of the jth component. For τ_j^{rot} values, dependent only on the system temperature T_*, there is a simple solution to (5.47):

$$T_j^{\text{rot}}(t) - T_* = [T_j^{\text{rot}}(t=0) - T_*]\exp(-t/\tau_j^{\text{rot}}). \tag{5.48}$$

This solution indicates an exponential behavior of the rotational temperature T_j^{rot} with respect to T_*.

With the exception of hydrogen and its mixtures, it is commonly accepted that under standard conditions all values of τ_j^{rot} overlap a range where very little difference exists between T_j^{rot} and T_* for nearly all times t and initial values T_j^{rot} (t = 0). This fact, indeed, allows us to determine experimentally the local gas temperature T_* by contactless methods such as Raman spectroscopic measurements. The same methods may be applied to the vibrational spectra of fluids, although at present it is hard to use the available data processing methods for complex molecular constituents.

In the engineering sciences, the concept of relaxation times constitutes definitely the level of description for most applications connected with chemical reactions. This means that the standard case of chemically reacting fluids is classified by a set of *characteristic times* τ_r assigned to each reaction r that occurs simultaneously with all other reactions involved. For example, let us consider the case where R chemical reactions take place according to the *stoichiometric equations*

$$\sum_{j=1}^{J'} \nu_{jr}{}' \,\mathfrak{R}_j \underset{\kappa_r^b}{\overset{\kappa_r^f}{\rightleftarrows}} \sum_{j=J'+1}^{J} \nu_{jr}{}'' \,\mathfrak{R}_j ; \qquad r = 1(1)R. \tag{5.49}$$

Here ν_{jr}—denoted by a prime on the left-hand side and by a double prime on the right-hand side—marks the stoichiometric number of the jth component $\mathfrak{R}_j$ in the rth reaction. All $\nu_{jr}{}'$ ($j = 1(1)J'$) and $\nu_{jr}{}''$ ($j = J' + 1(1)J$) are integers that indicate how many molecules of the chemical species $\mathfrak{R}_j$ take part in either the forward or backward reactions. It follows that in the particle term $\Sigma_j \mu^t{}_{j*} dN_j$ the particle change dN_j must be proportional to the respective differences $(\nu_{jr}{}'' - \nu_{jr}{}')$, valid for each index j and r. By convention, the stoichiometric coefficients are assumed positive if the component appears in the second term of the equation (i.e., if it is a product of the reaction r) and negative if it appears in the first term (i.e., if it is a reactant in the reaction r). If a particular species j does not occur in the reaction r, the coefficients ν_{jr} will be zero.

Mathematically, the whole reaction mechanism is specified by the $J \times R$ array of the stoichiometric numbers ν_{jr}, which must fulfill element conservation. Since there are no more than three species on either side of an elementary reaction, a large number of the elements of the *stoichiometric matrix* are normally zero. In addition, the reaction mechanism under consideration is physically established by its functions $\kappa^f{}_r$ and $\kappa^b{}_r$, called the *forward* and the *backward rate coefficients* of reaction r, respectively. For a set of elementary reactions, these rate coefficients are functions of the system temperature T_* alone, provided that there is no vibrational relaxation. The *Arrhenius ansatz* may then be used along with the limiting law of chemical kinetics, viz.:

$$\kappa^f{}_r = A_r T^{n_r} \exp(-E_r/\mathcal{R}T_*); \qquad \kappa^b{}_r = K_r^{-1}(T_*)\kappa^f{}_r. \tag{5.50}$$

From statistical mechanics, the equilibrium constant $K_r(T_*)$ *is obtained for all equilibrium concentrations, symbolized by* $[X_j^{\text{equ}}]$ and expressed by the *equilibrium mass action law*

$$\prod_{j=1}^{J} [X_j^{\text{equ}}]^{\nu_{jr}{}'' - \nu_{jr}{}'} = K_r(T_*) ; \qquad r = 1(1)R^{\text{ind}}. \tag{5.51}$$

Since the element conservation at the baryon – lepton level must be considered, it follows that only R^{ind} independent equilibrium conditions (5.51) are needed to determine the complete set of species concentrations. If more than R^{ind} reactions occur, it can easily be shown that algebraic relationships exist between their equilibrium constants so that with an independent set of R^{ind} equilibrium constants the remaining equilibrium constants may be determined from them.

In contrast, a non-equilibrium chemical kinetic mechanism may involve any number of chemical reactions, which must be greater than or equal to R^{ind}. The problem—whether the chemical reactions are independent of each other or not—can be solved from the rank of the stoichiometric matrix noted above.

To describe chemical processes by means of Gibbs–Falkian dynamics, we will start with the Gibbs main equation (5.25) for any multicomponent single-phase body-field system

$$\rho de = \mathbf{v} \bullet \rho\, d\mathbf{i} - \rho \mathbf{f} \bullet d\mathbf{r} + T_* \rho\, ds + (p_*/\rho) d\rho + \sum_j \mu_{*j}\, \rho\, d\omega_j; \qquad j = 1(1)J.$$

Let us now define a new quantity λ_r by means of the stoichiometric coefficients v_{jr} measured in moles:

$$de = \cdots + T_*\, ds + \cdots + \sum_j \mu_{*j}\, d\omega_j$$

$$m^{B+L}\, d\omega_j := \sum_r M_j v_{jr}\, d\lambda_r, \tag{5.52}$$

$$de = \cdots + T_*\, ds + \cdots + \sum_r \sum_j \mu_{*j}\, v_{jr}\, d\lambda_r = \cdots + T_*\, ds + \cdots \sum_r a_r\, d\lambda_r$$

$$a_r := -\sum_j (m^{B+L})^{-1} M_j \mu_{*j}\, v_{jr}. \tag{5.53}$$

Because the v_{jr} is measured in moles, the molar masses M_j according to (5.17) are needed. The new independent and nondimensional variable λ_r introduced here is called the *progress variable* for the rth reaction. From the chemical point of view, the differential $d\lambda_r$ denotes a small number that measures the extent to which the reaction has taken place. The second new quantity is the specific *affinity* a_r of the rth reaction, first extensively applied by de Donder in 1922.

The integration of (5.52) for a single reaction ($R = 1$) yields an expression for the progress variable

$$\lambda = v_j^{-1}(\omega_j - \omega_j^0), \tag{5.54}$$

where ω_j^0 stands for the initial mass fraction of the jth component. Together with the inequality

$$0 \le \omega_j \le 1 \tag{5.55}$$

the scope of validity of λ can be determined. For the reactants (with $v_{jr} \le 0$) the interval becomes

$$\frac{1 - \omega_j^0}{v_j'} \le \lambda \le -\frac{\omega_j^0}{v_j'}; \qquad j = 1(1)j, \tag{5.56}$$

whereas for the products (with $v_{jr} > 0$) a symmetric solution results:

$$-\frac{\omega_j^0}{v_j''} \le \lambda \le \frac{1 - \omega_j^0}{v_j''}; \qquad j = J' + 1(1)J. \tag{5.57}$$

The value $\lambda = 0$ corresponds to the starting condition where only reactants exist with given initial values ω_j^0. The affinity A_r, related to the rth reaction, is a state variable that allows us to concisely formulate some important behaviors of chemical reactions. For instance, natural processes can simply be characterized by the inequality

$$A_r \, d\lambda_r > 0; \qquad r = 1(1)R. \tag{5.58}$$

In practice, the identity

$$A_r \equiv 0; \qquad r = 1(1)R \tag{5.59}$$

is of great relevance: It describes the thermodynamic equilibrium state defined by the maximum of entropy with respect to the variable λ_r (i.e., $(\partial s/\partial\lambda_r)_{e,\,\ldots,\,\lambda \neq 0} \equiv 0$) and fixed by the complete set of the R reactions.

Of course, the case of non-equilibrium reactions is of special interest. It is advisable to quantify such non-equilibrium processes by referring to the actually observable effect of any reaction. Clarke and McChesney (1976, p. 80) suggest using the net rate of production of any given species j in the mixture. Their result concerns the *mass rate of production* Γ_{jr} of the jth component $\mathfrak{R}_j$ in the rth reaction and is given as follows:

$$\Gamma_{jr} = \rho M_j \left[\nu_{jr}'' - \nu_{jr}' \right] \left\{ \frac{1}{\tau_r^f} \prod_{j=1}^{J} \omega_j^{\nu_{jr}''} - \frac{1}{\tau_r^b} \prod_{j=1}^{J} \omega_j^{\nu_{jr}''} \right\}. \tag{5.60}$$

The relaxation times τ_r^f and τ_r^b concerning the rth forward reaction and the backward reaction, respectively, are of special interest:

$$\frac{1}{\tau_r^f} := \rho^{-1} \left[\kappa_r^f \prod_{j=1}^{J} (\rho M_j^{-1})^{\nu_{jr}'} \right]; \qquad \frac{1}{\tau_r^b} := \rho^{-1} \left[\kappa_r^b \prod_{j=1}^{J} (\rho M_j^{-1})^{\nu_{jr}''} \right]. \tag{5.61}$$

For any given pair of values of temperature and pressure both characteristic chemical times may easily be determined, provided that the set of functions κ^f_r and κ^b_r—the forward and the backward rate coefficients of reaction r—are known. Note that in (5.60) the mass fractions ω_j are used instead of concentrations $[X_j]$ of the mixture in units of moles per unit volume. The conversion follows from the simple formula

$$[X_j] = M_j^{-1} \rho \, \omega_j; \qquad j = 1(1)J, \tag{5.62}$$

where ρ and M_j are the mass density of the mixture and the molar mass of the jth species, respectively.

Assuming that the fluid mixture is in chemical equilibrium, there will be no net rate of production of any particular species. This means that forward and backward reactions are dynamically balanced and each production Γ_{jr} vanishes. Taking the rth reaction as a typical one, the ratio of the corresponding equilibrium characteristic chemical times

$$\frac{\tau_r^{b,\,\mathrm{equ}}}{\tau_r^{f,\,\mathrm{equ}}} = \prod_{j=1}^{J} (\omega_j^{\mathrm{equ}})^{\nu_{jr}'' - \nu_{jr}'}; \qquad r = 1(1)R \tag{5.63}$$

immediately follows from Equation (5.60) in principle. Its right-hand side becomes admissible by (5.51). This limiting case serves as a suitable reference state provided that in practice the mass fraction ω_j^{equ} can be calculated by any of the very efficient computing techniques now available (see, e.g., Gordon and McBride, 1976).

Equations (5.60) – (5.63) may now be handled to show that, under non-equilibrium conditions, there exists the following set of interesting relations

$$\frac{\kappa_r^f}{\kappa_r^b} = \left[\frac{\tau_r^b \tau_r^{f,\,equ}}{\tau_r^f \tau_r^{b,\,equ}}\right] K_r \,(\rho/\rho^{equ})^{\sum_j (\nu_{jr''} - \nu_{jr'})} \tag{5.64}$$

between the forward and the backward rate coefficients κ_r^f and κ_r^b of reaction r and the corresponding relaxation times τ_r^f and τ_r^b.

Be aware that Equations (5.60)–(5.64) touch the limits of knowledge concerning the connection between theory and experiment in non-equilibrium chemistry. In general, the function κ_r^f is found exclusively by experiments for a relatively narrow range of temperature and pressure for plots using the Arrhenius ansatz (5.50). Hence, for known κ_r^f values the corresponding κ_r^b values assumed to be valid for the same, but reverse, reaction are calculated with the assigned equilibrium constant K_r. Consequently, this approximation leads from (5.64) to a direct statement about the ratio τ_r^b/τ_r^f:

$$\frac{\tau_r^f}{\tau_r^b} = \frac{\tau_r^{f,\,equ}}{\tau_r^{b,\,equ}}\left(\frac{\rho}{\rho^{equ}}\right)^{\sum_j (\nu_{jr''} - \nu_{jr'})} \quad . \tag{5.65}$$

This refers to the two characteristic chemical times proved to be valid for the rth non-equilibrium reaction. It is noteworthy that any deviation from the corresponding equilibrium state only depends on the mixture density as one of the local non-equilibrium variables. This result is slightly incorrect for physical reasons because the very special case $\rho = \rho^{equ}$ cannot yield a solution represented by values calculable for well-known equilibrium states. Neither can this solution be used without considering the actual prevailing conditions. Unfortunately, it is hard to find well-founded and mutually independent data that allow us to evaluate Equation (5.61) in a precise way.

Nevertheless, it should be emphasized that a relation such as (5.60) generally applies to non-equilibrium conditions. This means that for the particular values of, say, ρ and T_* located at some chosen place and time, the mass-fractional composition at this point in space-time will not adequately be described by equation (5.63) with T^{equ} and ρ^{equ}, equated just by T_* and ρ, respectively. If we were to do so, we would get a mass-fractional composition that would undoubtedly differ from the actual composition.

We cannot discuss the details of chemical reactions involved in any non-equilibrium processes here. Yet it should be stressed that, generally, we cannot construct an arrangement of the characteristic chemical times τ_r^b and τ_r^f as in (5.36) with the intention of eliminating some reactions for the theoretical description of the whole non-equilibrium process in question. In reality, chemical kinetics must only be considered as a coherent subsystem whose mathematical structure is chiefly coined by the circumstances due to the constraints of the entire process. Thus, chemical kinetics become especially important in view of the very short residence times of combustible gases within the combustion chamber of propulsion systems like scramjets. Here the circumstances differ essentially from those in a chamber with subsonic combustion, where flames can easily be bound on flame holders that only produce recirculation to increase that residence time. When the flame is ignited locally, it will at once spread widely into all such regions where a reactive mixture exists.

In supersonic combustion chambers, other conditions prevail. Save for the mixing at different speeds, the dominating reaction conditions in supersonic regions decide whether the reactions are fast enough to complete the combustion within the chamber or whether the engine exhaust gas will spill rather large portions of reactive species whose chemical energy is lost for pure thrust generation. With proper adverse reaction conditions, it is also possible to prohibit the flame propagation into the core flow along the chamber. For this reason, a realistic description of these phenomena must consider the detailed reaction of the kinetic process within the field. This means that the complete reaction schemes of several technically relevant combustion processes, like the hydrogen-air combustion, have to be taken into account. Such a scheme sometimes consists of dozens of reactions with more than ten chemical species. Some of these reactions only take part in certain sequences of the whole mechanisms. For instance, when reactions are starting, no free atoms or radicals will exist in the combustible mixture. These can only be formed in reactions between the stable molecules of the initial gas. For the correct description of the process in time and space the pertaining reaction scheme must include such start reactions, even though they do not play a role during the steady course of the process. The same is true for a considerable number of other reactions, such as some chain branching mechanisms or chain termination reactions. Obviously, each case requires individual attention.

In summary, three kinds of time quantities prove to be relevant for any system $\{\wp; \mathcal{O}\}$ and could be identified (the word system here means $\wp$ observed by $\mathcal{O}$). The first time quantity, the observer's time scale $\mathcal{T}$, furnishes a subjective element of the system under consideration. Exemplarily, this time $\mathcal{T}$ may be expressed by the residence time noted above and defined by the ratio of $L : \mathcal{V}$, where L is a characteristic length and $\mathcal{V}$ is a characteristic velocity. Keep in mind the extension of the combustion chamber as L and the average gas flow velocity (derived from the steady mass flow rate) as $\mathcal{V}$.

The second time quantity refers to the internal processes taking place within the system. This time defines characteristic relaxation phenomena with emphasis on vibrational relaxation and, particularly, on non-equilibrium chemical reactions. Although it is difficult to give any general recommendation, these sets of relaxation times may be employed to select the appropriate system variables according to the rules of order in (5.35) and (5.36).

The last time quantity, relevant for a mathematically adequate mapping of any macroscopic system, is the curve parameter t. Processes near equilibrium have a time scale dt of the same order of magnitude as the observer's time scale $\mathcal{T}$, and it is this latter scale that determines the coordinates of the observer's state-space $\chi(\mathcal{T})$. Because dt and $\mathcal{T}$ are comparable in most cases, the Gibbs main equation (5.43), modified by the definition of affinity A_r (5.53), may be split up for $\chi(\mathcal{T})$ by dt. This conversion leads to the following rate equation

$$\frac{de}{dt} = \mathbf{v}\cdot\frac{d\mathbf{i}}{dt} - \mathbf{f}\cdot\frac{d\mathbf{r}}{dt} + T_*\frac{ds}{dt} - p_*\frac{d\rho^{-1}}{dt} - \sum_{j=1}^{J}\omega_j\sum_{\nu=2}^{\aleph}\left(\left(\frac{T_*}{T_{j\nu}}\right)-1\right)\frac{de_{j\nu}}{dt} - \sum_{r=1}^{R}A_r\frac{d\lambda_r}{dt}, \quad (5.66)$$

where all fast relaxation processes with $\tau_r/\mathcal{T} << 1$ are omitted from the last summation term.

It is a crucial point that the element dt cannot be made arbitrarily small. In other words, because we want to describe the process entirely in $\chi(\mathcal{T})$-space, there is some limit to the rates of change appearing in (5.66). Higher rates of change demand the inclusion of additional terms on its right-hand side. As processes are frequently initiated in a system $\{\wp; \mathcal{O}\}$ by some changes at its boundaries, the longest response time throughout $\{\wp; \mathcal{O}\}$ of the affected thermodynamic variables imposes an obvious lower limit on dt for $\{\wp; \mathcal{O}\}$ to remain uniform. Of course, extended systems cannot behave uniformly on short time scales. Consequently, $\{\wp; \mathcal{O}\}$ needs to be divided into a large number of phases. The consequent hypothetical experiment, however, has to correspond to real experience containing the actual fluid velocities within $\{\wp; \mathcal{O}\}$. From this idea a convenient thermodynamic subsystem may be introduced. This so-called *fluid particle* is located by position vector $\mathbf{r}$ at time t. It may be treated as a convected fluid element of macroscopically infinitesimal volume $d\tau$, distinguished by a typical dimension $|d\mathbf{r}|$ much larger than a mean free path.

In sharp contrast with the common theory, Equation (5.66) takes into account exactly this fundamental mathematical concept. This Gibbs main equation contains the *total* energy (not the internal energy only) and the (linear) momentum as an obvious manifestation of the fluid particle's motion. There is no need to fall back on the hypothesis regarding the fluid particle as a quasi-mass-point. This model, thoroughly discussed in Chapter 3 and disproved in Chapter 6, presupposes the element $d\tau$ to be equipped with all thermostatic properties and moved as a whole according to the laws of classical mechanics.

Two other constraints often claimed apodictically are also questionable: Introducing the longest response time $\tau_\wp$ for any signal to transport information across the system, the constraints

$$dt \approx \mathcal{T} \geq \tau_\wp, \quad (5.67)$$

are believed to ensure that processes remain in temporal equilibrium. Allegedly, this means that during their changes the system under consideration should maintain nearly uniform properties. There is no question that such a statement is correct, as long as it expresses the idea of the fluid particle in only a formalized manner or in words different from those used above. But the constraints (5.67) are erroneously interpreted if "their imposition is usually termed the hypothesis of local thermodynamic equilibrium" (Woods, 1975, p. 69). Such an overall conclusion will have to fail, particularly with regard to the gradation rule (5.36).

Note that the operator d/dt is no more than the time derivative in any convenient reference frame. Within its coordinates the motion of each fluid particle is described by the independent variables of (5.66), which are now functions of $\mathbf{r}$ and t.

Equation (5.66) exhibits two further features to be accentuated:

- Specific variables are adopted in place of the total quantities generally appearing in Falk's dynamics.
- The complete interaction between $d\tau$ and its environment can explicitly be described by means of the First Law in its version valid for continuum physics and adapted in an appropriate way to (4.53).

The latter point offers the option of dealing with open systems and is the topic of the next section. To conclude these ideas another point should be noted: "the error committed in the Gibbs relation [(5.66)] by ignoring a relaxation process that is not quite fast enough to be relaxed, is only second-order in the displacement of the progress variable λ from its equilibrium value" Woods (1975, p. 69).

5.4 Balance Equations

5.4.1 *IS THERMODYNAMICS MERELY A USEFUL BODY OF IDEAS?*

In the former chapters manifold use has been made of the knowledge concerning facts and methods available at the microscopic level. This information, however, should be regarded as motivation for and illustration of the ideas hitherto presented, not as a substantial part of a mathematical theory of macroscopic phenomena, which is the main subject of this book. Indeed, the incorporation of the Gibbs–Falkian concept into a mathematical field description by means of partial differential equations with respect to time- and space-coordinates should be regarded as a completely independent step of the theory. This point of view is controversial; there is violent disagreement about the epistemological value of certain macroscopic theories compared with the standard theories, such as the theories of quantum magic and gravity, which internationally dominate education in physics. Particularly with regard to several modern trends in physics concerning mainly nonlinear phenomena, the following remarks might be useful.

In the prevailing philosophy of continuum physics, it is common to disregard the knowledge of the microscopic world in the mathematical presentation of macroscopic fundamentals. In fact this is admissible because modern mathematics, with the help of some physical basics, can furnish efficient tools for the reliable description of real phenomena. Furthermore, it must be acknowledged that the numerous relationships between field theories relying on the continuity axioms and the particle theories are much more complicated than originally thought. For this reason it is fairly hard to understand Penrose's view of the theoretical status of thermodynamics compared with other main branches of physics. In his brilliant book *The Emperor's New Mind,* this prominent author casually remarks that

> thermodynamics, as it is normally understood, being something that applies only to averages, and not to the constituents of a system—and being partly a deduction from other theories—is not quite a physical theory in the sense that I mean it here (the same applies to the underlying mathematical framework of statistical mechanics). (Penrose, 1990, p. 221)

Perhaps Penrose is right about thermodynamics, for he understands it "as a merely useful body of ideas," but he is certainly in error in view of theories such as the Gibbs–Falkian concept being executed in engineering practice in favor of the real world of central heatings, airplanes, or dialyzers. He takes the widespread view among the scientific community, where many members use the amenities of modern technology, like combustion and heat techniques,[11] but do not realize that these amenities could never have been successful by means of a theory such as Penrose quite obviously envisions. A physical theory in Penrose's sense may serve to discover black holes or imaginary times, even at the risk of being forced to theoretically change the black holes into yellow ones. (By the way, statistical mechanics has never been a mandatory prerequisite for all sections of technical thermodynamics and process engineering, for example, as applied in combustion technology, heat and mass transfer, or multiphase flows.)

Nevertheless, all information may be useful concerning reliable empirical facts and theories that prove to be applicable to the microscopic and even to the mesoscopic level of description. This is true regarding the use (at least in principle) of kinetic theories for the determination of all parameters defined by the macroscopic theories under consideration and normally determined by experiment. In addition, to this day Truesdell's following statement is an appropriate description of the integration procedures in the kinetic theories: "The whole problem of the existence and status of dominant solutions in general remains, more than a century after the first groping toward such solutions by Maxwell and Boltzmann themselves, covered by

[11] Thus, for instance, in his inaugural lecture, a prominent German professor in theoretical physics answered the question, "What heat does really mean?" with an amazing statement concerning sensation of heat and coldness: " ... that they cannot be fully objectified ... and forbid being precisely specified. In a word: They are inaccessible to the physical method" (author's translation; cf. Straub, 1990, p. 171). Similar ways of looking at things seem to lead some theoretical physicists to a mathematical representation of work and heat that has been proven false. Fortunately, this defect is irrelevant because, as a rule, such a representation does not affect the other topics of the respective treatment (e.g., Schlögl, 1989, p. 60).

mystery and approached through a delicate combination of guesswork and staggering formal calculations" (Truesdell, 1984, p. 422).

Clearly, this means that formulas for viscosity, heat or electrical conductivity, and so on are approximations based on special molecular models of matter. This is not trivial, for these very approximations offer no obligatory access to thermodynamics. For instance, the thermodynamic temperature is ambiguously connected with the concept of kinetic temperature, "because in the theory of monatomic gases it is simply a multiple of the energetic. We wished to leave nobody a chance to miss the strictly mechanical character of the kinetic theory" (Truesdell, 1984, p. 405).

Some attempts to derive the form of macroscopic constitutive equations directly from the Boltzmann equation are of a more recent date. The respective dominant solutions may or may not conform with the idea of irreversibility in continuum physics, but certainly they do obey specific constitutive relations appropriate to a certain kind of nonsimple material. As a matter of fact there is not a single solution believed to be valid for any real transfer phenomenon. In practice, all results are either based on a purely empirical foundation or their theoretical concept concerning any model fluids demands approximations of such a drastic degree for a solution of the Boltzmann equation that it is no longer possible to obtain a reliable statement about the mathematical structure of the transfer laws in the sense of Newton, Fourier, or Fick.

These experiences were well confirmed by the investigations of Carleman and McKean; it was their ambitious objective, regardless of costs, to derive an *exact* solution of the Boltzmann equation for an inhomogeneous temporal and spatial domain. To achieve this, they even allow their model fluid to correspond to a caricature of a substance rather than to a real one. Nonnenmacher, as well as Straub et al. (1987), have obtained further exact solutions of the Boltzmann equation, particularly for the diffusional current density.

In view of macrophysical phenomena there are of course other important reasons to consider statistical and kinetic theory on a molecular basis. Due to the recent evolution in physics, quantum mechanics in its more interesting parts now deals with the description of unstable particles and their mutual conversions. Relativity, originated as a geometric theory, at present is mainly associated with cosmological items such as the history of the universe or the search operation for any dark matter.

Notwithstanding, remember that quantum theory is well-founded, for instance, on the linear and reversible Schrödinger equation. For this reason, most physicists believe that the particle world is ruled by the principle of microphysical reversibility. Perhaps such a belief has been reinforced by the fact that the classical branches such as flow mechanics, thermodynamics, and electrodynamics had a great part in all advanced technologies of the last seventy years. On the other hand, theoretical physics, with only a few important exceptions, furnished these fields neither with fundamentals nor with results relevant for applications. All textbooks of today's theoretical physics will confirm this statement, contrasted with modern textbooks in the engineering sciences.

To grasp this remarkable development, it is certainly helpful to remember that since the 1890s nonlinear continuum physics and nonlinear thermodynamics have continuously advanced. This progress was not restricted to technical inventions or to the great number of contributions in process engineering. Up to the final victory of quantum mechanics during the late 1920s, many physicists and mathematicians made many important contributions to the understanding of these traditional branches of nonlinear physics. Poincaré should be mentioned with emphasis because of his brilliant achievements in the field of nonlinear mathematics. His work, published in the last decade of the nineteenth century, has recently been rediscovered as a pathway of a "new" field of modern science—nonlinear physics. He set the stage where, since the beginning of this century, four actors have been playing their double matches: reversible versus irreversible events and deterministic versus stochastic processes.

It is remarkable that such a development could occur over only two decades. Undoubtedly triggered by some unexpected discoveries, quite a new view of matter arose associated with spontaneous activity and complicated phenomena running far from equilibrium.

This development has led to an interesting occurrence: Nowadays one finds a flood of books in nearly all bookshops entitled with at least one of the typical key terms of modern physics and aimed at the university-educated amateur believed to be well-versed in a number of subjects. The level of these publications ranges from philosophical studies to popularized books, where even purely esoteric brochures may be encountered. Each title contains one or two scientific standard expressions that may evoke daily associations normally far from any notations in mathematical physics. But the fascination of words like catastrophe (Thom, 1984), chaos (Hao, 1989), complexity (Nicolis and Prigogine, 1989), synchronicity (Peat, 1989), synergetics (Haken, 1978), and so on obviously provokes curiosity and today even produces bestsellers.

Most of these books, when looked at closely, are designed along a recurrent pattern: Beginning with the general introduction of the key term in question (e.g., complexity) and its vocabulary (e.g., dissipation, nonlinearity, stability, symmetry breaking, correlations), the authors (e.g., Nicolis and Prigogine, 1989) offer some special items assumed to be characteristic for the topic (e.g., complexity and dynamics, complexity and randomness, complexity and knowledge). As a rule, the reader will find little mathematically coherent theory, but there are always some selected examples for the theme under consideration (e.g., chaotic behavior as a prototype of special dynamics). The mathematics presented is focused on a very special case [e.g., Lorenz's strange attractor (Lorenz, 1963, p. 137)] without any commentary on the constraints of the equations used. But the latter are often weighty, particularly compared with the sometimes far-reaching inferences [e.g., "The model gives rise to an attractor that captures the principal features of turbulent convection" (Nicolis and Prigogine, 1989, p. 124)] drawn from the exposed phenomenon and its mathematical modeling.

Strictly speaking, this method—explaining new and sometimes questionable effects—seems more appropriate for experts than for amateur readers. However, the usual selection of simple but relevant examples, supplemented with the corresponding mathematical tools, might be pedagogically opportune, provided that the reader is able to understand any new physical effects. Nevertheless, the authors, especially many scientific journalists, deal with problems as though they were of the simplest kind. This invariably implies that their solution might easily be generalized to complex problems of reality. Unfortunately, such a procedure is common in the natural sciences. An example is the classical case of the one-body-field interaction: Its simple solution cannot be regarded as a special case of the so-called two-body problem, which is neither solvable nor even exactly describable as a mathematical problem.

Notably, all these publications may help to gain insights into problems that in the past were found intractable. But they should not be accepted as common textbooks in the sense that they can enable the reader to employ them for any concrete problems in science and technology. They cannot, aside from a few applications of the chaos theory. Even from that theory there have been, in spite of all opposite affirmations, no substantial contributions to any investigation on turbulence phenomena for instance, seen at least in comparison with the decades of research on this field in the engineering sciences. The situation seems to be similar to Heisenberg's famous contribution to turbulence theory, which is always cited but is only applied in rare cases.

5.4.2 BALANCE EQUATIONS

In Section 5.3 we studied some essential prerequisites and barriers for an appropriate mathematical field theory adequate for continuum models of physical systems. Such continuum theories are also concerned with phases, whose relevant properties were discussed in Section 5.2. Let us now stress the decisive difference between a *field* and a *phase*. By definition, the latter depends only on the values of the local variables assigned to diverse volume elements of the fluid. Furthermore, its energy is invariant against any metric distortions. In contrast, the variables of a field may be distinguished by an inherent coherence length that indicates the distance where effects of their local changes can no longer be ascertained. For a sufficiently small coherence length, any field may be treated as a phase. Strictly speaking, a phase is a limiting case in physics that may even simultaneously occur along with fields. Critical phenomena and phase transitions of the second order are relevant examples for some spatially extended systems, where the manifold of states appears partly as a phase and partly as a field.

The main aim of the classical field theory is the investigation of those partial differential equations that are valid in three-dimensional Euclidean space for the mechanical, electromagnetic, and thermal state parameters that depend on space and time in such a way that their evolution is represented by space-time functions. There are some excellent monographs in which the systematic and exact development of classical field theories can be studied.

Let us now briefly outline the derivation of the conservation laws for any fluid particle introduced in the last section. It is located by position vector $\mathbf{r}$ at time t and may be treated as a convected fluid element of a macroscopically infinitesimal volume $d\tau$, distinguished by a typical dimension $|d\mathbf{r}|$ much larger than a mean free path. The fluid particle is moving with the kinetic velocity $\mathbf{v}^{\#}$ relative to a fixed coordinate frame, where $\partial/\partial t$ denotes the partial time derivative. Conversely, in a frame convected with the same fluid particle, the time derivative will now be defined by

$$D := \frac{\partial}{\partial t} + v^{\#} \bullet \nabla \ , \tag{5.68}$$

where ∇ is the gradient vector operator. The operator D is known as the *material* or *substantial derivative.*

Equation (5.68) refers to the well-known alternative possibility of describing processes by means of two different coordinate systems in space and time. "The Lagrangian coordinates are fixed in the material and move with it at its center of mass (barycentric) velocity. Thus Lagrangian coordinates constitute a thermodynamic *closed system* or *control mass analysis* frame in which the conservation of mass law has a convenient representation. Eulerian coordinates are fixed in space and thus constitute thermodynamic *open system* or *control volume analysis* frame. To distinguish between the time derivatives in these frames, the symbol D/Dt is used for the Lagrangian frame differential (also called the material or substantial derivative), whereas the usual differential symbols are used for the Eulerian frame differentials" (Balmer, 1983, p. 19). In this sense, the left-hand side of Equation (5.68) represents the Lagrangian frame and the right-hand side stands for the Eulerian frame.

All classical field theories are primarily based on kinematics. Each material domain may be seen as an ensemble of infinitesimal subsystems assigned to the volume elements $d\tau$. Geometric space is needed to enumerate the subsystems by the space-coordinates. The meaning of this numbering makes clear the very intrinsic neighborhood affinities or, mathematically expressed, the topologic relations of the domain taken as a whole. This point of view reduces dynamics to the solution of two isolated problems: the description of the subsystems themselves and the interactions occurring between adjoining volume elements only.

Whereas these items lay down the foundations for the weighty position of kinematics in conventional theories, Falk's dynamics postulates an inverse idea: Metric space properties like the notions of distances, angles, and even velocities should primarily follow from dynamics. For this reason, the usual method propagated by textbooks is questionable, in so far as it claims to derive fundamental relations of continuum physics by changing any properties along with attending changes of infinitesimal volume elements, regardless of the possible changes of other physical properties.

Note that this criticism contradicts Euler's famous *cut principle,* which follows the dictum "isolate the system." But, in view of modern terms, constitutive relationships, and material models, this cut principle creates some subtle confusion

> in that it rests upon a purely geometrical statement, namely, that the traction is normal to the surface on which it acts. Moreover, the confusion is deep, for, first, it obscures the tensorial character of the stress, as expressed by Cauchy's fundamental theorem, and, second, it renders the principles of linear momentum and moment of momentum equivalent, as they are for many degenerate systems, e.g., for a punctual system subject to pairwise equilibrated central forces. (Truesdell, 1968, p. 193)

According to the basic axiom of kinematics, the kinetic velocity $\mathbf{v}^{\#}$ follows from the superposition of the translational velocity of the continuous flow medium, defined by its elasticity rate in the direction of the principal axes and from the rotational velocity around them. It is a crucial point that this result is independent of the special nature physically relevant for the fluid particle system in question. Following the reasoning stated above, it is mandatory that a connection be established by the *kinematic network condition*

$$\left(\frac{\partial E}{\partial \mathbf{P}}\right)_{\mathbf{r}, S, V, \ldots, N_j} = \mathbf{v} \Leftrightarrow \mathbf{v}^{\#} = \frac{d\mathbf{r}}{dt} \tag{5.69}$$

valid for all body-field systems as defined by equations like (5.9). In fact, the important equivalence (5.69) guarantees that with the help of the (dynamical) velocity $\mathbf{v}$ all the individual properties of the system in question will automatically be considered.

A further conclusion from the basic axiom of kinematics leads to a substantial relation between the material derivative and the macroscopically infinitesimal volume $d\tau$, according to which

$$D(d\tau) = (d\tau)\, \nabla \bullet \mathbf{v} \tag{5.70}$$

holds (Spurk, 1989, p. 25). Here Equation (5.70) is introduced as a definition of $d\tau$.

Virtually all modern non-equilibrium thermodynamics texts (viz. Balmer, 1983, p. 19) present local equilibrium field theories in essentially the same way. Their objective is to obtain a rate balance equation applied to an infinitesimal volume, $d\tau$, which is subject to the local equilibrium hypothesis concerning the purely thermodynamic properties of $d\tau$. For instance, this is true for the *extended irreversible thermodynamics*: "Imagine a system which is not in equilibrium with its surroundings. We may assume that the system may be subdivided into a large number of subsystems, small enough so that their volume (or extension) is small compared with the system's volume but still containing a large number of molecules to be regarded as a macroscopic system. The local equilibrium assumption now asserts that each of these larger number of subsystems may be regarded as being in thermodynamic equilibrium" (Garcia-Colin and Uribe, 1991, p. 108).

Suppose that $Z_V(\mathbf{r}, t)$ denotes some physical quantity of the system per volume unit. Then the rate of change $\mathscr{S}_Z$ of this quantity in a frame moving with the fluid particle may be defined by

$$(d\boldsymbol{\tau})^{-1} D(Z_V \, d\boldsymbol{\tau}) := \mathscr{S}_Z = DZ_V + Z_V \nabla \bullet \mathbf{v} = \partial_t Z_V + \nabla \bullet (Z_V \mathbf{v}), \tag{5.71}$$

where the abbreviation ∂_t is used for the partial time derivative $\partial/\partial t$.

According to postulate (5.20) of the baryon–lepton constancy, we let $Z_V(\mathbf{r}, t)$ first be the mass density ρ. Then, as the mass $\rho \, d\boldsymbol{\tau}$ of the fluid particle is constant, the source term $\mathscr{S}_Z$ vanishes, and (5.71) yields the various forms

$$\begin{aligned} D(\rho \, d\boldsymbol{\tau}) &= 0, \qquad & \rho D\rho^{-1} &= \nabla \bullet \mathbf{v} \\ D\rho + \rho \nabla \bullet \mathbf{v} &= 0, \qquad & \partial_t \rho + \nabla \bullet \rho \mathbf{v} &= 0 \end{aligned} \tag{5.72}$$

of the law of mass conservation. Introducing now the specific quantity $z := \rho^{-1} Z_V$, Equation (5.71) may easily be changed by means of the conservation law (5.72) into a form useful in practice:

$$\rho \, Dz = \partial_t \rho z + \nabla \bullet (\rho z \, \mathbf{v}); \qquad Z_V := \rho \, z. \, . \tag{5.73}$$

In a continuum consisting of J components the barycentric velocity of the mixture may be defined by

$$\mathbf{v} := \sum_j \rho_j \mathbf{v}_j; \qquad j = 1(1)J, \tag{5.74}$$

where the individual velocities $\mathbf{v}_j$ of the different components will macroscopically result in the diffusion phenomena. For this reason we introduce the *diffusion velocity* $\mathbf{u}_j$ and the *diffusion current density* $\mathbf{j}_j$ of the jth component. We will define both these diffusion properties with respect to the barycentric velocity $\mathbf{v}$ as follows:

$$\mathbf{u}_j := \mathbf{v}_j - \mathbf{v}; \qquad \mathbf{j}_j := \rho_j \, \mathbf{u}_j; \qquad j = 1(1)J. \tag{5.75}$$

It follows immediately from (5.74) and (5.75) that the diffusion current densities are not independent, because their number is reduced to $J - 1$ by the summing of (5.75):

$$\sum_j \mathbf{j}_j = \sum_j \rho_j \mathbf{u}_j = \sum_j \rho_j (\mathbf{v}_j - \mathbf{v}) = 0. \tag{5.76}$$

If we let z be any arbitrary specific quantity referred to unit mass, then the corresponding extensive state variable is defined by

$$Z := \int_V \rho j z d\boldsymbol{\tau}, \tag{5.77}$$

where the integration should be carried out over the volume elements $d\boldsymbol{\tau}$ from which the whole domain V under consideration is composed. The course of time defined by

$$\frac{dZ}{dt} = \frac{d}{dt} \int_V \rho z \, d\boldsymbol{\tau} \tag{5.78}$$

can generally arise for the following two reasons only:

1. Due to the flux into and out of Z through the boundary surface ∂V of the volume V.
2. Due to the annihilation or the production of the quantity Z within the volume V that is the consequence of sinks or sources for the quantity Z, present in any macroscopically infinitesimal volume $d\tau$ of the continuum.

This is a pure reasoning of common logic characterized by its basic axiom expressed traditionally in a Latin version, "tertium non datur." The formulation of the general balance equations is effected on the basis of these points. Translated into mathematics, points (1) and (2) may be expressed with regard to (5.78) as follows:

$$\frac{dZ}{dt} = \frac{d}{dt}\int_{V} \rho z \, d\tau := -\oint_{\partial V} \mathbf{j}_z \bullet \mathbf{n} d\Omega + \int_{V} \sigma_z \, d\tau. \tag{5.79}$$

The current density $\mathbf{j}_z$ is the quantity of Z crossing a plane of unit area in unit time, provided that the plane is oriented normal to $\mathbf{j}_Z$. Here $d\mathbf{\Omega} = \mathbf{n}\ d\Omega$ (is a vectorial surface element whose magnitude is $d\Omega$, and its direction is determined by the outward normal $\mathbf{n}$. The integrand σ_z of the last volume integral denotes the density of the internal sink or source of Z within $d\tau$.

Balance equations like (5.79) cannot be confirmed directly for any quantity. In principle, the integral expression (5.79) is nothing but a simultaneous definition for the quantities involved. On the whole, its significance lies in the fact that given two quantities, any of the other quantities can be determined. But to do so, it is paramount that the balance be further specified based on whether the local or the material description is the primary one.

In the latter case, general balance equations may be easily obtained as the selected volume element $d\tau$ moves together with the medium which in turn will move with a velocity $\mathbf{v}$ in the frame of coordinates x_k ($k = 1, 2, 3$) fixed to span the space. Hence, the integration of Equation (5.73) has to be extended to the time-dependent volume $V(t)$ moving with the continuum. It becomes

$$\frac{d}{dt}\int_{V(t)} \rho z \, d\tau = \int_{V(t)} \rho D z d\tau. \tag{5.80}$$

because the substantial differentiation with respect to time only affects the quantity z. Hence, by application of the Gauss theorem, the combination of the last two equations yields

$$\rho D z + \nabla \bullet \mathbf{j}_z = \sigma_z, \tag{5.81}$$

which is the differential form of the material balance. Bear in mind that Equation (5.81) is an "empty formula" that introduces two quantities defined qualitatively by quite distinct references to the system variable z. Complementary quantitative expressions for $\mathbf{j}_z$ and σ_z are the subject matter of several theories commonly characterized as constitutive. In principle, all macroscopic parameters defined by the corresponding constitutive equations have to be determined experimentally. But there are some approximate methods by which those transport coefficients may also be derived theoretically, at least for relatively simple fluids. As a rule, such mathematically ambitious procedures are established at a mesoscopic level using statistical elements for the respective theory. Kinetic theories of gases founded on the Maxwell–Boltzmann equation are by far the best-known approaches.

A further generally valid rule should be mentioned: The quantity $\mathbf{j}_z$ has a vectorial order that is always increased by one degree compared with the identical order of the two quantities z and σ_z. Adding the terms of zero sum $-\nabla \bullet (\rho z\, \mathbf{v}) + \nabla \bullet (\rho z\, \mathbf{v})$ to (5.81), the equation

$$[\, \rho\, Dz - \nabla \bullet (\rho z\, \mathbf{v})] + \nabla \bullet \mathbf{j}_z + \nabla \bullet (\rho z\, \mathbf{v}) = \sigma_z = \partial_t\, \rho z + \nabla \bullet (\, \mathbf{j}_z + \rho z\, \mathbf{v}) \tag{5.82}$$

arises, which, using (5.73) and the definition

$$\mathbf{j}_z{}^0 := \mathbf{j}_z + \rho z\, \mathbf{v}, \tag{5.83}$$

will yield the differential form of the local balance

$$\partial_t(\rho z) + \nabla \bullet \mathbf{j}_z{}^0 = \sigma_z. \tag{5.84}$$

Here $\mathbf{j}_z{}^0$ denotes the local current density of the field quantity z.

The source density σ_z of arbitrary field quantities z appears as a direct consequence of internal and external influences manifested by all their irreversible processes. Gradients of the system's local velocity, temperature, chemical potential, and so on with respect to the corresponding values of its surroundings represent inhomogeneities in the interior of the system under consideration. The current densities caused by these inhomogeneities are the factors that determine the contributions pertaining to the source density σ_z.

External contributions stem from the long-range action of external forces acting on the interior of the system. In the sense of body-field systems discussed above, gravitational or electromagnetic fields acting on the material of the system always exist. In this context, the question of the possible conservation of some field quantity is again raised. Recall that the conservation behavior of a system quantity is not a property in itself, but rather a relation defined by special conditions on each quantity in question. Thus, for instance, the conservation of mass is related to a special level of description of certain processes fixed by the constancy of definite particle numbers, like those of leptons and baryons in the case of chemical reactions. Undoubtedly, as the following comparison demonstrates, the most prominent examples for conservation properties are immediately related to the time- and space-coordinates.

- spatial infinitesimal translation → conservation of linear momentum
- temporal infinitesimal translation → conservation of energy
- spatial rotation → conservation of angular momentum
- uniform motion → conservation of center of mass

Further conservation quantities such as the electrical charge are well-known.

In Section 5.1, we saw that the conservation of the linear momentum can be mathematically expressed by (5.4). This leads to the remarkable conclusion that the equation of motion, valid only for bodies in any body-field system, is characterized by a nonvanishing source density $\sigma_{\mathbf{p}}$. Such a result agrees with the balance of specific entropy with its "half" conservation law $\sigma_s \geq 0$, but is apparently inconsistent with the usual practice, according to which the general condition

$$\sigma_z \equiv 0 \tag{5.85}$$

is presumed to be valid for each conserved field quantity z. Outstanding examples for (5.85) are the balances of mass and energy. Whereas the former is given by (5.72), the latter has to be established by means of the differential form (5.81) in connection with the First Law and the Noether theorem.

Let us start with (5.81) applied to the specific energy e:

$$\rho De + \nabla \cdot \mathbf{j}_e \equiv 0. \tag{5.86}$$

In accordance with the First Law, the substantial energy current density $\mathbf{j}_e$ needs to be decomposed into two vectors, corresponding to the rule noted above, in connection with the interpretation of the general balance (5.81). The first vector $\mathring{\mathbf{q}}_*$ refers to the heating Q and the second vector $\mathring{\mathbf{w}}_*$ to the work rate W. Hence, the formal expression of this decomposition is

$$\mathbf{j}_e := \mathring{\mathbf{q}}_* + \mathring{\mathbf{w}}_*. \tag{5.87}$$

As we will see in the next chapters, $\mathbf{j}_e$ is part of the mathematical theory to substantiate both these vectors.

For the remaining explanations, the question arises as to whether there is a rule that helps us decide how many balances should be obtained and, particularly, which special balances should be considered. It is hard to answer this question from the vantage point of the formal theory of balances alone because it is a set of definitions settled by means of mathematics and logic. In my opinion, however, the solution of this problem is rendered simple by the use of the Gibbs–Falkian dynamics extended to a continuum description by the transition from the Gibbs space to the configuration space. For instance, if Equation (5.66) is applied to a material or local description, each additional term of this Gibbs main equation may be substituted by its own balance equation. In this way we can determine the number and kind of the balances involved. The inferences of this procedure will be presented in the following chapters.

It should be noted that even now electromagnetic influences on any body-field system are allowed to be considered in the same way as described above. Due to the well-known fact that all electromagnetic quantities are mutually related by Maxwell's equations, it is hard to understand how this set of quantities could be connected with the balance equations at all and, especially, incorporated into the formalism of Falk's dynamics. This problem will be solved in the last chapter of this book.

Chapter 6 Non-equilibrium Processes

"The customer cannot win at this game; this is the first law.
In fact, the customer is likely to lose; this is the second law."
—***Book of instructions for croupiers at the roulette table***

6.1 Dissipation Velocity

From Einstein's mechanics (discussed in the Sections 3.3 and 3.4) along with Planck's modification, two conclusions may be drawn with regard to the distinction between the state of motion and the state at rest of an energy–momentum transport.

1. In pressurized systems, the energy E as the dependent variable has to be replaced by a quantity $E_*^{[V]}$ resulting from a Legendre transformation with respect to the volume variable V. Hence, all independent variables associated with the rest enthalpy H_0 are Lorentz invariants.
2. According to Planck's M–G function (3.36) of pseudo-particles, motion is associated with the linear momentum, whereas the state at rest is defined by the limit $\mathbf{P} \rightarrow \mathbf{0}$ and quantified by the rest enthalpy H_0 depending on the variables S, p_*, and N.

To discuss these statements with regard to any multicomponent single-phase system, classified as a body-field system (BFS) by its Pfaffian

$$dE = \mathbf{v} \cdot d\mathbf{P} - \mathbf{F} \cdot d\mathbf{r} + T_* \, dS - p_* \, dV + \sum_{j=1}^{J} \mu_{*j}^{t} \, dN_j, \tag{6.1}$$

let us first perform the corresponding Legendre transformation with respect to V. Using Equation (2.75), generally valid for the m-fold Legendre transformation, the new quantity

$$E_*^{[V]} = E - V\left(\frac{\partial E}{\partial V}\right)_{\mathbf{P},\,\mathbf{r},\,S,\,N} = E + Vp_* \tag{6.2}$$

leads to the M–G function $E_*^{[V]}(\mathbf{P}; \mathbf{r}; S; p_*; N_j)$ of the system in question as well as to its corresponding Gibbs main equation

$$dE_*^{[V]} = \mathbf{v} \cdot d\mathbf{P} - \mathbf{F} \cdot d\mathbf{r} + T_* \, dS + V \, dp_* + \sum_{j=1}^{J} \mu_{*j}^{t} \, dN_j. \tag{6.3}$$

Compare this Pfaffian with the results presented in Section 3.4.2 for a system consisting of pseudo-particles. This special case is of great interest because, for instance, Equations (3.37) immediately furnish the general rules to determine the conjugate variables **v**, **F**, T_*, V, and μ^t_{*j}. Furthermore, Equations (3.37) yield some additional information about those variables, provided the M–G function $E_*^{[V]}(\mathbf{P}; \mathbf{r}; S; p_*; N_j)$ is known in all details (such as those of pseudo-particles). Forming the double limit $\mathbf{P} \to \mathbf{0}$ and $\mathbf{r} := \mathbf{r}_\# \equiv$ constant, assumed to be true for all changes of the other system variables, we obtain from (6.3)

$$dH = Tds + Vdp + \sum_{j=1}^{J} \mu_j^t \, dN_j, \tag{6.4}$$

where a new symbol H is introduced for $E_*^{[V]}$ if this double limit is actually affected. Of course, H means the well-known enthalpy depending on the variables S, p, and N_j each. By dropping the asterisks in (6.4), an essential result is anticipated that will be proven in this chapter: The M–G function $H(S; p; N_j)$ exists only for the absolute state at rest defined by the double limit $\mathbf{P} \to \mathbf{0}$ and $\mathbf{r} := \mathbf{r}_\# \equiv$ constant. This case may be referred to as the common state of thermodynamic equilibrium. In the Gibbs space any moving multicomponent single-phase BFS, defined by its Gibbs–Euler function (3.8), has to be described explicitly by (6.3). Its mapping onto the configuration space by means of parameterization, using time as a curve parameter, leads to an expression similar to Equation (5.66) that can be combined with balance equations of the form (5.81) or (5.84), respectively. This will be demonstrated below.

First of all, however, we should discuss a new way of discerning more precisely non-equilibrium states from any equilibria. The basic idea is simple: Provided that only systems that can generally be characterized as nonrelativistic ones are considered, we can introduce the notion of kinetic energy as usual. This procedure follows Einstein's mechanics and leads to the separation of the total energy E from the rest energy E_0:

$$E_{\text{kin}} := E - E_0 \tag{6.5}$$

Be aware that with regard to Gibbs–Falkian theory the additional introduction of the term *kinetic energy,* according to the equation

$$E_{\text{kin}} := \frac{1}{2} m^{B+L} \mathbf{v}^2, \tag{6.6}$$

is a mere definition, introduced with reference to Leibniz's *vis viva* $m\mathbf{v}^2$ (cf. Mirowski, 1990, p. 19). E_{kin} is associated with the constant mass m^{B+L} of all baryons and leptons involved [see Equation (5.23)]. Certainly, there are some well-founded arguments to employ this definition in practice, but there is no theoretical need to do so. For this reason it is necessary to accept definitions (6.5) and (6.6) as not being intrinsically inconsistent with the Gibbs–Euler function of the system under consideration and its mathematical derivatives. To fulfill this condition for all real and virtual changes of state, an interesting differential equation[12] can be derived as follows: Starting with the kinetic energy E_{kin}, defined by (6.5), Equation (6.6) can be written

$$\frac{E_{kin}}{(1/2)\,m^{B+L}} = \mathbf{v}^2 = v_x^2 + v_y^2 + v_z^2$$
$$= \left(\frac{\partial E_{kin}}{\partial P_x}\right)^2_{\mathbf{r},S,V,N_j} + \left(\frac{\partial E_{kin}}{\partial P_y}\right)^2_{\mathbf{r},S,V,N_j} + \left(\frac{\partial E_{kin}}{\partial P_z}\right)^2_{\mathbf{r},S,V,N_j}, \tag{6.7}$$

where the generally valid derivative $\mathbf{v} = (\partial E/\partial \mathbf{P})_{...}$ can be replaced by $\mathbf{v} = (\partial E_{kin}/\partial \mathbf{P})_{...}$, because $E_0(\mathbf{r}; S; V; N_j)$ is independent of $\mathbf{P}$. Equation (6.7) is a nonlinear partial differential equation for E_{kin} with respect to the three Euclidean components of the momentum $\mathbf{P}$. By means of the vectorial form

$$\frac{2E_{kin}}{m^{B+L}} = \left(\frac{\partial E_{kin}}{\partial \mathbf{P}}\right)^2_{\mathbf{r},S,V,N_j}, \tag{6.8}$$

Equation (6.7) is written in a more compact manner. A general solution of Equation (6.8) is easily found by differentiating and inserting the formula

$$m^{B+L}E_{kin} = \frac{1}{2}[\mathbf{P} + \boldsymbol{\phi}]^2; \qquad \boldsymbol{\phi} = \boldsymbol{\phi}(\mathbf{r}; S; V; N_j) \tag{6.9}$$

into the differential equation (6.8). The vector $\boldsymbol{\phi}$ follows from the integration and depends on the system variables assumed to be constant for the purpose of integration. This is a typical example for the so-called *principle of equipresence* "as a rule to guide us when ... we set up constitutive equations" (Truesdell, 1984, p. 300). According to this principle, it is reasonable to assume that any constitutive quantity depends on the complete set of the independent state variables, always provided that there are no reasons to exclude some of them as is done with the momentum in $(6.9)_2$.

It is, however, more illustrative to use specific quantities: with $e_{kin} = E_{kin}/m^{B+L} = 1/2\,\mathbf{v}^2$, Equation (6.9) immediately becomes:

$$\mathbf{v} = \mathbf{i} + \boldsymbol{\varphi}; \qquad \boldsymbol{\varphi} = \boldsymbol{\varphi}(\mathbf{r}; s; \rho^{-1}; \omega_j). \tag{6.10}$$

Just like the flow velocity $\mathbf{v}$ and the specific momentum $\mathbf{i}$, the vectorial quantity $\boldsymbol{\varphi}$ also has the unit of a velocity variable. This is one reason we call the quantity $\boldsymbol{\varphi}$ a *dissipation velocity.*

Note that Equation (6.10) is immediately derived from two "icons" of classical physics: the decomposition of energy (6.5) along with the familiar one-half *vis viva*

[12]A similar differential equation may be deduced for the motion of an electrically charged particle within a magnetic field (Falk and Ruppel, 1983, p. 182). This problem concerns a simple, but important, system of the second kind; that is, its velocity does not only depend on the conjugated variable $\mathbf{P}$, but also on the field coordinate $\mathbf{r}$.

(6.6), always provided that a proper Gibbs fundamental equation $\Gamma(E, \mathbf{P}, \mathbf{r}, \ldots) \equiv 0$ is accepted. From this point of view it is puzzling that the scientific community adheres to the very special solution $\varphi(\ldots) \equiv 0$ for nearly all actual cases, without serious theoretical or experimental necessity. Concerning the general solution of any mathematical problem, each kind of discussion is actually a pure truism for every mathematician or theoretical physicist. Therefore, it is hard to grasp why a dispute about a possibly more general expression than $\mathbf{v} = \mathbf{i}$ has never taken place. Perhaps the force of habit, along with the predominance of mechanical paradigms over all branches of physics are responsible for this omission. These rather psychological influences seem to have prevented traditional science from asking for the consequences of a fundamental relation $\Gamma(E, \mathbf{P}, \mathbf{r}, \ldots) \equiv 0$ with respect to the universally accepted terms of kinetic energy and momentum originally founded only for Newton–Eulerian mass-point mechanics.

As a first result it is remarkable that the velocity φ simply explains an important fact normally experienced in all real flows along any given walls. That is, on account of the *wall adhesion* of any viscous flow, *maximum dissipation* occurs at the wall if the flow velocity $\mathbf{v}$ tends toward zero. In this case the corresponding specific momentum $\mathbf{i}$ locally equals φ, though with opposite signs. In my opinion, this important example convincingly proves the conclusion that, unlike traditional physics would have it, there is a physically fundamental difference between the two key terms *linear momentum* $\mathbf{P}$ and the property $m_{\#}\mathbf{v}$, originally labeled as *quantitas motus* by Isaac Newton in 1687. Very recently, Sieniutycz and Berry (1993, p. 1772) obtained a similar result by means of a Lagrangian with dissipative potentials. A thermal momentum density appears, which equals $\delta\ell$.

We cannot offer detailed information here about the diverse properties of the dissipation velocity φ. With regard to the mathematical structure of the dependency on its variables, this holds especially true. Therefore, let us take a first step in this direction by some respective preparations.

Starting with the total derivative of the specific kinetic energy e_{kin}, according to definition (6.6), we could combine de_{kin} with the total derivative $d\varphi$ of the dissipation velocity φ. In particular, we obtain the following formulas:

$$de_{\text{kin}} = \mathbf{v} \cdot d\mathbf{v} = \mathbf{v} \cdot d\mathbf{i} + \mathbf{v} \cdot d\varphi = \mathbf{v} \cdot d\mathbf{i} + \mathbf{i} \cdot d\varphi + \varphi \cdot d\varphi \tag{6.11}$$

$$d\varphi = \left(\frac{\partial \varphi}{\partial \mathbf{r}}\right)_{s,\,\rho,\,\omega_j} \cdot d\mathbf{r} + \left(\frac{\partial \varphi}{\partial s}\right)_{\mathbf{r},\,\rho,\,\omega_j} ds + \left(\frac{\partial \varphi}{\partial \rho}\right)_{s,\,\mathbf{r},\,\omega_j} d\rho + \sum_{j=1}^{J} \left(\frac{\partial \varphi}{\partial \omega_j}\right)_{s,\,\rho,\,\mathbf{r},\,\omega_{\neq j}} d\omega_j \tag{6.12}$$

This is reminiscent of the Gibbs main equation (5.25)

$$\rho\, de = \mathbf{v} \cdot \rho\, d\mathbf{i} - \rho \mathbf{f} \cdot d\mathbf{r} + T_* \rho\, ds + (p_* \rho^{-1}) d\rho + \textstyle\sum_j \mu_{*j}\, \rho\, d\omega_j; \qquad j = 1(1)J,$$

which is presumed to be valid for any multicomponent single-phase BFS and represented by specific variables. Inserting now the energy form of motion $\mathbf{v} \cdot d\mathbf{i}$ given by (6.11) and (6.12) and using some simple algebraic manipulations, the Pfaffian can be transformed into

$$de = de_{\text{kin}} - \boldsymbol{\varphi} \bullet d\boldsymbol{\varphi} - [\mathbf{f} - \mathbf{i} \bullet (\partial\boldsymbol{\varphi}/\partial\mathbf{r})_{s,\rho,\omega j}] \bullet d\mathbf{r} + [T_* - \mathbf{i} \bullet (\partial\boldsymbol{\varphi}/\partial s)_{\mathbf{r},\rho,\omega j}]ds$$
$$+ \rho^{-2}[p_* - \rho^2 \mathbf{i} \bullet (\partial\boldsymbol{\varphi}/\partial\rho)_{\mathbf{r},s,\omega j}]d\rho + \sum\nolimits_j [\mu_{*j} - \mathbf{i} \bullet (\partial\boldsymbol{\varphi}/\partial\omega j)_{\mathbf{r},\rho,s,\omega \neq j}]d\omega_j. \tag{6.13}$$

For simplicity, a possible dependency of $\boldsymbol{\varphi}$ on the field vector $\mathbf{r}$ may be dropped in the following part. Furthermore, the specific field force $\mathbf{f}$ is presupposed to be conservative. This means that $\mathbf{f}$ depends only on the conjugated variable $\mathbf{r}$, which is equivalent to the fact that there exists a total derivative

$$de_{\text{pot}} := -\mathbf{f}(\mathbf{r}) \bullet d\mathbf{r} \tag{6.14}$$

called the *specific potential energy.*[13] A simplification may be obtained by partial integration of the energy form $-\boldsymbol{\varphi} \bullet d\boldsymbol{\varphi}$. The result is termed the *specific dissipation energy*:

$$e_{\varphi} := -\frac{1}{2}\boldsymbol{\varphi}^2. \tag{6.15}$$

Looking back at Equation (6.13), let us build in the last two definitions. Hence, the following set of relationships will arise:

$$e - e_{\#} := u + e_{\text{kin}} + e_{\text{pot}} + e_{\varphi}; \tag{6.16}$$

$$du = T\,ds + p\rho^{-2}\,d\rho + \sum\nolimits_j \mu_j\,d\omega_j; \tag{6.17}$$

$$T := T_* - \mathbf{i}\bullet\left(\frac{\partial\boldsymbol{\varphi}}{\partial s}\right)_{\rho,\,\omega_j}; \quad p := p_* - \rho^2\mathbf{i}\bullet\left(\frac{\partial\boldsymbol{\varphi}}{\partial\rho}\right)_{s,\,\omega_j}; \quad \mu_j := \mu_{*j} - \mathbf{i}\bullet\left(\frac{\partial\boldsymbol{\varphi}}{\partial\omega_j}\right)_{s,\,\rho,\,\omega_{\neq j}} \tag{6.18}$$

There is, of course, more in this set than mere abbreviations; the set is essentially identical with (6.13). For this reason we cannot identify (6.17) with the usual form used in conventional physics. This is also evident from (6.16), by which the common notion of specific internal energy may be defined, but a significant divergence from the norm is indicated. In other words, the classical result only appears under the condition that the vanishing of the dissipation velocity may be justified physically. However, the general case implies that moving systems, which do not consist of Eulerian mass-points, are characterized by an additional kind of energy, viz. e_{φ}, provided that the specific kinetic energy e_{kin} of those systems is introduced by definition. Very recently Sieniutycz and Berry (1993, p. 1773) published a similar result

[13]The existence of conservative body forces is needed, provided a viscous flow within a corresponding force field could be described by Euler's law of motion in the frictionless limiting case. This case is theoretically consistent with the thermodynamic principles only if local equilibrium prevails. Such a condition can be satisfied by (6.14) (see Sommerfeld, 1964, p. 36) and the derivation of the Bernoulli equation shown.

arising from the Lagrange multiplier of the entropy generation balance for consistent treatment of irreversible processes via a variational formalism.

The remarkable triplet (6.18) intensifies the latter statement. Indeed, there are two ways to identify each of the state quantities with its associate marked by an asterisk. Theoretically, both options may be realized by a zero vector either of the specific momentum **i** or the dissipation velocity $\boldsymbol{\varphi}$. For this reason, we introduce symbolically the two items

$$\begin{array}{ll} \textit{State at rest:} & \mathbf{i} \to 0 \Leftrightarrow \begin{cases} \boldsymbol{\varphi} \to \mathbf{0} \\ \mathbf{v} \to \mathbf{0} \end{cases} \\ \textit{Reversible process:} & \boldsymbol{\varphi} \to \mathbf{0} \Leftrightarrow \mathbf{i} \to \mathbf{v} \end{array} \tag{6.19}$$

A more profound understanding of the above properties and statements is the subject of the subsequent sections. Nevertheless, both items help us begin to discern in a basic way equilibrium states or processes from non-equilibrium phenomena. The main argument rests on the **general axiom** that equilibrium is always connected with reversibility, whether the system is in motion or not. Consequently, equilibrium exists only under the condition that $(6.19)_2$ is fulfilled. The reverse is also assumed to be true: If $(6.19)_2$ is not satisfied, the system is in a non-equilibrium state described by (6.13) for the case of any multicomponent single-phase body-field system.

6.2 Kinetic Equilibrium in Fields

There are only a few ambitious treatises on rational hydrodynamics that deal with perfect fluids defined as frictionless (i.e., reversible) flows. It is characteristic for those mathematical theories to look upon any dissipationless flow as one of the various paradoxes of fluid dynamics (see, e.g., Birkhoff, 1960). In one of his famous Princeton lectures Weyl, the great mathematician and natural philosopher, outlined this prejudice in a few words: "The following paradox arises in connection with viscous fluids: If a perfect fluid is to be regarded as the limit of a viscous one as the viscosity $\eta \to 0$, then how is it that the viscous fluid sticks to the boundary, even with small η, while the perfect fluid glides smoothly?" (Weyl, 1942, p. 61). Unfortunately, little else can be found on this topic in modern textbooks. Of course, Weyl's question is part of the fundamental problem of what extent Euler's equation of motion may be considered as a limiting case of the Navier–Stokes equation of motion. This is particularly true with regard to the fact that the viscosity of a fluid as well as its heat capacity is unanimously treated either as a constant or as a material function only depending on the local state variables. Under such a premise, however, the limit $\eta \to 0$ may perhaps be justifiable mathematically, but physically it is inadmissible in principle.

To my knowledge, Maxwell published the first, and up to now the only, relevant study about the problem. In his treatise concerning the trend toward equilibrium, he introduced in an original manner the notion of kinetic equilibrium for flows described by the Boltzmann equation: This special kind of equilibrium always prevails provided that the *collision integral* $I(f, f)$ of the Boltzmann equation vanishes identically and even independently of its residual terms. Those terms, however, lead directly to the balance equations of a frictionless flow, subject to any reversible changes of local equilibrium states. Of course, this far-reaching consequence is due to the equilibrium distribution function resulting from $I(f, f) \equiv 0$ and escorting via zero-valued transfer such properties as heat conduction and viscous shearing (Truesdell, 1984, p. 414). Such an equilibrium solution refers to time- and space-dependent velocity fields, wherein some gradients such as those of local pressure and flow velocity will exist. This result is inconsistent with the postulates preferred by some adherents of rational mechanics and exemplified as follows: "By equilibrium is meant a steady state with time independent variables and no temperature gradient, no symmetric velocity gradient and no dissipative fluxes" (Boukary and Lebon, 1987, p. 106). With reference to Maxwell's work the following two items may be stated:

1. Despite definitely existing gradients, an equilibrium flow of real fluids may occur. This matter of fact is proved phenomenologically by Euler's equation of motion as a prototype of kinetic equilibria.
2. Kinetic equilibrium may be realized as the reversible limiting case of every real, and this can only mean dissipative flow.

This nontrivial definition, founded on the kinetic theory of gases, furnishes proof of the fact that, now as before, classical hydrodynamics lacks modern designations for the comprehension of basic terms such as equilibrium, steady state, dissipation, reversibility, and so on. Yet it allows a transparent design of a mathematically consistent field theory that avoids paradoxes similar to those that we have discussed.

To consider equilibria in extended and continuous regions, we need only regard the limiting case of field quantities, that is, infinitesimal phases according to the rules discussed in Subsection 5.4.2. For this purpose all extensive quantities ξ_k have to be replaced by the pertaining densities $\xi_{V,j}$ ($j = 0, 1, \ldots, r$), which may be treated as functions of space and time. Given the M–G function $\xi_{V,0} = G_V(\xi_{V,1}(\mathbf{r}), \ldots, \xi_{V,r}(\mathbf{r}), \mathbf{r})$, assumed to be substantial for any infinitesimal volume element $d\tau$ of the region $\mathscr{B}$, the dependent quantity $\xi_{V,0}$ may be explicitly independent of the position vector $\mathbf{r}$. In this special case the density function G_V will describe the same infinitesimal system at all places of $\mathscr{B}$. Consequently, $\xi_{V,0}$ only depends implicitly on the $\mathbf{r}$-functions of the other variables. By integration of each density $\xi_{V,j}$ over the whole region $\mathscr{B}$,

$$\xi_j = \int_{\mathscr{B}} \xi_{v,\,j}\, d\tau; \qquad j = 0(1)r, \tag{6.20}$$

the extensive quantities ξ_j assigned to $\mathscr{B}$ will result. Within this region exchange processes occurring between infinitesimal phases tend toward phase equilibria defined by the general variational condition

$$\delta \int_{\mathcal{B}} \xi_{V,j}\, d\tau \equiv 0; \qquad j = 0(1)r, \tag{6.21}$$

which expresses the mutually unrestrained interaction of the jth density $\xi_{V,j}$ between the volume elements $d\boldsymbol{\tau}$. If we let $\xi_{V,0}$ be the energy density, then the total energy E determined by (6.20) has to be minimized according to the basic problem of the variational principle:

$$\delta E = \int_{\mathcal{B}} \delta E_V [\xi_1(\mathbf{r}), \ldots, \xi_r(\mathbf{r})]\, d\tau \equiv 0. \tag{6.22}$$

By means of Lagrange's method of multipliers λ_j, it is easy to incorporate the finite number of supplementary conditions (6.21) into the optimization

$$\int_{\mathcal{B}} \left\{ \left(\frac{\partial E_V}{\partial \xi_{V,1}} - \lambda_1 \right) \delta\xi_{V,1} + \ldots + \left(\frac{\delta E_V}{\partial \xi_{V,r}} - \lambda_r \right) \delta\xi_{V,r} \right\} d\tau \equiv 0 \tag{6.23}$$

yielding the solutions

$$\frac{\delta E_V}{\delta \xi_{V,j}} = \lambda_j; \qquad j = 1(1)r \tag{6.24}$$

which are, in accordance with Equation (2.73), the corresponding intensive variables.

This means that for equilibrium of region $\mathcal{B}$ the conjugates of the variables $\xi_j(\mathbf{r})$ with respect to the total energy E have an identical value λ_j ($j = 1, \ldots, r$) for all $\mathbf{r}$ in $\mathcal{B}$. The resulting statement is very important and will be demonstrated below:

> Inside any spatially extended region $\mathcal{B}$ each intensive variable $\boldsymbol{\tau}_j$ that is conjugated to the density $\xi_{V,j}$ has the same value everywhere, provided that kinetic equilibrium prevails in $\mathcal{B}$.

It is remarkable that this statement completely agrees with the findings from the Boltzmann equation mentioned above. Thus, in kinetic equilibrium, for example, the existence of an entropy density enforces an isothermal region $\mathcal{B}$ for which local heat fluxes are suppressed. Hence, the derived assertion about kinetic equilibrium is a limiting value theorem concerning non-equilibrium phenomena.

6.3 Three Additional Theorems Concerning Non-equilibrium

6.3.1 *DIVERGENCE THEOREM*

Since further inferences from the Gibbs relations of the system in question will have to be drawn, it seems inappropriate to expound on the physical meaning of the dissipation velocity. Therefore, let us turn back to the Legendre-transformed Gibbs main equation (6.3) valid for a multicomponent single-phase system and classified as a body-field system. Two significant modifications of Equation (6.3) should be carried out first using specific quantities instead of the original system variables and specifically relating those quantities to time as a curve parameter as discussed in Section 5.3.

Dividing (6.3) by the constant baryon–lepton mass m^{B+L} corresponding to (5.23), we get

$$\rho\, de_*^{[\rho]} = \rho\mathbf{v}\bullet d\mathbf{i} - \rho\mathbf{f}\bullet d\mathbf{r} + T_*\rho ds + dp_* + \sum_{j=1}^{J} \mu_{*j}\rho d\omega_j , \tag{6.25}$$

where, aside from the Legendre-transformed specific energy $e_*^{[\rho]}$, the same quantities are employed as in Equation (5.25).

Note that the Pfaffian (6.25) now contains an energy form without a conjugate variable. As this differential form also holds true for any constant mass density ρ, Equation (6.25) may thus be expressed by densities $\xi_{V,j}$ as variables. Hence, the boxed statement of the previous section is inapplicable for the pressure p_*. But it may be applied the other way around: Assuming an explicit dependency of the Legendre-transformed energy density $E_{*V}{}^{[V]} = \rho e_*{}^{[\rho]}$ on the position vector **r**, the pressure p_* is a generally well-defined and nonconstant function of **r**, provided that kinetic equilibrium prevails in $\mathscr{B}$. Simultaneously, a second energy form loses its conjugate variable. According to (6.19) the specific momentum **i** equals the flow velocity **v** under the condition that kinetic equilibrium may be identified with a reversible process. Then, again for constant density ρ, the energy form of motion will change from $\rho\mathbf{v} \bullet d\mathbf{i}$ to $dE_{\text{kin},V}$, with the consequence that this special kind of energy density also is a well-defined and nonconstant function of **r**. For this reason, in a region $\mathscr{B}$ presumed to be subject to kinetic equilibrium, there may be simultaneously existing gradients of pressure and velocities. Naturally, they cause motion of the kind described by the commonly named Eulerian equation.

A special case of this equation may be derived at once from Equation (6.25). From a Legendre transformation of (6.3) with respect to the entropy S we obtain

$$E_*^{[V,S]} = E_*^{[V]} - \left(\frac{\partial E_*^{[V]}}{\partial S}\right)_{\mathbf{P},\mathbf{r},p_*,N_j} \bullet S = E_*^{[V]} - T_*S . \tag{6.26}$$

After division by the constant baryon–lepton mass m^{B+L}, the following Gibbs main equation arises

$$\rho\, de_*^{[\rho,s]} = \mathbf{v}\bullet\rho\, d\mathbf{i} - \rho\mathbf{f}\bullet d\mathbf{r} - \rho s\, dT_* + dp_* + \sum_{j=1}^{J} \mu_{*j}\rho\, d\omega_j . \tag{6.27}$$

We assume a constant mass density ρ_o as well as a potential approach for the field force density $\mathbf{F}_V = \rho\mathbf{f}$, according to $\rho\mathbf{f} \bullet d\mathbf{r} := -dE_{\text{pot},V}$. Then a considerable simplification of (6.27) may be performed for the limiting case of kinetic equilibrium, provided that the consequences of the boxed statement are reconsidered as follows:

$$d\rho_o\, e_*^{[\rho,s]} \equiv 0; \qquad dT_* \equiv 0; \qquad d\omega_j \equiv 0 \text{ for all } j. \tag{6.28}$$

In addition, the dissipation velocity φ vanishes by definition (6.19), with the result that the energy form $\mathbf{v} \bullet \rho d\mathbf{i}$ may be replaced by $dE_{\text{kin},V}$. Furthermore, the asterisks with T_* and p_* are then dropped. Hence, Equation (6.27) takes on the form of a total derivative

$$d(E_{\text{kin},V} + E_{\text{pot},V} + p) \equiv 0 \tag{6.29}$$

with the simple solution

$$E_{\text{pot},\,V} + \frac{1}{2}\rho_0\mathbf{v}^2 + p \;=\; \text{constant}. \tag{6.30}$$

This is the well-known *Bernoulli equation,* proven to be valid for a special class of body-field systems from a few general principles of Falk's dynamics. It is a useful relationship in many applications and also demonstrates the theoretical relevance of the limiting value theorem concerning non-equilibrium phenomena as discussed above.

Let us turn back to Equation (6.25), where time as the curve parameter t may be introduced by means of the operator D denoted as the material derivative and defined by (5.68). The resulting *Gibbs rate equation,* valid for a multicomponent single-phase system and classified as a body-field system,

$$\boxed{\rho De_*^{[\rho]} \;=\; \mathbf{v}\bullet\rho D\mathbf{i} - \mathbf{v}\bullet\rho\mathbf{f} + T_*\rho Ds + D_{p*} + \sum_{j=1}^{J}\mu_{*j}\rho D\omega_j\,,} \tag{6.31}$$

is of such a form, useful for all investigations on other types of systems in this text. Modifications solely concern the energy forms that have to be considered additionally, such as the three components of the angular momentum vector, or which term should be dropped (like the sum over the species terms). For this reason, the results may indeed be regarded as an example for the method to be proposed. Aside from the pressure term, it is evident that each expression in Equation (6.31) can be replaced by the corresponding balance equation settled in (5.81). Moreover, and in accordance with the postulate (5.69) of concurrent flow velocities in kinematics and dynamics, the common kinematic form $d\mathbf{r}/dt$ is replaced by the dynamic flow velocity $\mathbf{v} = (\partial e_*^{[\rho]}/\partial\mathbf{i})_{\mathbf{r},s,p_*,\omega j}$ of the body under consideration.

To substitute the various balances, we must above all convert the energy balance (5.86), supplemented by Equation (5.87), into a balance equation of the Legendre-transformed specific energy $e_*^{[\rho]} = e + \rho^{-1}p_*$. The derivation is subsequently given as

$$\begin{aligned} \rho De + \nabla\bullet\mathbf{j}_e \equiv 0 \Rightarrow \rho D\left(e_*^{[\rho]} - \rho^{-1}p_*\right) + \nabla\bullet\mathbf{j}_e \equiv 0 \\ \rho D\, e_*^{[\rho]} - Dp_* + \frac{p_*}{\rho}De\ \frac{p_*}{S}DS + \nabla\bullet\mathbf{j}_e \equiv 0 \end{aligned} \tag{6.32}$$

$$\rho D\, e_*^{[\rho]} - \partial_t p_* - \mathbf{v} \bullet \nabla p_* - p_* \nabla \bullet \mathbf{v} + \nabla \bullet \mathbf{j}_* \equiv 0$$

$$\Rightarrow \rho D e_*^{[\rho]} + \nabla \bullet [\mathbf{j}_e - p_* \mathbf{v}] = \partial_t p_* .$$

In reference to the First Law of Thermodynamics, the energy flux density $\mathbf{j}_e$ consists of two vectorial parts. According to equation (5.87), the heat flux vector $\mathring{\mathbf{q}}_*$ as well as the work rate vector $\mathring{\mathbf{w}}_*$ are now introduced. Later on their properties will be determined consistently by the theory developed in this and the following sections. Thus, the two definitions

$$\mathbf{j}_e := \mathring{\mathbf{q}}_* + \mathring{\mathbf{w}}_* \tag{6.33}$$

$$\mathbf{j}_{e*}^{[\rho]} := \mathbf{j}_e - p_* \mathbf{v} = \mathring{\mathbf{q}}_* + \mathring{\mathbf{w}}_* - p_* \mathbf{v} \tag{6.34}$$

appear to be of advantage, particularly for the description of various transfer processes such as friction and heat conduction within the dissipative flowing fluid.

Let us insert all the relevant balance equations into the general equation (6.31). It is helpful to use Table 6.1 to observe all quantities in question at a glance, such as the convection term ρDz, the flux density $\mathbf{j}_z$, and the production density σ_z according to (5.81).

It is noteworthy that the use of some special symbols and words for the various densities $\mathbf{j}_z$ and σ_z is common practice. Thus, for instance, the two vectors $\mathbf{j}_s$ and $\mathbf{j}_j$ denote the *entropy flux density* and the *diffusional flux density,* whereas the *momentum flux density* $\mathbf{JI}$ is a second-order tensor. Above all, however, the following list "constitutes an empty scheme" (Cercignani, 1988, p. 85) as long as the fluxes and production densities are physically unspecified quantities.

A peculiarity of mass conservation should be emphasized: The set of *component continuity equations* for the mass fraction ω_j, $j = 1(1)J$,

$$\rho\, D\omega_j + \nabla \bullet \mathbf{j}_j = \Gamma_j \tag{6.35}$$

leads to a closure condition for the *production density* Γ_j of the jth species. By summing these component balances

$$\rho D \sum_j \omega_j + \nabla \bullet \sum_j \mathbf{j}_j = \sum_j \Gamma_j$$

Table 6.1 Balance Equations for Energy, Momentum, Entropy, and Mass Fraction

Balance equation	*Convection term*	*Flux density*	*Production density*
Legendre-transformed energy	$\rho\, De_*^{[\rho]}$	$\mathbf{j}_{e*}^{[\rho]}$	$\partial_t p_*$
linear momentum	$\rho\, D\mathbf{i}$	$\mathbf{JI}$	$\sigma_\mathbf{P}$
entropy	$\rho\, Ds$	$\mathbf{j}_s$	σ
mass fraction	$\rho\, D\omega_j$	$\mathbf{j}_j$	Γ_j

and taking into consideration the closure condition (5.76) for the diffusion flux vectors $\mathbf{j}_j$ as well as the rule $\Sigma_j\omega_j = 1$, the restriction

$$\sum_{j=1}^{J} \Gamma_j \equiv 0 \tag{6.36}$$

results. If no chemical reactions take place between the components, all source terms in the component balances vanish identically, that is, $\Gamma_j \equiv 0$ for every j. Whereas chemical reactions do occur, the Γ_j source densities of the components do not all vanish.

Let us insert all balances into the Gibbs rate equation (6.31). Then, the result

$$\begin{aligned}\partial_t p_* - \nabla \cdot [\mathring{\mathbf{q}}_* + \mathring{\mathbf{w}}_* - p_*\mathbf{v}] &= [\sigma_\mathbf{P} - \nabla \cdot \mathbf{JI} - \rho\mathbf{f}] \cdot \mathbf{v} \\ &+ \partial_t p_* + [\nabla p_*] \cdot \mathbf{v} \\ &+ T_*\sigma - T_*\nabla \cdot \mathbf{j}_s + \sum_{j=1}^{J} \mu_{*j}\Gamma_j - \sum_{j=1}^{J} \mu_{*j}\nabla \cdot \mathbf{j}_j\end{aligned} \tag{6.37}$$

leads to some remarkable consequences, provided that the following *tensor identities* are used (Bird, Stewart, and Lightfoot, 1960, p. 731).

$$\begin{gathered}[\nabla \cdot p_*\mathbf{1}] = \nabla p_* \\ [\nabla p_*] \cdot \mathbf{v} = \nabla \cdot [p_*\mathbf{v}] - p_* \nabla \cdot \mathbf{v} \\ (p_*\mathbf{1} : \nabla \mathbf{v}) = p_* (\nabla \cdot \mathbf{v}); \qquad (\mathbf{JI} : [\nabla \mathbf{v}] = (\nabla[\mathbf{JI} \cdot \mathbf{v}]) - (\mathbf{v} \cdot [\nabla \cdot \mathbf{JI}]) .\end{gathered} \tag{6.38}$$

Of course, these important formulas of the tensor analysis (compiled in Appendix 2.2) are also valid for other scalars (used here: the pressure p_*), vectors (flow velocity $\mathbf{v}$), and second-order tensors (momentum flux density $\mathbf{JI}$).

The unit tensor $\mathbf{1}$ is one whose diagonal elements are unity and whose nondiagonal elements are zero:

$$1 := \begin{bmatrix} 1 & 0 & 0 \\ 0 & 1 & 0 \\ 0 & 0 & 1 \end{bmatrix} . \tag{6.39}$$

Some manipulations and the use of (6.38) will transform (6.37) into the expression

$$\begin{aligned}&\nabla \cdot \left[\mathring{\mathbf{q}}_* + \mathring{\mathbf{w}}_* - \mathbf{JI} \cdot \mathbf{v} - T_*\mathbf{j}_s - \sum_{j=1}^{J} \mu_{*j}\mathbf{j}_j \right] + [\sigma_\mathbf{P} - \rho\mathbf{f}] \cdot \mathbf{v} \\ &+ \left[-p_* (\nabla \cdot \mathbf{v}) + \mathbf{JI} : \nabla\mathbf{v} + T_*\sigma + \mathbf{j}_s \cdot \nabla T_* + \sum_{j=1}^{J} \mu_{*j}\Gamma_j + \sum_{j=1}^{J} \mathbf{j}_j \cdot \nabla\mu_{*j} \right] \equiv 0 .\end{aligned} \tag{6.40}$$

This identity should hold for all possibilities of any admissible process realization. Thus, for the purpose of a mathematical theory, it seems sufficient to settle on some demands for the basic properties of body-field systems. For instance, it is reasonable to assume that the second term of (6.40) vanishes independently of the other two terms. It can be dropped for all local flow velocities $\mathbf{v} \neq \mathbf{0}$, but, aside from some pathological cases (e.g., $[\sigma_\mathbf{P} - \rho\mathbf{f}] \perp \mathbf{v}$), only under the condition that the vectorial *momentum production density* σ_P equals the field force density $\mathbf{F}_V := \rho\mathbf{f}$. That is,

$$\sigma_\mathbf{P} = \mathbf{F}_V. \tag{6.41}$$

A second aspect should be taken into account: The first term contains the energy flux vector $\mathbf{j}_{e*}{}^{[\rho]} := \mathring{\mathbf{q}}_* + \mathring{\mathbf{w}}_* - p_*\mathbf{v}$ assigned to (6.40). Because the second-order tensor $\mathbf{\Pi}$ is not specified in more detail for the moment, we may decompose $\mathbf{\Pi}$ into two parts, according to

$$\mathbf{\Pi} := p_*\mathbf{1} + \boldsymbol{\tau}_*. \tag{6.42}$$

Hence, the expression $\mathring{\mathbf{q}}_* + \mathring{\mathbf{w}}_* - \mathbf{\Pi} \bullet \mathbf{v}$, first occurring in Equation (6.40), becomes

$$\mathring{\mathbf{q}}_* + \mathring{\mathbf{w}}_* - \mathbf{\Pi} \bullet \mathbf{v} = \mathbf{j}_{e*}{}^{[\rho]} + p_*\mathbf{v} - \mathbf{\Pi} \bullet \mathbf{v} = \mathbf{j}_{e*}{}^{[\rho]} + p_*\mathbf{v} - p_*\mathbf{v} - \boldsymbol{\tau}_* \bullet \mathbf{v}.$$

This means that for a vanishing divergence term of (6.40) the energy flux vector $\mathbf{j}_{e*}{}^{[\rho]}$ is established by the relationship

$$\mathbf{j}_{e*}{}^{[\rho]} = \boldsymbol{\tau}_* \bullet \mathbf{v} + T_* \mathbf{j}_s + \sum_j \mu_{*j}\, \mathbf{j}_j, \tag{6.43}$$

where $\boldsymbol{\tau}^*$ is called *viscous pressure tensor.*

From a theoretical point of view it seems more reasonable to expand this result into a theorem concerning the flux densities of all balances involved:

$$\mathbf{j}_e - \mathbf{\Pi} \bullet \mathbf{v} - T_*\mathbf{j}_s - \sum_{j=1}^{J} \mu_{*j}\mathbf{j}_j \equiv 0. \tag{6.44}$$

The importance of this so-called *divergence theorem* extends far beyond its direct application to multicomponent single-phase body-field systems. This is true because the structure of (6.44) indicates that for other classes of systems only the pertaining flow densities will have to be included.

6.3.2 DISSIPATION THEOREM

In Section 4.4.3 we discussed several reasons for splitting the energy flux density $\mathbf{j}_e$ into the two vectors $\mathring{\mathbf{q}}_*$ and $\mathring{\mathbf{w}}_*$. Under the condition that even in continuum physics the notion of work rate $\mathring{\mathbf{w}}_*$ refers to any change of information content as a characteristic of matter, $\mathring{\mathbf{w}}_*$ may be measured by the change of shape encountered by each infinitesimal volume element $d\tau$. Such an effect is induced by all interactions between $d\tau$ and its surroundings, but primarily by flow influences. In particular, the local pressure as the natural conjugate property of $d\tau$ is the driving force for shape variations.

In this context, it should be stressed that the pressure p_* in question must not be taken for an equilibrium quantity of state. In addition, the volume element $d\tau$ is also subject to dissipative shearing stresses conventionally expressed by local velocity gradients.

There cannot be any doubt that irreversible phenomena control details of flow patterns. The same is true for the specially prescribed boundary conditions. Still, since it is proved even in the limiting case of Euler's equation of motion, the main influence on flow field topology stems from the spatial distributions of local pressure and flow velocity as well as their mutual interaction.

For this reason it is opportune to decompose (6.44) into two parts

$$\overset{\circ}{\mathbf{w}}_* := \mathbf{JI} \cdot \mathbf{v}, \tag{6.45}$$

$$\overset{\circ}{\mathbf{q}}_* := T_* \mathbf{j}_s + \sum_j \mu_{*j} \mathbf{j}_j, \tag{6.46}$$

where, according to (6.42), the work rate definition includes the non-equilibrium pressure p_*. We may provisionally imagine the pressure p^* as a quantity that does not obey the equation of state commonly used in practice.

The introduction of a viscous pressure tensor $\boldsymbol{\tau}_*$, defined by reference to the momentum flux density $\mathbf{JI}$ and pressure p_*, leads to an important conclusion from (6.40): Both the quantities $\mathbf{JI}$ and p_* appear in two additive terms in the third bracket of this equation term, which may be modified by means of the tensor identities (6.38) as follows:

$$\begin{aligned} -p_*(\nabla \cdot \mathbf{v}) + \mathbf{JI} : \nabla\mathbf{v} &= -p_*(\nabla \cdot \mathbf{v}) + p_*\mathbf{1} : \nabla\mathbf{v} + \boldsymbol{\tau}_* : \nabla\mathbf{v} \\ &= -p_*(\nabla \cdot \mathbf{v}) + p_*(\nabla \cdot \mathbf{v}) + \boldsymbol{\tau}_* : \nabla\mathbf{v}. \end{aligned}$$

This result, together with (6.41) and (6.44), leads to the identity

$$\boldsymbol{\tau}_*:\nabla\mathbf{v} + \mathbf{j}_s \cdot \nabla T_* + \sum_{j=1}^{J} \mathbf{j}_j \cdot \nabla\mu_{*j} + T_*\sigma + \sum_{j=1}^{J} \mu_{*j}\Gamma_j \equiv 0 \; . \tag{6.47}$$

This so-called *dissipation theorem* also extends beyond its application to multi-component single-phase body-field systems, because the structure of (6.4) indicates that for other classes of systems only the pertaining production densities have to be built in.

This theorem summarizes in a characteristic manner all flux and production densities arising in the relevant balance equation. But the most striking result clearly concerns the fact that all these characteristics of non-equilibrium phenomena produce a zero sum. The theorem obviously refers to two kinds of dissipative effects. The first considers the occurring flux densities $\mathbf{j}_z$ along with the corresponding gradients $\nabla\zeta_z$, where ζ_z stands for the intensive variables assigned to z and distinguished by the condition $\zeta_z \geq 0$. The second effect comprises all actually occurring production densities σ_z. From a mathematical point of view, it is reasonable to consider three notable arguments:

1. Avoiding any chemical reactions, diffusion, or dissipative flows, the system in question may be reduced in such a way that either all terms in (6.47) will vanish identically, or at least two terms distinguished by opposite signs will remain.
2. There is at least one element of (6.47) that can never be switched off by admissible procedures of process realization without simultaneously dropping the other existing elements.
3. In addition to the definitions of the two kinds of dissipative effects given above, it is beneficial to distinguish them by different signs.

With respect to the Gibbs main equation $\rho De_*^{[\rho]} = Dp_* - \rho\mathbf{f} \bullet \mathbf{v} + \Sigma^r_{j=3} \zeta_{z,j}\, \rho\, Dz_j$, let us formalize argument (3) by the convention

$$\sum_{j=4}^{r} \zeta_{z,j}\sigma_{z,j} \geq 0; \quad \mathbf{j}_{z,j} \bullet \nabla\zeta_{z,j} \leq 0 \qquad j = 3\,(1)\,r, \tag{6.48}$$

which is assumed to be valid for all specific variables z under consideration. The subscript j refers to r variables and allows us to distinguish two types of elements concerning dissipation for each indicated z. Every first-type element of (6.48) needs the summation over the subscript j; the second type does not. The values $j = 1$ and $j = 2$ of the subscript relate to the pressure p_* and the specific field force $\mathbf{f}$, which do not explicitly furnish any contribution to dissipation.

There is an important example for the first-type element of $(6.48)_1$, well-known as *De Donder's inequality* (Prigogine and Defay, 1962, p. 71). For the special case of a closed system realized by vanishing diffusional flux densities $\mathbf{j}_j$, it becomes

$$\sum_{r=1}^{\rho} A_{V,r}\mathbb{V}^r \geq 0, \tag{6.49}$$

where $\mathbb{V}^r$ is the velocity of the rth reaction. De Donder defined this quantity for the first time in reference to the progress variable λ_r for the rth reaction:

$$\mathbb{V}^r := \frac{d}{dt}\lambda_r; \qquad r = 1\,(1)\,\rho. \tag{6.50}$$

According to Equation (6.35), the chemical production density Γ_j for the jth species agrees with its convection term $\rho D\omega_j$ if the system is closed. By means of Equations (5.52) and (5.53), the direct relation

$$\sum_{j=1}^{J} \mu_j\Gamma_j = \sum_{r=1}^{\rho} A_{V,r}\mathbb{V}^r \geq 0 \tag{6.51}$$

may then be obtained between Γ_j for all j and the affinity density $A_{V,r} := \rho a_r$ for all r.

The total production, caused by simultaneously occurring chemical reactions per time unit, equals the sum over the products, which are each formed from the velocity of the rth reaction and the assigned affinity density. The same is true for the sum over all involved components consisting of the products, each of them now made up from the chemical production density of the *j*th component and the assigned chemical potential per mass unit. Reaction coupling along with feedback events may happen and are mathematically represented by the possibility that some of the sum terms are singled out by opposite signs as compared to (6.51).

There is an alternative possibility in studying details of chemical reaction dynamics. Guldberg and Waage found by an empirical analysis that the concentrations of reacting species can be related to the corresponding reaction velocities. "Even when applied to complex reactions involving intermediate steps and transient species this empirical approach so excels in the elucidation of reaction mechanisms that Mass Action might be termed the First Law of Chemical Kinetics" (Garfinkle, 1992, p. 282).

Recently, a promising method, the so-called *natural path approach,* has been developed by Garfinkle, who studied the progress of a homogeneous stoichiometric chemical reaction in a closed isothermal system. He worked out chemical thermodynamic formalisms in a manner consistent with the laws of thermodynamics, and therefore, independently of mechanistic considerations. Reactions and their velocities are now successfully described in terms of the rate of change of a proper thermodynamic function, *the affinity decay rate* $\mathring{A}_{TV}$ at uniform temperature and fixed volume. In response to Garfinkle's work a critical review was prepared by Hjelmfelt, Brauman, and Ross in 1990. They found that the affinity decay rate $\mathring{A}_{TV}$ is directly dependent on the respective reaction mechanisms, contrary to Garfinkle's observation based on direct empirical analysis (Garfinkle, 1992, p. 283). Summarizing his extensive studies, Garfinkle confirmed his assertion that there exists a unique reaction path over the entire range of empirical observation independent of any reaction mechanisms. Moreover, he stated that "The excellent data correlation observed permits a thermodynamic reaction velocity to be computed that corresponds to the mechanistic reaction velocity determined by kinetics based on an understanding of reaction mechanism. Excellent agreement between these reaction velocities is achieved over the range of experimental observations" (Garfinkle, 1992, p. 299).

Statements like (6.48) or (6.51) are metaphysical inasmuch as they based on a universal theorem that can never be proven by a finite number of data. The same is true for the case where inequalities (6.48) are assumed to formalize the Second Law of Thermodynamics. In this context it is remarkable that the second part of (6.48) indicates the well-known relationship between a prescribed direction of any flux density $\mathbf{j}_z$ and the subsequent direction of the assigned gradient $\nabla\zeta_z$. This is the deeper meaning of the general observation that heat fluxes can never flow in the direction of increasing temperature. Curiously, this observation is rarely mentioned in connection with the mathematical formulation of the Second Law.

Although the entropy production density σ obeys the same inequality $(6.48)_1$ as the other pertaining production densities, there is still an essential difference with respect to argument (2) listed above: In agreement with experience, the entropy production density σ is presumed to be the one production density that must always exist in real processes. All others may be manipulated, at least in principle, in such a way that their respective values tend to zero. However, we must pay attention to various possibilities of creating dissipation through some hidden processes. Frozen chemical reactions are a prominent example.

To quantify the special role of the entropy production density σ, a further rule should supplement the dissipation theorem with its parts (6.47) and (6.48):

$$\lim_{\sigma \to 0} \begin{cases} \sum_{k=1}^{r} \sum_{j=4}^{r} \delta_j^k \zeta_{z,j} \sigma_{z,j} = 0 \\ \sum_{k=1}^{r} \delta_j^k \mathbf{j}_{z,j} \bullet \nabla \zeta_{z,j} = 0; \quad j = 3\,(1)\,r \quad \text{for all } z_j. \end{cases} \tag{6.52}$$

This limiting law is indeed an essential part of the mathematical theory presented here and refers to all dissipation terms except the term $T_*\sigma$ itself. The primary significance of (6.52) is that we may couple *states* of equilibrium to those *processes* assumed to be dissipationless. This is particularly relevant for both the special case of kinetic equilibrium and the state at rest defined by (6.19).

The two theorems (6.44) and (6.47) derived from the general axioms of the Alternative Theory deserve comment particularly with regard to the *extended irreversible thermodynamics* (EIT). The EIT is characterized by a rate equation commonly dedicated to Gibbs,

$$\mathring{U} = T\mathring{S} - p\mathring{V} + \sum_i \mu_i{}' \mathring{N}_i - \sum_i \sum_\alpha X_i^\alpha \bullet (d\Phi_i^\alpha / dt).$$

It is assumed to be valid for an irreversible multicomponent single-phase system (Eu, 1986, p. 217). The corresponding generic function $U(S, V, N_i, \phi_i^\alpha)$ refers to the internal energy U by definition, following an idea of Meixner (cf. Garcia-Colin and Uribe, 1991, p. 111). The respective relation between the entropy production density σ and the dissipative contributions represented by the flux ϕ_i^α of species i and its potential conjugate $X^\alpha{}_i = T(\partial S/\partial \Phi_i^\alpha)$ is given by the expression

$$\sigma = \sum_i \sum_\alpha T^{-1} X_i \bullet \Lambda_i^\alpha,$$

where the tensorial properties Λ_i^α are the representative dissipative terms of the system in question. The sum over the integers α (running from 1 to 4) relates to an ordering directed to the following characteristic gradients:

$$\alpha = 1 \Rightarrow \nabla \mathbf{v};\ \alpha = 2 \Rightarrow \nabla \bullet \mathbf{v};\ \alpha = 3 \Rightarrow \nabla T;\ \alpha = 4 \Rightarrow \rho^{-1}\nabla p - \nabla_T \mu_i.$$

Although there is a formal similarity between the expression for α and Equation (6.47), it is obvious that the conceptual differences are significant. The two theorems (6.44) and (6.47) are derived from Equation (4.54) applied to a multicomponent single-phase body-field system. This Pfaffian results from the Gibbs fundamental equation $\Gamma(E, \mathbf{P}, \mathbf{r}, S, V, N_i) = 0$ and refers to the total energy E of the system under consideration. The internal energy U may be formally determined by Equation (6.16), provided that the complete solution via $\Gamma(E, \mathbf{P}, \mathbf{r}, S, V, N_i) = 0$ is available. Thus, the Alternative Theory leads to a result that is inconsistent with the two relationships given above as representative statements of the EIT.

In the EIT formalism, the theoretical approach is not related to the total energy E. Under the condition that the EIT refers to Gibbsian ideas and notions, E is exclusively established by the complete set of extensive variables, which work as coordinates of the corresponding Gibbs space. The common practice of separating the contributions of kinetic and potential energies from E implies the same conclusion for the internal energy U as for E. Assuming U as an M–G function of the system, then U cannot depend on properties such as the moments ϕ_i^{α}, which do not belong to the extensive state variables by definition (cf. Eu, 1986, p. 215).

6.3.3 GENERAL EQUATION OF MOTION

The various results summarized by the relations quantifying both the divergence and the dissipation theorems may be applied together with (6.41) to derive a more useful version of the momentum balance equation of the body-field system under consideration. Using the definition of the momentum flux density—that is, the second-order tensor $\mathbf{JI} := p_* \mathbf{1} + \boldsymbol{\tau}_*$—it is easy to write

$$\rho\, D\mathbf{i} + \nabla \cdot \{p_* \mathbf{1} + \boldsymbol{\tau}_*\} = \rho \mathbf{f}, \tag{6.53}$$

where, aside from the viscous pressure tensor $\boldsymbol{\tau}_*$, all relevant quantities are either part of the balance frame or furnished by the corresponding Gibbs rate equation. The problem is how to transform this balance equation into an equation of motion, in other words, a relation concerning the flow velocity $\mathbf{v}$ instead of the specific linear momentum $\mathbf{i}$.

In traditional physics *Cauchy's two laws of motion* are used. They are given by the expressions

$$\rho\, D\mathbf{v} - \nabla \cdot \mathbf{T} = \rho \mathbf{f}; \qquad \mathbf{T} = \mathbf{T}^t; \qquad \mathrm{T}(\mathrm{r};t) \equiv T_{\alpha\beta} = \begin{pmatrix} T_{11} & T_{21} & T_{31} \\ T_{12} & T_{22} & T_{32} \\ T_{13} & T_{23} & T_{33} \end{pmatrix}, \tag{6.54}$$

where at a fixed time the tensor field $\mathbf{T}$ is always differentiable, and the field force $\mathbf{f}$ and the mass density ρ are assumed to be continuous. The existence of the stress tensor $\mathbf{T}$ (along with its tensor divergence $\nabla \cdot \mathbf{T}$) is often brought into connection with *Cauchy's fundamental theorem*. Equation $(6.54)_2$ formalizes the symmetry of $\mathbf{T}$ by its equivalence with the transposed second-order tensor $\mathbf{T}^t$.

On September 30, 1822, Cauchy announced the *stress principle*, which has been the foundation of the rational mechanics of continua and of modern material theories ever since. This principle asserts that upon any smooth, closed, and orientable surface $\partial\mathcal{B}$—be it an imagined surface within the body $\mathcal{B}$ or the bounding surface of the body itself—there exists an integrable field of traction vectors presumed to depend on the surface $\partial\mathcal{B}$ only through the normal $\mathbf{n}$. Therefore, the occurrence of $\mathbf{t_n}$ (leading to the total force and the total torque acting upon the matter in the region $\mathcal{B}$) implies that there exists a linear transformation $\mathbf{T}$ over the vector space. This "law," $\mathbf{t_n} = \mathbf{T} \cdot \mathbf{n}$, has had a long history (see Truesdell, 1968, p. 194). Cauchy's first law claims that the resultant contact force per unit volume is identical with $\nabla \cdot \mathbf{T}$, while the second law lays down the necessary and sufficient condition that the resultant contact torque be the moment of the resultant contact force. "The rational mechanics of materials is the theory of contact forces" (Truesdell, 1968, p. 186).

Cauchy's two laws are sometimes insufficiently general for modern continuum mechanics. This is especially true for shock waves or so-called oriented materials, for which it may be necessary to introduce higher-order stresses, spin angular momentum, and so on. Nevertheless, $(6.54)_1$ seems to provide a sound basis for an extended class of materials, even though it rests exclusively on Newton–Eulerian force ideas. In other words, in sharp contrast to Equation (6.53) Cauchy's deduction is mainly based on geometric arguments; it does not include explicitly any property that one could connect with fractional or other dissipation effects. This is made clear with regard to Euler's hydrodynamics, which was the first correct three-dimensional theory of a deformable material.

To obtain Euler's from Cauchy's theory, it is sufficient to assume the traction $\mathbf{t_n}$ to be normal to the surface on which it acts. The resulting scalar field p_E defined by $\mathbf{t_n} := p_E\mathbf{n}$ leads to the simplification of Cauchy's transformation rule; that is, $\mathbf{T} = -p_E\mathbf{1}$, where once again $\mathbf{1}$ is the unit tensor. For this reason, it seems adequate to decompose $\mathbf{T}$ into two parts

$$T := -p_E\mathbf{1} + \tau_C, \tag{6.55}$$

where now the Euler pressure p_E and the (Cauchy-) viscous pressure tensor τ_C appears without abandoning the symmetry relation $(6.54)_2$. Although for the moment (6.55) bears only a formal character, the negative stress tensor $\mathbf{T}$ may be related to the momentum flux density $\mathbf{JI} := p_*\mathbf{1} + \tau_*$, provided that the convection term of Cauchy's first law is assumed to be $\rho\, D\mathbf{i}$. Of course, this assumption is justified because Cauchy's reasoning closely followed Newton's basic force law, according to which all forces involved equal the respective time rate of the linear momentum. Nevertheless, this a priori identity $\mathbf{v} = \mathbf{i}$ is a far-reaching simplification with respect to the facts and concepts presented and discussed in Chapters 3 to 5. Furthermore, neither the terms for the respective pressures defined by (6.42) and (6.55) nor the respective viscous pressure tensors need detailed explanation. Despite these objections, the question arises whether there are theoretical reasons to support Cauchy's motion laws by means of the results presented above as the hard core of the Alternative Theory. To do so, it is convenient to interpret the pressure tensor $\mathbf{\Pi}$ as the negative of the stress tensor $\mathbf{T}$:

$$\mathbf{\Pi} \equiv -\mathbf{T}. \tag{6.56}$$

The question posed above will be answered by combining the dissipation velocity $\varphi = \mathbf{v} - \mathbf{i}$ with the differential balance equation of momentum (6.53). It is easy to verify that the resulting relationship, with the help of (5.73), becomes:

$$\rho\, D\mathbf{v} - \rho\, D\boldsymbol{\varphi} + \nabla \bullet \{p_*\mathbf{1} + \boldsymbol{\tau}_*\} = \rho\mathbf{f} = \rho\, D\mathbf{v} - \partial_t \rho\boldsymbol{\varphi} + \nabla \bullet \{p_*\mathbf{1} + \boldsymbol{\tau}_* - \rho\mathbf{v}\boldsymbol{\varphi}\}.$$

Assuming now that the bracket term $\{p_*\mathbf{1} + \boldsymbol{\tau}_* - \rho\mathbf{v}\boldsymbol{\varphi}\}$ has the same mathematical structure as the one for the pressure tensor Π in connection with Equations (6.55), we can relate the unknown second-order tensors $\boldsymbol{\tau}_*$ and $\boldsymbol{\varphi}$ to each other:

$$\Pi:\ =\ p_* 1 - \tau_*; \qquad \tau_*:\ =\ \frac{1}{2}\rho\mathbf{v}\boldsymbol{\varphi}. \tag{6.57}$$

This result is especially remarkable with respect to (6.47): The dissipation velocity $\boldsymbol{\varphi}$ is obviously a measure of the irreversibility attributed to the dissipation theorem.

From this conclusion we may derive the following limiting laws by using the momentum balance (6.53) and the relations $(6.18)_2$, which connect the non-equilibrium pressure p_* with the pressure p of the state at rest:

$$\boldsymbol{\varphi} \to \mathbf{0} \Rightarrow \mathbf{i} \Rightarrow \mathbf{v} \Rightarrow \begin{cases} p_* \to p \\ \\ \boldsymbol{\tau}_* \to \mathbf{0} \end{cases} \Rightarrow \rho D\mathbf{v} + \nabla p = \rho\mathbf{f} \tag{6.58}$$

According to this so-called *reversibility theorem,* the Euler equation of motion results without using paradoxically vanishing viscosity parameters. Clearly, Equation (6.58) also confirms the theoretically essential connotation between kinetic equilibrium and reversibility. Yet for the mathematical theory of non-equilibrium phenomena the following "limes law"

$$\lim_{\sigma \to 0} \boldsymbol{\varphi}\ =\ \mathbf{0} \tag{6.59}$$

should be noted. Accordingly, dissipationless processes are simply defined by the simultaneously vanishing quantities $\boldsymbol{\varphi}$ and σ, especially with regard to (6.52).

It is easy to transform the momentum balance Equation (6.53) into the equation of motion expressed for Newton's "*quantitas motus*" ($= m^{B+L}\mathbf{v}$). Considering (6.57), the following relationship

$$\rho\, D\mathbf{v} = \rho\mathbf{f} - \nabla \bullet \boldsymbol{\Pi} + \partial_t \rho\boldsymbol{\varphi}; \qquad \boldsymbol{\Pi} := p_*\mathbf{1} - \boldsymbol{\tau}_* \tag{6.60}$$

appears, which formally resembles Cauchy's equation of motion except for the unsteady term $\partial_t\rho\boldsymbol{\varphi}$. Note that this term implicitly contains the partial time derivatives of all system variables except the local momentum. The viscous pressure tensor $\boldsymbol{\tau}_* = \frac{1}{2}\rho\mathbf{v}\boldsymbol{\varphi}$ stands for a dyadic product whose components may be arrayed as follows:

$$\tau_* = \frac{1}{2}\rho \begin{bmatrix} v_1\varphi_1 & v_2\varphi_1 & v_3\varphi_1 \\ v_1\varphi_2 & v_2\varphi_2 & v_3\varphi_2 \\ v_1\varphi_3 & v_2\varphi_3 & v_3\varphi_3 \end{bmatrix} \tag{6.61}$$

This expression demonstrates what the notion of non-equilibrium really means mathematically, namely, motion represented by a velocity component v_k, but weighted by the difference between the local contributions of the velocity component v_ℓ and the assigned momentum component i_ℓ; that is, $\varphi_\ell = v_\ell - i_\ell$. Thus, local velocity gradients do not play a primary role. The more interesting contribution to any non-equilibrium phenomenon stems from the pressure term. The pressure p_* deviates considerably from the pertinent local equilibrium value, even for relatively small values of flow velocity. This statement corresponds to the fact that p_* cannot be described normally by the use of the respective thermal equation of state, although this is commonly practiced and euphemistically paraphrased as the Duhem–Hadamard conjecture (see Truesdell, 1984, p. 60). But it should also be stressed that this convention contradicts the Gibbs–Duhem relation offered in its general form (2.76), by which a dependency on the flow velocity is established. The relevance of the dissipation velocity vector $\boldsymbol{\varphi}$ concerning the viscous pressure tensor $\boldsymbol{\tau}_*$ brings up another crucial point of the theory presented in this text. Therefore, we will discuss further some important properties of $\boldsymbol{\varphi}$.

The special structure of the pressure tensor $\boldsymbol{\Pi}$ clearly indicates that for every no-slip condition $\mathbf{v} = \mathbf{0}$ the tensor $\boldsymbol{\tau}_*$ will vanish. Consequently, all corresponding dissipative effects are now included in the non-equilibrium pressure p_*. This fact can be easily recognized by direct integration of Equation $(6.18)_2$

$$p_* = p + \rho^2 \left(\frac{\partial e_\varphi}{\partial \rho} \right)_{\mathbf{r},\, s,\, \omega j}, \tag{6.62}$$

where the equivalence of $\mathbf{i}$ and $\boldsymbol{\varphi}$ for $\mathbf{v} = \mathbf{0}$ is considered. This result reveals that the influence of any local dissipation manifests itself by means of the specific dissipation energy e_φ introduced in Equation (6.15) ($e_\varphi = -\frac{1}{2}\varphi^2$).

With respect to the basic relationship between the three motion quantities $\mathbf{i}$, as well as $\mathbf{v}$ and $\boldsymbol{\varphi}$, it is apparent that two further energies can be attained, by scalar multiplication of that relationship, that also manifest local irreversibility:

$$\mathbf{v} = \mathbf{i} + \boldsymbol{\varphi} \Rightarrow -\frac{1}{2}\boldsymbol{\varphi} \cdot \mathbf{v} = -\frac{1}{2}\boldsymbol{\varphi} \cdot \mathbf{i} - \frac{1}{2}\boldsymbol{\varphi} \cdot \boldsymbol{\varphi} \Rightarrow e_\tau = e_{\mathbf{i}} + e_\varphi. \tag{6.63}$$

We may call the first two specific energies concerning dissipations, e_τ and e_i, the (specific energy of) stress dissipation and the (specific energy of) momentum dissipation, respectively. An examination of the stress dissipation e_τ using several basic rules of vector analysis leads to some interesting conclusions.

Starting from the definition we obtain directly

$$\begin{aligned} e_\tau &:= -\frac{1}{2}\mathbf{v} \cdot \boldsymbol{\varphi} = -\frac{1}{2}\mathbf{v}^2 + \frac{1}{2}\mathbf{v} \cdot \mathbf{i} \\ &\qquad\qquad \mathbf{v} \cdot \mathbf{i} = |\mathbf{v}||\mathbf{i}| \cos(\mathbf{v}, \mathbf{i}) \\ &= \frac{1}{2}\mathbf{v}^2 [\chi \cos(\mathbf{v}, \mathbf{i}) - 1], \end{aligned} \tag{6.64}$$

where the factor χ is introduced to measure the amount of $\mathbf{i}$ as the multiple of $\mathbf{v}$. The angle between the two vectors $\mathbf{v}$ and $\mathbf{i}$ denoted by $(\mathbf{v}, \mathbf{i})$ follows from the definition of their scalar product. The second step concerns the foundation of the theory presented here. If the stress dissipation e_τ is interpreted as a manifestation of irreversibility, then e_τ should display significant properties indicating its most relevant characteristics. These characteristics may be furnished by means of *transport coefficients* valid for the special fluid under consideration and of a property that is true for all real substances under non-equilibrium conditions and for the equilibrium limit. Following the traditional view (established mainly by Maxwell), those transport coefficients should be represented by characteristic relaxation times (discussed in Section 5.3). In principle, these internal times may be presumed to be a positive definite function, each of which depends on the same variables as the dissipation velocity φ itself. Within the scope of a phenomenological theory, such material properties must be determined by proper experiments.

Another quantity that influences e_τ in a more general manner is the local entropy production σ. It is noteworthy that, as a consequence of this disposition, the entropy balance equation will be built directly into the mathematical apparatus of the theory. Hence, the ansatz

$$\rho e_\tau := t_{\tau *} T_* \sigma \Rightarrow \begin{cases} t_{\tau *} \geq 0 \\ T_* \geq 0 \quad 0 \Rightarrow e_\tau \geq 0 \\ \sigma \geq 0 \end{cases} \tag{6.65}$$

defines the stress dissipation e_τ with factors that are always zero-point bounded and positive valued for all admitted processes. Moreover, the previously noted reversibility theorem in its special form (6.59) is considered.

By regarding (6.65) with respect to Equation $(6.64)_3$, an interesting inference is that the basic ratio between the amounts of the two vectors **i** and **v**, exhibited by the coefficient χ, is evidently constituted by an algebraic expression for the angle $(\mathbf{v}, \mathbf{i})$ formed by those two vectors:

$$\chi \geq \cos^{-1}(\mathbf{v}, \mathbf{i}). \tag{6.66}$$

This remarkable result indicates that dissipation in a flow is induced primarily by a characteristic drift between the local vectors of flow velocity and specific momentum. Accordingly, angles of more than 90° are irregular. Thus, an obtuse angle does not occur because the amount of φ would surpass that of **i**. On the other side, that drift vanishes only in the limiting case of kinetic equilibrium which is equivalent to φ values tending toward zero. As for (6.65), an obvious conclusion is to prefer the same mathematical structure for the other specific energies e_i and e_φ arising in Equation $(6.63)_3$. The result will be a simple linear relation between the corresponding relaxation times $t_{\mathrm{i}*}$, $t_{\varphi *}$, and $t_{\tau *}$.

However, we should note that the specific dissipation energy $e_\varphi := -\frac{1}{2}\varphi^2$ always takes on negative values by definition. Consequently, the corresponding approach

$$\rho e_\varphi := -t_{\varphi *} T_* \sigma \Rightarrow \begin{cases} t_{\varphi *} \geq 0 \\ T_* \geq 0 \quad \Rightarrow e_\varphi \leq 0 \\ \sigma \geq 0 \end{cases} \tag{6.67}$$

differs from (6.65) by the negative sign although all factors are also positive, just as their behavior indicates in (6.65). Hence, the linear connection between the three relaxation times becomes

$$t_{\tau*} + t_{\varphi*} = t_{i*} \tag{6.68}$$

where the third relaxation time t_{i*} is deduced with the knowledge of the other two.

6.4 Hypothetical State at Rest

6.4.1 *BASIC PROBLEMS REGARDING INFORMATION FOR NON-EQUILIBRIUM PHENOMENA*

The elementary yet rigorous deduction of the equation of motion establishes some basic relationships necessary for a deeper understanding of non-equilibrium phenomena in physics. We now present the following proofs utilizing the reference properties of those variables marked by an asterisk—pressure, temperature, and the ρ–Legendre-transformed energy $e_*^{[\rho]}$. These quantities have been identified already by the definitions (6.19) and the reversibility theorem [Equations (6.58) and (6.59)] as manifestations of a dissipationless equilibrium state at rest, since they play a distinguished role in material science. They are mutually related by the so-called equations of state, which may be determined experimentally and, in part, theoretically with high precision.

Thus, for instance, the equilibrium pressure p, related to the non-equilibrium pressure p_* by Equation $(6.18)_2$, is well-known for numerous gases and liquids as a characteristic material function of mass density ρ and temperature T, as well as of mass fractions in the case of mixtures. Each function has its individual mathematical structure, mainly influenced by the chemical constitution of the fluid under consideration, but it is subject to some universal stability conditions [as indicated by Equation (2.35)] and limiting laws like the ideal gas equation of state. In summary, every equation of state can only be precisely determined for the state at rest.

In other words, as a rule relations between non-equilibrium quantities do exist theoretically. In practice, however, they cannot be determined with sufficient accuracy. Therefore, it seems reasonable to include the equations of state into the whole network of any mathematical theory of non-equilibrium phenomena. Nearly all commonly used theories do this, but they do so in a more heuristic way, either founded on the principle of local equilibrium or the Duhem–Hadamard conjecture. Such arguments can easily lead to misunderstandings and paradoxes because of the risk of confusing approximations with the more rigorous and stringent theoretical elements of a theory. Moreover, kinetics traditionally dominate classical mechanics and fluid dynamics, as well as the so-called theories of nonuniform gases. Thus kinetics overestimates the theoretical weight and influence of space and time variables, compared with the physically original events running dynamically in the phase space. This will be clarified below and illustrated fully by means of a few examples concerning transport phenomena and their characteristic coefficients. Strikingly, the results obtained demonstrate such a degree of mathematical complexity, even for simple cases of heat conductivity or flow friction, that we should not wonder about the comparatively meager results attained from the Maxwell–Boltzmann equation. In particular, this refers to the significant difference between the two kinds of solutions, aptly called *general* and *dominant* by Truesdell.

General solutions correspond to an initial-value problem and yield gross quantities that obey the field equations of continuum mechanics. "But in general gas flows there is no reason to think that either the irreversibility or any constitutive equation of continuum thermomechanics is satisfied" (Truesdell, 1984, p. 451). As for the corresponding transport coefficients, this kind of solution always presumes the explicit knowledge of the formally mathematical relation between the flux density under consideration and its respective variables. Fourier's and Fick's constitutive laws of heat conduction and diffusion are prominent examples.

Dominant solutions, already discussed in Section 5.4, have been conjectured to exist. Formal procedures have been devised to find successive approximations to any property that may be assigned to specific constitutive relations appropriate to a certain kind of material. In other words, the primary aim of these solutions is to confirm and apply those constitutive laws as response functions. Unfortunately, there is no single solution that may be regarded as representative for real systems, that is, valid for physically founded molecular density functions. But it is perhaps more peculiar that even artificial molecular models of extreme simplicity lead to rather complicated response functions, compared to traditional laws like those of Fourier's and Fick's.

6.4.2 *FIELD EQUATION OF LEGENDRE-TRANSFORMED ENERGY*

Let us first carry out a Legendre transformation of the specific internal energy u with respect to the variable ρ^{-1}. By means of the transformation rule

$$u^{[\rho^{-1}]}(s; p; \omega_j) = u(s; \rho^{-1}; \omega_j) - \frac{\partial u(s; \rho^{-1}; \omega_j)}{\partial \rho^{-1}} := u + \frac{p}{\rho} = h, \tag{6.69}$$

the corresponding Gibbs main Equation (6.17) becomes a Pfaffian now expressed in the variables s, p, and ω_j, for a multicomponent single-phase body-field system

$$\rho dh = T\rho ds + dp + \sum_{j=1}^{J} \mu_j \rho d\omega_j \, . \tag{6.70}$$

Beginning with (6.16), the specific enthalpy h consequently leads to the following expression by summing up all energy species involved:

$$e + p\,\rho^{-1} - e_{\#} := u + p\,\rho^{-1} + e_{\text{kin}} + e_{\text{pot}} + e_{\varphi} \tag{6.71}$$

$$e^{[\rho]} - e_{\#} := h + e_{\text{kin}} + e_{\text{pot}} + e_{\varphi}.$$

The density-transformed energy $e^{[\rho]}$ (without an asterisk) may be formally assigned to an energy balance equation

$$\rho De^{[\rho]} + \nabla \bullet \mathbf{j}_e^{[\rho]} = \partial_t p, \tag{6.72}$$

which is related to Equation (6.32) with its respective star-marked quantities by the formula

$$\mathbf{j}_e^{[\rho]} = \mathbf{j}_{e*}^{[\rho]} + (p_* - p)\mathbf{v}. \tag{6.73}$$

Now, using the connection (6.34) between the energy flux densities $\mathbf{j}_e^{[\rho]}$ and $\mathbf{j}_e$, along with their reference (6.45) to the heat flux vector $\mathring{\mathbf{q}}_*$ and work rate $\mathring{\mathbf{w}}_* := \mathbf{\Pi} \bullet \mathbf{v}$, (6.73) becomes

$$\mathbf{j}_e^{[\rho]} = \mathring{\mathbf{q}}_* + \mathbf{\Pi} \bullet \mathbf{v} - p_*\mathbf{v} + (p_* - p)\mathbf{v} = \mathring{\mathbf{q}}_* + \boldsymbol{\tau}_* \bullet \mathbf{v} + (p_* - p)\mathbf{v}. \tag{6.74}$$

This relation will play a significant role in the formulation of the constitutive laws derived in this section. Meanwhile, note that the energy flux density $\mathbf{j}_e^{[\rho]}$ is assigned to the set of variables s, p, and ω_j, where the non-equilibrium effects are implicitly taken into account by the balance of the energy species (6.71) as well as the star-marked viscous pressure tensor and the pressure alone. This remark will be evident immediately from the deduction of the enthalpy balance equation as follows: Equation of motion (6.60), scalar multiplied by the local flow velocity, for the moment yields

$$\rho\mathbf{v} \bullet D\mathbf{v} = \rho\mathbf{v} \bullet \mathbf{f} - \mathbf{v} \bullet \nabla \bullet \mathbf{\Pi} + \mathbf{v} \bullet \partial_t\rho\boldsymbol{\varphi}; \qquad \mathbf{v} \bullet \mathbf{\Pi} := p_*\mathbf{v} \bullet \mathbf{1} - \mathbf{v} \bullet \boldsymbol{\tau}_*. \tag{6.75}$$

Assume again a potential approach for the field force density $\mathbf{F}_V = \rho\mathbf{f}$, according to $\rho\mathbf{f} \bullet \mathbf{v} := -\rho De_{\text{pot}} + \rho\partial_t e_{\text{pot}}$. Then a considerable simplification of (6.75) may be achieved, if the kinetic energy per mass unit is introduced by $\mathbf{v} \bullet D\mathbf{v} = De_{\text{kin}}$ and the balance of the energy species (6.71) is considered. Then, by elimination of the potential and kinetic energies, we first change Equation (6.75) into the following form

$$\rho D\left(e^{[\rho]} - h - e_\varphi\right) = -\mathbf{v} \bullet \nabla p_* + \mathbf{v} \bullet \nabla \bullet \boldsymbol{\tau}_* + \mathbf{v} \bullet \partial_t\rho\boldsymbol{\varphi} + \rho\partial_t e_{\text{pot}}.$$

Next, we substitute the two substantial derivatives De_φ and $De^{[\rho]}$ by means of Equations (5.73) and (6.72). Furthermore, we must apply the tensor rules given by (6.38). Hence, the latter equation turns into the expression

$$\begin{aligned} \rho Dh = {} & -\nabla \bullet \mathbf{j}_e^{[\rho]} + \mathbf{v} \bullet \nabla p_* + \boldsymbol{\tau}_* : \nabla\mathbf{v} - \nabla \bullet [\mathbf{v} \bullet \boldsymbol{\tau}_* + \rho e_\varphi \mathbf{v}] \\ & -\mathbf{v} \bullet \partial_t\rho\boldsymbol{\varphi} - \partial_t\rho e_\varphi - \rho\partial_t e_{\text{pot}} + \partial_t p. \end{aligned} \tag{6.76}$$

The numerous terms representing unsteady changes attract attention, as do the problematical mixture of quantities belonging to two different sets of system variables. However, a fortunate circumstance allows a remarkable transformation of (6.76) to a mathematical structure that contains quantities related only to the reference state discussed above.

The following tensor rules (Bird, Stewart, and Lightfoot, 1960, p. 730, see also Appendix 2.2)

$$\mathbf{v}(\boldsymbol{\varphi} \bullet \mathbf{v}) = \{\mathbf{v}\boldsymbol{\varphi}\} \bullet \mathbf{v}; \qquad \mathbf{v} \bullet \{\mathbf{v}\boldsymbol{\varphi}\} = (\mathbf{v} \bullet \mathbf{v})\boldsymbol{\varphi} = \boldsymbol{\varphi}(\mathbf{v} \bullet \mathbf{v}) = \{\boldsymbol{\varphi}\mathbf{v}\} \bullet \mathbf{v} \tag{6.77}$$

represent dyads {..} and scalar products (.·.), respectively, combined with one vector each. Hence, the divergence term $\nabla \bullet [\ldots]$ may be altered as follows:

$$\nabla \bullet [\mathbf{v} \bullet \boldsymbol{\tau}_* + \rho e_\varphi \mathbf{v}] = \nabla \bullet \left[\frac{1}{2}\rho \mathbf{v} \bullet \{\mathbf{v}\boldsymbol{\varphi}\} - \frac{1}{2}\rho\,(\boldsymbol{\varphi} \bullet \boldsymbol{\varphi})\,\mathbf{v}\right]. \tag{6.78}$$

This expression may be manipulated considerably by a condition assumed to be essential for the theory presented. In fluid dynamics that condition refers to a Cauchy theorem that asserts that the stress tensor **T** is symmetric. By means of the nomenclature used here, it is easy to formulate this *symmetry condition* by the tensor relation

$$\{\mathbf{v}\boldsymbol{\varphi}\} = \{\boldsymbol{\varphi}\mathbf{v}\}, \tag{6.79}$$

which refers to the viscous pressure tensor $\boldsymbol{\tau}_*$ only, apparently quite different from Cauchy's result. For flows with internal rotation, (6.79) is not true (see Section 7.4).

Inserting (6.79) into (6.78), a significant simplification may be obtained:

$$\begin{aligned}\nabla \bullet [\mathbf{v} \bullet \boldsymbol{\tau}_* + \rho e_\varphi \mathbf{v}] &= \nabla \bullet \left[\frac{1}{2}\rho \mathbf{v} \bullet \{\boldsymbol{\varphi}\mathbf{v}\} - \frac{1}{2}\rho\,\{\mathbf{v}\boldsymbol{\varphi}\} \bullet \boldsymbol{\varphi}\right] \\ &= \nabla \bullet \left[\frac{1}{2}\rho\,\{\mathbf{v}\boldsymbol{\varphi}\} \bullet \mathbf{i}\right] = \nabla \bullet \left[\left(\frac{1}{2}\rho\boldsymbol{\varphi} \bullet \mathbf{i}\right)\mathbf{v}\right] \\ &= \mathbf{v} \bullet \nabla\left(\frac{1}{2}\rho\boldsymbol{\varphi} \bullet \mathbf{i}\right) + \left\{\left(\frac{1}{2}\rho\boldsymbol{\varphi} \bullet \mathbf{i}\right)\mathbf{1}\right\} : \nabla\mathbf{v}.\end{aligned} \tag{6.80}$$

The last equation is the key to resolving the indefiniteness contained in (6.76) by using two sets of variables. If the left-hand side of that equation is substituted for the same expression in (6.76), then it will be evident with regard to (6.80) that the first term of its right-hand side may be formally assigned to the pressure term $\mathbf{v} \bullet \nabla p_*$ of (6.76). The corresponding assignment to the viscous pressure term $\boldsymbol{\tau}_* : \nabla\mathbf{v}$ of (6.76) will be done with the second term. Hence, Equation (6.76) turns into the form

$$\begin{aligned}\rho Dh = &- \nabla \bullet \mathbf{j}_e^{[\rho]} + \mathbf{v} \bullet \nabla p + \boldsymbol{\tau} : \nabla\mathbf{v} \\ &- \mathbf{v} \bullet \partial_t \rho\boldsymbol{\varphi} - \partial_t \rho e_\varphi - \rho\partial_t e_{\text{pot}} + \partial_t p,\end{aligned} \tag{6.81}$$

using

$$p := p_* - \frac{1}{2}\rho\,(\boldsymbol{\varphi} \bullet \mathbf{i}); \qquad \boldsymbol{\tau} := \boldsymbol{\tau}_* - \frac{1}{2}\rho\,(\boldsymbol{\varphi} \bullet \mathbf{i})\,\mathbf{1} \tag{6.82}$$

by definition.

Combining (6.81) and (6.82) into a single expression, the pressure terms of Equation (6.81) as well as its two terms concerning unsteady changes of the dissipation velocity $\boldsymbol{\varphi}$, will lead to

$$
\begin{aligned}
\rho Dh = & -\nabla \bullet \mathbf{j}_e^{[\rho]} + Dp + \boldsymbol{\tau}:\nabla \mathbf{v} \\
& -\mathbf{v} \bullet \partial_t \rho \boldsymbol{\varphi} - \partial_t \rho e_{\varphi} - \rho \partial_t e_{\text{pot}}.
\end{aligned} \tag{6.83}
$$

This field equation of specific enthalpy h seems to be quite similar to the common representations given in modern textbooks of thermofluid dynamics. But there are some significant differences as to the meaning of the various terms used. Thus, Equation (6.83) describes the energetic part of non-equilibrium phenomena by means of a set of variables that allow us to consider additional information only available for equilibrium states. As a matter of fact, this information concerns all equations of state characterizing the moving matter in question. It should be stressed that the deduction of (6.83) does not contradict any axioms and theorems previously presented.

The utilized equations of state $p(T,\rho,\omega_j)$ and $h(T,\rho,\omega_j)$ refer to a hypothetical state at rest, the properties of which do not agree with the actual values of the local non-equilibrium state variables. The hypothetical state at rest defines a certain kind of accompanying process for the real process as it was formerly discussed by Muschik (1988, p. 11). Indeed, this must be viewed as an essential discrepancy with conventional practice, for which there is no well-founded method available to gain the correct non-equilibrium values T_* and p_*. In contrast to this, the Alternative Theory constructs a mathematical procedure to determine the field variables in their local non-equilibrium states. Note that Equation (6.83) also includes some non-equilibrium quantities (such as the dissipation velocity $\boldsymbol{\varphi}$) that only vanish simultaneously with the trend toward any local equilibrium (i.e., with $\mathbf{i} = \mathbf{0}$). But before we tackle these quantities, we need to settle two questions regarding whether the definitions (6.82)—as an immediate consequence of the symmetry condition (6.79)—are contradictory to other known relationships.

The first question refers to $(6.82)_1$ and asks for the condition by which an identical difference $p_* - p$ in the pressures involved might be induced, using either Equation $(6.82)_1$ or Equation $(6.18)_2$. The second question focuses on a basic problem of transfer processes concerning constitutive equations and their physical parameterization by means of transport coefficients. The answers are difficult, for they touch the fundamentals of each physically based continuum theory in a very different way.

As discussed above in connection with (6.74), theoretical considerations prompt us to introduce a flux density such as $\mathbf{j}_e^{[\rho]}$, presumed to be dependent on the set of variables that characterize the hypothetical state at rest. Each related flux density $\mathbf{j}_z$ is also defined to be a part of the pertaining balance equation in such a way that $\mathbf{j}_z$ cannot vanish at any position with values $\mathbf{v} = \mathbf{0}$ for the flow velocity. This problem will be dealt with in the next chapter in view of the viscous pressure tensor $\boldsymbol{\tau}$, defined by $(6.82)_2$ and assigned to the hypothetical state at rest.

In a next step an extended form of the field Equation (6.83) will be derived, thereby pointing out its validity for any multicomponent single-phase system. For this purpose it is first convenient to express the left-hand side of (6.83) by its total differential with respect to the variables T, p, and ω_j related to the corresponding hypothetical state at rest

$$
\begin{aligned}
Dh &= \left(\frac{\partial h}{\partial T}\right)_{p,\,\omega j} DT + \left(\frac{\partial h}{\partial p}\right)_{T,\,\omega j} Dp + \sum_{j=1}^{J} \left(\frac{\partial h}{\partial \omega_j}\right)_{p,\,T,\,\omega_{\neq j}} D\omega_j \\
&= c_{p,\,\mathrm{mix}} DT + h_{p,\,\mathrm{mix}} Dp + \sum_{j=1}^{J} h_j\, D\omega_j .
\end{aligned}
\tag{6.84}
$$

Two characteristic properties of the mixture composed of all components $j = 1(1)J$ are introduced: the specific heat $c_{p,\mathrm{mix}}$ and the pressure coefficient $h_{p,\mathrm{mix}}$. In addition, the so-called partial specific enthalpy h_j of the jth component is defined. All properties are assumed to be available from the known caloric equation of state $h(T, p, \omega_j)$ given for the mixture. Note that for a mixture consisting of some ideal gases, the function $h_{p,\mathrm{mix}}$ vanishes identically and h_j depends only on temperature.

Multiplying (6.84) by the mass density ρ of the mixture, its resulting last term may be substituted by the set of component continuity equations for the mass fraction ω_j, as introduced by (6.35). Thereby, (6.83) may be written in its final form:

$$
\begin{aligned}
\rho c_{p,\,\mathrm{mix}} DT = &-\nabla\bullet\mathbf{j}_e^{[\rho]} + (1 - h_{p,\,\mathrm{mix}})\, Dp + \boldsymbol{\tau}:\nabla\mathbf{v} - \sum_{j=1}^{J} h_j\Gamma_j - \sum_{j=1}^{J} h_j\nabla\bullet\mathbf{j}_j \\
&-\mathbf{v}\bullet\partial_t\rho\boldsymbol{\varphi} - \partial_t\rho e_{\varphi} - \rho\partial_t e_{\mathrm{pot}} .
\end{aligned}
\tag{6.85}
$$

Beyond the frame covered by Equation (6.83), this field equation of the temperature T explicitly reveals the influence of any chemical reactions and diffusional processes represented by the chemical production densities Γ_j and the diffusion fluxes $\mathbf{j}_j$ of the jth component.

6.5 Constitutive Properties of Matter

6.5.1 *VACUUM THEOREM*

The first question posed in the previous section concerns a comparison between the two previously stated relations

$$
p := p_* - \rho^2 i\bullet\left(\frac{\partial\boldsymbol{\varphi}}{\partial\rho}\right)_{\mathbf{r},\,s,\,\omega_j} \tag{6.18$_2$}
$$

$$
p := p_* - \frac{1}{2}\rho\,(\boldsymbol{\varphi}\bullet\mathbf{i}) \tag{6.82$_1$}
$$

with the aim to clarify the conditions for which they are identical. A solution of the relation

$$
\rho^2\mathbf{i}\bullet\left(\frac{\partial\boldsymbol{\varphi}}{\partial\rho}\right)_{\mathbf{r},\,s,\,\omega_j} = \frac{1}{2}\rho\,(\mathbf{i}\bullet\boldsymbol{\varphi})
$$

can easily be obtained by integration. Multiplication of both sides with $\boldsymbol{\varphi}$ from the right eliminates the specific momentum **i**. This first step furnishes a result with a dyadic product on each side. These tensors may then be resolved in a second step as follows:

$$\rho\left\{\left(\frac{\partial\boldsymbol{\varphi}}{\partial\rho}\right)_{\mathbf{r},\,s,\,\omega_j}\boldsymbol{\varphi}\right\} = \frac{1}{2}\{\boldsymbol{\varphi}\boldsymbol{\varphi}\} \Rightarrow \rho\boldsymbol{\varphi}\bullet\left\{\left(\frac{\partial\boldsymbol{\varphi}}{\partial\rho}\right)_{\mathbf{r},\,s,\,\omega_j}\boldsymbol{\varphi}\right\} = \frac{1}{2}\boldsymbol{\varphi}\bullet\{\boldsymbol{\varphi}\boldsymbol{\varphi}\}\ .$$

Therein the second expression may be simplified as

$$\rho\boldsymbol{\varphi}\left\{\left(\frac{\partial\boldsymbol{\varphi}}{\partial\rho}\right)_{\mathbf{r},\,s,\,\omega_j}\boldsymbol{\varphi}\right\} = \frac{1}{2}\boldsymbol{\varphi}\bullet\{\boldsymbol{\varphi}\boldsymbol{\varphi}\} \Rightarrow -\rho\left(\frac{\partial e_{\varphi}}{\partial\rho}\right)_{\mathbf{r},\,s,\,\omega_j} = -e_{\varphi}.$$

The specific dissipation energy e_φ is now assumed to be functionally structured by a linear relationship with the mass density ρ, according to $e_\varphi = \in(\mathbf{r}, s, \omega_j)\rho$ + constant, where the auxiliary function $\in(\mathbf{r}, s, \omega_j)$ is independent of ρ. This result may be summarized by the following statement, called the vacuum theorem.

> The mathematical concept of non-equilibrium processes presented here is based on a matter model for which the proportionality
>
> $$\boldsymbol{\varphi}^2 \sim \rho$$
>
> is presumed to be true. Motion and matter are macroscopically constituted by inrreversability.

This means that dissipation does not occur under vacuum conditions, in other words, if $\rho \to 0$. This result seems reasonable and may be accepted as physically founded within a broad range of applications. It may even be used as a definition of a *macrophysical vacuum.*

An essential consequence follows directly from $(6.82)_2$, along with the equivalence of both $(6.18)_2$ and $(6.82)_1$. If we form the pressure tensor **Π** by the use of non-equilibrium quantities as well as by means of adequate properties defined only for the hypothetical state at rest, the result

$$\mathbf{\Pi} := p_*\mathbf{1} - \boldsymbol{\tau}_* = p\mathbf{1} - \boldsymbol{\tau} \tag{6.86}$$

demonstrates the invariance of **Π** with respect to the appropriate levels of description. This statement confirms the influential and far-reaching conjecture of Duhem and Hadamard according to which the pressure p may be substituted by the respective equations of state. However, one should realize that this proof is of value only with the complete design and frame of the Alternative Theory along with Falk's dynamics. The common method, widely used in flow mechanics to equate the pressure with the equilibrium equation of state, cannot be justified either by the theoretical results given above or by means of experiments. Neumaier (1996) has very recently reported on p, ρ, T, ω_j - experiments with various gases and fluid mixtures. As he has demonstrated by some extensive series of highly accurate measurements concerning the trend toward equilibrium, there is no agreement between local pressure, temperature, and density according to the respective equation of state. His experimental program has been performed for propane (C_3H_8) within a considerable range of various temperatures and pressures with the help of a new setup that allows process realizations via variations of all three state variables. Neumaier's analysis proves that even small deviations from equilibrium conditions immediately lead to motion of the fluid. Only within narrow bounds of pressure and density variations do gas flows seem to exist for which kinetic equilibrium prevails (as it is described in Subsection 6.2) for flows obeying the Euler equation of motion. Outside these bounds significant discrepancies between the experimentally determined state parameters and their assigned equation of state have been observed in connection with the time behavior of the parameters in the course of their trend toward equilibrium.

With respect to (6.86) and the constitutive Equation (6.74) concerning $\mathbf{j}_e^{[\rho]}$, let us compile some information as to the values of $\mathbf{\Pi}$ and $\mathbf{j}_e^{[\rho]}$ for three limiting cases:

state at rest $\mathbf{i} \to 0$ no-slip condition $\mathbf{v} \to 0$ dissipationless $\boldsymbol{\varphi} \to 0$

$$\lim \mathbf{\Pi} = \begin{cases} p & for\ \mathbf{i} \to 0 \\ p_* & for\ \mathbf{v} \to 0; \\ p & for\ \boldsymbol{\varphi} \to 0 \end{cases} \qquad \lim \mathrm{j}_e^{[\rho]} = \begin{cases} 0 & for\ \mathbf{i} \to 0 \\ \overset{\circ}{q}_* & for\ \mathbf{v} \to 0 \\ 0 & for\ \boldsymbol{\varphi} \to 0 \end{cases} \tag{6.87}$$

Considering the invariance relation (6.86), the limiting case $\lim\mathbf{\Pi} = p_*$ for stagnation conditions $\mathbf{v} \to \mathbf{0}$ may be alternately expressed either by $\lim\mathbf{\Pi} = p_{\mathrm{v}=0}\mathbf{1} - \boldsymbol{\tau}_{\mathrm{v}=0}$ (where $p_{\mathrm{v}=0}$ and $\boldsymbol{\tau}_{\mathrm{v}=0}$ denote the pressure and the viscous pressure tensor for $\mathbf{v} \to \mathbf{0}$) or by the hypothetical state at rest. The question arises how to calculate $\mathbf{j}_e^{[\rho]}$ and $\boldsymbol{\tau}$.

Before giving an answer (in Subsection 6.5.2), a short remark seems appropriate with regard to the relevance of the state at rest. In practice, this limiting case belongs to a special area of material science occupied with the task of determining very precisely the thermal equations of state as well as the caloric equations of pure substances and multicomponent mixtures. Corresponding data sets and algebraic formulas are the result of research commonly performed by means of special theoretical methods and certain kinds of experimental devices. Most of their theoretical framework is elaborated worldwide either in phenomenological or in molecular and statistical thermodynamics. The textbooks of Prausnitz (1969) and Lucas (1990) offer competent examples concerning the essence of the problem, which is how to obtain a precise equation of state $p(T, \rho, \omega_k)$ for an extended range of the fluid behavior under consideration. The high efficiency of modern methods to establish such an equation is illustrated by a Helmholtz equation $A(T, \rho)$ proven to be valid for water and its vapor. This new 56-coefficient fundamental equation is true in the entire fluid region covered by the selected data set from the melting line to about 1000°C at pressures up to about 1000 MPa. The selected data set includes the following properties of the fluid in its real and ideal gas states: $p\rho T$-data, thermal properties at saturation, isochoric heat capacity, speed of sound, isobaric heat capacity, enthalpy, internal energy at saturation, Joule–Thomson coefficient, and isothermal throttling coefficient.

When approaching the critical point, the description shows a singular behavior with regard to the heat capacity and the speed of sound. The overall accuracy of this new "water equation" satisfies most requirements for all extremes of today's applications. The selected data set, composed of more than 6400 values, is bounded by their experimental uncertainty. Values of the relative density, for instance, can be reproduced with an accuracy better than 0.0001% for ideal gas conditions and 1% for 1000 K and 1000 MPa (cf. Pruß and Wagner, 1996).

6.5.2 HEAT FLUX DENSITY

The following consideration touches the core of the Alternative Theory since the option of dealing with the theory of constitutive equations is uncommon. This section will be restricted to investigations concerning the energy flux density $\mathbf{j}_e^{[\rho]}$ of any one-component single-phase system; the determination of the viscous pressure tensor $\boldsymbol{\tau}$ is reserved for the next chapter.

The simplest starting point is offered by the equation

$$\mathbf{j}_e^{[\rho]} = \overset{\circ}{\mathbf{q}}_* + \mathbf{JI} \bullet \mathbf{v} - p_* \mathbf{v} + (p_* - p)\mathbf{v} = \overset{\circ}{\mathbf{q}}_* + \boldsymbol{\tau}_* \bullet \mathbf{v} + (p_* - p)\mathbf{v}, \tag{6.74}$$

which may be changed with the help of $(6.82)_1$. The following sequence of relations results from the application of the pertaining definitions of $\boldsymbol{\tau}_* := \frac{1}{2}\rho\mathbf{v}\boldsymbol{\varphi}$ and $e_\mathbf{i} = \frac{1}{2}(\mathbf{i} \bullet \boldsymbol{\varphi})$ as well as from using the tensor rules (6.77):

$$\begin{aligned}\mathbf{j}_e^{[\rho]} &= \overset{\circ}{\mathbf{q}}_* + \boldsymbol{\tau} \bullet \mathbf{v} + (p_* - p)\,\mathbf{v} = \overset{\circ}{\mathbf{q}}_* + \boldsymbol{\tau}_* \bullet \mathbf{v} + \frac{1}{2}\rho\,(\mathbf{i} \bullet \boldsymbol{\varphi}) \\ &= \overset{\circ}{\mathbf{q}}_* + \frac{1}{2}\rho\,\{\boldsymbol{\varphi}\mathbf{v}\} \bullet \mathbf{v} - \rho e_\mathbf{i}\mathbf{v} = \overset{\circ}{\mathbf{q}}_* + \frac{1}{2}\rho\mathbf{v}\,(\boldsymbol{\varphi} \bullet \mathbf{v}) + \frac{1}{2}\rho\mathbf{v}\,(\boldsymbol{\varphi} \bullet \mathbf{i}) \\ &= \overset{\circ}{\mathbf{q}}_* + \frac{1}{2}\rho\mathbf{v}\,([\mathbf{v} - \mathbf{i}] \bullet [\mathbf{v} + \mathbf{i}]) = \overset{\circ}{\mathbf{q}}_* + \frac{1}{2}\rho\mathbf{v}\,(\mathbf{v}^2 - \mathbf{i}^2)\end{aligned} \tag{6.88}$$

The last relation yields the key to the problem: Defining a further dissipation energy in analogy to the three various quantities introduced by (6.63), the following expression, which may be called specific heat energy,

$$e_\mathbf{q} := -\frac{1}{2}(\mathbf{v}^2 - \mathbf{i}^2) \tag{6.89}$$

arises. Obviously, the last term of (6.88) represents a correction like that of (6.82) concerning $\boldsymbol{\tau}$ and $\boldsymbol{\tau}_*$, if $\boldsymbol{\tau}_*$ is accepted as the viscous pressure tensor of any non-equilibrium process. Then $\boldsymbol{\tau}$ reflects the respective viscous pressure tensor that is thought to be definitely assigned to the corresponding hypothetical state at rest. It seems reasonable to assume that the same is true for both the vectors $\overset{\circ}{\mathbf{q}}_*$ and $\mathbf{j}_e^{[\rho]}$. Thereafter, the solution is simple, particularly if we consider an idea realized in gas kinetics that defines the heating flux vector $\mathbf{q}$ by an expression of the form (Cercignani, 1988, p. 83)

$$\mathbf{q} \equiv \rho \frac{1}{2} \overline{\mathbf{c}^2 \mathbf{c}}. \tag{6.90}$$

Here **c** is the so-called *peculiar velocity* that describes the random deviation of the molecular velocity from the ordered motion with velocity **v**. The bar indicates, as usual, an arithmetic mean value. "The heating flux vector is nothing more than a gross manifestation of transfer of molecular kinetic energy" (Truesdell, 1984, p. 408). Assuming now that the specific momentum **i** is the relevant source of the local motion, then the ansatz

$$\mathring{\mathbf{q}}_* := \rho e_{\mathbf{q}} \mathbf{i} \tag{6.91}$$

for the heat flux density $\mathring{\mathbf{q}}_*$ immediately leads to an important intermediate result,

$$\mathbf{j}_e^{[\rho]} = \rho \frac{1}{2}\left(\mathbf{v}^2 - \mathbf{i}^2\right)\left[\mathbf{v} - \mathbf{i}\right] = -\rho e_{\mathbf{q}} \boldsymbol{\varphi}, \tag{6.92}$$

for the energy flux density $\mathbf{j}_e^{[\rho]}$. The structural similarity between Equations (6.92) and (6.90) is striking. Clearly, all limiting cases (6.87) are covered, provided that the dissipation is taken into account explicitly. This should be done according to the linear relationships (6.65) and (6.67) discussed before. Let us, therefore, define such a relation

$$\rho e_{\mathbf{q}} := t_{\mathbf{q}*} T_* \sigma \tag{6.93}$$

that reduces the specific heat energy $e_{\mathbf{q}}$ to the entropy production density σ by means of a transport coefficient represented by a characteristic relaxation time $t_{\mathbf{q}*}$.

The evaluation of (6.93) offers some characteristics of the Alternative Theory, but also, above all, several notable insights into the general ideas of transport phenomena. Starting with Equation (6.92) and using the entropy balance equation according to (5.81), an important expression is obtained:

$$\mathbf{j}_e^{[\rho]} = -t_{\mathbf{q}*} T_* \sigma \boldsymbol{\varphi} = -t_{\mathbf{q}} T \boldsymbol{\varphi} [\rho \mathbf{v} \bullet \nabla s + \rho \partial_t s + \nabla \bullet \mathbf{j}_s]. \tag{6.94}$$

In the last part of (6.94), the term $t_{\mathbf{q}*} T_*$ is replaced by $t_{\mathbf{q}} T$, indicating that the characteristic time $t_{\mathbf{q}}$ is presumed to be also related to the hypothetical state at rest.

Formula (6.94) is crucial for the theory of constitutive relations inasmuch as it demands the specification of the class of systems to be treated. The last term of the right-hand side contains the entropy flux density, which through the divergence theorem (6.44) links all other relevant flux densities of the system. Thus, both pure fluids and mixtures formally obey the same Equation (6.94). However, the main difference consists of different equations of state valid for the entropy and the thermal properties of state and implicitly depends on mass fractions in the case of a mixture. As to the relaxation time $t_{\mathbf{q}*}$, special dependencies are needed in addition. Above all, however, the influence of the heat flux density as well as that of the diffusional flux vectors has to be considered. Expressed in mathematical notation, this relation concerning fluxes becomes

$$\mathring{\mathbf{q}}_* := T_* \mathbf{j}_s + \sum_j \mu_{*j}\, \mathbf{j}_j, \tag{6.46}$$

where a closure condition holds for the diffusional flux densities $\mathbf{j}_j$.

To underscore this deduction, a one-component single-phase system will be treated first. On one hand, the problem of determining $\mathbf{j}_e^{[\rho]}$ by means of (6.94) is strongly influenced by the divergence term $\nabla \bullet (\mathring{\mathbf{q}}_*/T_*)$ for diffusion assumed to be absent. On the other hand, the heat flux density $\mathring{\mathbf{q}}_*$ may be evaluated correctly, at least in principle, following its definition (6.91). Inserting again the entropy balance equation for σ, the following equation results:

$$\mathring{\mathbf{q}}_* = t_{\mathbf{q}*} T_* \sigma \mathbf{i} = t_{\mathbf{q}*} T_* \mathbf{i}\left[\rho \mathbf{v} \bullet \nabla s + \rho \partial_t s + \nabla \bullet \frac{\mathring{\mathbf{q}}_*}{T_*}\right]. \tag{6.95}$$

For the sake of simplicity, we obtain (6.97) from (6.95) using the vector rule

$$\nabla \bullet \left[\mathring{\mathbf{q}}_* T_*^{-1}\right] = T_*^{-1} \nabla \bullet \mathring{\mathbf{q}}_* - T_*^{-2} \mathring{\mathbf{q}}_* \bullet \nabla T_* \tag{6.96}$$

together with (6.91):

$$\begin{gathered} \mathring{\mathbf{q}}_* - t_{\mathbf{q}*} \mathbf{i} \nabla \bullet \mathring{\mathbf{q}}_* = -\lambda_* \nabla T_* + t_{\mathbf{q}*} \rho \{\mathbf{i}\mathbf{v}\} \bullet T_* \nabla s + t_{\mathbf{q}*} \rho \mathbf{i} T_* \partial_* s \\ \lambda_* := t_{\mathbf{q}*} T_*^{-1} \rho e_{\mathbf{q}} \mathbf{i}^2. \end{gathered} \tag{6.97}$$

This result, gained without disregarding any relations, is highly significant. Transport phenomena are plainly proven to be a part of Gibbs–Falkian dynamics by mapping the description of the system in question from the Gibbs space onto a space of events where processes may be controlled by means of time- and space-coordinates.

The structure of both Equations (6.97) is very complex and reflects several physically different effects for which a simple and visual explanation seems illusionary. However, we can simplify Equations (6.97) drastically for an important application in practice: Supposing that the flow velocity $\mathbf{v}$ may be ignored due to conditions concerning creeping flows or near no-slip walls, then the equations

$$\begin{gathered} \mathring{\mathbf{q}}_* + t_{\mathbf{q}*} \varphi \nabla \bullet \mathring{\mathbf{q}}_* = -\lambda_* \nabla T_* - t_{\mathbf{q}*} \rho \varphi T_* \partial_t s \\ \lambda_* := \frac{1}{2} t_{q*} T_*^{-1} \rho \end{gathered} \tag{6.98}$$

follow approximately from (6.97). We may easily prove that $(6.98)_2$ is the well-known *heat conductivity*. In agreement with one of the main principles of gas kinetics, λ_* is positive for all real cases, provided that $t_{\mathbf{q}*} > 0$ holds for all values of the relaxation time.

To conclude this chapter, we should mention that Equation (6.98) does not confirm Fourier's classical law of heat conduction in solids. Also, note that the mathematical structure of (6.98) is inconsistent with that of the prominent phenomenological relationship proposed by Cattaneo in 1948 in terms of elementary kinetic considerations, viz.

$$\overset{\circ}{\mathbf{q}} + \theta\, \partial_t \overset{\circ}{\mathbf{q}} = -\, k\, \nabla T, \tag{6.99}$$

where θ denotes the relaxation time of the heat flux vector $\overset{\circ}{\mathbf{q}}$ and the corresponding value of the thermal conductivity is symbolized by k (Astarita, 1990, p. 185). Furthermore, the more general expression (6.97) takes into account some cross-effects between flow stress and energy fluxes that are approximately considered in Grad's famous kinetic theory of nonuniform gases.

However, there is a fundamental discrepancy between the classical results and this new relationship (6.98) that is fully associated with the non-equilibrium level of the processes under investigation. This level cannot be reduced further. Therefore, Equation (6.98) must be used for an iteration procedure, starting with proper approximations. It is true, nevertheless, that this method is required to determine the entropy flux vector $\mathbf{j}_s$. But in connection with any local energy transfer, the latter quantity appears only as a part of the relevant energy flux density $\mathbf{j}_e^{[\rho]}$ according to Equation (6.94). This energy flux vector may be related to the actual energy flux density $\mathbf{j}_{e*}^{[\rho]}$ with Equation (6.73) given above

$$\mathbf{j}_e^{[\rho]} = \mathbf{j}_{e*}^{[\rho]} + (p_* - p)\mathbf{v} \tag{6.73}$$

Using now Equation $(6.82)_1$, the pressure difference will be eliminated to furnish the relationship

$$\mathbf{j}_e^{[\rho]} = \mathbf{j}_{e*}^{[\rho]} + \boldsymbol{\tau}_* \bullet \mathbf{i}, \tag{6.100}$$

which is comparable with the structure of Equation (6.18) attributing the non-equilibrium quantities p_* and T_* respectively to their properties p and T of the corresponding hypothetical state at rest. This means that $\mathbf{j}_e^{[\rho]}$ would only be realized, if, at least in principle, the identity could be achieved for the limiting case $\mathbf{i} \rightarrow \mathbf{0}$.

Chapter 7 | General Equation of Motion and Its Approximations

"Everybody believes it completely because mathematicians think it is an experimental fact, and experimentalists think it is a theorem in mathematics."—***H. Poincaré***

7.1 Elementary Picture of Dissipation

In the preceding chapter we established a substructure on which to build a realistic theory of continuous bodies. Founded on Falk's concept of a general theory of systems, the main principles and axioms of the Alternative Theory lead to a set of theorems referring to the theory as a whole as well as to some special cases of constitutive equations. Let us now continue with the analysis of these equations, through which the flow friction problem should be treated. In short, the question is how to determine the viscous pressure tensor $\boldsymbol{\tau}$ introduced by Equation $(6.82)_2$.

Some of the relationships derived in Chapter 6 pertain and are compiled again for convenience here.

$$\left.\begin{aligned} &\rho\, D\mathbf{v} = \rho\mathbf{f} - \nabla \cdot \Pi + \partial_t \rho \boldsymbol{\varphi} && (6.60) \\ &\boldsymbol{\Pi} := p_* \mathbf{1} - \boldsymbol{\tau}_* = p\mathbf{1} - \boldsymbol{\tau} && (6.86) \\ &p := p_* - \frac{1}{2}\rho\,(\boldsymbol{\varphi} \cdot \mathbf{i}); \qquad \boldsymbol{\tau} := \boldsymbol{\tau}_* - \frac{1}{2}\rho\,(\boldsymbol{\varphi} \cdot \mathbf{i})\mathbf{1} && (6.82) \\ &\boldsymbol{\tau}_* := \frac{1}{2}\rho\, \mathbf{v}\boldsymbol{\varphi}. && (6.57) \end{aligned}\right\} \qquad (7.1)$$

In Sections (6.1) and (6.3) we explained two properties of the dissipation velocity $\boldsymbol{\varphi}$: one refers to the mere formal aspects of $\boldsymbol{\varphi}$, whereas the other relates to its geometric connections with $\mathbf{i}$ and $\mathbf{v}$ [as given by (6.64)]. The compilation in (7.1) points out a third property. This compares with ideas applied to the kinetic theory of gases.

Aside from the two pressure terms, p_* and p, and the mass density, ρ, the set (7.1) contains three different quantities that use the same unit of velocity. They enter into the Alternative Theory through the constitutive properties discussed in the last section. Within that framework, a characteristic correspondence is observed between the propounded concept of flux densities like $\mathbf{j}_e^{[\rho]}$ and some well-known integrals defining the same fluxes in any kinetic theory of gases. Those integrals also work with three properties, each of which have the same unit of velocity. An example using the heating flux vector is given by Equation (6.90), which concerns only two physically divergent terms of velocity: an ordered one and a random one. The third kind is simply formed by the difference of the other two.

It seems reasonable to pursue a similar interpretation for the three quantities **i**, **v**, and $\boldsymbol{\varphi}$. Surely it is true that statistical arguments cannot be used in macroscopic theory. Thus, when attempting to expound actually occurring processes, the notion of dissipation is preferred over an explanation by random events. Notwithstanding, there seems to be a grave discrepancy between the two concepts caused mostly by the a priori assumption of an unrealistic model of matter consisting only of Eulerian mass-points (like billiard balls). In such a model there is no difference between the specific momentum of a particle and its molecular velocity. But within the framework of the Alternative Theory this difference is essential because of the prominent role of the momentum variable for all systems in motion. This is true especially with regard to the notion of work and its connection with the idea of information as interpreted by Stonier. His theses were discussed extensively in Section 4.4, so we will simply summarize them here with the following quotation.

> To obtain useful work, the applied energy must either contain information itself, or act on some organized object or device which acts as an "energy transducer" (or both). Such a device may be a fairly "simple" and inert body, such as a dugout canoe with a sail being propelled by the wind (although both the canoe and the sail are highly organized structures containing an enormous amount of information). Or, by contemporary standards, it may be highly complex, such as a nuclear reactor generating high pressure steam. In all cases, energy transducers possess organization—whether they be an individual atom, a protein molecule, a photovoltaic cell, a membrane, a cell, a battery, a steam engine, a nuclear reactor, or whatever. All exhibit organization. All possess structural information without which they could not act as energy transducers. (Stonier, 1990, p. 96)

Since every volume element of any flow is axiomatically presumed to be an open system that is able to interact with its surroundings by heating and work rates, Stonier's conjecture refers mostly to the fluid itself. In other words, the application of work rates causes particles to be bound into fixed patterns and ordered motion, whereas the application of heat to a system brings about a local randomization of the flow region—it produces disorder. This descriptive interpretation is supported by the formulas (6.45) and (6.97–6.98), whereby work can directly be identified as a process property controlled by the locally acting momentum flux density $\boldsymbol{\Pi} := p_* \mathbf{1} + \boldsymbol{\tau}_*$. It is noteworthy that $\boldsymbol{\Pi}$ does not vanish in the case of kinetic equilibrium. Thus, useful information stored in the fluid can be locally exchanged with the immediate surroundings by the flow pressure triggered by the respective momentum alone. In sharp contrast to this, the heat flux density $\mathring{q}_*$ completely tends toward zero values if dissipationless flows are permitted to be approximately realized.

The exchange of information by work is supposed to be executed by an interplay between the two leading contributions of the momentum balance, viz. the local pressure gradient and the momentum rate. This operation, however, is permanently influenced by many specific intermolecular interactions that mainly depend on the density of the fluid and manifest themselves by the equations of state used. Furthermore, every well-posed problem of non-equilibrium physics is specified by some boundary conditions affecting the flow as a whole. Other local influences follow from additional variables such as mass fractions or temperature.

These various influences induce a characteristic drift between the local velocity $\mathbf{v}$ and the specific momentum $\mathbf{i}$. Note that $\mathbf{v}$ itself is immediately coupled to $\mathbf{i}$ according to $\mathbf{v} = (\partial \mathbf{e}/\partial \mathbf{i})_{\mathbf{r},s,\rho,\omega,\,\dots}$. The classical identity $\mathbf{v} = \mathbf{i}$ will only arise if those various influences do not exist or cancel. It is exactly this case that is equivalent to the reversible limit. Real events, however, can never compensate this difference $\mathbf{v} - \mathbf{i}$, which forms a third velocity $\boldsymbol{\varphi}$ in quite a natural way. Obviously, an average molecular velocity explained by random motion in gas kinetics finds its actual counterpart in the macroscopic flow velocity $\boldsymbol{\varphi}$, mastered in suitable proportions by various kinds of dissipative contributions. The latter manifest themselves by this drift $\boldsymbol{\varphi}$, which reveals the genuine mechanism of dissipation in a very transparent manner.

To examine its basic phenomenon more closely, let us look at a somewhat oversimplified example: A car gets its momentum from the driving power commonly transmitted from the motor to the rear wheels. Usually, the direction of the momentum coincides with the line of motion determined by the speed of the car and fixed by the current adjustment of the front wheels. But in special situations like aquaplaning, both directions diverge. The resulting effect from both directions can instantaneously cause quite an unexpected behavior of motion. Such consequences, always unwanted for the driver, correspond to the widespread prejudice to which dissipation is commonly exposed. But for this example the limits have been reached. Many applications follow from the striking fact that dissipation is indeed often responsible for new and even innovative developments occurring in non-equilibrium events.

7.2 General Equation of Motion

Let us start with the deduction of the pressure tensor in more common terms such as velocity gradients and so on. The first step concerns the non-equilibrium viscous pressure tensor $\boldsymbol{\tau}_*$ defined by Equation (6.57). However, to determine $\boldsymbol{\tau}_*$ exactly, we can use (the specific energy of) the stress dissipation e_τ defined by (6.65) as follows:

$$e_\tau := -\frac{1}{2}(\mathbf{v}\bullet\boldsymbol{\varphi}); \qquad \rho e_\tau := t_{\tau *} T_* \sigma .$$

Combining both definitions and inserting the entropy balance equation for the entropy production density σ, we first obtain

$$-\frac{1}{2}\rho\,(\boldsymbol{\varphi}\bullet\mathbf{v}) = t_{\tau *} T_* \rho\,(\nabla s\bullet\mathbf{v} + \rho^{-1}\nabla\bullet\mathbf{j}_s + \partial_t s)\,. \tag{7.2}$$

Then, left-side multiplication by $\mathbf{v}$ alters (7.2) to

$$-\frac{1}{2}\rho \mathbf{v}\,(\boldsymbol{\varphi}\bullet\mathbf{v}) \;=\; t_{\tau*}T_{*}\rho\,[\,\mathbf{v}\,(\nabla s\bullet\mathbf{v}) + \mathbf{v}\,(\rho^{-1}\nabla\bullet\mathbf{j}_s + \partial_t s)\,]\,,$$

which leads to the relationship

$$-\frac{1}{2}\rho\,\{\mathbf{v}\boldsymbol{\varphi}\}\bullet\mathbf{v} \;=\; -t_{\tau*}T_{*}\rho\,[\,\{\mathbf{v}\nabla s\}\bullet\mathbf{v} + (\rho^{-1}\nabla\bullet\mathbf{j}_s + \partial_t s)\,\mathbf{1}\bullet\mathbf{v}]\,.$$

This expression contains three scalar products for which the elementary tensor rules $\mathbf{v} = \mathbf{1}\bullet\mathbf{v} = \mathbf{v}\bullet\mathbf{1}$ and $\mathbf{v}(\boldsymbol{\varphi}\bullet\mathbf{v}) = \{\mathbf{v}\boldsymbol{\varphi}\}\bullet\mathbf{v}$, laid down in (6.77), have been applied.

By canceling the factor $\mathbf{v}$ on both sides of the last expression, the desired result will appear on the left side:

$$\boldsymbol{\tau}_{*}^{\text{tot}} \;=\; -t_{\tau*}T_{*}\rho\left[\;\{\mathbf{v}\nabla s\} + \left\{\left(\frac{1}{\rho}\nabla\bullet\mathbf{j}_s + \partial_t s\right)\mathbf{1}\right\}\right]. \tag{7.3}$$

In considering this result, bear in mind the following comments.

1. In accordance with the basic assumptions of the Alternative Theory, Equation (7.3) represents a general relationship for the non-equilibrium viscous pressure valid for nonrelativistic systems. Differences between various classes of systems may be found in different expressions for the entropy flux density $\mathbf{j}_s$.
2. The deduction of (7.3) does not assume the symmetry condition (6.79) concerning $\boldsymbol{\tau}_{*}$. Therefore, Equation (7.3) comprises the complete contribution of the entropy production σ to the viscous pressure tensor. For this reason, the superscript "tot" indicates this general case.
3. Due to the no-slip condition $\mathbf{v} = \mathbf{0}$ for which $\boldsymbol{\tau}_{*}$ vanishes by definition, the empirical function $t_{\tau*}$ is assumed to be proportional to the flow velocity $\mathbf{v}$.
4. Flow friction requires symmetry of $\boldsymbol{\tau}_{*}$ as an immediate consequence of (7.3). This may be realized by the symmetric dyad

$$\{\mathbf{v}\nabla s\} \;=\; \{\mathbf{v}\nabla s\}^{T} \;=\; \begin{bmatrix} v_1\partial_1 s & v_1\partial_2 s & v_1\partial_3 s \\ v_2\partial_2 s & v_2\partial_2 s & v_2\partial_3 s \\ v_3\partial_3 s & v_3\partial_3 s & v_3\partial_3 s \end{bmatrix} \tag{7.4}$$

representing an ensemble of six quantities. Each of its elements marks a velocity component, weighted with the locally assigned gradient $\partial_r s$ of the specific entropy ($r = 1, 2, 3$). This result clearly demonstrates the close connection between motion and its special structure organized by the entropy variable of the system. Primarily, entropy gradients dominate local flow friction; velocity gradients do not.

Nevertheless, it should be stated that (7.4) is a rather formal result. However, the question arises how to split the second-order tensor $\boldsymbol{\tau}_{*}^{\text{tot}}$ into its symmetric part $\boldsymbol{\tau}_{*}$ and its antisymmetric part $\boldsymbol{\tau}_{*}^{a}$. Are there any mathematical conditions to accomplish this ? And if so, what would be their physical meaning?

The occurrence of the entropy flux vector $\mathbf{j}_s$ as part of a second-order tensor reminds one of a conjecture ascribed to Pierre Curie in connection with his work on the piezoelectricity of crystals. This conjecture deals with the so-called Curie theorem, sometimes called the Curie principle (even though the words *theorem* and *principle* are not synonyms). [However, it is interesting that, if one follows the Curie principle through the chain of literature references back to its origin, one will finally find the report by Curie in which nothing even remotely similar to that principle is stated (see Truesdell, 1984, p. 389).]

Two different formulations of the principle may be found (Astarita, 1990, p. 194):

> *Strong form:* "Fluxes and thermodynamic forces of different tensorial character do not interfere."
>
> *Weak form:* "Forces and fluxes that correspond to a coupling of tensors whose orders differ by an odd number do not occur."

Trivial counterexamples to both statements can be found. Thus, Truesdell (1984, p. 388) recommended considering the isotropic form of Fourier's law, with a thermal conductivity that depends linearly on temperature corresponding to

$$\mathbf{q} = k_0\,(1 + k_1 T)\,\nabla T,$$

where the heat flux vector $\mathbf{q}$ (i.e., an order-one tensor) is coupled with the scalar (zero-order tensor) quantity, viz. temperature T. The selected material is characterized by the two constant parameters k_0 and k_1.

The crucial point is that there is no such thing as a Curie theorem in algebra. But in physics, too, it is hard to justify this principle by means of any experiences or mathematical theorems referring to symmetry relations. For this reason it may safely be disregarded in this treatise.

Let us continue to express the viscous pressure tensor $\boldsymbol{\tau}_*$ in a proper form that allows the application of the properties of the hypothetical state at rest. This option is based on $(6.82)_2$ and leads, along with (6.63), to the formula

$$\boldsymbol{\tau} = \boldsymbol{\tau}_* - \frac{1}{2}\rho\,(\mathbf{i}\bullet\boldsymbol{\varphi})\,\mathbf{1} = \boldsymbol{\tau}_* + \rho e_{\mathbf{i}}\mathbf{1} = \boldsymbol{\tau}_* + \rho e_{\tau}\mathbf{1} - \rho e_{\varphi}\mathbf{1}. \tag{7.5}$$

Using the definitions of the dissipation energies e_τ and e_φ, (7.5) will then become

$$\boldsymbol{\tau} = \boldsymbol{\tau}_* - \frac{1}{2}\rho\,(\mathbf{i}\bullet\boldsymbol{\varphi})\,\mathbf{1} = \boldsymbol{\tau}_* + \rho e_{\mathbf{i}}\mathbf{1} = \boldsymbol{\tau}_* - \rho\frac{1}{2}\,(\mathbf{v}\bullet\boldsymbol{\varphi})\,\mathbf{1} + \rho\frac{1}{2}\,(\boldsymbol{\varphi}\bullet\boldsymbol{\varphi})\,\mathbf{1} \tag{7.6}$$

where the first and second term on the right-hand side are given by Equations (7.3) and (7.2), and the last term may be substituted by the identity $\frac{1}{2}\rho\varphi^2 \equiv t_{\varphi*}T_*\sigma$. Thus, (7.6) becomes

$$\begin{aligned} \boldsymbol{\tau} = {} & -t_\tau T\rho\left[\,\{\mathbf{v}\nabla s\}^T - \{\,(\mathbf{v}\bullet\nabla s)\,\mathbf{1}\}\,\right] \\ & + t_\varphi T\rho\left\{\left(\rho^{-1}\nabla\bullet\mathbf{j}_s + \mathbf{v}\bullet\nabla s + \partial_t s\right)\mathbf{1}\right\}. \end{aligned} \tag{7.7}$$

Note that the substitution $t_* T_* := t\,T$ is assumed to be valid for both relaxation times $t_{\varphi *}$ and $t_{\tau *}$.The required symmetry of $\boldsymbol{\tau}_*$ is indicated by using the transposed tensor $\{\mathbf{v}\nabla s\}^T$ according to (7.4). However, we still need to consider the antisymmetric part $\boldsymbol{\tau}_*{}^a$.

The relationship (7.7), proven to be appropriate for the hypothetical state at rest, is equipped with some remarkable properties:

1. The viscous pressure tensor $\boldsymbol{\tau}$ does not vanish for no-slip conditions $\mathbf{v} = \mathbf{0}$: The *wall shear stress* τ_w appears.
2. Aside from a term representing unsteady entropy effects, the scalar τ_w is mainly determined by the local entropy flux density $\mathbf{j}_s$, which is given by (6.98) for a one-component single-phase system.
3. There exist two transport coefficients substantiated by the times $t_{\varphi *}$ and $t_{\tau *}$:

$$\beta_\tau := t_\tau T \rho s; \qquad \beta_\varphi := t_\varphi T \rho s. \tag{7.8}$$

These coefficients may be considered viscosities like the ones introduced by Maxwell more than a hundred years ago. However, Maxwell preferred the definition $\beta_M := t_M p$ (Truesdell, 1984, p. 453). The definitions (7.8) agree with Maxwell's concept for the case where the specific entropy s is replaced by the gas constant $\check{R}$ of the fluid, believed to be an ideal gas subject to the thermal equation of state: $p = \check{R}\; T\, \rho$. Note that both material parameters β_τ and β_φ are proven to be functions of state depending only on the local values of the thermal variables. The most remarkable result is the interpretation of the viscosity term as a special measure of the corresponding absolute entropy density ρs. Furthermore, Equation (7.7) reveals that flow friction and heat conductivity as well as diffusion are mutually joined by the actual entropy flux density $\mathbf{j}_s$, which must always be considered. Although the viscous pressure tensor depends in a distinctive way on the local gradients of the specific entropy s, it still seems useful to underline the explicit dependency of the tensor $\boldsymbol{\tau}$ on the pertaining velocity gradients. Indeed, such a representation corresponds with tradition and also permits a detailed comparison with other results. For this reason, the dyad $\{\mathbf{v}\,\nabla s\}$ in (7.7) will be expanded according to Equation (A.21) of Appendix 2.2, viz.

$$\{\mathbf{v}\nabla s\} = -s\,\{\nabla \mathbf{v}\} + \{\nabla s \mathbf{v}\} + \{\,[\nabla s \times \mathbf{v}] \times \mathbf{1}\} \tag{7.9}$$

where $\nabla\mathbf{v}$ denotes the velocity gradient. Here two terms appear in the theory: The term $\{\nabla s\mathbf{v}\}$ is a second-order tensor, indicating gradients of entropy-weighted velocity components. The last term of the identity (7.9) refers to the *axial vector* $\{\nabla s \times \mathbf{v}\}$ and must be dropped to guarantee the symmetry of the non-equilibrium viscous pressure tensor $\boldsymbol{\tau}_*$. This is correct, because an axial vector can never be part of any symmetric tensor. Conversely, only the symmetric part includes scalar properties expressed by one-third of the trace of the complete tensor $\boldsymbol{\tau}_*{}^{\text{tot}}$ that must be subdivided. As a consequence, an important condition for the antisymmetric part $\boldsymbol{\tau}_*^{\text{a}}$ of $\boldsymbol{\tau}_*{}^{\text{tot}}$ results:

$$\boldsymbol{\tau}_*^{a} = t_\tau T \rho\{[\nabla s \times \mathbf{v}] \times \mathbf{1}\} \equiv \mathbf{0}. \tag{7.10}$$

The corresponding symmetry of $\boldsymbol{\tau}_*^{tot}$ is well-founded by (6.79) and will be discussed in Section 7.4 with regard to the conservation of angular momentum.

Equation (A.15) of Appendix 2.2 yields a tensor rule for the scalar product $(\mathbf{v} \bullet \nabla s)$:

$$(\mathbf{v} \bullet \nabla s) = (\nabla \bullet s\,\mathbf{v}) - (s \nabla \bullet \mathbf{v}).$$

Inserting the last two identities into (7.7), the viscous pressure tensor $\boldsymbol{\tau}$ may then be calculated by the relationship

$$\begin{aligned}\boldsymbol{\tau} = {} & \beta_\tau \langle \{\nabla \mathbf{v}\} - s^{-1}\{\nabla s \mathbf{v}\}\rangle + \left(\beta_\tau + \beta_\varphi\right)\left\{\left(s^{-1}(\nabla \bullet s\mathbf{v}) - (\nabla \bullet \mathbf{v})\right)\mathbf{1}\right\} \\ & + \beta_\varphi \{((\rho s)^{-1} \nabla \bullet \mathbf{j}_s + \partial_t \ell n s)\,\mathbf{1}\},\end{aligned} \tag{7.11}$$

which includes the Maxwell viscosities β_τ and β_φ defined by (7.8). Equation (7.11) deserves further comment: Note that once more $\boldsymbol{\tau}$ vanishes for frictionless flow expressed by constant specific entropy s prevailing everywhere. This is true independent of the fact that the viscosities β remain finite. Note also that the mathematical structure of $\boldsymbol{\tau}$ reveals three dominating effects: (1) the potency of the velocity gradients, (2) the influence of compressibility, and (3) several transfer mechanisms subsumed under the local entropy flux density $\mathbf{j}_S$. Certain similarities to relationships commonly used in today's physics cannot be overlooked. Notwithstanding this fact, the obvious differences lead to two notable inferences:

1. All non-classical terms of Equation (7.11) are formulated by means of thermodynamic notions completely unknown in Cauchy's time.
2. For the first time, the material parameters β_τ and β_φ may be represented as individual properties proven to be proportional to the existing absolute entropy.

Combining Equations (6.60) with (6.86) yields the relationship

$$\rho D\mathbf{v} = \rho \mathbf{f} - \nabla p + \nabla \bullet \boldsymbol{\tau} + \partial_t \rho \boldsymbol{\varphi}, \tag{7.12}$$

which in connection with the solution (7.11) may be called the *Navier–Saint Venant equation of motion,* following the well-founded recommendation of Szabó (1987, p. 267). In 1834, Saint Venant submitted his "Mémoire sur la dynamique des fluides" to the French Academy of Science. His main thesis helped to propagate the idea that flow friction is due to relative velocities of neighboring fluid elements. Along with Cauchy's term of stress continuity, Saint Venant's construction of the stress tensor gave a realistic description of viscous flows for the first time (i.e., eleven years before Stokes' publication on internal friction of fluids in motion). The crucial point was the mathematically correct partition of the stress tensor into a hydrostatic pressure and an additional second-order tensor presuming the shear components to be proportional to the local slip velocities. Compared with the common equations of motion, (7.12) together with the $\boldsymbol{\tau}$-relation (7.11) is obviously much more complicated. Both equations not only contain additional quantities such as s or $\boldsymbol{\varphi}$, but also mathematically special structures with some interesting physical consequences. Those structures can be more easily explained with the $\boldsymbol{\tau}$-relation in its version (7.7), because we can distinguish between two important limiting cases:

no-slip condition:

$$\mathbf{v} \equiv \mathbf{0} \Rightarrow \tau_w = \beta_\varphi \left\{ \left((\rho s)^{-1} \nabla \bullet \mathbf{j}_s + \partial_t \ell n\, s \right) \mathbf{1} \right\} \Big|_w$$

isentropic limit: (7.13)

$$Ds \equiv 0 \Rightarrow \boldsymbol{\tau} = 0.$$

Both cases yield new results. The first case connects friction with the heat flux vector $\mathring{\mathbf{q}}_*$ and the non-equilibrium temperature T_* via the divergence theorem (6.46), based on the relation $\mathbf{j}_s = \mathring{\mathbf{q}}_* / T_*$ for existing wall constraints. The second case is traditionally concerned with the unrealistic assumption of a vanishing viscosity, whereas (7.13) results from the isentropic condition joined with the mathematically special structure of the $\boldsymbol{\tau}$-relation. But the most interesting aspect deals with the "in-between-case": Many flow fields show patterns that may be illustrated by a *near-wall region* and a *core region* far from any wall-like influences. Equation $(7.13)_1$ is approximately valid for the near-wall region.

Conversely, the core region, as part of dissipative flow phenomena, is never well described by $(7.13)_2$, particularly for turbulent flows. For many applications the core region may be defined by local values of the specific entropy s, which expands around a nearly constant value s_0.

The mathematical structure of (7.11) reveals then that the first two terms mutually cancel for a zero approximation. Furthermore, the unsteady change of entropy may be dropped. Consequently, the viscous stress is determined by the local entropy flux density $\mathbf{j}_s$ corresponding to

$$\tau_{\text{bulk}} = \beta_\varphi \left\{ \left((\rho s_0)^{-1} \nabla \bullet \mathbf{j}_s \right) \mathbf{1} \right\} \Big|_{\text{bulk}} . \tag{7.14}$$

Provided that this flow pattern legitimates both conclusions, typical fluid friction is then decisively constituted by local dependencies on temperature gradients and heat flux divergencies. Therefore, it is questionable (in my opinion) whether the commonly used turbulence models, established chiefly by means of velocity effects, are physically justified. This seems especially true for the popular theories of turbulence applied to incompressible fluids. This argument affects the foundation of theories in natural science. Based on the principles of the Alternative Theory, there is no reason to accept any matter model believed to be incompressible in every classical sense. But this is from a purely physical standpoint and is focused on the purpose of mathematically describing certain kinds of natural phenomena by means of a finite set of general axioms and proven theorems.

Within the range of purely mathematical interests such physical reasoning can sometimes seem irrelevant. Physical inconsistencies, often called paradoxes, are sometimes approved with a good chance of success when accompanied by a proper proof of any theorem concerning typically mathematical issues like existence and uniqueness. In this context it seems appropriate to refer to Lakatos' warning about the uncritical assessment of any experience with regard to theoretical considerations. In accordance with Popper, Lakatos claimed that it is impossible to prove any statement concerning facts by experiments. Such statements can only be derived from verbally expressed logic or mathematically proven theorems; they do not result from facts. "Experience can no more prove statement than can a blow with the fist on the table." (Lakatos, 1974, p. 97; author's translation).

Nevertheless, experiences are the natural source of every theory in physics. But in fact they enter such theories only in a rather indirect and abstract way. For instance, one main principle of this text is the presupposition that every mathematical theory in part and as a whole (1) has to be based on a finite number of data, and (2) should avoid the use of so-called metaphysical elements like the First and Second Law of Thermodynamics that in principle would demand an infinitive number of data for their verification. Even that would not suffice if the nature of the First and Second Laws were axiomatic only.

A further point aims at another class of problems of special interest for understanding continuum physics. For an initial boundary problem described by partial differential equations, the convention is to regard solutions in space and time as a way to predict the future evolution of the system in question. Computer-aided models for the evaluation of climatic evolutions are a well-known example. Such scenarios may be useful, especially for purposes of determining safety measures. But we cannot expect any concrete forecast from adequate solutions of partial differential equations that are only assumed to be true for natural events.

More than two hundred years ago, David Hume spoke quite openly about the human ability to make statements regarding future events by the principle of induction: He categorically denied such an ability. This inability concerns deterministic processes as well as stochastic ones (see Straub, 1990, p. 140). In the case of field theories, Humes's skeptical inference is easy to comprehend. Every problem in continuum physics is allegedly so well-posed that consistent boundary conditions can be prescribed. But for a solution in space and time these boundary values themselves can never be given in their actual dependencies on space and time. This is especially evident for long time scales and spatially complex evolutions. Of course, approximations by means of linear extrapolation are always justified for short time scales and simple space configurations of the boundaries. But for this crucial simplification it is often hard to procure reliable information about all values needed for the boundary conditions.

To illustrate this situation, let us compile all relevant quantities and equations for a complete field description of any non-equilibrium flow, where the fluid is assumed to locally be a one-component single-phase body-field system. This represents by far the simplest case imaginable.

Four kinds of quantities must be considered. The classification follows primarily from the equations used with preference for each quantity in question.

Field quantities: 5 velocity $\mathbf{v}$; mass density ρ; temperature T; pressure p; dissipation velocity $\boldsymbol{\varphi}$

Constitutive quantities: 8 viscous pressure tensor $\boldsymbol{\tau}$; energy flux density $\mathbf{j}_e^{[\rho]}$; entropy flux density $\mathbf{j}_s$; heat flux vector $\mathring{\mathbf{q}}_*$; relaxation times t_τ, t_φ, t_q; heat conductivity λ_*.

Non-equilibrium quantities: 7 non-equilibrium temperature T_*; heat dissipation e_q; specific momentum $\mathbf{i}$; dissipation energy e_φ; Maxwell viscosities β_φ, β_τ; entropy production density σ.

Thermal and caloric equations of state: 4 pressure $p(T, \rho)$; specific entropy $s(T, \rho)$; specific heat capacity $c_p(T, \rho)$; pressure coefficient $h_p(T, \rho)$

Because there are twenty-four relevant quantities, twenty-four relations in form of algebraic or differential equations are needed. In the preceding chapters all these relationships have been either derived from the basic principles without any approximation or expounded (like the equations of state). Additionally, a set of initial and boundary conditions have to be prescribed along with the knowledge of the occurring specific field force $\mathbf{f}$. For example, the following constraints may exist.

$$\begin{aligned} &\textit{Initial conditions:} && \mathbf{v}(\mathbf{r}; t=0) = \mathbf{v}_0(\mathbf{r}); \quad p(\mathbf{r}; t=0) = p_0(\mathbf{r}); \\ & && T(\mathbf{r}; t=0) = T_0(\mathbf{r}) \\ &\textit{Boundary conditions:} && \mathbf{v}(\mathbf{r} = \mathbf{r}_{\mathcal{B}}, t) = \mathbf{v}_{\mathcal{B}}(t); \quad p(\mathbf{r} = \mathbf{r}_{\mathcal{B}}, t) = p_{\mathcal{B}}(t); \\ & && T(\mathbf{r} = \mathbf{r}_{\mathcal{B}}, t) = T_{\mathcal{B}}(t) \end{aligned} \tag{7.15}$$

Appendix 3 contains a complete compilation of all relevant equations to be solved simultaneously for the system under consideration. Unfortunately, it seems unlikely that precise numerical solutions can be determined at present, even with the help of very fast and efficient computers and highly sophisticated iteration procedures.

We need not be pessimistic, but we cannot ignore the fact that the equations presented in Appendix 3 clearly contradict the prominent doctrine that things and events in this world are simple and straightforward. Contrary to such expectations even this simplest case of non-equilibrium flow processes is highly complex. Simple answers to questions about nature are an exception. In reality, there is no place for paradigms concerning physics of mass-points, harmonic oscillators, two-body motions, incompressible fluids, and quantum mechanics of integrable and reversible systems without the "scandalous" problem of potential energy (in de Broglie's sense), and so forth.

Of course, it is justifiable to use approximations and idealizations of theoretical concepts and mathematical models in physics based on general axioms and proven theorems. But they will inevitably lead to paradoxes if they are extended to conclusions generalized in a way that is inconsistent with their own principles. The "treatment of pressures in a moving Newtonian fluid" is an example discussed by Lu (1970). He pointed out that the hydrostatic pressure simply does not exist at points with deformation and that, similarly, the thermostatic pressure does not exist for the degenerate case of incompressible fluids. The latter point raises some intricate questions about the correct handling of the pressure term in flows. The usual extension of Gibbs's thermostatics to thermodynamics is another example. In this text the opposite way is proven to work with no inconsistencies, but at the cost of considerably more mathematical complexity and notational abstraction. Based on this discussion, we may make some inferences with regard to one of the most relevant, yet unsolved, problems in physics and technology: turbulence.

7.3 Some Remarks on Turbulent Flows

Following the traditional point of view, worldwide research on turbulent flows is mainly based on the Navier–Stokes equation of motion, applied to incompressible flows. In spite of great success in the understanding of turbulent patterns by special mathematical and experimental techniques, there is no indication that a generally accepted theory of turbulence may be expected in the near future. From the viewpoint of the set of equations given in Appendix 3, such a status quo is not surprising. It seems hard to imagine physically convincing reduction of this set to an equation of motion, along with the continuity equation of incompressible flows only.

This statement covers the primary conclusion drawn by Häfele and Schönauer from their exhaustive study on a transitional low Reynolds number turbulence model. The authors summarized their theoretical and numerical results as follows: "The model contains not enough physics to be extrapolated to another flow situation apart from the scaling flow. ... Remember once again that the ambition of the model was to predict automatically also the transition" (Häfele and Schönauer, 1985, pp. 6, 7). Note that Häfele's and Schönauer's study represents all basic research on turbulent flows that uses *computational fluid dynamics*. These authors first developed a voluminous program package for the calculation of three-dimensional laminar boundary layer equations in streamline coordinates (Schönauer and Häfele, 1987). This was adapted to incompressible turbulent boundary layers on three-dimensional configurations, which are given pointwise. Then, they applied their solver to a three-dimensional test configuration of an inclined 6:1 prolate spheroid for which precise measurements of the wall shear stress were available. The numerical calculations were performed with the help of a two-equation transitional low Reynolds number model proposed recently by Wolfshtein for turbulent three-dimensional flows. The model was scaled at a 0° angle of attack of the configuration; in other words, suitable results were obtained in comparison with adequate reference data published formerly by Rotta. But the application to a 10° body in three-dimensional flow did not produce satisfying physical results. In spite of numerically enormous efforts, the authors could not avoid the occurrence of negative values of the turbulence energy near the front stagnation point. Moreover, the intended transition phenomena did not evolve. It is noteworthy that this crucial result does not depend only on the turbulence model applied. Indeed, tests with the three-dimensional version of the low Reynolds number k-ϵ model of Lam, Bremhorst, and Rodi showed a similar behavior; that is, again the effect of negative turbulence energy occurred (Haas and Schönauer, 1987).

Häfele and Schönauer gave a simple but plausible explanation for this failure: Even very sophisticated turbulence models that should include transition, only work as an "interpolating" tool fixed between scaled measurements. The application to three-dimensional flows does not work when each streamline is characterized by its own history. Therefore, models of Wolfshtein's kind can actually find some sort of transition from laminar to turbulent two-dimensional flow only if it is properly scaled. To summarize, it is suspected that all models of this kind are unable to describe adequately the viscous sublayer that is believed to be the real source of turbulence. Notwithstanding, further progress of CFD will modify the present view particularly with respect to special applications (cf. O'Shea and Fletcher, 1994, p. 496).

In 1985, the same year that Schönauer and his collaborators published the first part of their treatise, a workshop on the state of the art of turbulence research was held in Mountain View, California. The group consisted of the ten leading turbulence researchers in the UK, the U.S., and France. The workshop was organized around three formal talks designed to stimulate discussion. The first focused on experimental methods, the second on turbulence modeling, and the last on numerical simulations. The report from that workshop is a serious reference for experimental work, theoretical descriptions of turbulent flows and their modeling, numerical simulations, and future research.

The main criterion used in the evaluation of a turbulence model is whether it is able to give adequate engineering accuracy for a large class of problems without any change in the constants. In view of these ambitious demands on engineering models, Schönauer's "disappointing numerical tests" confirmed the doubts of the workshop members about the commonly used k-ϵ models (Nixon, 1986, p. 16) by concrete experiences: Two of the most prominent turbulence models are in error as a rule and can only be accepted for the scaled reference case. Furthermore, his study offers some surmise how to proceed methodologically in using the numerical and experimental contributions to the analysis. Employing only such numerical and experimental data whose boundary conditions are consistent with each other is not recommended. This condition is commonly ignored by the "modelers," as Rotta (1984, p. 14) deplored, while the experts required it in their final report of the workshop mentioned above (Nixon, 1986, p. 24). Unfortunately, this depreciation of experiments in basic turbulence research has increased during the past decade.

Therefore, theoretical and numerical studies based on appropriate observations will be increasingly fewer. Obviously, this is true for separated flows, for which characteristic vortex and turbulence structures behind two-dimensional drag bodies can be demonstrated by reliable pressure measurements and, especially, precise LDA measurements (Leder, 1983). Founded on the time-averaged approach for large Reynolds numbers, the typical pressure distributions that are always observed in *dead water* can be explained by the characteristic vortex pair. This pair may be theoretically interpreted as a virtual body, as panel calculations of potential theory prove. Moreover, the zero-streamline enveloping the vortex pair can be understood as a displacement contour for the external flow. Because of limited physical knowledge about the structure of turbulent motions in dead water, physical information in the form of empirical laws are quite helpful for turbulence modeling. Thus, for example, the experiments reveal a typical pressure distribution with a pronounced pressure minimum that does not agree with any alleged pressure constancy. Currently, similar distributions of turbulence energy are also observed experimentally. They are anisotropic and characterized by some extrema. [By the way, the dead water turbulence structures behind a sharp-edged cylinder are quite similar to those of flat plates with correspondingly definite angles of attack (Geropp and Leder, 1984).]

Unfortunately, these profound experiments are mainly executed for the clear demonstration of phenomena that were of common interest in theories concerning so-called incompressible turbulence. The actual understanding of *compressible turbulence* is still quite limited, especially in comparison to what is known of turbulence in incompressible fluids. Thus, for instance, experiments with turbulent flows in compressible media "suffer from a lack of adequate facilities and instrumentation" (Friedrich, 1993, p. 15-1).

In free shear flows, there are no complicating effects due to solid walls such as boundary layers. These flows, therefore, are appropriate to use in the study of turbulence because they allow us to focus on the details of the energy transfer along the whole spectrum of space scales without external complications. The largest flow structures, or eddies, can be in the order of the extension of the flow regime. Conversely, the smallest meaningful scales for the flow are in the order of the *Kolmogorov length,* where molecular transfer effects like the viscosity of the fluid dominate its convection within an extension of microns or millimeters.

For convenience, to study these complex phenomena scientists designed *model eddies* in an incompressible fluid. Neglecting, for instance, the dynamics of the eddy ends or the process of eddy breakup, the local interplay between potential and kinetic energy can be quantified to obtain the order of magnitude of coherence effects occurring in a Navier–Stokes fluid. Details about *Rankine* and *Oseen* model eddies can be found in a paper by Ahlborn et al. (1991). The authors investigate what happens in eddy-eddy collisions and in interactions between eddies and flow.

We have strongly emphasized experiments in the field of turbulence research for the simple reason that the field should not be dogmatically determined by the Navier–Stokes equations alone (Oran and Boris, 1993, p. 34-2). In general, research on incompressible turbulence is founded exclusively on the Navier–Stokes equation of motion along with the incompressibility condition. Compressible turbulence is, strictly speaking, based on another set of notions that is not expressed only by a greater number of equations and an extended set of variables. Indeed, the main dissension may be found on the different meaning of all the comparable quantities used in both sets of Navier–Stokes equations. Thus, the direct comparison "reveals the classic problem regarding the pressure: In compressible flow it is a state variable, in incompressible flow it is a dynamic variable which enforces the incompressibility constraint" (Friedrich, 1993, p. 15–7). The same is true for energy and even for the flow velocity. Whereas in classical continuum mechanics the pressure is derived from the notable fact that classical mechanics operates with an incomplete notion of energy, internal energy is considered as an independent field quantity for compressible turbulence theory. For this reason the dynamical reference of the local flow velocity to the total energy e and the linear momentum $\mathbf{i}$ (i.e., $\mathbf{v} = (\partial \mathbf{e}/\partial \mathbf{i})_{\ldots}$ is also caused by the different energy terms.

To my knowledge, the consequences of these fundamental discrepancies have never been discussed with regard to the various peculiarities of any turbulence theory, such as *Reynolds* or *Favre averaging* or *Reynolds stress transport* and *turbulent flux equations*. Evidently, this widespread use of physical quantities sharply contradicts the principles of Falk's dynamics concerning the notion of universally physical quantities (see Sections 2.2 and 2.4).These remarks should be considered within the computation scheme presented in Appendix 3 and derived for non-equilibrium flow phenomena of simple fluids. They imply that, without uttering any criticism, a set of equations is declared a truism and believed to be a mathematically well-founded image of nature. Yet, the expositions in the preceding chapters disclosed the suspicion that the Navier–Stokes equations (at least for incompressible turbulence) were originally based on questionable mechanical principles due to the lack of the modern apparatus of notions like energy, enthalpy, entropy, equation of state, absolute temperature, Mach and mole numbers, or even the First and Second Law. No serious attempts have been made to develop an alternative idea or to formulate other theories of turbulence phenomena.

Let us illustrate this dogmatic kind of reasoning about the role of the Navier–Stokes equations by the sober statements given on the workshop mentioned above:

> The numerical simulations are either Large Eddy Simulations (LES) in which only the small eddies are modeled or Direct Numerical Simulations (DNS) in which the Navier–Stokes equations are solved numerically with no modeling[14] ... Turbulence research experiments face the serious shortcomings of the relative sparseness of data points and an inability to measure pressure fluctuations. With the development of more sophisticated, computer-controlled measuring techniques, experiments may be able to produce larger volumes of data, but can never be expected to yield simultaneous data at as many points as simulations. (Nixon, 1986, p. 2)

[14]Although modeling is needless, DNS require some specific conditions in the form of certain symmetry constraints along the flow boundaries. But sometimes such studies will give the impression that these intrinsically mathematical requirements are yet another kind of modeling.

Some results of the thorough workshop/discussion about the training and ability necessary to conduct good research in turbulence cannot be misunderstood. The main result was undoubtedly the conjecture that "turbulence research may be getting too goal-oriented, and this may not be productive in the long term" (Nixon, 1986, p. 32). Concern was also expressed that the brightest students, necessary to conduct turbulence research, are not entering the engineering sciences. But also "the blasé attitude of some of the physicists who worked in turbulence in the past is not to be encouraged" (Nixon, 1986, p. 31). Yet, the conclusion quite plainly drawn by the participants is striking and, above all, self-revealing: "The suggestion was made that the best turbulence research might be done by microelectronic engineers" (Nixon, 1986, p. 32).

Such recommendations reflect the state of the art during the 1980s. There is no doubt about their intentions, especially regarding the actual development in this field:

1. The relevance of key experiments declines rapidly when they are applied to decisively support theoretical considerations in turbulence research.
2. The Navier–Stokes equations are believed to be a mathematically correct mapping of natural phenomena in the sense that their solutions yield results within a broad range of parameter variations and of boundary conditions. Therefore, one has the impression that expensive test runs in wind tunnels may be substituted by DNS in abundance.

Of course, the last point has increased in significance during the 1990s. This success revealed, with the help of some ingenious and innovative studies, that certain patterns of turbulence structures are distinguished by certain characteristics that sometimes allow great simplifications of turbulence properties. For instance, these experiences have explained why it does not seem to be necessary to resolve the eddy structure down to the Kolmogorov scale and even below. Furthermore, new massive parallel computers allow very large and fast computations. Allegedly, the possibility now exists of doing LES or DNS even for large reactive systems (Oran and Boris, 1993, p. 34-12).

Nevertheless, it appears that all these approaches may in the end produce qualitatively acceptable but not quantitatively correct results: They may be faulty in some one-dimensional test problems or certain types of flow. But they often hit the mark with regard to multidimensional experiments or more complex types of turbulent flows. The question remains whether the "microelectronic engineers" will be able to analyze various possibilities of adequate corrections and manipulations incorporating all relevant theoretical, experimental, and numerical constraints. All things considered, it seems advisable to examine only the factual scope of validity, which is assigned a priori to the vectorial Navier–Stokes equation of motion. The problem refers to its true role as a mathematical basis, either of an appropriate tool for engineering applications or of a basic property of actually moving non-equilibrium flows. Surely the former purpose is purely pragmatic, whereas the latter aspect mainly aims at the fundamentals of physical understanding.

On the whole, it should be emphasized that the turbulence problem is not limited to questions concerning the Navier–Stokes equation. Turbulence models, as well as chaos-oriented interpretations, do have thoroughly independent meanings. This is not only true in face of new theoretical considerations and experimental research programs, but also with respect to the great efforts spent in computational fluid dynamics. For this reason, it is sufficient to note that the Alternative Theory presented in this book can only offer some specific contributions to the whole cosmos of modern turbulence research. Bear in mind that there are some topics involved that do not belong to the physical part of the turbulence problem. They can obviously be reduced to various sources of errors associated with every proper procedure of numerical mathematics. This holds a fortiori for the formulations of initial and boundary conditions, as well as symmetry constraints applied to the flow field under consideration. Unfortunately, it is difficult to assign the diverse errors of a numerical solution to their original sources, particularly since compensation effects may occur. Nevertheless, there is no denying that the physical fundamentals should be consistent with the generally accepted principles in physics.

7.4 Conservation of Angular Momentum

Falk and Ruppel (1983, p. 225) dealt at great length with this difficult problem in reference to the principles presented in the preceding chapters. For this reason it seems sufficient to focus the following study on the *angular momentum* and its conservation property applied to continuum physics. Let us first remember that the conservation of linear momentum is equivalent to the invariance of the field against any linear affine displacement (see Section 4.3). A similar significance has to be assigned to the conservation of the angular momentum as characteristic of the invariance of any field against rotational motions. As a consequence, every volume element of a fluid that can carry out either a translational or a rotational nonuniform motion will remain in the state of virtual and actual material stability. This necessary and sufficient condition is satisfied if the result of all the forces acting upon any volume, as well as of the moments of these forces, is zero for all events. The pertaining mathematical formulation will be presented with two common tensor notations that refer to an appropriate index notation or to the shorthand symbolic formalism usually practiced in this text.

As usual, the *total angular momentum* **J** per unit mass is divided into the *external* and *internal* angular momentum **L** and Ω:

$$J_{\alpha\beta} = \mathrm{L}_{\alpha\beta} + \Omega_{\alpha\beta}; \qquad \alpha, \beta = 1, 2, 3. \tag{7.16}$$

The external angular momentum $\mathbf{L} := \mathbf{r} \times \mathbf{i}$ of the unit mass is defined as the vector product of the specific momentum $\mathbf{i}$ multiplied by the position vector $\mathbf{r}$ with respect to the spatial system of coordinates:

$$\mathrm{L}_{\alpha\beta} := \mathrm{r}_\alpha \mathrm{i}_\beta - \mathrm{r}_\beta \mathrm{i}_\alpha. \tag{7.17}$$

Hence, the substantial time derivative is $\overset{\circ}{\mathbf{L}} := \mathbf{r} \times D\mathbf{i} - \mathring{\mathbf{r}} \times \boldsymbol{\varphi}$, using the basic relation (6.10), $\mathbf{i} = \mathbf{v} - \boldsymbol{\varphi}$. Because $\mathring{\mathbf{r}} = \mathrm{v}$, the cross product $\mathring{\mathbf{r}} \times \mathbf{v}$ vanishes, but at first glance $\mathbf{v} \times \boldsymbol{\varphi}$ does not. Such an inference, however, is premature: In accordance with Equation (A.19) of Appendix 2.2, valid for dyads, the cross product $\mathbf{v} \times \boldsymbol{\varphi}$ is directly proportional to $\boldsymbol{\tau}_*^{\mathrm{a}}$, that is, the antisymmetric part of the total non-equilibrium viscous pressure tensor $\boldsymbol{\tau}_*^{\mathrm{tot}}$. For this reason this cross product

$$\boldsymbol{\tau}_*^a = \frac{1}{2}\rho\,[\mathbf{v} \times \boldsymbol{\varphi}] \times \mathbf{1} \tag{7.18}$$

vanishes under the symmetry condition $\boldsymbol{\tau}_*^{\mathrm{a}} = \mathbf{0}$. Equation (7.18) is equivalent to the notable statement that the vectors $\mathbf{v}$ and $\boldsymbol{\varphi}$ are aligned in parallel. Evidently, this result confirms Equation (7.10), where the local flow velocity $\mathbf{v}$ is parallel to the respective entropy gradient ∇s. Of course, this fact indicates certain material constraints that decisively influence the physical behavior of the flow, subject to specific initial and boundary conditions. An immediate consequence of (7.10), using the tensor rule (A.22) of Appendix 2.2, is found with regard to a class of motions that is of central significance for classical fluid dynamics: flows distinguished by *irrational motion* (see Lamb, 1879, p. 32). This consequence is equivalent to the property of some portions of matter determined by means of a characteristic *velocity potential* (see Lamb, 1879, p. 18).

The consequence addressed above undoubtedly states that real flows never experience irrational motion. In mathematical terms the condition

$$\nabla \times \mathbf{v} \neq \mathbf{0} \tag{7.19}$$

holds for all fluid flows defined by M–G functions like Equation (6.31). We may presume that this equation constitutes most of the systems believed to be relevant in practice. It should be stressed also that (7.19) is proven without any exception to the axioms and theorems of the mathematical theory presented in this text.

Let us continue the analysis of the problem concerning the conservation of angular momentum. Right-side multiplication by $\mathbf{r}$ alters the momentum balance Equation (6.53) to

$$\rho\mathbf{L}^{\circ} = \mathbf{r} \times (\rho\mathbf{f} - \nabla \bullet \mathbf{JI}), \tag{7.20}$$

where $\mathbf{JI}$ stands for the momentum flux density and ρ for the mass density. Making use of the tensor rule in the symbolic form

$$\mathbf{r} \times \nabla \bullet \mathbf{JI} = \nabla \bullet (\mathbf{r} \times \mathbf{JI}) + \mathbf{JI}^T - \mathbf{JI}, \tag{7.21}$$

and of the index notation

$$r_\alpha\left(\partial \mathbf{JI}_{\beta\gamma}/\partial r_\gamma\right) - r_\beta\left(\partial \mathbf{JI}_{\alpha\gamma}/\partial r_\gamma\right) = \frac{\partial}{\partial \mathbf{r}_\gamma}\left[r_\alpha \mathbf{JI}_{\beta\gamma} - r_\beta \mathbf{JI}_{\alpha\gamma}\right] \tag{7.22}$$
$$+ \mathbf{JI}_{\beta\alpha} - \mathbf{JI}_{\alpha\beta}; \quad \alpha, \beta, \gamma = 1, 2, 3,$$

the rate equation of the external angular momentum (7.20) becomes:

$$\rho\mathbf{L}^{\circ} + \nabla \bullet (\mathbf{r} \times \mathbf{JI}) = \mathbf{r} \times \rho\mathbf{f} + \mathbf{JI} - \mathbf{JI}^T, \tag{7.23}$$

where the superscript T denotes the respective transposed tensor.

A relevant phenomenological flow property is the angular velocity $\boldsymbol{\omega}$ of the mass elements classified by a second characteristic fluid parameter. This average moment of inertia per unit mass θ is used to represent so-called *micromorphic continua* (with the Cosserat-continuum as a special material class). Occasionally, their affine microstructures are discussed (Truesdell, 1984, p. 169). The axial vector $\boldsymbol{\Omega} := \theta\boldsymbol{\omega}$ combines these two quantities by definition. This specific internal angular momentum is the local contribution of the fluid to the total angular momentum $\mathbf{J}$. The respective rate equation for $\boldsymbol{\Omega}$ may be expressed in the generally valid form (5.81),

$$\rho\overset{\circ}{\boldsymbol{\Omega}} + \nabla \bullet \mathbf{\Pi}_{\Omega} = \boldsymbol{\kappa}, \tag{7.24}$$

where the axial vector $\boldsymbol{\kappa}$ denotes the local production term; the angular flux density $\mathbf{\Pi}_{\Omega}$ is a second-order tensor.

The balance equation for the total angular momentum $\mathbf{J}$ is obtained if the balance of the internal angular momentum (7.24) is added to (7.23). Anticipating

$$\mathbf{\Pi} - \mathbf{\Pi}^T + \boldsymbol{\kappa} \equiv \mathbf{0}, \tag{7.25}$$

as conservation condition for $\mathbf{J}$, the following expression results:

$$\rho\overset{\circ}{J}_{\alpha\beta} + \frac{\partial}{\partial r_\gamma}\left[r_\alpha \Pi_{\beta\gamma} - r_\beta \Pi_{a\gamma} + \Pi_{\Omega, a\gamma}\right] = \rho\,(r_\alpha f_\beta - r_\beta f_\alpha)\,; \qquad \alpha, \beta, \gamma = 1, 2, 3. \tag{7.26}$$

The index notation, along with Einstein's summation convention, is used for the presentation of $\mathbf{J}$ as an antisymmetric tensor. Three aspects relating to the resulting sum are of interest:

1. The rate equation for $\mathbf{J}$ must correspond with the mathematical form (5.81) of every conservation law:

$$\rho\overset{\circ}{\mathbf{J}} + \nabla \bullet \mathbf{j}_{\mathbf{J}^\circ} = b_{\mathbf{J}^\circ}. \tag{7.27}$$

   In other words, the internal source density $b_{\mathbf{J}^\circ}$ vanishes. Since the field force $\rho\mathbf{f}$ must be interpreted as an external source acting on the continuum, the principle of the total angular momentum conservation can only be observed locally by taking into consideration that the condition (7.25) is satisfied. Hence, the production vector $\mathbf{k}$ may be expressed in index notation:

$$\kappa_\gamma = \Pi_{\beta\alpha} - \Pi_{\alpha\beta}\,; \qquad \alpha, \beta = 1, 2, 3, \quad \text{cyclic.} \tag{7.28}$$

2. To satisfy (7.25), the quantity $\boldsymbol{\kappa}$ must be identified with an axial vector that is proportional to the antisymmetric part $\mathbf{\Pi}^a$ of the momentum flux density $\mathbf{\Pi}$. The superscript a denotes this antisymmetric tensor. Corresponding to the tensor rule

$$\mathbf{\Pi} - \mathbf{\Pi}^T = 2\,\mathbf{\Pi}^a, \tag{7.29}$$

   the simple relation $\boldsymbol{\kappa} = -2\,\mathbf{\Pi}^a$ must hold.
3. According to Equation (6.53) the pressure terms contained in the difference $\mathbf{\Pi} - \mathbf{\Pi}^T$ cancel each other. Hence, the identity $\mathbf{\Pi} - \mathbf{\Pi}^T = \boldsymbol{\tau}_* - \boldsymbol{\tau}_*^T$ holds. Using

the dyad rule corresponding to Equations (A.19) and (A.20) of Appendix 2.2 for the two polar vectors **v** and **w** (viz. $\{\mathbf{1} \bullet \mathbf{vw} - \mathbf{wv}\} = \mathbf{v} \times \mathbf{w} \times \mathbf{1}$), the relationship

$$\mathbf{1} \bullet \{\boldsymbol{\tau}_* - \boldsymbol{\tau}_*^T\} = \frac{1}{2}\rho\, \mathbf{1} \bullet \{\mathbf{v}\boldsymbol{\varphi} - \boldsymbol{\varphi}\mathbf{v}\} = \frac{1}{2}\rho\, [\mathbf{v} \times \boldsymbol{\varphi}] \times \mathbf{1} = \boldsymbol{\kappa} \bullet \mathbf{1} \qquad (7.30)$$

then results as an expression of far-reaching relevance. It identifies the production density $\boldsymbol{\kappa}$ of the balance (7.22) with the two flow quantities **v** and $\boldsymbol{\varphi}$, which are determined by an actual non-equilibrium behavior. Note that $\boldsymbol{\kappa} \bullet \mathbf{1} = \boldsymbol{\kappa}$ is valid.

Obviously, this connection between **v** and $\boldsymbol{\varphi}$ reveals that internal rotation is conditioned not only by fluid properties like the material parameter θ but also by the local dissipation velocity $\boldsymbol{\varphi}$. The inverse inference seems even more interesting: With $\theta \neq 0$, real flows are never isentropic or frictionless; that is, $\boldsymbol{\varphi}$ is always a finite function for which the cross product $[\mathbf{v} \times \boldsymbol{\varphi}]$ does not disappear. In this "normal" case, the commonly applied symmetry condition $\boldsymbol{\tau}_*^a = \mathbf{0}$ is faulty so that both vectors **v** and **i** are pointing in different directions.

Furthermore, two limiting cases must be noted with regard to balance (7.24):

1. For no-slip conditions $\mathbf{v} = \mathbf{0}$, the divergence term of the reduced rate equation for $\boldsymbol{\Omega}$(viz. $\rho\overset{\circ}{\boldsymbol{\Omega}} + \nabla \bullet \mathbf{JI}_{\Omega} = \mathbf{0}$) contains a dissipative contribution to the wall friction caused by angular momentum.
2. For dissipationless flow, that is with $\boldsymbol{\varphi} \equiv \mathbf{0}$, the balance of the internal rotation $\rho\overset{\circ}{\boldsymbol{\Omega}} + \nabla \bullet \mathbf{JI}_{\Omega} = \mathbf{0}$ describes a special kind of kinetic equilibrium exhibited in Euler's equation of motion.

The crucial point concerns a special system variable responsible for the internal fluid rotation. That relationship is given by the respective M–G function with the angular momentum as an independent variable. In principle, its conjugate quantity depends also on the other system variables. The inverse is true as well: All field variables depend on the angular momentum **J**. For this reason, the dissipation velocity $\boldsymbol{\varphi}$ also becomes a function of **J**. Furthermore, a difference arises between $\{\boldsymbol{\tau}_* - \boldsymbol{\tau}_*^T\}$ and $\frac{1}{2}\rho\{\mathbf{v}\boldsymbol{\varphi}' - \boldsymbol{\varphi}'\,\mathbf{v}\}$, where $\boldsymbol{\varphi}'$ denotes the part of $\boldsymbol{\varphi}$ that is independent of **J** and fulfills the symmetry condition (6.79). Hence, the vector product $[\mathbf{v} \times \boldsymbol{\varphi}]$ does not vanish and the production term $\boldsymbol{\kappa} = -2\mathbf{JI}^a$ remains finite in accordance with (7.30). This result indicates a drift between the specific momentum **i** and the flow velocity **v**, as described in Section 6.3. This aspect cannot be pursued any further, yet its deduction is straightforward and does not need to be covered here.

To close this section, a remark should be made concerning the true meaning of the total angular momentum **J**. It is evident that Equation (7.27) mainly introduces a formal condition to satisfy a conservation law. As far as the real angular momentum is concerned, this conservation rule is not required for the internal angular momentum $\boldsymbol{\Omega}$. More interesting is the fact that the quantity **L**, defined by Equation (7.17), includes an additional property of the (specific) linear momentum **i** rather than its own physical quantity. Therefore, a mathematical connection between the internal and the linear momentum appears explicitly. In other words, the vector **L**, above all and along with the true angular momentum Ω, enables us to comprehend the obligatory invariance of the configuration space of events toward any infinitesimal rotation.

7.5 Navier–Stokes–Fourier Fluids

The viscous pressure tensor of the Navier–St. Venant equation of motion may easily be adapted to certain model fluids. This procedure reveals some physical constraints, according to which quite a number of characteristic mathematical conditions must be considered. The desired adaptation is always distinguished by the same mathematical structure as defined for the description of the model fluid.

Let us take as a first example the *Navier–Stokes–Fourier fluids*. They are defined by the Navier–Stokes relation (Truesdell, 1984, p. 426)

$$\mathbf{\Pi}_{\mathrm{NS}} := \varpi(\rho; T)\mathbf{1} - \chi(\rho; T)\,(tr\,\mathbf{D})\mathbf{1} - 2\mu(\rho; T)\,\mathbf{D}, \tag{7.31}$$

where the meaning of the symbols $\mathbf{\Pi}$, ρ, and T are known beforehand. The viscous pressure tensor

$$\boldsymbol{\tau}_{\mathrm{NS}} := 2\mu(\rho; T)\,\mathbf{D} + \chi(\rho; T)\,(tr\,\mathbf{D})\mathbf{1}, \tag{7.32}$$

is not only a symmetric tensor but also one of a very special kind. The symmetry follows from the stretching $\mathbf{D}$, which is determined from a velocity field $\mathbf{v}(\mathbf{r}; t)$ by the dyad rule

$$\mathbf{D} := \frac{1}{2}\{\nabla\mathbf{v} + \mathbf{v}\nabla\} = \frac{1}{2}\{\nabla\mathbf{v} + \nabla\mathbf{v}^T\} \equiv \nabla\mathbf{v}^s. \tag{7.33}$$

It is evident that the tensor $\mathbf{D}$ is identical with the symmetric velocity gradient $\nabla\mathbf{v}^s$ in accordance with Equations (A.18) and (A.19) of Appendix 2.2. The trace of $\mathbf{D}$ ($= tr\,\mathbf{D}$) is simply the velocity divergence $\nabla \bullet \mathbf{v}$. The symmetric velocity gradient $\nabla\mathbf{v}^s$ becomes, along with the antisymmetric velocity gradient $\nabla\mathbf{v}^a$, the common velocity gradient, that is,

$$\nabla\mathbf{v} = \nabla\mathbf{v}^s + \nabla\mathbf{v}^a = \nabla\mathbf{v}^s + \frac{1}{2}\{[\nabla\times\mathbf{v}]\times\mathbf{1}\}, \tag{7.34}$$

which is usually not identical with the transposed velocity gradient $\mathbf{v}\nabla = \nabla\mathbf{v}^T$.

The last term of Equation (7.34) can be related to the *vorticity vector* ς by

$$\varsigma := \frac{1}{2}\nabla\times\mathbf{v}; \qquad \varsigma_\lambda := \frac{1}{2}\{\nabla\mathbf{v}^a\}_{\beta\alpha}\varepsilon_{\alpha\beta\gamma}, \tag{7.35}$$

where the epsilon tensor $\varepsilon_{\alpha\beta\gamma}$ is used for the index notation (Bird, Stewart, and Lightfoot, 1960, p. 720).

> The scalar functions ϖ, χ, and μ are assigned by hypothesis or determined from the results of experiment. We may regard ρ, T, and $\mathbf{D}$ as arbitrarily variable fields subject only to conditions of smoothness, sign, and symmetry. In particular, the field of body force plays no part. Whatever it may be, it imposes no restriction upon ρ, T, and $\mathbf{D}$ at this stage. (Truesdell, 1984, p. 426)

Two aspects concerning the definition (7.31) should be added: First, definition (7.31) is to be supplemented by a constitutive equation for the heat flux vector $\mathring{\mathbf{q}}$ according to Fourier's first law of heat conduction

$$\mathring{\mathbf{q}} := -k\,\nabla T, \tag{7.36}$$

where k denotes the thermal conductivity (Bird, Stewart, and Lightfoot, 1960, p. 245). Second, from definition (7.31) alone, we can make no comment regarding the rotational motion of either the antisymmetric part $\mathbf{\Pi}^a{}_{\mathrm{NS}}$ of the pressure tensor or any properties of Navier–Stokes–Fourier fluids. The inclusion of Fourier's law is motivated by experience with kinetic theories of gases, where an immediate relation exists between the thermal conductivity k and the viscosity μ of perfect gases. The second aspect mentioned is likely to result due to historical reasons. Except for some textbooks concerned with linear irreversible thermodynamics, there are no serious discussions about influences of rotational motions to $\mathbf{\Pi}^a{}_{\mathrm{NS}}$ (see de Groot and Mazur, 1974, p. 125; Gyarmati, 1970, p. 150; Woods, 1975, p. 191). However, even studies that cover some basic properties of $\mathbf{\Pi}^a{}_{\mathrm{NS}}$ are bound to explain the limiting case $\mathbf{\Pi}^a{}_{\mathrm{NS}} \to \mathbf{0}$ either by a frictionless mechanism of rotation or by a vanishing rotational viscosity μ_r. (By the way, reliable values of this coefficient μ_r are not known.)

In this section we will try to discern the deduction of the mathematical structure of (7.31) from the Alternative Theory represented by the Navier–St. Venant differential equations. Then, we will discuss the pertaining mathematical conditions that imply certain physical consequences of chief interest.

First, we will compare the pressure tensor $\mathbf{\Pi} = p(T, \rho)\,\mathbf{1} - \boldsymbol{\tau}$ with definition (7.31): According to the result for the hypothetical state at rest, the function $\varpi(T, \rho)$ "assigned by hypothesis" can be identified with the thermal equation of state $p(T, \rho)$ of the fluid. This option rests on the pertaining theorem derived for $p(T, \rho)$ in Section 6.4. Next, we will consider the viscous pressure tensor $\boldsymbol{\tau}$ that is specialized from (7.11) to the expression

$$\boldsymbol{\tau}_{\mathrm{AT}} = \beta_\tau\{\nabla\mathbf{v}\}^s + \beta_0\{(\nabla \bullet \mathbf{v})\mathbf{1}\}. \tag{7.37}$$

Obviously, Equation (7.37) exhibits the same mathematical structure as (7.32), if the relationship (7.33) between the symmetric dyad $\{\nabla\mathbf{v}\}^s$ and the stretching $\mathbf{D}$ is considered. Both the coefficients β_τ and β_0 relate to the shear viscosity μ and the second velocity χ defined by the Navier–Stokes equation of motion. Beginning with the identity

$$\begin{gathered}\boldsymbol{\tau} = \left\langle \beta_\tau \frac{1}{2}\{\nabla\mathbf{v}\} - s^{-1}\{\nabla s\mathbf{v}\} + s^{-1}(\mathbf{v}\bullet\nabla s)\,\mathbf{1}\right\rangle + \beta_\varphi\{((\rho s)^{-1}\nabla\bullet\mathbf{j}_s + s^{-1}(\mathbf{v}\bullet\nabla s) \\ \partial_t \ell n\, s)\mathbf{1}\} = \boldsymbol{\tau}_{AT} = \beta_\tau\{\nabla\mathbf{v}\}^s + \beta_0(\nabla\bullet\mathbf{v})\,\mathbf{1}\end{gathered} \tag{7.38}$$

based on (7.11), the velocity gradient $\nabla\mathbf{v}$ will first be substituted by the symmetric part $\{\nabla\mathbf{v}\}^s$ according to (7.34). Eliminating $\beta_\tau\{\nabla\mathbf{v}\}^s$ on both sides, the following condition must then be fulfilled for the structural equivalence to be proven:

$$\beta_\tau \left\langle \frac{1}{2}\{\,[\nabla \times \mathbf{v}] \times \mathbf{1}\} - s^{-1}\{\nabla s\mathbf{v}\} + s^{-1}(\mathbf{v}\bullet\nabla s)\,\mathbf{1} \right\rangle$$
$$+ \beta_\varphi\{\,(\,(\rho s)^{-1}\nabla\bullet\mathbf{j}_s + s^{-1}(\mathbf{v}\bullet\nabla s) + \partial_t \ell n\; s)\,\mathbf{1}\} \;=\; \beta_0(\nabla\bullet\mathbf{v})\,1. \tag{7.39}$$

Evaluation (1) of Appendix 2.2 is now inserted for a third of the term $s^{-1}(\mathbf{v} \cdot \nabla s)$ enclosed in the first brackets$\langle ..\rangle$, which now becomes

$$\beta_\tau \left\langle \frac{1}{2}\{\,[\nabla \times \mathbf{v}] \times \mathbf{1}\} - s^{-1}\{\nabla s\mathbf{v} - \frac{1}{3}(\nabla\bullet s\mathbf{v})\,\mathbf{1}\} + s^{-1}\,\frac{2}{3}(\mathbf{v}\bullet\nabla s)\,\mathbf{1} - \frac{1}{3}(\nabla\bullet\mathbf{v})\,\mathbf{1} \right\rangle$$
$$+ \beta_\varphi\{\ldots\} \;=\; \beta_0(\nabla\bullet\mathbf{v})\,\mathbf{1}. \tag{7.40}$$

This expression may be transposed into another form using a second-order tensor with *zero trace* (called *deviator* and marked by the superscript °); by definition we obtain

$$\beta_\tau \left\langle \frac{1}{2}\{\,[\nabla \times \mathbf{v}] \times \mathbf{1}\} - s^{-1}\{\nabla \mathring{s}\,\mathbf{v}\} + s^{-1}\,\frac{2}{3}(\mathbf{v}\bullet\nabla s)\,\mathbf{1} \right\rangle + \beta_\varphi\{\cdots + s^{-1}(\mathbf{v}\bullet\nabla s)\,\mathbf{1}\ldots\}$$
$$- \left(\left(\beta_0 + \frac{1}{3}\beta_\tau\right)\nabla\bullet\mathbf{v}\right)1 \equiv \mathbf{0}. \tag{7.41}$$

Equation (7.41) is the starting point for several interesting conclusions derived in a mathematically rigorous manner. It is divided into two parts that are completely different in their mathematical structure. Whereas the first bracket term $\langle ..\rangle$ contains only second-order tensors each distinguished by a zero trace, the inverse is true for the two other bracket terms $\{\ldots\}$ and $(\ldots)$. That is, every term to be summed is a second-order tensor, the secondary diagonals of which are zeros, and the principal diagonal is occupied by tensor elements forming a finite value of its trace. In accordance with the tensor rules, there is no interaction between the two parts. Consequently, both parts must vanish independently of each other:

$$\left\langle \frac{1}{2}\{\,[\nabla \times \mathbf{v}] \times \mathbf{1}\} - s^{-1}\{\nabla \mathring{s}\,\mathbf{v}\} \right\rangle = \mathbf{0};$$
$$\beta_\varphi\left\{\frac{1}{\rho s}\nabla\bullet\mathbf{j}_s + \partial_t \ell n\; s\right\} + \left(\frac{2}{3}\beta_\tau + \beta_\varphi\right)s^{-1}\mathbf{v}\bullet\nabla s - \left(\left(\beta_0 + \frac{1}{3}\beta_\tau\right)\nabla\bullet\mathbf{v}\right) \equiv 0 \tag{7.42}$$

The first part results in a tensorial condition, whereas the second part can be reduced to an algebraic relationship with exclusively scalar factors and sum terms.

Some solutions, especially those of (7.42), may be equally relevant for science and practice. Since the details of their mathematical proofs are rather cumbersome and not at all transparent (even for experts in fluid dynamics), we present them thoroughly in Appendix 2.3 rather than here. Note that the proofs there are attained with no mathematical approximation.

The theorem, accurately derived from $(7.41)_1$, demonstrates that very strict conditions must be observed if the full viscous pressure tensor (7.11) is used in its simplified structure (7.37). Both results

$$\{\nabla s\mathbf{v}\} = \begin{bmatrix} \partial_1 s v_1 & 0 & 0 \\ 0 & \partial_2 s v_2 & 0 \\ 0 & 0 & \partial_3 s v_3 \end{bmatrix}; \qquad \nabla\times\mathbf{v} \equiv \mathbf{0} \tag{7.43}$$

are highly restrictive for the flow field in question. The first expression requires a so-called spherelike structure of the dyad $\{\nabla s\mathbf{v}\}$ such that apart from the prime diagonal all other elements are zero. In this context, the second-order tensor $\{\nabla s\mathbf{v}\}$ is believed to represent the principal characteristic of a real, that is, compressible and frictional flow.

The second result, however, is much more surprising: The identity $(7.43)_2$ states that flows of any fluid, subject to the mathematically special structure (7.37) and its frictional behavior, are irrotational. Such a statement contradicts the theorem discussed in connection with Equation (7.19).

Let us now continue with $(7.42)_2$ to find some adequate theorems for the various viscosity parameters β. Multiplication by ρs and then division by β_φ transform Equation (7.42) into

$$\frac{1}{\beta_\varphi}\left(\left(\beta_0+\frac{1}{3}\beta_\tau\right)\rho s\nabla\bullet\mathbf{v}\right)-\frac{2}{3}\frac{\beta_\tau}{\beta_\varphi}\rho\mathbf{v}\bullet\nabla s = \sigma, \tag{7.44}$$

where the balance equation of the specific entropy s is used. Hence σ, the entropy production density, appears. By means of the so-called *bulk viscosity* β_v,

$$-\frac{2}{3}\beta_v := \beta_0+\frac{1}{3}\beta_\tau, \tag{7.45}$$

Equation (7.44) turns into the simplified version

$$-\frac{2}{3}\frac{\beta_v}{\beta_\varphi}\rho s\nabla\bullet\mathbf{v}-\frac{2}{3}\frac{\beta_\tau}{\beta_\varphi}\rho\mathbf{v}\bullet\nabla s = \sigma. \tag{7.46}$$

Furthermore, the basic Equation (7.37) of the Navier–Stokes type may then be written

$$\tau_{\mathrm{AT}} = \beta_\tau\left\{\overset{\circ}{\nabla\mathbf{v}}\right\}^s-\frac{2}{3}\beta_v\,(\nabla\bullet\mathbf{v})\,\mathbf{1}, \tag{7.47}$$

where the symmetric velocity gradient is replaced by its deviator. This manipulation follows from the so-called Stokes' relation, which asserts that the bulk viscosity should vanish (see Truesdell, 1984, p. 409).

Equation (7.46) leads to two important conclusions with respect to Equations (7.47) and (7.37):

- Incompressibility, globally defined by $\nabla\bullet\mathbf{v}\equiv 0$, is an ill-posed condition. In this case, the local no-slip condition $\mathbf{v}=\mathbf{0}$ would result in a value $\sigma=0$, although, in general, such wall effects are extraordinarily dissipative.
- The Stokes relation does not hold. For the identity $\beta_v\equiv 0$, the same argument stated above applies.

It is remarkable that the unconstrained selection of a certain value of $\beta_v\neq 0$ yields a simple expression for the entropy production density σ connected with a definite sign control: Assuming $\beta_v := \beta_\tau$ followed by $\beta_0=\beta_\tau$, (7.46) turns into the relationship

$$-\frac{2}{3}\frac{\beta_\tau}{\beta_\varphi}\,\rho\nabla\bullet s\mathbf{v} = \sigma\geq 0, \tag{7.48}$$

which implies negative divergences of the specific entropy rate $s\mathbf{v}$, provided that both Maxwellian viscosities β_φ and β_τ are positive definite functions. Experience will prove whether the relation $\beta_\nu := \beta_\tau$ will hold for a wide class of materials or only for some special cases under certain boundary conditions.

The results of the analysis presented above are striking. Nevertheless, each commentary for those results should consider the complete set of premises forming the basis of the Alternative Theory. All theorems used above have been derived from a few axioms and principles including a complete system theory. The latter is based on Falk's dynamics expounded in Chapter 2 and extended mainly in the last two chapters. On the whole, the network of all the introduced relations, proven relevant for the system under consideration, allows the inclusion of every interaction between the pertaining quantities of motion, the mechanical properties, and the thermodynamical properties. This scientific approach differs essentially from the theoretical concepts employed in rational mechanics. Thus, for instance, one of the fundamentals in modern continuum mechanics, the so-called *representation theorem,* is applicable only to quite simple functions between two second-order tensors. More complex systems, consisting of several tensorial quantities, cannot be adequately described by this mathematical tool (see Becker and Bürger, 1975, pp. 141, 165).

Another point is even more crucial: Any comparison between the Navier–Stokes theory commonly used in practice and the results presented here is only admissible for such cases where thermodynamic state variables, like entropy and temperature, are incorporated in the set of Navier–Stokes equations. If circumstances are reversed—if only mechanical quantities are applied to any flow system—then there is no reason to consider the Alternative Theory. This is particularly true for the traditional Navier–Stokes theory consisting of the ordinary equation of motion, the condition of incompressibility $\nabla \cdot \mathbf{v} = 0$, and the set of initial and boundary constraints. The classical mathematical questions connected with these relationships lead to theorems concerning purely mathematical answers in view of four field quantities of a fluid, viz. the three velocity components v_1, v_2, v_3, and the pressure function p/ρ, whereas the kinematic viscosity $\nu = \mu/\rho$ as a single parameter identifies the fluid.

All modern textbooks on this topic contain a description of some theorems concerning the existence, uniqueness, and, in a few cases, the regularity of solutions for the linear and nonlinear case, but they also include the steady and time-dependent cases. Adequate approximations of these problems by discrete methods are also addressed. Moreover, questions of stability and convergence of various numerical procedures are treated. Problems regarding the notions of weak and strong solutions and their relations to classical solutions are studied. Even fractal and Hausdorff dimensions of a universal attractor are estimated, involving global Lyapunov exponents as known from chaos theory (see, e.g., Temam, 1979; or Constantin and Foias, 1988).

Yet nowhere does this mathematical research reflect the Navier–Stokes equation of motion itself. Real physical aspects seem to be only of secondary interest. For this reason, all arguments are irrelevant in reference to, for example, conservation of angular momentum and its influence on the mathematical structure of the Navier–Stokes equation. However, such a point of view cannot be sustained if physical reasoning prevails and technological applications are required. Thus, our decision is indispensable: you might either reject the weighty restrictions (7.43) as well as the conclusions obtained from (7.46), or you might prefer to deal thoroughly with the theoretical background of the theorems proved above.

Of course, the statement $(7.43)_2$ is highly controversial with regard to the dogmatic role assigned to the Navier–Stokes equations in turbulence research. Undoubtedly, the theorem that states that Navier–Stokes–Fourier flows are irrotational at all times fails to agree with reality, although it is true for fluids modeled by the Navier–Stokes equations. Hence, it is hard to appraise the consequences of Equation $(7.43)_2$ for certain approximations that lead to vortex flow descriptions regardless of the condition $\nabla \times \mathbf{v} \equiv \mathbf{0}$ being considered for the correct use of the Navier–Stokes equations. Boundary layer equations are well-known examples, as are equations used in turbulence theories such as Reynolds stress transport, pressure variance, and turbulent flux of internal energy (see, e.g., Friedrich, 1993, p. 15–6). It seems possible that in some cases the model equations, particularly as used in turbulence research, have long lost their physical origin from the Navier–Stokes equations. Due to the purely formal modeling of the field quantities alone and the introduction of certain questionable closure conditions, the fundamentals of turbulence research may indeed be founded on convention rather than on the original Navier–Stokes equations. Henri Poincaré voiced this suspicion nearly a hundred years ago even in view of Newton's basic laws. His credo is very outspoken: "The principles of mechanics are nothing but conventions and disguised definitions" (Poincaré, 1906, p. 140; author's translation). Hence, it is not unlikely that with continuously increasing degrees of complexity any affected scientific branches will degenerate to pure conventionalism.

It should be mentioned also that the condition $\nabla \times \mathbf{v} \equiv \mathbf{0}$ was first asserted by Domingos, who offers a partial differential equation for the velocity potential of a Navier–Stokes–Fourier fluid. This equation responds to a polytropic change of state along the flow path and is consistent with his hypothesis "...that a velocity potential always exists" (Domingos, 1984, p. 2). He emphasized that the correct deduction of his scalar equation of motion "was concerned with the instantaneous velocity field without any separation between mean and fluctuating quantities" (Domingos, 1984, p. 9). One of the crucial points of Domingos's argument refers to a basic item of fluid dynamics: He concluded that incompressibility is physically untenable. Indeed, it is easy to prove that every Navier–Stokes–Fourier fluid with an irrotational and incompressible flow will obey an equation of motion for which viscous effects cancel automatically.

Domingos's remarkable treatise sheds light on the seemingly widespread behavior of the scientific community when faced with unusual facts or conjectures: Domingos is never quoted, his results are ignored, no scientific dispute takes place. This is in accord with Friedrich Nietzsche's famous statement: "People are only mediocre egoists; if necessary, the wisest man prefers his habits even at the risk of his own advantage" (author's translation). An important example of this lack of regard is Ahmadi's work on turbulence models of compressible flows. His approach laid claim to review the thermodynamics of turbulence based on the averaged Clausius–Duhem entropy inequality. By means of his thermodynamic presuppositions Ahmadi allegedly proved a rate-dependent turbulence model for incompressible Navier–Stokes fluids (Ahmadi, 1990, p. 88). Of course, such a limiting behavior is inconsistent with Domingos's proof according to which incompressibility is physically incompatible with thermodynamics. The same conclusions follow from the Alternative Theory. A sore point of Ahmadi's theory is the lack of transparency of his assumptions along with the estimation of a lot of constant parameters without sufficient reference to his own thermodynamic framework. For instance, Ahmadi's turbulent flows are assumed without comment to be in local equilibrium and subject to Stoke's viscosity law. Thus, it is hard to pronounce judgment on the true value of Ahmadi's turbulence theory.

Some other consequences of the condition $\nabla \times \mathbf{v} \equiv \mathbf{0}$ for Navier–Stokes–Fourier fluids are discussed by Straub and Lauster (1994), which especially concerns the theories of irreversible thermodynamics and gas kinetics.

A concluding remark might be devoted to Sommerfeld's (1964, pp. VII, 261) notorious dilemma: He stopped working on his theory of turbulence the moment a sharp contrast with experiments occurred. He identified two elements possibly responsible for this discrepancy: Either the trustworthy method of small oscillations or the Navier–Stokes equations were failing. Sommerfeld decided not to chose one against the other. Later on he hoped for better concepts that are based either on Tollmien's stability criterion or on von Weizsäcker's and Heisenberg's ideas of a statistical interpretation of turbulent phenomena. It is notable that both approaches assume the Navier–Stokes equations to be the true base and, again, the most relevant input. You may draw your own conclusion.

7.6 Simplified Models of Dissipative Flows

After the surprising analytical results with the Navier–Stokes equation of motion, we will use the same procedure to study the simplest approach for the viscous pressure tensor $\boldsymbol{\tau}$. It may become part of a well-defined reference model for flow descriptions. In the next chapter, we will conduct some special tests with this model to compare their solutions with well-known results from the Navier–Stokes equations and from experiments. Defining the simplest $\boldsymbol{\tau}$-model by

$$\boldsymbol{\tau}_{\mathrm{AT}} := \beta_0\{(\nabla \bullet \mathbf{v})\mathbf{1}\}, \tag{7.49}$$

the corresponding restrictions with respect to the Navier–St. Venant Equation (7.7) can be analyzed from the tensor equation

$$\beta_\tau\, s^{-1}\Big[-\{\mathbf{v}\nabla s\}^T + \{(\mathbf{v}\bullet\nabla s)\,\mathbf{1}\}\Big] + \beta_\varphi\left\{\left(\frac{\sigma}{\rho s}\right)\mathbf{1}\right\} = \beta_0\{(\nabla\bullet\mathbf{v})\,1\}\,. \tag{7.50}$$

Except for the first expression all other summation terms consist of diagonal dyads. For this reason, the complete set of elements placed in the secondary diagonals of $\{\mathbf{v}\nabla s\}^T$ must vanish. The respective identity

$$\{\mathbf{v}\overset{\circ}{\nabla} s\}^T \equiv 0 \tag{7.51}$$

can only be satisfied either by the trivial case $\mathbf{v} \equiv \mathbf{0}$ or by the entropy condition

$$\nabla s \equiv 0, \tag{7.52}$$

which excludes spatial entropy differences along with flow events. In other words, due to local entropy production, entropy fluxes will occur but convection does not contribute to this mechanism. Hence, the direct connection

$$\sigma = \nabla \cdot \mathbf{j}_s \tag{7.53}$$

holds for all events. Considering (7.51), Equation (7.50) may be simplified to the scalar expression

$$\sigma = \frac{\beta_0}{\beta_\varphi} \rho s \nabla \cdot \mathbf{v}, \tag{7.54}$$

which reveals that the two quantities (β_0/β_φ) and $\nabla \cdot \mathbf{v}$ always have the same signs due to the exclusively positive values of the entropy density ρs and the entropy production σ. Of course, it is trivial that incompressibility turns out to be a physically inadmissible property.

To establish the desired flow model, we must agree on some properties of matter. These properties belong to the simplest matter model, that of the elementary gas, assumed to be a special case of an *ideal gas* but equipped only with constant material parameters. General rules for ideal gases were discussed in Section 3.4.3. Therefore, we will compile here all properties relevant to the primitive model in question.

$$\begin{aligned}
&\textit{Thermal equation of state:} && \Omega = RT; \qquad \Omega := \frac{p}{\rho} \text{ (pressure function)} \\
&\textit{Caloric equation of state:} && u = \frac{\Omega}{\kappa - 1}; \qquad \kappa := \frac{c_p}{c_v} \text{ (heat capacity ratio)} \\
&\textit{Heat conductivity–viscosity ratio:} && \frac{\lambda}{\beta_\tau} = \frac{f^{\#} R}{\kappa - 1} \\
&\textit{Kinematic Maxwell viscosity:} && \overset{*}{v} = \frac{\beta_\tau}{\rho} = \frac{t_\tau \rho T s}{\rho} = t_\tau \Omega \frac{s}{R}
\end{aligned} \tag{7.55}$$

Some remarks about the relationships in (7.55) are in order. Note that the pressure function Ω is proportional to the temperature T, but only in the case of an ideal gas. Unlike this, real pure substances obey functions p/ρ, which depend in turn on temperature T and the mass density ρ. Whereas Ω may be an absolute quantity, the specific internal energy u contains an arbitrary reference value that is commonly suppressed, since usually only energy differences Δu are relevant. For the sake of simplicity only gases with nonrotating spherically symmetrical molecules are considered. Accordingly, the heat capacities c_p and c_v are also constants. In fact, this is true for monatomic gases, such as the noble gases, for which the specific heat c_v may be expressed by $c_\mathrm{v} = R(\kappa - 1)^{-1}$, with the isentropic index κ equal to 5/3. For other ideal gases, the latter properties should be regarded as a more or less appropriate approximation.

The ratio λ/β_τ relates to an equivalent heat conductivity-shear viscosity ratio k/μ, which is well-known in the kinetic theory of transport properties. For gases with nonrotating spherically symmetrical molecules the ratio $k/(\mu c_v)$ is 5/2 (Chapman-Cowling, 1970, p. 160). Opposed to that the heat conductivity λ is defined by Equation (6.97) and approximated for the hypothetical state at rest. This is also true for the Maxwell viscosity

β_τ established by Equation (7.8). Nevertheless, the ratio λ/β_τ cannot be thought of as a constant. For this reason, the respective equation of the set (7.55) should be interpreted as a definition for the material function $f^\#$ rather than an otherwise derived relationship. Of course, for a first approach we are justified in considering $f^\#$ as a pure number of the same order of magnitude as the one given in the classical theories.

The last equation of the set (7.55) reveals that the basic assumption of an ensemble of constant parameters must be classified as a rough approximation at best. Evidently, the kinematic Maxwell viscosity instead depends in a complicated way on temperature and in a weak way on density via the specific entropy. Thus, the selection of any averaged but constant value of the kinematic viscosity is only justified for field studies concerning mainly qualitative scenarios. As will be proved in the next chapter, the parameter ν^* sometimes diverges significantly from any comparable value of the traditional kinematic viscosity $\nu = \mu/\rho$. Further simplifications for the envisioned primitive flow model might be reasonable with regard to the qualitative scenarios to be studied. Only for this purpose will all partial time derivatives appearing explicitly in the balance equations be dropped.

Consequently, the field equation of motion as well as that of the specific enthalpy (6.83) of any one-component single-phase system can be simplified considerably. Moreover, the specific enthalpy h will be substituted by the specific internal energy u via the well-known formula $u = h - p/\rho$. If the term ρDh is displaced by the expression $\rho Du + Dp - \rho^{-1}\, D\rho$, where the last term is to be replaced by $\nabla \bullet \mathbf{v}$, then the original balance equation

$$\begin{aligned}\rho Dh &= -\nabla \bullet \mathbf{j}_e^{[\rho]} + Dp + \boldsymbol{\tau} : \nabla \mathbf{v} \\ &\quad -\mathbf{v} \bullet \partial_t \rho \varphi - \partial_t \rho e_\varphi - \rho \partial_t e_{\text{pot}}\end{aligned} \tag{6.83}$$

is reduced to

$$\rho Du = -\nabla \bullet \mathbf{j}_e^{[\rho]} + \boldsymbol{\tau} : \nabla \mathbf{v} - \nabla \bullet \mathbf{v}. \tag{7.56}$$

Except for the shock wave structure problem, all other cases described by means of this primitive flow model neglect the dissipation function $\boldsymbol{\tau} : \nabla \mathbf{v}$ and approximate the energy current density $\mathbf{j}_e^{[\rho]}$ by the formula

$$\mathbf{j}_e^{[\rho]} = -t_q \frac{1}{T_*} \{\boldsymbol{\varphi} \nabla T_*\} \bullet \mathring{\mathbf{q}}_* \approx -\frac{1}{2}\left(\frac{t_q^2}{T^2}\right) \rho \varphi^4 (\boldsymbol{\varphi} \bullet \nabla T) \nabla T = -\lambda \nabla T. \tag{7.57}$$

Here, Equations (6.94), (6.97), (7.52), and (6.98) have been combined along with the approximation $T \approx T_*$. The full tensor $\{\boldsymbol{\varphi} \nabla T_*\}$ is approximated by its trace $(\boldsymbol{\varphi} \bullet \nabla T_*)$. Using (7.55), the field equation of the specific internal energy becomes

$$D\,\Omega = f^\# \nu^* \nabla \bullet \nabla \Omega - (\kappa - 1)\Omega \nabla \bullet \mathbf{v}, \tag{7.58}$$

which applies the Laplacian operator $\nabla^2 := \nabla \bullet \nabla$ to the scalar Ω. Equation (7.58) contains the two field variables Ω and $\mathbf{v}$. Now the vectorial equation of motion, simplified by the compiled model assumptions, will be used. There are two options: The first refers to the Navier–Stokes approach (7.37) discussed in Section 7.5; the second concerns the simplest $\boldsymbol{\tau}$-model (7.49). Considering the well-known tensor rule (Bird, Stewart, and Lightfoot, 1960, p. 725)

$$\nabla^2 \mathbf{v} := \nabla(\nabla \bullet \mathbf{v}) - [\nabla \times [\nabla \times \mathbf{v}]], \tag{7.59}$$

which defines the Laplacian operator ∇^2 applied in rectangular coordinates to a vector field $\mathbf{v}$, the two options may be quantitatively represented as follows:

Simplest $\boldsymbol{\tau}$-model:

$$\begin{aligned} \rho D\mathbf{v} &= \beta_0 \nabla(\nabla \bullet \mathbf{v}) - \nabla p \\ &= \beta_0 \nabla^2 \mathbf{v} + \beta_0 \nabla \times [\nabla \times \mathbf{v}] - \nabla p \end{aligned} \tag{7.60}$$

Navier–Stokes model:

$$\begin{aligned} \rho D\mathbf{v} &= \beta_\tau \nabla \bullet \{\nabla \mathbf{v}\} + \beta_0 \nabla(\nabla \bullet \mathbf{v}) - \nabla p \\ &= (\beta_0 + \beta_\tau)\nabla^2 \mathbf{v} - \nabla p. \end{aligned} \tag{7.61}$$

We should mention that the simplest $\boldsymbol{\tau}$-model permits the description of vorticity flows, whereas for the Navier–Stokes model the condition $[\nabla \times \mathbf{v}] \equiv \mathbf{0}$ holds according to Theorem (7.43). To introduce the pressure function $\Omega := p/\rho$ into Equations (7.60) and (7.61), two possibilities are available:

1. *Boussinesq-approximation:* Under the condition $\nabla\rho/\rho << 1$, the term $\rho^{-1}\nabla p$ may be replaced by $\nabla\Omega$ in accordance with

$$\nabla \frac{p}{\rho} = \nabla\Omega = \frac{p}{\rho}\left(\frac{\nabla p}{p} - \frac{\nabla \rho}{\rho}\right) \approx \frac{1}{\rho}\nabla p. \tag{7.62}$$

2. *Polytropic process:* Assuming a polytropic realization of a flow process, according to the relationship

$$\frac{p}{p_\#} = \left(\frac{\rho}{\rho_\#}\right)^n \tag{7.63}$$

(see, e.g., Landsberg, 1961. p. 205), it is easy to prove by serial expansion that the relation

$$\frac{1}{p}\nabla p = \frac{n}{n-1}\nabla\Omega = \nabla\left(\frac{n}{n-1}\Omega\right) := \nabla\Xi \tag{7.64}$$

approximately holds. The subscript # denotes an arbitrary standard state, and the exponent n is the polytropic index presumed to be constant.

Since both equations of motion (7.60) and (7.61) form the expression $\rho^{-1}\nabla p$, it seems reasonable to use a common symbol Ω_N, indicating the alternatives (7.62) or (7.64) by $\Omega_N = \Omega$ or $\Omega_N \equiv \Xi$. Thereby two sets of balance equations describing flow phenomena can be presented, resulting from the energy relationship (7.58) together with Equations (7.60) and (7.61), which are concerned with the continuous motion of pure fluids.

$$D\mathbf{v} = \nu_0 \nabla(\nabla \bullet \mathbf{v}) - \nabla\Omega_N; \qquad D\Omega_N = f_0^{\#}\nu_0 \nabla^2\Omega_N - (\kappa - 1)\,\Omega_N \nabla \bullet \mathbf{v} \tag{7.65}$$

$$D\mathbf{v} = \nu_\tau \nabla^2 \mathbf{v} - \nabla\Omega_N; \qquad D\Omega_N = f_\tau^{\#}\nu_\tau \nabla^2\Omega_N - (\kappa - 1)\,\Omega_N \nabla \bullet \mathbf{v} \tag{7.66}$$

The parameters are defined by Equations (7.55) and expressed in terms of $(\beta_0 + \beta_\tau)$ or β_0.

Nehring published two sets of balance equations whose mathematical structures are similar to that of Equations (7.65) and (7.66). He preferred purely mechanical arguments and established his equations more heuristically than systematically, intending to study some characteristic features of viscous flows in an exclusively qualitative way (Nehring, 1983 and 1993). Certain symmetries of these pairs of equations are of special interest, as may easily be pointed out by inserting (7.59) into Equation (7.65). The resulting relationship,

$$D\mathbf{v} = \nu_0 \nabla^2 \mathbf{v} - \nabla \Omega_N + \nu_0 \nabla \times \nabla \times \mathbf{v}; \qquad D\Omega_N = f_0{}^{\#} \nu_0 \nabla^2 \Omega_N - (\kappa - 1)\, \Omega_N \nabla \cdot \mathbf{v}, \tag{7.67}$$

demonstrates that, aside from the rot-rot term of $(7.67)_1$, the $\nabla^2 - \nabla$ symmetry of (7.66) and (7.67) is quite obvious. For convenience, we will give Equations (7.65) and (7.66) a name: Equation (7.65) will be called the *Original Nehring Equation* (ONE) and Equation (7.66) will be called the *Modified Nehring Equation* (MNE). Of course, a comparison of the classical and alternative fluid equations with exact solutions is of interest. Using symmetry methods, pertaining studies on ONE and MNE provide some remarkable results, especially in comparison with the classical (i.e., incompressible) Navier–Stokes equations.

Employing symmetry groups and reductions, *solution families* may be constructed for systems of nonlinear partial differential equations, provided that initial or boundary conditions are disregarded. These families of solutions depend on coefficients in the partial differential equations and on free coefficients generated as part of their symmetry construction.

The Lie symmetry groups, and the associated algebra of generators, are already known (see Rogers and Ames, 1989) for the Navier–Stokes equations. An appropriate exposition of this theory was recently presented for comparison with the symmetries of ONE and MNE (Ames et al., 1995). They were efficiently calculated, using an interactive symbolic computer package. The results concern generators of Lie algebra in terms of mathematical operations of invariance. All sets of equations under consideration possess generators representing translations, rotations, dilation, and Galilean boosts. Thus, they are compatible with Newtonian mechanics at least. An essential difference exists between the alternative sets and the classical Navier–Stokes equations: The generators of the latter form a Lie algebra that is of infinite dimension because of several arbitrary functions. The alternative systems ONE and MNE both have a finitely dimensional Lie algebra of dimension eleven.

The main discrepancy, however, concerns the notion of pressure. A very special generator is only assigned to the Navier–Stokes equations and works with a pressure term presumed to be a quantity measured on an interval scale. In contrast, the thermodynamic pressure is endowed with an absolute zero-point and hence a ratio-scaled quantity. Examinations of the respective variables of ONE and MNE, as well as of the classical Navier–Stokes equations, suggested a search for "traveling waves" (with not necessarily oscillatory solutions). The analysis led to some quite different results for ONE and MNE than for the classical Navier–Stokes equations. For instance, MNE possesses traveling wave solutions, for which equilibrium states exist.

The paper published by Ames et al. furnishes many further results allowing conclusions regarding the occurrence of bounded or unbounded solutions, the dependency of the wave speed on transport coefficients, and the limiting case of the vanishing of a transport coefficient.

As mentioned above, the Lie group method is restricted to problems for which initial or boundary conditions are disregarded. Therefore, we will conclude this section with some remarks about representative boundary conditions, which are useful for several flow events. Whereas it is easy to formulate the no-slip condition

$$\mathbf{v}\big|_{\text{wall}} \equiv \mathbf{v}_{\text{wall}} = \begin{cases} \mathbf{0} & \text{for fixed walls} \\ \mathbf{v}_0 & \text{for moved walls} \end{cases} \tag{7.68}$$

as an adequate boundary condition concerning the local flow velocity v, it is more difficult to find a correct boundary condition for the pressure function Ω. Starting with the specific Legendre-transformed energy $e^{[\rho]}$ (cf. Equation $(6.71)_2$).

$$e^{[\rho]} - e_{\#} = h + e_{\text{kin}} + e_{\text{pot}} + e_{\varphi},$$

its balance equation

$$\rho\, De^{[\rho]} + \nabla \bullet \mathbf{j}_e{}^{[\rho]} = \partial_t p \tag{6.72}$$

may be considerably simplified by means of some elementary manipulations with respect to the special internal energy $u = h - p/\rho$ and the inequality.

$$u + e_{\text{kin}} >> e_{\text{pot}} + e_{\varphi}, \tag{7.69}$$

which is assumed to be valid near any wall, along with the three steady-state conditions

$$\partial_t e^{[\rho]} \equiv 0; \qquad \partial_t p \equiv 0; \qquad \partial_t \rho \equiv 0. \tag{7.70}$$

Hence, the resulting balance equation for $e^{[\rho]}$,

$$\nabla \bullet [\rho\mathbf{v}(u + e_{\text{kin}}) + \mathbf{j}_e{}^{[\rho]}] \equiv 0, \tag{7.71}$$

furnishes the desired boundary condition in the form of two options for *fixed walls*:

with the respect to *local heat transfer*: $\nabla \bullet \mathbf{j}_e{}^{[\rho]} = -\lambda\nabla^2 T = -(\lambda/R)\, \nabla^2\Omega \equiv 0$;
with respect to the *adiabatic case*: $\nabla\upsilon = (\kappa - 1)^{-1}\nabla\Omega \equiv 0$.

In both applications the Ω function is subject to the *wall condition*

$$\nabla^2\Omega \equiv 0, \tag{7.72}$$

on which the first integration yields either a constant or zero.

With Equations (7.68) and (7.72), two mathematical expressions may serve as constraints for the solutions of ONE and MNE. Note that the continuity equation, usually used in conventional fluid dynamics as a member of the set of conservation laws, does not explicitly appear with ONE or MNE. But, if solutions of ONE or MNE for the field quantities $\mathbf{v}$ and Ω are available, the density field may be determined by integration of the continuity equation.

Chapter 8 | Paradigmata Are the Winners' Dogmata

"Don't believe anything merely because you have been told it, or because it is traditional, or because you have imagined it."
—***Gautama Buddha***

Chapter 8 was written in cooperation with Michael Lauster. Its title is a phrase coined by Threv S. W. Salomon.

8.1 Selection of Paragons

The Alternative Theory offers some approximations with physical assumptions that may on one hand be precisely deduced from its general axioms. On the other hand, those simplified results concern two reference cases that are of special interest in practice. The first case—the *Original Nehring Equations* (ONE)—is based on the simplest approach imaginable for the viscous pressure tensor $\boldsymbol{\tau}$. The second case—the *Modified Nehring Equations* (MNE)—refers to the Navier–Stokes model (7.61) with its second-order tensor $\boldsymbol{\tau}$ as well.

Of course, the question arises whether these two approximations, ONE and MNE, work well in comparison with experiments and with other purely mathematical relations. Some answers will be given by means of a few examples of flow phenomena, chosen to cover a broad range of relevant applications.

First of all, a paradigmatic case is presented by the Hagen–Poiseuille flow, which is well-known both as an allegedly exact solution of the Navier–Stokes equation of motion and as an efficient tool for experimentally determining the shear viscosity of the fluid in question. Second, for an apparently perfect field of simulation, the Lorenz equations may be used to study several kinds of nonlinear and even chaotic behavior. Amazingly, the deduction of this set of equations is motivated by a well-posed experimental setup constructed primarily with intent to study special patterns of compressible flows. Third, the behavior of *boundary layer flows* is of special interest with respect to characteristic techniques for approximate solutions of the Navier–Stokes equations. A certain kind of problem leads to the *Falkner–Skan equation*, for which a corresponding expression will be derived from MNE.

Fourth, structures of a *normal shock wave* in perfect gases take up a key position as to their description by means of differential equations of different types, such as the Navier–Stokes or the Nehring equations. Past experience has shown that there is usually no agreement between experiments and calculations, so a negative test is to be expected.

The fifth example concerns a basic problem of thermodynamics: *Gay-Lussac's second law* (1807), which is equivalent to Joule's law (1845), which states that the internal energy of an ideal gas is independent of volume (see, e.g., Landsberg, 1961, p. 190). Curiously, up to now there exist only incomplete and physically unsatisfactory solutions of this irreversible flow problem. By means of MNE, we intend to grasp this process in detail and to obtain quantitative results concerning relaxation times to be compared with experimental data sets.

Last but not least, a *complex flow configuration* will be studied: Along a channel, gaseous matter flows around a vertically installed cylinder that is able to rotate and may have its surface heated. At the 1985 workshop on turbulence, the team of experts (see Chapter 7) recommended just this configuration for research as one of two key problems of modern fluid dynamics, "although the numerical details ... are difficult" (Nixon, 1986, p. 25).

Fortunately, we may refer to Lauster's extensive studies concerning non-equilibrium events in compressible fluids based on Gibbs–Falkian thermodynamics. In the course of these studies, he dealt with many theoretical problems that are also relevant in practice. All six examples outlined above are part of his studies. Therefore, we may peruse the main results of these six paragons as to both their physical aspects and their mathematical peculiarities. For all six, you will find details of the mathematically relevant deductions in Lauster's work (see e.g., Lauster, 1995 and 1996).

Before presenting the results for the six selected cases, let us discuss some general aspects of the first three examples, which refer to the significance of the vorticity term $\varsigma := \nabla \times \mathbf{v}$ for some theoretical conclusions to be drawn from the Navier–Stokes equations, especially.

With regard to the physical relevance of the vorticity notion for current turbulence theories (see, e.g. Rotta, 1972, p. 54), the logical incompatibility of the well-known *Navier–Stokes vorticity equation*

$$\mathring{\varsigma} = \varsigma\nabla\mathbf{v} + \nu\nabla^2\varsigma; \qquad \nu := \frac{\mu}{\rho} \tag{8.1}$$

with the theorem $\varsigma := \nabla \times \mathbf{v} \equiv \mathbf{0}$ provokes a serious dispute about the true relationship between theory and experiment in fluid dynamics.

One example will do: The mathematical description of the incompressible Hagen–Poiseuille flow has the reputation of being an exact integration of the Navier–Stokes equation of motion. In addition, there is much experimental data that seemingly confirm the numerical computations. According to Lamb, who cited the authority of Reynolds "this hypothesis has been put to a very severe test in the experiments of Poiseuille and others, [so that] we can hardly hesitate to accept the equations as a complete statement of the law of viscosity" (Truesdell, 1974, p. 111).

In reality it is easy to verify that relevant assumptions for the integration procedure mentioned are more sophisticated than physically justified. The usual way to describe the pressure drop as constant along the flow path is normally an ill-posed condition for most tube flows in practice. The problem discussed here refers to (1) the compatibility between definitions (as for the Navier–Stokes–Fourier fluid) and first principles (as to the conservation laws), and (2) the need for an engineer to work with correct approximations. Thus, for example, both sets of Nehring equations may be used and have numerical solutions that demonstrate the well-known parabolic curve in the core of a laminar tube flow consistent with empirical evidence. However, a thorough examination will reveal a higher-order curve determined by conditions for the chosen process realization.

Moreover, the Hagen–Poiseuille mass flow equation of the fourth power (Spurk, 1989, p. 179) only results if the viscosity function μ degenerates to a constant. According to any concurrent flow conditions (and only then) the velocity profile is parabolic. "Thus the experiments, to which Lamb and Reynolds referred, are far from sufficient to establish the validity of the Navier–Stokes equation in general" (Truesdell, 1974, p. 119).

Another consequence refers to applications of the actual *theory of deterministic chaos*: First, the foundation of the Lorenz equations cannot be justified by the Navier–Stokes equations as is common practice. Therefore, we cannot make any realistic prediction for dynamical evolution problems [or for a Rayleigh parameter far from unity (Sparrow, 1982, p. 3)] by integration of this set of ordinary differential equations (Sparrow, p. 1). Second, the first tentative attempts to find a quantitative explanation of the turbulent flow mechanism are in vain. Using the conventional incompressible Navier–Stokes equation of motion, some authors construct a structural function $D(r)$, together with the so-called Lagrange relaxation rate $\gamma(r)$, for which the variable r refers to the selected flow field level to be analyzed. For different flow patterns, $D(r)$ describes how the velocities will vary between two arbitrary field points separated by the distance r. By this means, a connection can be made with the rate of energy dissipation supposed to be responsible for the structure of turbulent flows.

The differential–integral equation for $D(r)$ results directly from the Laplace operator ∇^2 acting on $\mathbf{v}$ (Grossmann, 1990, p. 5) in the pressure tensor divergence of the Cauchy equation [(6.54)] applied to incompressible flows. With both the restrictions $\nabla \times \mathbf{v} = \mathbf{0}$ and $\nabla \bullet \mathbf{v} \equiv 0$, the tensor rule (7.59) leads to the identity $\nabla^2\mathbf{v} \equiv \mathbf{0}$; hence this theoretical approach quantified by $D(r)$ breaks down. This conclusion is correct, provided that Cauchy's first law [(6.54)] does not tend for an asymptotically vanishing shear viscosity to any universal turbulence structures that are mainly dependent on the nonlinear convective term and independent of the friction term chosen. Note that although some experiences might indicate such a behavior, it seems that there is then, for instance, no realistic intermittency. As an unexpected result, therefore, the implied self-similar mathematical structure of the flow velocity field, suggested to be chaotic (Grossmann, 1990, p. 5), cannot be proved by means of the Navier–Stokes equation of motion. For this reason, some special knowledge that seems assured becomes fairly questionable (cf., for example, Ebeling and Feistel, 1992, p. 305).

A third class of problems is formed by the boundary layer flows. These are limited to small flow regimes, distinguished chiefly by steep gradients of flow velocities. Their mathematical description takes place by the boundary layer equations (BLE), derived from the Navier–Stokes equations and combined with some characteristic boundary conditions. On the one hand, BLE are subject to some significant changes in their mathematical structure compared with the original partial differential equations. But, on the other hand, BLE also gain new mathematical dimensions insofar as it is easy to prove that they now describe flows with vorticity that is no longer zero. Perhaps BLE should be defined as a tool for the appropriate description of boundary layers, completely independent of the Navier–Stokes equations.

8.2 Vorticity-influenced Flows

8.2.1 HAGEN–POISEUILLE FLOW

A highly informative example of classical ideas and methods, the Hagen–Poiseuille flow appears in every engineering, physics, and medical (e.g., circulation) textbook. It is a steady-state, laminar flow of any viscous fluid along a horizontally arranged circular cylinder tube. Its theoretical relevance is in the fact that it can be described by an allegedly exact solution of the Navier–Stokes equation of motion. It is also significant for measuring shear viscosities of any fluid by the experimental determination of mass flow rates and pressure drops (see, e.g., Spurk, 1989, p. 160). Due to this well-known application, some authors recommend confining this flow description to capillary-like tubes, along which laminar flow patterns might be realized. However, such a statement is only justified by empirical arguments because the conventional deduction is valid for all diameter–length ratios.

In Figure 8.1, a laminar velocity profile at a fixed position along the tube is sketched together with the velocity components u, v, and w assigned to the cylindrical coordinates r, φ, and z used.

Let us investigate the Hagen–Poiseuille flow by means of the Modified Nehring Equations (MNE) proven in Chapter 7 to be closely related to the Navier–Stokes equation (NSE). If the assumptions are valid for NSE, a contradiction arises at once: Note a steady-state flow according to $\partial_t(\ldots) \equiv 0$, which is taken to be symmetrical with respect to the angle-coordinate, viz. $\partial(\ldots)/\partial\varphi = 0$. This flow will then degenerate as to the velocity components u and v (i.e., $u = 0$, $v = 0$). Finally, driven by a constant pressure drop in the z-direction, it will yield the relation $\partial\Omega/\partial z = \text{constant}$.

Hence, we can draw a remarkable conclusion: These assumptions are only consistent with MNE if the conditions $\Omega = \text{constant}$ and $w = 0$ were presupposed, even

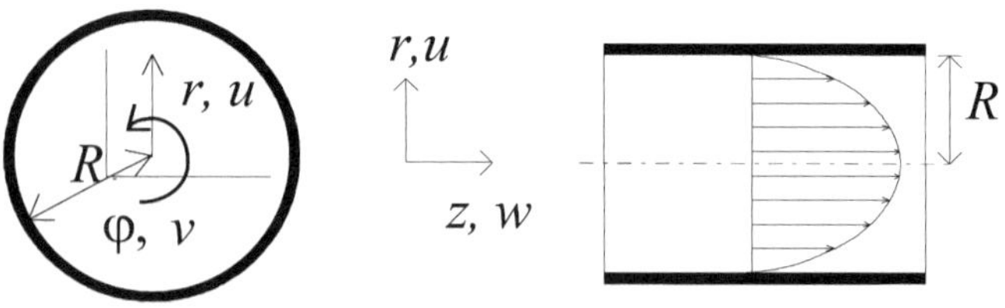

Figure 8.1 Space and velocity coordinates of the Hagen–Poiseuille flow.

though the momentum equation $(7.66)_1$ yields the well-known parabolic velocity profile

$$w(r) = -\frac{1}{\nu_\tau}\frac{\partial\Omega_N}{\partial z}\frac{R^2}{4}\left(1-\frac{r^2}{R^2}\right), \tag{8.2}$$

this time in complete agreement with the NSE result.

Another picture is obtained by neglecting the u-component in the r-direction, in comparison with the main stream component w, whereas the corresponding derivative $\partial u/\partial r$ does not vanish. In other words, the constraints

$$u \ll w \quad \text{and} \quad \frac{\partial u}{\partial r} \neq 0 \tag{8.3}$$

lead, additionally, to a distribution function for the velocity component u, as well as to the pressure function $\Omega_N(r)$, viz.

$$u(r) = \frac{R^2}{4\nu_\tau\sqrt{\kappa-1}}\frac{\partial\Omega_N}{\partial z}\left\langle \frac{r}{R}\left[\frac{\pi}{2}-\sqrt{1-\frac{r^2}{R^2}}\right]-\arcsin\left(\frac{r}{R}\right)\right\rangle \tag{8.4}$$

$$\Omega_N(r)-\Omega_{N_0} = -\frac{R}{2\nu_\tau\sqrt{\kappa-1}}\frac{\partial\Omega_N}{\partial z}\left\langle\sqrt{1-\frac{r^2}{R^2}}-1\right\rangle \tag{8.5}$$

where the index 0 refers to the values valid for the core of the flow at position $r = 0$. The graphs of both functions are given in Figure 8.2.

From the solutions (8.2), (8.4), and (8.5) of the set of equations (7.66), we make the following assertions. First, comparing (8.4) with (8.2), it becomes evident that the Hagen–Poiseuille flow runs only along tubes whose radii R are very small compared to the tube lengths L. In all other cases the velocity component u of the flow in the r-direction obtains values of the order of magnitude assigned to the value m of the axial component w. Of course, this behavior contradicts inequality (8.3).

Second, it is physically untenable to foster the premise of laminar flow along any tube when that flow type is in a first step defined by streamlines running parallel to the axis of the tube (Sommerfeld, 1964, p. 69). Every motion of a fluid in the direction of this axis is accompanied by a radial velocity unequal to zero. Following from Equation (8.4), the gradient $\partial u/\partial r$, viz.

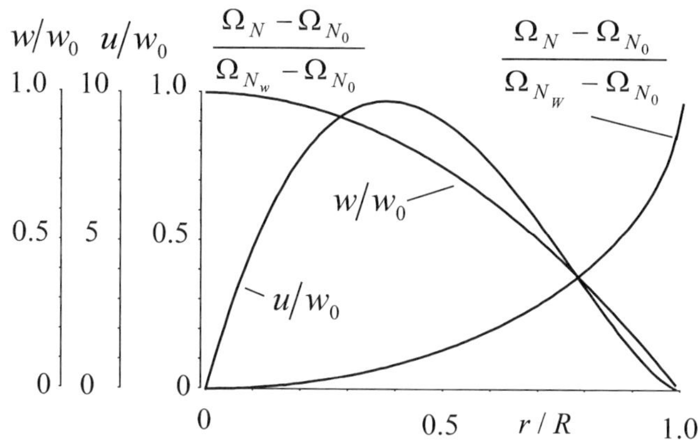

Figure 8.2 Course of pressure and velocity functions for the radial direction of the Hagen–Poiseuille flow.

$$\frac{\partial u\,(r)}{\partial r} = \frac{R}{2v_\tau\sqrt{\kappa-1}}\frac{\partial \Omega_N}{\partial z}\left[\frac{\pi}{4} - \sqrt{1-\frac{r^2}{R^2}}\right], \tag{8.6}$$

is characterized by a single-valued maximum of the radial flow velocity at position $r/R = [1 - (\pi/4)^2]^{1/2}$. Additionally, quite in agreement with (8.5), radial motion always occurs from the wall of the tube toward the core of the flow.

Third, the experimental determination of the kinematic viscosity ν_τ of any (nearly) perfect gas by the use of Hagen–Poiseuille flows is reduced to measurements of a pair of flow velocity and temperature values as well as the actual temperature drop $(\partial T/\partial z)_L$ along the tube of length L. The resulting expression

$$\nu_\tau = \frac{\left[8\sqrt{\kappa-1}\left(T_W - T_0\right) - \pi\left(\frac{\partial T}{\partial z}\right)_L R\right]\Re\left(\frac{\partial T}{\partial z}\right)_L R^2}{8\left[4\sqrt{\kappa-1}\,T_0 - \pi\left(\frac{\partial T}{\partial z}\right)_L R\right]w_0} \tag{8.7}$$

contains averaged temperature values of the gas (characterized by the specific gas constant $\Re$ and the isentropic index κ) for wall (i.e., $r = R$) and core (i.e., $r = 0$) conditions of the flow with its mass flow rate $\overset{\circ}{m} \propto w_0$. In contrast to the classical solution, the compressibility behavior explicitly influences the shear viscosity of the fluid $\mu = \nu_\tau\rho = \nu_\tau(p/\Re T)_{\text{inlet}}$, whereas the incompressible Hagen–Poiseuille flow knows temperature only as a fluid index.

Likewise, the Original Nehring Equations (ONE) (7.65) lead to trivial solutions if the common premises of classical NSE are used. However, considering the changed conditions (8.3) we obtain a new set of differential ONE

$$\frac{r}{2\nu_0}w\frac{\partial \Omega_N}{\partial z} = \nu_0\left(\frac{\partial^2 u}{\partial r^2} + \frac{1}{r}\frac{\partial u}{\partial r}\right) - \frac{\partial \Omega_N}{\partial r}; \tag{8.8}$$

$$0 = \nu_0\left(\frac{\partial^2 u}{\partial r \partial z} + \frac{1}{r}\frac{\partial u}{\partial z}\right) - \frac{\partial \Omega_N}{\partial z}; \tag{8.9}$$

$$w\frac{\partial \Omega_N}{\partial z} = f_0 \nu_0\left(\frac{\partial^2 \Omega_N}{\partial r^2} + \frac{1}{r}\frac{\partial \Omega_N}{\partial r}\right) - (\kappa - 1)\,\Omega_N \frac{\partial u}{\partial r}, \tag{8.10}$$

where no differential condition for the flow component w along the tube is included. In other words, for ONE the velocity distribution $w(r)$ of the main stream direction only plays the role of a curve parameter. Exactly for this reason, and considering the experimental facts, it seems appropriate to declare the parabolic $w(r)$ profile (8.2) also obligatory for ONE in spite of some legitimate objections against this profile. After replacing ν_τ by the corresponding quantity ν_0, Equation (8.2) can be inserted into Equation (8.8). Subsequently, the resulting differential equation can be integrated along with Equation (8.9). The solution,

$$\Omega_N - \Omega_{N0} = \nu_0 \int_{r^*=0}^{r^*=r} \frac{1}{r^*}\frac{\partial}{\partial r^*}\left(r^*\frac{\partial u}{\partial r^*}\right)\partial r^* + \frac{1}{2\nu_0^2}\left(\frac{\partial \Omega_N}{\partial z}\right)^2 \frac{R^4}{4}\left(1 - \frac{r^2}{2R^2}\right)\frac{r^2}{2R^2}, \tag{8.11}$$

represents the pressure function. Neglecting the second summand of the right side for capillary-like tubes and small temperature gradients in the z-direction, we obtain the same results as given by (8.5) for MNE.

8.2.2 *LORENZ EQUATIONS*

For modern chaos theory, one of the first respective set of equations—the so-called Lorenz equations—was derived from NSE. Lorenz, an American meteorologist interested in climate forecasting, was motivated by the nowadays well-known Rayleigh–Bénard convection (Moore, 1964, p. 619). This flow configuration can be realized within an infinitely extended thin liquid layer locked up between the two horizontal and free boundaries called the bottom and top surfaces. Due to the difference ΔT at these plain boundaries, a temperature gradient is maintained parallel to the vertically directed field vector of gravity. Beyond a critical difference $(\Delta T)_{cr}$ the resulting heat transfer processes, initially caused by heat conduction only, are always superimposed by increasingly dominating convective motion. Divergences in local mass densities give rise to local flows transporting hot liquid from below and turning cold fluid over at the top. Typical patterns in the form of *steady-state liquid rolls* occur and form the characteristic flow picture of the steady-state Rayleigh–Bénard convection (see, e.g., Kreuzer, 1983, p. 111).

Lorenz based his work on Saltzman's solution of this type of flow, the steady state of which can be described by means of Fourier series. Considering only three primary modes, Lorenz simplified Saltzman's set of partial differential equations—originally modeled for two time-dependent, two-dimensional flow velocity fields and one temperature field—to a set of only time-dependent ordinary differential equations. This set becomes

$$\begin{aligned} \dot{X}(\tau) &= \quad -Pr \bullet X + Pr \bullet Y \\ \dot{Y}(\tau) &= \quad -X \bullet Z + r \bullet X - Y \\ \dot{Z}(\tau) &= X \bullet Y \quad -b \bullet Z, \end{aligned} \tag{8.12}$$

where the quantities X, Y, and Z denote *amplitude functions*. Abbreviations are used as follows:

Prandtl Number: $\quad Pr := \mu\lambda^{-1}f^{\#}\Re(\kappa - 1)^{-1}$, according to (7.55); (8.13)

Rayleigh Number: $$Ra := \frac{gH^3\rho\gamma\Delta T}{\mu\alpha}; \qquad Ra_{cr} = \frac{\pi^4(1+a^2)^3}{a^2}; \tag{8.14}$$

Dimensionless process time: $$\tau := \alpha\pi^2\frac{1+a^2}{H^2}t; \qquad b := \frac{4}{1+a^2} \in [0;4]. \tag{8.15}$$

The most important parameter of (8.12) is $r := Ra/Ra_{cr}$, where the Rayleigh number Ra is defined by external quantities such as the constant temperature difference ΔT, the height H of the fluid layer, and the gravitational acceleration g, as well as some fluid parameters such as the thermal diffusivity α and the coefficient of volume expansion γ. However, its critical value Ra_{cr} may only be expressed by a single process property—the spatial extension a of one of the fluid cells—which is the most noticeable quality of the Rayleigh–Bénard convection. Note that this length a can normally be obtained only by experiment.

The solutions of the Lorenz Equations (8.12) contain the first "strange attractor" ever known that has become an emblem of chaos theory. The properties of these equations, as well as the very different dependencies on the three parameters Pr, r, and b, are discussed in full by Sparrow (1982). For certain values of these parameters, the typical form of the well-known Lorenz attractor is illustrated in Figure 8.3.

Let us now deduce a set of equations that may be regarded as analogous relations to the Lorenz Equations (8.12). These new differential equations refer to the physical background of (8.12) and will be derived from MNE as well as from ONE, using the Saltzman–Lorenz method. Assuming a gaseous layer arranged horizontally and extended arbitrarily in the x- and y-direction, the field force vector $\mathbf{f}$, along with the gradient of the pressure function Ω (i.e., the gas temperature), is vertically directed opposite to the z-direction. This arrangement is sketched in Figure 8.4, wherein the two velocity components u and w are also plotted together with the surface values of Ω.

The following analysis will be executed for a sectional area with y = constant. To include the field force $\mathbf{f}$, we must first extend MNE and ONE by the well-known Archimedean buoyant forces expressed by the term

$$\rho\mathbf{f} = -\rho g\left(\frac{\rho_H^0 - \rho}{\rho}\right). \tag{8.16}$$

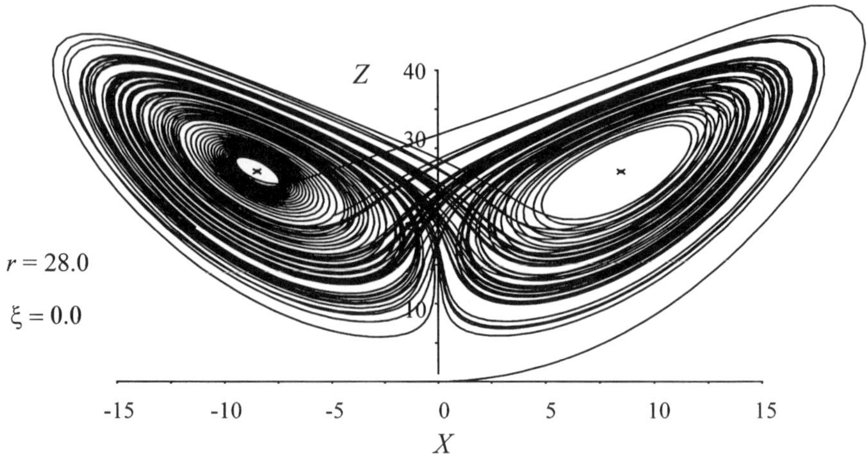

Figure 8.3 Original Lorenz attractor for characteristic values $Pr = 10$, $b = \frac{8}{3}$, and $r = 28$.

By crosswise multiplication and subsequent subtraction of the two equations of motion for the velocity components u and w, we obtain the *vorticity equation of motion*

$$D\omega + \omega\nabla \bullet \mathbf{v} = \nu_\tau \Delta\omega + g\frac{\partial}{\partial x}\left(\frac{\rho_H - \rho}{\rho}\right) \tag{8.17}$$

for MNE first, where the single existing component ω of the vorticity vector ς is given by

$$\omega = \frac{\partial u}{\partial z} - \frac{\partial w}{\partial x}. \tag{8.18}$$

Then, the vorticity equation of motion for ONE becomes

$$D\omega + \omega\ \nabla \bullet \mathbf{v} = g\frac{\partial}{\partial x}\left(\frac{\rho_H - \rho}{\rho}\right), \tag{8.19}$$

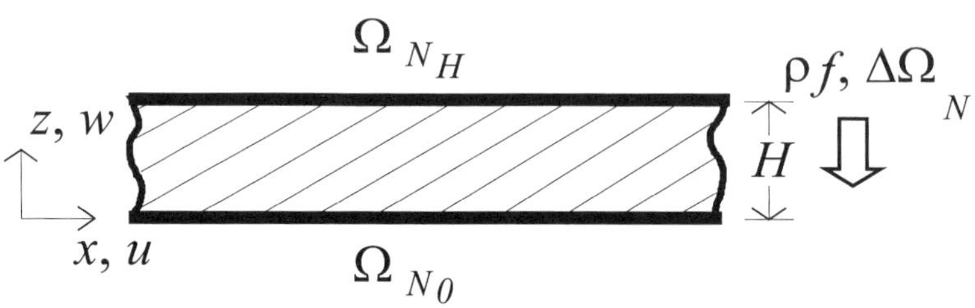

Figure 8.4 Basic configuration of Rayleigh–Bénard convection with free surfaces.

where, as before, ρ_H is the value of the mass density at the unheated upper surface ($z = H$) of the fluid layer.

The pressure function is represented by the superimposition of the linear Ω-distribution, valid for pure heat conduction, with an interference function Θ according to the relation

$$\frac{\Omega_N(x, z, t)}{\Omega_{NH}} = 1 + \frac{\Omega_{N0} - \Omega_{NH}}{\Omega_{NH}}\left(1 - \frac{z}{H}\right) + \Theta(x, z, t)\,. \tag{8.20}$$

This will lead to a change of both vorticity equations of motion by means of a polytropic relation between the density ρ and the gas temperature $\Re T = \Omega$ as well as an adequate series expansion.

For MNE and ONE, the vorticity equations now become

$$D\omega + \omega\,\nabla\cdot\mathbf{v} = \nu_\tau \Delta\,\omega - \frac{1}{n-1} g \frac{\partial\Theta}{\partial x} \tag{8.21}$$

and

$$D\omega + \omega\,\nabla\cdot\mathbf{v} = -\frac{1}{n-1} g \frac{\partial\Theta}{\partial x} \tag{8.22}$$

respectively. The energy equation, aside from the different viscosity parameters for MNE and ONE, can be transformed into a differential relation for the interference function Θ:

$$D\Theta - \frac{\Delta\Omega_N}{\Omega_{NH}}\frac{1}{H} w = f_\ell \nu_\ell\,\Delta\Theta - (\kappa - 1)\,\Theta\,\nabla\cdot\mathbf{v}; \qquad \ell = 0, \tau. \tag{8.23}$$

Following Lorenz's approach, first the velocities, then the vorticity, and finally the Θ-function are each approximated by a degenerate Fourier series expansion, for which only the first harmonics with the real wave number k are considered, viz.

$$\begin{aligned}
u(x, z, t) &= -\frac{\alpha}{H} X(t) \sin(k\pi x) \cos(\pi z) \\
w(x, z, t) &= \frac{\alpha}{H} Y(t) \sin(k\pi x) \cos(\pi z) \\
\omega(x, z, t) &= \frac{\alpha}{H^2} \pi (k+1) X(t) \sin(k\pi x) \cos(\pi z) \\
\Theta(x, z, t) &= \frac{\nu_\ell \alpha}{g H^3} (n-1)\,\pi^2 \frac{k+1}{k} (k^2+1)\,[Y(t) \cos(k\pi x) \sin(\pi z) - Z(t) \sin(2\pi z)]\,.
\end{aligned} \tag{8.24}$$

Inserting these expansions into Equations (8.18), (8.21), or (8.22), respectively, and then also into Equation (8.23), we can use the resulting relations, after some algebraic manipulations, to derive a set of differential equations. The solutions of these differential equations refer to the dimensionless amplitudes X, Y, and Z as mere algebraic functions of the dimensionless time coordinate, defined as follows:

$$\hat{\tau} := \left(k^2 + 1\right)\left(\frac{\pi}{H}\right)^2 \kappa\alpha\, t. \tag{8.25}$$

The resulting set of ordinary nonlinear differential equations

$$\begin{aligned}\dot{X}(\hat{\tau}) &= \xi XX \qquad \kappa^{-1}PrX + \kappa^{-1}PrY \\ \dot{Y}(\hat{\tau}) &= \qquad \psi XZ + rX - \qquad Y \\ \dot{Z}(\hat{\tau}) &= \varsigma XY + \frac{\kappa-1}{2}\xi XZ \qquad -bZ\end{aligned} \tag{8.26}$$

is characterized by two fluid parameters, the Prandtl number Pr and the isentropic index κ, as well as two constants, ς and ψ, following from the approximations used (see Lauster, 1995, p. 114). The constants also determine the coefficient ξ, by which two nonlinear terms in (8.26) appear. Within the framework of the theory, these nonlinear terms are related to the divergence $\nabla \bullet \mathbf{v}$ in such a way that they vanish for $\nabla \bullet \mathbf{v} \equiv 0$, in other words, exclusively for incompressible fluids. For this reason, the coefficient ξ may be thought to be an apt measure of compressibility influences and therefore can be varied within certain bounds.

The set (8.26) formally embraces the Lorenz equations as a limiting case for the values $\xi \equiv 0$, $\varsigma \equiv 1$, and $\psi \equiv -1$. Notwithstanding, there are some characteristic differences between the numerical results computed by means of the two Lorenz sets (8.12) and (8.26), although they both originate from the closely related NSE and MNE. Figures 8.5 through 8.8 show the compressibility effect by a sequence of representative plots presented for the Lorenz set (8.12)—if for example values $\varsigma \equiv 1$, and $\psi \equiv -1$ as well as $Pr = 10\kappa$ are selected for (8.26)—with the exception that the coefficient ξ will have to be varied. Starting from Lorenz's strange attractor (Figure 8.3), a stable fixed point appears by varying ξ values. For $\xi < 0$, this fixed point is located in the right square of the Z-X plane. It will arrive at its final position after a finite number of windings, the amount of which decreases with increasing values of ξ.

A second example refers to a solution of the original Lorenz equations that involves the stable, symmetric, and periodic orbit shown in Figure 8.9 for $r = 160$. This orbit remains (approximately) stable in the interval $154.4 < r < 166.07$ (see

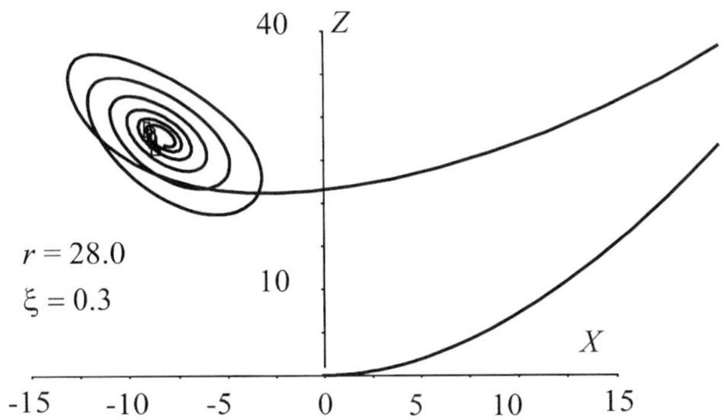

Figure 8.5 Stationary point for the Lorenz system based on MNE, $\xi = 0.3$.

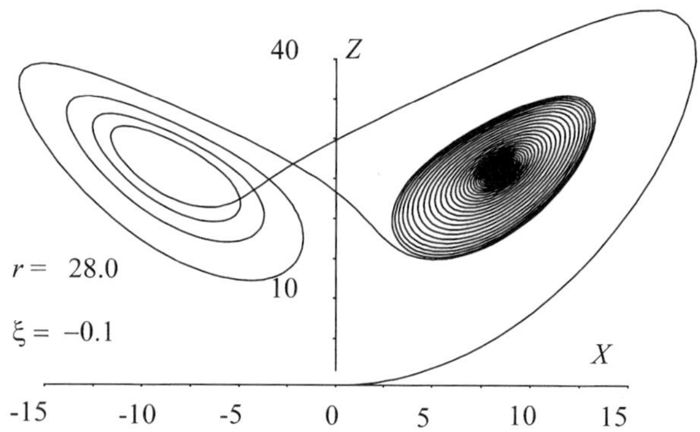

Figure 8.6 Stationary point for the Lorenz system based on MNE, $\xi = 0.1$.

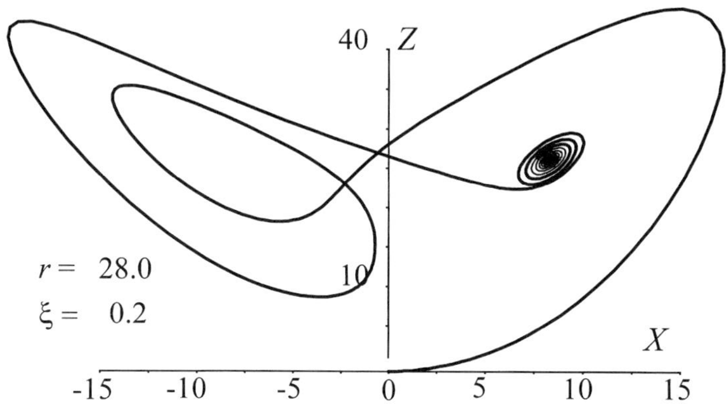

Figure 8.7 Stationary point for the Lorenz system based on MNE, $\xi = 0.2$.

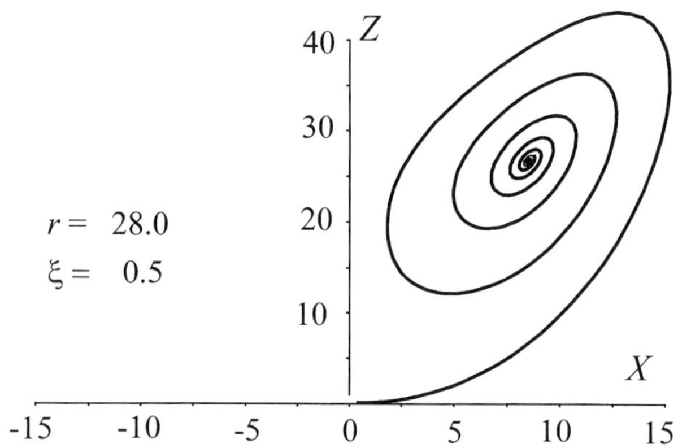

Figure 8.8 Stationary point for the Lorenz system based on MNE, $\xi = 0.5$.

Sparrow, 1982, p. 59). The Lorenz system based on MNE for $r = 160$ confirms this stable orbit even for increasing values of the compressibility coefficient ξ. But Figures 8.10 and 8.11 reveal that there are some characteristic differences from the reference case in Figure 8.9. First, the symmetry is lost. Moreover, there exists a special kind of orbit ruptured into fibers and forming noisy periodicity (Sparrow, 1982, p. 69). Surprisingly, the course of the orbit becomes more simple again for higher ξ values.

Some concluding remarks concerning MNE for Lorenz's problem may be useful:

- It may be proven that the coefficient ξ turns out to be a measure that gives information on the fluid compressibility at the actual position of any orbit point regarded. Thus, it can be shown that values $\xi < 0$ relate to points either of the right square with an increasing mass density or the left square with decreasing mass density. The reverse is true for $\xi > 0$.
- For small amounts of ξ, "chaotic solutions" (as for example the strange attractor for $r = 28$) become stable solutions that either run along stable orbits or tend toward certain stationary points.
- Periodic solutions along stable orbits split up in such a way that fibers appear forming certain bands within which some trajectories oscillate. Anomalous periodic orbits do occur for small ξ values, but for higher ones the orbit loses its complicated pattern, even compared with the original orbit (that is, for $\xi \equiv 0$, see Figure 8.9).

The solutions to the Lorenz problem have entirely different representations when ONE is used. Due to the different vorticity equation (8.22) for ONE, the Lorenz set (8.26) has to be replaced by the set

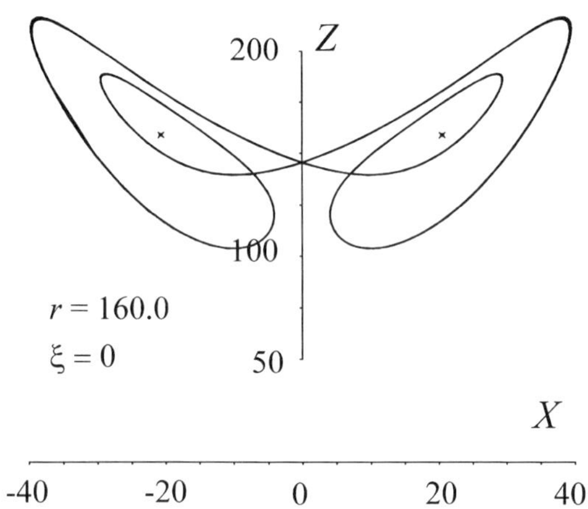

Figure 8.9 Stable orbit for the Lorenz system based on MNE, $\xi = 0$.

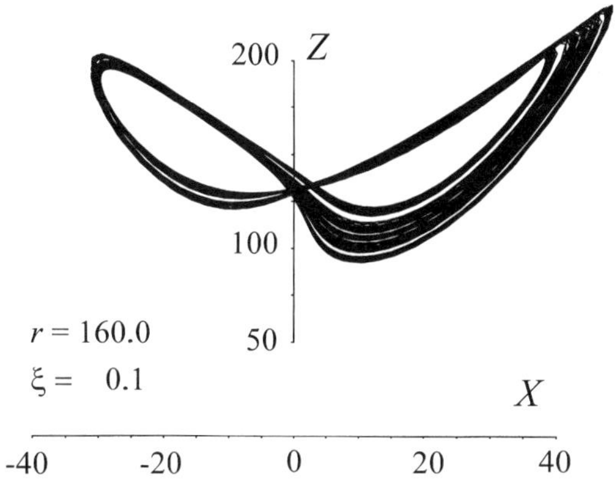

Figure 8.10 Stable orbit for the Lorenz system based on MNE, $\xi = 0.1$.

$$\begin{aligned}
\dot{X}(\hat{\tau}) &= \xi XX + \kappa^{-1} PrY \\
\dot{Y}(\hat{\tau}) &= \psi XZ + rX - Y \\
\dot{Z}(\hat{\tau}) &= \varsigma\, XY + \frac{\kappa - 1}{2}\xi XZ - bZ,
\end{aligned} \tag{8.27}$$

where only the second right-hand term of $(8.26)_1$ does not enter. This apparently small deviation causes quite different behavior in the solution than with either the NSE and MNE systems.

For parameters like those of the original Lorenz equations, the set (8.27) possesses only a single stationary point; this also holds for arbitrarily selected Rayleigh

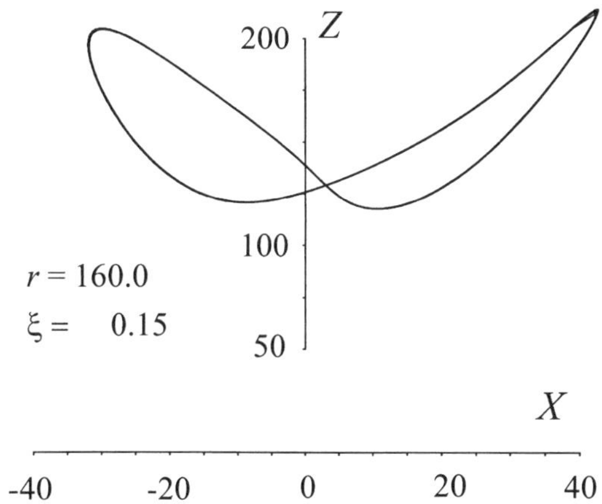

Figure 8.11 Stable orbit for Equations (8.26).

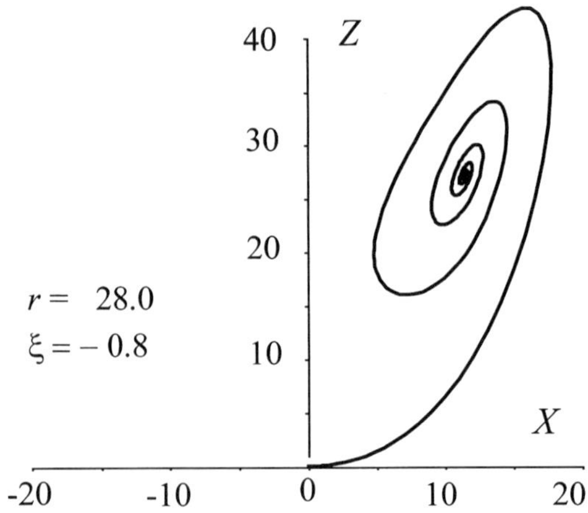

Figure 8.12 Stationary point for Equations (8.27).

ratios r. This means that pure heat conduction prevails even for high temperature differences. But, indeed, this distinguished solution is very susceptible to the slightest deflection from the neutral position.

With values $\xi < 0$ this sensitivity can be eliminated. Figures 8.12 to 8.14 show the change of behavior with increasing amounts of ξ. Additionally, the values of the ratio r will increase as the admissible ξ-intervals become ever smaller.

Concerning the Lorenz equation set (8.12) and the Lorenz-like sets (8.26) and (8.27), note that although the original Lorenz set is of great interest for the theory of deterministic chaos, its physical relevance is questionable, particularly with respect

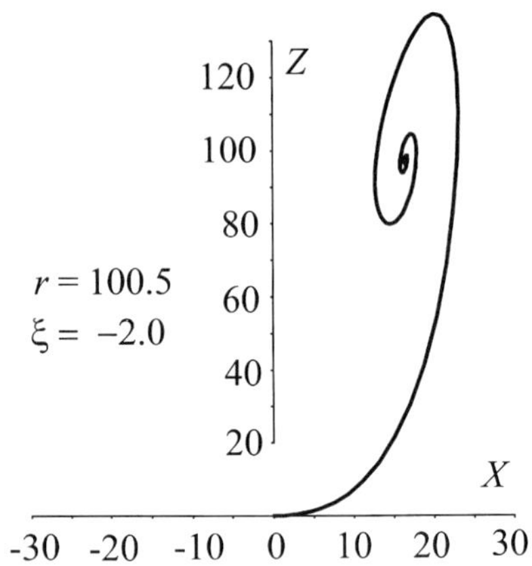

Figure 8.13 Fixed points for the Lorenz system based on ONE; $r = 100, 5, \xi = -2.0$.

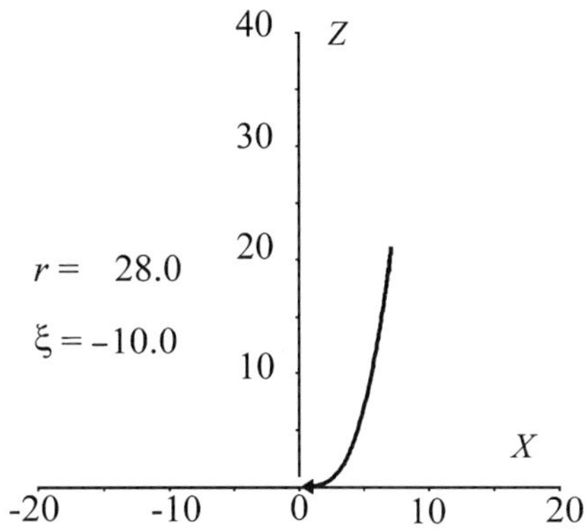

Figure 8.14 Fixed points for the Lorenz system based on ONE; $r = 28.0$, $\xi = -10.0$.

to Lorenz's intention to apply it to weather forecasting or even to turbulence research. Gumowski concisely commented on the latter point as follows: "The Lorenz turbulent flow thus makes no contribution to the understanding of physically turbulent fluid flow" (1989, p. 6).

In conclusion, four statements should be sufficient:

- Lorenz's original parameters, especially the Prandtl number $Pr = 10$, prevents any conclusions relevant for the actual behavior of atmospheric motion.
- All three Lorenz sets presented above describe nothing but some fluctuations of velocity components and the gas temperature around their mean values. The latter are focused on a single space point. The details of the deduction reveal that in fact this unrealistic result is only due to the diverse approximations commonly used.
- The boundary value problem of the partial Navier–Stokes differential equations is converted into the initial problem of Lorenz's ordinary differential equations by arbitrarily truncating Ritz–Galerkin series expansion. It is a verifiable fact that all the mathematical restrictions and physical approximations used for the deduction of the Lorenz equations were of such a kind that every similarity, or even identity, between any solutions of Lorenz's equations and the original NSE can be safely excluded.
- Ames proved that neither the Lorenz equations nor the Lorenz-like equations following from MNE display any chaotic behavior for certain combinations of parameters (cf. Lauster, 1995, p. 115). It is remarkable that fluids with $Pr \approx 1$ are affected. For this special case, one should expect to lose all the interesting large r behavior (Sparrow, 1982, p. 184). However, any numbers $Pr < 10$ seem to be consistent only with the main assumptions of the Lorenz equations, if compressibility is considered. The corresponding experiences with the Lorenz sets derived from MNE and ONE suggest that the famous chaotic behavior of

the Lorenz equations chiefly follows from the incorporated incompressibility condition.

Still, further research is needed, provided that knowledge of real flow patterns are of any interest in the physical background of phenomena in which trajectories oscillate in a pseudo-random way for long periods of time before finally settling down to stable stationary or stable periodic behavior. The same is true when information about dynamical events is demanded. Certainly, it is of interest when trajectories alternate irregularly between chaotic and apparently stable periodic behaviors and when trajectories appear to be chaotic even though they stay very close to a nonstable periodic orbit.

8.2.3 BOUNDARY LAYER FLOWS

Since boundary layer flows are of great interest in practice, we will relate this concept to the Nehring approximations (7.65) and (7.66). It does not make sense to transform ONE into their boundary layer representation: Prandtl's idea falls short of them in that their decomposition of the flow field into a near-wall regime and a frictionless region of the so-called potential flow will not work. The ONE set offers an integrated concept, that is, a rough approximation of the field equations without abandoning effects that in reality simultaneously dominate the fluid behavior near the wall and along the bulk flow.

Another situation is given for MNE that are structurally comparable with NSE. The usual procedure for finding their adequate boundary layer representations first includes the reference of all field variables and parameters of MNE to an appropriate flow quantity, such as the free flow velocity U_∞ and a characteristic length L for the whole flow configuration under consideration.

The following set of references may be regarded as representative for MNE:

$$\begin{aligned} x &:= L\hat{x}; \qquad Re := \frac{U_\infty L}{\nu_\tau} & u &:= U_\infty \hat{u}; \\ y &:= \frac{L}{\sqrt{Re}}\hat{y}; & v &:= \frac{U_\infty}{\sqrt{Re}}\hat{v}; \\ z &:= L\hat{z}; & w &:= U_\infty \hat{w}; \\ t &:= \frac{L}{U_\infty}; & \Omega_N &:= U_\infty^2 \hat{\Omega}_N \end{aligned} \tag{8.28}$$

The passage to the limit $Re \to \infty$ for the Reynolds number leads to the nondimensional form of the boundary layer equations. After introducing the desired units by means of (8.28), we obtain the dimensionalized boundary layer equations derived from MNE:

$$\begin{aligned} \frac{\partial u}{\partial t} + u\frac{\partial u}{\partial x} + v\frac{\partial u}{\partial y} + w\frac{\partial u}{\partial z} &= \nu_\tau \frac{\partial^2 u}{\partial y^2} - \frac{\partial \Omega_N}{\partial x}; \\ 0 &= \frac{\partial \Omega_N}{\partial y}; \end{aligned}$$

$$\frac{\partial w}{\partial t} + u\frac{\partial w}{\partial x} + v\frac{\partial w}{\partial y} + w\frac{\partial w}{\partial z} = \nu_\tau \frac{\partial^2 w}{\partial y^2} - \frac{\partial \Omega_N}{\partial x};$$
$$\frac{\partial \Omega_N}{\partial t} + u\frac{\partial \Omega_N}{\partial x} + w\frac{\partial \Omega_N}{\partial z} = -(\kappa - 1)\,\Omega_N\left(\frac{\partial u}{\partial x} + \frac{\partial v}{\partial y} + \frac{\partial w}{\partial z}\right). \tag{8.29}$$

To check the capability of (8.29) it obviously suffices to study a boundary layer problem that is generally accepted as characteristic in a particular way for the boundary layer concept. Undoubtedly, an adequate example is given by the ordinary differential equation derived in 1930 by Falkner and Skan and validated for so-called self-similar solutions of steady-state, two-dimensional boundary layer equations. These similar solutions concern such velocity profiles as $u(x, y)$, which differ among themselves at various positions x only by a scale factor in the coordinates x and y.

The general assumptions for finding scaling generators for differential equations are elaborated, for example, by Rogers and Ames (1989). Ames et al. (1995) have also recently published results of an analysis that includes certain scaling properties of MNE. Applied to the Falkner–Skan equation, its derivation is confined to the boundary layer on a semi-infinite wedge at zero angle of attack, that is, to a special kind of potential flows (Moore, 1964, p. 116) dealing with an external velocity distribution given by

$$U(x) := U_\infty x^m. \tag{8.30}$$

For negative values of m, the pressure gradient in this distribution is adverse, and separation may be expected. The velocity U_∞ refers to a constant initial value, and the exponent m relates to the wedge angle ($-0.199 < \beta < 2$) as follows:

$$m := \frac{\beta}{2-\beta}. \tag{8.31}$$

Choosing appropriate values for β it is possible to describe some other flow forms, such as a flow toward a flat plate normal to the stream ($\beta = \pi$).

Introducing the similarity variable

$$\eta := y\sqrt{\frac{m+1}{2}\frac{U_\infty}{\nu_\tau}}\,x^{\frac{m-1}{2}} \tag{8.32}$$

and a scalar stream function

$$\psi := \rho_\infty \sqrt{\frac{2}{m+1}\nu_\tau U_\infty}\,x^{\frac{m+1}{2} - \frac{2m}{1-n}} f(n), \tag{8.33}$$

here defined for compressible flows by the relationships

$$\rho u := \frac{\partial \psi}{\partial y}; \qquad \rho v := -\frac{\partial \psi}{\partial x}, \tag{8.34}$$

the steady-state continuity equation is identically fulfilled first. Characteristic consequences of both the definitions (8.32) and (8.33) concern the following similarity property: If b is an arbitrary constant, the system is invariant under the mappings

$$y \Rightarrow by; \qquad x \Rightarrow b^2x; \qquad \psi \Rightarrow b\psi.$$

Inserting now (8.32) and (8.33) into the set (8.29), the latter can be reduced to a single fourth-order ordinary differential equation

$$f''' + \left(1 - \frac{2\beta}{n-1}\right) ff'' + \beta\,(1 - f'^2) = 0 \tag{8.35}$$

with boundary conditions

$$f(0) = f'(0) \equiv 0; \qquad f'(\infty) = 1 \tag{8.36}$$

for the nondimensional stream function $f(\eta)$ and its derivative f' with respect to the similarity variable η (Lauster, 1995, p. 132).

Compared with the Falkner–Skan equation, the only additional factor in (8.35) is the factor $\langle 1 - 2\beta\,(n-1)^{-1}\rangle$. For this reason, some interesting special cases of the Falkner–Skan theory may easily be settled, provided that the polytropic index n is presumed to be infinite. This limiting case is identical to the assumption of an incompressible flow, for which the Falkner–Skan equation is valid. For $\beta \equiv 0$ it reduces to the well-known *Blasius equation.* With $\beta \equiv 1$, we can obtain the boundary layer solution for two-dimensional flow toward a flat plate, which is also a solution of the full Navier–Stokes equations (the exact solution). The same is true for $\beta \equiv \frac{1}{2}$, for which the exact solution for an axially symmetric flow toward a plane is formally identical to the boundary layer solution for a rectangular wedge (cf. Moore, 1964, p. 118).

Taking into account these three standard cases, along with the case of flow separation ($\beta \equiv -0.199$), the numerical solutions of the set (8.35–8.36) allow the study of the influence of the factor $\langle 1 - 2\beta\,(n-1)^{-1}\rangle$ on the behavior of certain boundary layers. In Figure 8.15 some results are sketched, for which a realistic index n—that is, an admissible value of the wedge angle β—has been selected.

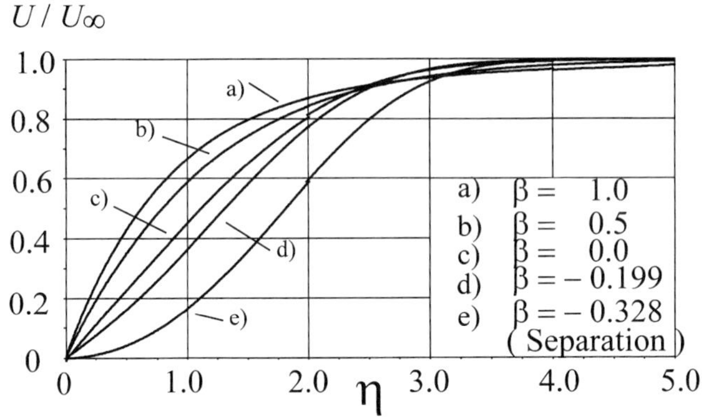

Figure 8.15 Falkner–Skan boundary layers based on MNE.

Whereas for $\beta \equiv 0$ identical results follow from (8.35) and the Falkner–Skan equation, the parameter n strongly influences the courses of the solutions characterized by the other values of β.The essentials of this behavior may be summarized by two items:

1. Values $\beta > 0$ represent compressible wedge flows forming boundary layers that thicken continuously with increasing length x, whereby each local value of the boundary layer thickness is larger, compared with the corresponding value of incompressible fluids.
2. Separation happens once and for all with β amounts that differ considerably from the well-known result of the Falkner–Skan theory. That theory offers only a single value indicating separation, whereas Equation (8.35) describes a behavior of separation depending on the polytropic index n. Since values $\beta < 0$ represent decelerated flows, the result $\beta = -0.328$ along with $n = 2$ manifests the fact that compressible boundary layers are by far more insensitive to increase in pressure than the respective incompressible flow patterns. This general behavior also confirms some experiences with turbulent boundary layers subject to a considerable increase in pressure (cf. Walz, 1966, p. 180).

We should appraise these statements with regard to the remarkable meaning of the Falkner–Skan equation for all approximations applied to any boundary layer theory (see, e.g., Walz, 1966, p. 34). A more detailed analysis of (8.35) indicates that the polytropic index n strongly influences the course of the velocity profiles. It should be borne in mind that this parameter n has a double quality. First, it is an index assumed to be discretely assigned to some thermodynamic standard cases:

$n \equiv 0 \Rightarrow$ isobaric changes of state

$n \equiv 1 \Rightarrow$ isothermal changes of state

$n \equiv \kappa \Rightarrow$ isentropic changes of state

$n \equiv \infty \Rightarrow$ incompressible changes of state

Although these four conventional changes of state manifest a special kind of process idealization, it is evident that they also indirectly express a certain measure of compressibility, provided that the concept of polytropic changes of state is accepted. Hence, in comparison with the highest possible value $n = 5/3$ for any isentropic changes of a monatomic gas, an arbitrary value of about $n = 2$ represents weakly compressible changes of state at best.

The second quality refers to the possibility of using n as a real number assumed to be continuously varied within certain bounds. Such an option is common practice in every applied boundary layer theory, where n plays the role of a so-called *form parameter* (Walz, 1966, p. 3) of any velocity profile. That parameter is then treated as one of the unknown quantities of any approximate theory. In this second case the index n does not only represent the degree of compressibility; it also implicitly comprises the dissipative dynamics of the boundary layer flow and its special dependency on the whole flow configuration, particularly defined by the body shape under consideration.

Altogether, MNE seem to be an appropriate tool for certain applications in boundary layer problems. This is especially true for approximate boundary layer theories related to the Falkner–Skan concept, whose physical meaning for the understanding of flow behavior near walls cannot be underrated (cf. Moore, 1964, p. 120). For such practice, MNE even extend the possibilities of the Navier–Stokes equations applied approximately to more complicated boundary layer flows. Thus, for instance, a more detailed analysis (Lauster, 1995, p. 59) indicates that for a wedge angle $\beta = 1$ along with an index $n = 1.2$, which might be realistic for any adiabatic process of a polyatomic gas, boundary layer flow is rather unlikely compared with the classical Falkner–Skan results.

8.3 Basic Applications of Gasdynamics

8.3.1 THE STRUCTURE OF SHOCK WAVES

Equations like the two sets (7.65) and (7.66) that point out the compressibility of real fluids are expected to be useful in practice for problems typical in gasdynamics. That such an expectancy may sometimes fail to be satisfied will be shown in the light of steady shock waves in gases. In the following exposition, we intend to explicate the decisive role of dissipation for the spatial structures of these shock waves.

A shock wave implicates the process of changing from a uniform upstream flow to a uniform downstream flow. In a shock wave the changes of velocity, temperature, and so on occur in the direction of motion. Thus, a shock wave is a longitudinal wave. It is one-dimensional by definition because there are no flow gradients in directions parallel to the plane of the wave. As a frame of reference, a zero velocity in this plane is assumed. A *shock Mach number* Ma_s is defined as the ratio of the speed of the wave (relative to the upstream flow) to the speed of sound in the gas.

Every shock wave is stationary with respect to the frame of reference. Consequently, the shock Mach number may be identified with the actual flow Mach number of the upstream gas. Furthermore, this stationary property allows us to fix the pressure, temperature, density, and steady stream velocity ratios across the wave by means of the *Rankine–Hugoniot jump relations* (cf. Vincenti and Kruger, 1967, p. 413). These relations are proper adaptions to the general conservation laws of mass, momentum, and energy; they depend only on the specific heat ratio and the shock Mach number. As an essential property, the internal structure of the wave is, however, conditioned by viscous and heat conduction effects. It represents a flow situation that is strongly influenced by thermal non-equilibrium for large values of Ma_s, but does not involve the uncertainties associated with unknown boundary conditions. For this reason, studies on shock wave structures are accessible to reliable experimentation.

The correct theoretical understanding of these structures, particularly their mathematical description, is quite another matter. Such understanding requires above all quantitative rather than qualitative interpretation. Therefore, we cannot simply calculate a shock structure for which the profiles agree with the suspected course but are faulty if compared with experimental data. This statement is also relevant insofar as incorrect laws of material properties and scattered experimental data strongly influence the reliability of calculated shock structures even for small Mach numbers (cf. Heß, 1981, p. 68). In this context it is evident that details of the prevailing dissipative phenomena must be known. Hence, efficient kinetic theories of gases, like the Chapman–Enskog approach for transport properties of nonuniform gases, are useful. Since the Chapman–Enskog theory leads to the Navier–Stokes equations, the use of the latter is familiar via the continuum hypothesis. This far-reaching foundation of NSE is the alleged sore spot of its application to exactly such special flow situations of fluids that are subject to steep gradients of their local state quantities.

Although the continuum hypothesis has always burdened the shock wave theory, the items concerning the internal structure of shock waves have been studied by various investigators since the work of Rankine in 1870. In particular, the normal shock relations have the distinct advantage that they do not involve the complicating effect of molecular interactions with solid surfaces.

Some prominent authors claim that just the NSE, as the outstanding example of continuous flow dynamics, "give an accurate description of the structure of weak shock waves" (cf. Vincenti and Kruger, 1967, p. 413). Such statements disguise the facts because they do not reveal the assumptions needed to obtain the postulated agreement between theory and experiment. Unfortunately, evidence for measured steady shock wave profiles is given in two publications by Sherman and Talbot for helium (1955) and argon (1959) regarding upstream, supersonic conditions with Mach numbers little less than two.

A more detailed analysis reveals that a good adaption of the theoretical results to the experimental data may be achieved by the help of appropriate values of transport characteristics like viscosity and heat conductivity. However, such a method is merely empirical and does not give effectual insight into the transport mechanisms involved. A notable example will help us to see the difficulties in connection with the meaning of, say, viscosity with regard to the structure of shock waves in pure fluids. Viscosity of a gas is defined well within the scope of both the early kinetic theories and the Chapman–Enskog approach. Systematic expansion of the latter, from the first-order perturbation solution to the Burnett approximation, will lead to completely different experiences with regard to shock wave experiments (cf. Simon, 1976). Thus, for instance, Bird interpreted Sherman and Talbot's data to be "among the first results to cast doubt on the validity of the Burnett formulation" (1976, p. 134). As opposed to this, both Foch's and Simon's research indicate that with these higher-order hydrodynamic theories we can improve the mathematical description of shock profiles even for Mach numbers up to four (see, e.g., Heß, 1981, p. 38). Indeed, the experimental results gained by Alsmeyer (1974, p. 54) confirm this expectation with respect to the usual asymmetry of the density profile, but in no other way if the quantitative courses of the profiles themselves are evaluated.

Before we can check the efficiency of ONE and MNE for shock structure representations, the pertaining options of the Navier–Stokes equations (NSE) must be analyzed. Applied to the shock structure problem, they proceed to a set of ordinary differential equations including mass conservation, law of motion, and balance of specific enthalpy h:

$$\frac{d}{dx}(\rho u) = 0; \qquad \frac{d}{dx}(p + \rho u^2 - \tau_{xx}) = 0; \qquad \frac{d}{dx}\left(\rho u \langle h + \tfrac{1}{2}u^2 \rangle + q_x - u\tau_{xx}\right) = 0. \quad (8.37)$$

Set (8.37) is amenable to numerical solutions, always provided that appropriate relationships are explicitly available for the heat flux rate q_x and the viscous pressure tensor τ_{xx} with respect to the x-coordinate. For the Chapman–Enskog first-order perturbation solutions there exist the well-known relations

$$q_x = -k\frac{dT}{dx}; \qquad \tau_{xx} = \frac{4}{3}\mu\frac{du}{dx}, \quad (8.38)$$

where the heat conductivity k as well as the (dynamical) viscosity μ are defined for nonuniform gases, quite within the scope of the kinetic theory of gases. It is relevant for gas dynamics that within its range of validity this step of approximation furnishes a direct coupling between k and μ, as well as the dependency of μ on temperature, as follows:

$$k = \frac{3}{2}c_p\mu; \qquad c_p = \frac{5}{2}\frac{k_B}{m}; \qquad \mu = \frac{5}{16}\frac{\sqrt{\pi m k_B}}{\sigma^2}T^{\omega}. \quad (8.39)$$

For a kinetic model for monatomic gases consisting of spherical atoms with mass m and diameter σ, the power function $\mu \propto T^{\omega}$ (with the Boltzmann constant k_B) is settled by the index ω, defined by $\omega := \frac{1}{2} + \Lambda$. Theoretically, the factor Λ is assigned to the range of values $0 < \Lambda < \frac{1}{2}$, but practically, Λ is determined by numerical adaption to different experimental values of certain flow phenomena. Hagen–Poiseuille flows, but also shock wave experiments, are commonly believed to be appropriate for such an adaptive procedure. However, after the adaption by a free parameter like Λ, it is almost trivial that any agreement between some experimental data and the respective theoretical scheme seems coincidental rather than consistent with the theory applied to different data. Indeed, some authors imply the validity of NSE for shock structure problems from just such an agreement (see, e.g., Fizsdon et al., 1974).

Notwithstanding, a more detailed analysis in light of the extensive and reliable measurements performed by Schmidt (1969) and Alsmeyer (1974)[15] leads to quite another conclusion. Supported by an efficient setup, using density-dependent ab-

[15]It is strange that the experiments published by Schmidt in 1969 are well-known in literature and are often quoted, even in textbooks. In contrast, Alsmeyer's research is almost unknown, although it improved, carried on, and extended Schmidt's work (Alsmeyer, 1974, p. 27).

sorption laws deduced by electron beam scattering, Alsmeyer's shock tube experiments with argon and nitrogen cover a shock Mach number range from about $Ma_s =$ 1.5 to $Ma_s = 10$. It should be stressed that the density profiles are accurate within a tolerance of about 1% around their averaged courses (Alsmeyer, 1974, pp. 43–44).

Alsmeyer comprehensively compared his experimental data with profiles calculated from the NSE set (8.37–8.39) and adapted to the measured curves by appropriate values of the factor Λ. His comment leaves no doubt:

> It becomes apparent that even for a small Mach number of about $Ma_s = 1.55$, the Navier–Stokes approximation does not actually agree with the measurements for either the solid sphere model ($\mu \propto T^{1/2}$) or for Maxwell molecules ($\mu \approx T$). The same is true for more realistic models specified by "laws" $\mu \sim T^{\omega}$ ($0.5 < \omega < 1$).This theory provides gradients of the gas density that are too large. These salient divergences increase with rising shock Mach numbers. Such deviations of the Navier–Stokes shock profiles from the real ones were foreseen by Hicks et al. even for the very small Mach number $Ma_s = 1.2$. . . . These conclusions are very contradictory to the so-far predominant opinion according to which the Navier–Stokes approximation should be valid up to $Ma_s = 2$. (Alsmeyer, 1974, pp. 53–54; author's translation).

The well-known moment method applied by Mott-Smith in 1951 to shock wave structures reproduces the typical asymmetry of the shock profiles no better than does NSE. Nevertheless, the Mott-Smith solution is remarkably successful in predicting the thickness of very strong shock waves (see, e.g., Schmidt, 1969). Moreover, it is notable that Alsmeyer's research (1974, p. 50), as contrasted with the textbook knowledge (Bird, 1976, p. 137), confirms the utility of Mott-Smith's approach even for small Ma_s values. Mott-Smith assumed that the particle distribution function within the wave can be represented as a linear combination of the equilibrium functions that may be applied to the uniform upstream and downstream flows. We cannot discuss the details of that theory here, but it should be stressed that, based on fundamentals of continuum physics, the key to Mott-Smith's approach may be found in his preference for an adequate non-equilibrium concept. We will discuss that concept at the end of this section.

To attach ONE and MNE in an unambiguous way to the problem explained above, it seems reasonable to perform the comparison by means of an exact solution of the one-dimensional NSE (8.37–38). This method corresponds to a recommendation given by Vincenti and Kruger (1967, p. 415); it is designed to get closed algebraic solutions only for some special parameters indicating the gas under consideration.

Figure 8.16 illustrates the basic configuration of a normal shock wave along with its steep gradients of the common field variables. The one-dimensional, steady-state flow runs from left to right: A monatomic gas entering from the state $-\infty$ with hypersonic velocity is decelerated within a very thin regime to a subsonic velocity determined in the limit $+\infty$ by the Rankine–Hugoniot relations.

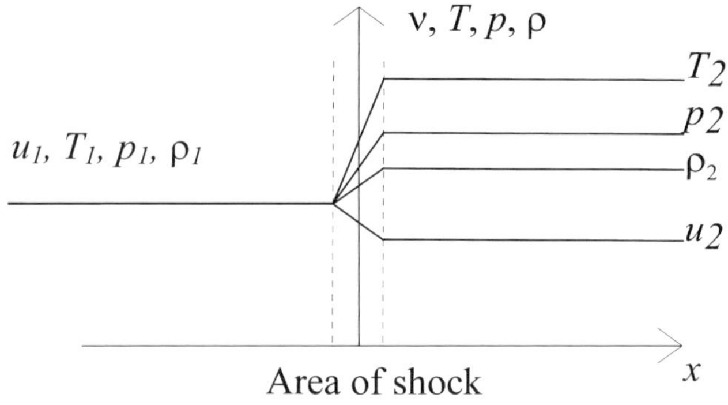

Figure 8.16 One-dimensional, steady-state shock wave structure.

Only for one-dimensional flows do both ONE and MNE possess identical equations of motion, viz.

$$u\frac{\partial u}{\partial x} = \nu\frac{\partial^2 u}{\partial x^2} - \frac{\partial \Omega_N}{\partial x}, \tag{8.40}$$

where ν stands for the two (normally different) kinematic viscosities of both ONE and MNE. Since both energy equations also agree formally, we may deal with either ONE (7.66) or MNE (7.67).

In connection with the energy Equation (7.56), note that the shock wave structure problem needs the consideration of the so-called dissipation function $\boldsymbol{\tau}:\nabla\mathbf{v}$ as a stringent consequence of the various steep gradients within the shock front. Consequently, both Nehring equations must be modified. For one-dimensional and stationary flow, the relationship

$$\rho u c_v \frac{\partial \Omega_N}{\partial x} = \frac{f\eta}{\Re}\frac{\partial^2 \Omega_N}{\partial x^2} + \rho u \frac{\partial \Omega_N}{\partial x} + \eta\left(\frac{\partial u}{\partial x}\right)^2 \tag{8.41}$$

results and is valid under the continuity restriction $\rho_1 u_1 = \rho_2 u_2 = \rho u$ for the constant mass flow. The subscripts 1 and 2 refer to the up- and downstream conditions, respectively (see Figure 8.16).

Let us now introduce a quantity that in my opinion contributes more irritation than profit for a deeper understanding of non-equilibrium phenomena in gasdynamics. It concerns the *effective mean free path* (mfp) ℓ, known from gas kinetics and given by the relationship

$$\ell = 3\frac{\eta}{\rho_1}\sqrt{\frac{n}{n-1}\,\frac{\pi}{8\Omega_{N_1}}}. \tag{8.42}$$

Allegedly, it is convenient to use ℓ as a microscopic length scale, regardless of the degree of non-equilibrium. That is, the mfp ℓ is subject to the convention defined by Equation (8.42). Thus, it is related to an effective intermolecular potential by means of the Chapman–Enskog first-order perturbation solution for the viscosity, regardless of an application of any eventual macroscopic relation between viscous stress and velocity gradients.

Following Lighthill's well-known deduction of an exact algebraic solution for the shock wave structure problem based on NSE, we easily obtain, after some mathematical manipulations (Lauster, 1995, p. 128), the nonlinear ordinary differential equation

$$\Phi\frac{d\Phi}{d\xi} = \beta Ma_1\,[2\,(\Phi-1)\,(\Phi-\alpha) - \gamma\,(\Phi^2-1)\,] \tag{8.43}$$

for the nondimensional gas velocity $\Phi := u/u_1$. This equation describes a function of the likewise nondimensional length coordinate $\xi := x/\ell$. An essential limitation of Equation (8.43) is that for the sake of transparency both ONE and MNE are here confined to elementary gases classified exactly by the constant Prandtl number $Pr = \frac{3}{4}$. It is found experimentally that for most gases at low density the Prandtl number varies between about 0.65 and 0.85, with the average quite close to 0.75. If $Pr = \frac{3}{4}$, we deduce from the set (8.37) that the stagnation enthalpy $h + \frac{1}{2}\,u^2$ is constant throughout the wave and, furthermore that $u\tau_{xx} = q_x$ for the x-derivatives. The constants α, β, and γ in (8.43) are given by the following definitions

$$\alpha := \frac{n-1}{n}\frac{\Omega_{N_1}}{u_1^2};$$

$$\beta := \frac{9}{8}\sqrt{\frac{n}{n-1}\frac{\pi\kappa}{8}};$$

$$\gamma := \frac{n-1}{n}\frac{\kappa-1}{\kappa}, \tag{8.44}$$

and the upstream supersonic Mach number Ma_1 becomes: $Ma_1 = u_1\,(\kappa\,\Omega_{N1})^{-1/2}$.

Before we present the analytical solution of (8.43), let us compare directly the differential Equation (8.43) with the original shock wave differential equation. This was first derived by Lighthill and is given here in the form later published by Bird, Stewart, and Lightfoot (1960, p. 335):

$$\Phi\frac{d\Phi}{d\xi} = 2\beta'\,Ma_1\,(\Phi-1)\,(\Phi-\alpha'); \tag{8.45}$$

$$\alpha' := 2\frac{\kappa}{\kappa+1}\frac{\Omega_{N_1}}{u_1^2} + \frac{\kappa-1}{\kappa+1}; \qquad \beta' := \frac{9}{16}\sqrt{\left(\frac{\kappa+1}{\kappa}\right)^2\frac{\pi\kappa}{8}}.$$

The exact solution of (8.45) can also be found in that textbook. It is included in the implicit solution of (8.43) for the nondimensional velocity Φ, viz.

$$\beta Ma_1\,(\xi-\xi_0) = \frac{1}{2\,(2-\gamma)}\ell n\left\{\left|\left[\frac{\Phi-\dfrac{2\alpha+\gamma}{2-\gamma}}{\Phi-1}\right]^{\frac{2\alpha+\gamma}{\alpha+\gamma-1}}(2-\gamma)\,(\Phi-1)^2\right|\right\} + C_{IV}. \tag{8.46}$$

The two constants of integration ξ_0 and C_{IV} may be chosen arbitrarily. Of course, it is useful to fix the ξ-origin with their help. If ξ_0 is selected for such a position, where the velocity Φ assumes the value $\Phi_0 = 0.75$, then it is indeed still possible to correctly describe even slight changes of velocity with weak shock waves.

Note that the upstream condition $\lim \Phi \Rightarrow 1$ for $\xi \Rightarrow -\infty$ can only be satisfied for the constraint

$$\frac{2-\gamma}{\alpha+\gamma-1} < 0, \tag{8.47}$$

which imposes the restriction $\alpha + \gamma < 1$ on the solution (8.46) because the values of γ cover the range $0 < \gamma < 1$ according to definition $(8.44)_3$. Clearly, this is equivalent to the upstream supersonic condition $Ma_1 > 1$ for apparently optional values of the parameters α and β. However, the last coefficient β immediately depends on the parameter γ, and the first parameter α is determined by the downstream condition for $\xi \Rightarrow \infty$. The latter refers to the downstream subsonic velocity Φ_2 as follows:

$$\Phi_2 = \frac{2\alpha+\gamma}{2-\gamma} = \frac{(\kappa-1)\, Ma_1^2 + 2}{\left(\frac{n+1}{n-1}\kappa + 1\right) Ma_1^2}. \tag{8.48}$$

Unlike this formula, Lighthill's classical solution does not contain a functional dependency on the index n. But it results as a limiting case of Equation (8.48) for $n \rightarrow \infty$; that is, the bracket of its denominator is replaced by $(\kappa + 1)$. Consequently, the following statement might be true: The velocity difference $\Phi_1 - \Phi_2$ along a stationary, one-dimensional shock wave is larger for a Nehring gas than for a Navier–Stokes–Fourier fluid. This is illustrated in Figure 8.17 for various upstream supersonic Mach numbers. Of course, the corresponding profiles of the gas temperature, pressure, and density are also steeper than those obtained by NSE. Both kinds of profiles agree for $n \rightarrow \infty$.

Regarding the shock thickness, recall that the solutions $u(x)$, $T(x)$, and so on are functions that asymptotically tend to the initial and final states $u = u_1$ and $u = u_2$, for

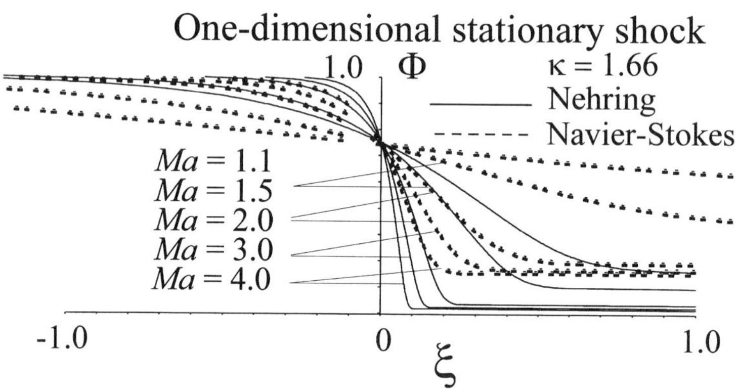

Figure 8.17 Shock wave profiles in monatomic gases for polytropic changes of state ($n = 2$).

which all dissipative events must vanish by definition. Both states are joined to each other by the Rankine–Hugoniot relations mentioned above. For ONE (and here also for MNE), they become

$$
\begin{aligned}
Ma_2^2 &= \frac{\left(\frac{n+1}{n-1}\kappa + 1\right) Ma_1^2}{(\kappa - 1)\, Ma_1^2 + 2} \\
\frac{\rho_2}{\rho_1} &= \frac{\left[2\frac{nk}{n-1} - (\kappa - 1)\right] Ma_1^2}{(\kappa - 1)\, Ma_1^2 + 2} \\
\frac{p_2}{p_1} &= \frac{(\kappa\, Ma_1^2 + 1)\left\{2\frac{nk}{n-1}\left[\frac{nk}{n-1} - (\kappa - 1)\right] Ma_1^2 - (\kappa - 1)\right\}}{\kappa\, Ma_1^2\left[(\kappa - 1) + 2\left(\frac{n}{n-1}\right)^2 \kappa - 2\frac{n}{n-1}(\kappa - 1)\right] + \kappa + 1} \\
\frac{\Omega_{N2}}{\Omega_{N1}} &= \frac{p_2}{p_1} \bullet \frac{\rho_1}{\rho_2}
\end{aligned}
\tag{8.49}
$$

They then convert into the classical relations for an infinite value of the index n.

Forming the shock wave thickness $\bar{x}_\rho$, defined as the largest gradient of the density profile

$$
\bar{x}_\rho := \frac{\rho_2 - \rho_1}{(d\rho/dx)_{\max}}, \tag{8.50}
$$

an indicator of the shock wave theory may be evaluated with the help of mfp ℓ, viz. the nondimensional reciprocal of the density-gradient shock wave thickness $\ell/\bar{x}_\rho$. Considering the continuity condition $\rho_1 u_1 = \rho_2 u_2 = \rho u = \text{constant}$, the thickness $\bar{x}_\rho$ can be reduced to an expression following from the differential equation of the velocity gradient (8.45) and its algebraic solution (8.46). Without going into details, the result of the evaluation affords the following view: For density profiles of Navier–Stokes–Fourier gases (i.e., for $n \to \infty$) the function $\ell/\bar{x}_\rho$, mainly depending on the shock wave Mach number $Ma_s \geq 1$, describes a monotonic increase from $\ell/\bar{x}_\rho = 0$ up to an asymptotic value slightly higher than 0.52 beyond $Ma_s \approx 5$. This is in sharp contrast to measured data, which tend to $\ell/\bar{x}_\rho \approx 0.3$ beyond $Ma_s \approx 3$ for argon (cf. Vincenti and Kruger, 1967, p. 421) and pass a flat maximum of about $\ell/\bar{x}_\rho \approx 0.37$ at nearly $Ma_s \approx 5.5$ for nitrogen (cf. Alsmeyer, 1974, p. 60).

These discrepancies are considerably augmented for decreasing values of the polytropic index n. Thus, particularly for strong shock waves, it might be that the continuum concept is now untenable. But, indeed, the enticement to draw such an inference is matched by one's own prejudice: Notions such as trajectory or mean free path are only meaningful within a purely mechanical point of view. In other words, those terms presume matter to be exclusively consisting of mass-point particles. For this reason, such models are admitted only for approximate calculations, but never for extreme cases like shock wave structures in hypersonic flows. Hence, even large deviations from experiments give no cause to reject the set of ONE,

MNE, or NSE due to doubts about their foundation by means of the continuum hypothesis. By the way, Mott-Smith's method, even though it agrees with experimental data up to high Mach numbers, cannot be used as a theory working with terms only defined on a molecular level. Yet this method indicates the primary problem of describing mathematically details of shock wave structures: the stationary *non-equilibrium* phenomena with variations along the flow paths and embedding between two asymptotically dissipationless reference states.

The solution (8.46–8.48) depends on the index n, which characterizes polytropic changes of state that represent the influence of compressibility and dissipation simultaneously along the path of the shock wave. Now, it is not trivial that on one hand this solution agrees with the one based on NSE for the hypothetical case $n \rightarrow \infty$. As discussed above, the NSE is in error with regard to measured shock wave profiles of virtually observed Mach numbers. Therefore, it may be said that NSE with constant transport properties furnish profiles that are too steep, but even equipped with adapted transport properties, NSE lead to erroneous results.

On the other hand, decreasing values of the index n in direction of realistic values will distort the results; in other words, deviations from experiments will clearly increase. These deviations are assumed to originate from a fundamentally wrong dissipation law applied to highly irreversible changes of state. Such a conclusion may be justified by the fact that it is surely the "most incompressible friction law" that provides the best (although wrong) agreement with experimental data—even for extremely compressible shock waves.

Indeed, the dissipation laws derived within the scope of the Alternative Theory and employed in the Navier–St. Venant equations (see Section 7.2 and Appendix 3) are extraordinarily complicated in comparison with the viscous pressure tensor and the heat flux vector used in NSE, ONE, and MNE. Recall that those laws are part of a theoretical concept that is based on non-equilibrium terms from the very beginning. For this reason, it seems inadmissible to insinuate that MNE, ONE, and even NSE are related in any hypothetical way to the principle of local equilibrium. This may be evident, provided that set of equations is accepted as mere approximations that can be systematically deduced from the Navier–St. Venant equations. Hence, the incomplete mathematical structure of the constitutive relationships used is likely to be responsible for the obvious divergence between the theoretical results available at present and the experimental data proved to be reliable for the structure of normal shock waves.

8.3.2 GAY-LUSSAC'S AND JOULE'S FREE EXPANSION EXPERIMENTS

There is some difference of opinion concerning the definition of an ideal gas. According to some authors (see, e.g., Landsberg, 1961, p. 191), it is equivalent to the single relation $p = \mathscr{R}T\rho$. This relation, however, is not satisfactory, since every real fluid obeys this thermal equation of state for the limit of vanishing density ρ. Moreover, for every real fluid there exists a curve called the *ideal gas curve,* on which each point is defined by $Z := p/\mathscr{R}T\rho \equiv 1$ and assigned to a definite ρ value peculiar to the fluid in question. For this reason, the following definition of an ideal gas may indeed be sufficient

$$Z \equiv 1; \qquad d\ell nZ \equiv 0 \tag{8.51}$$

because thereby the three derivatives $d\ell np$, $d\ell nT$, and $d\ell n\rho$ are joined in a quite characteristic manner. Definition (8.51) is equivalent to the description suggested by Zemansky (1968, p. 120).

Note that the property $\kappa := c_p/c_v \equiv$ constant for the isentropic index κ cannot be inferred from these definitions, but has to be stipulated separately. However, (8.51) does imply that the internal energy U of an ideal gas is independent of the volume V (or of pressure p). This fact follows immediately from a generally valid relationship given in every textbook of theoretical physics or thermodynamics:

$$\left(\frac{\partial U}{\partial V}\right)_T = T\left(\frac{\partial p}{\partial T}\right)_V - p = p\left(\frac{\partial \ell nZ}{\partial \ell nT}\right)_V. \tag{8.52}$$

It is noteworthy that Helmholtz boldly stated the reverse: "Now, it is of importance that the assumption $(\partial U/\partial V)_T = 0$ in connection with … equation [(8.52)]. …giving expression to both the First and Second Laws leads of necessity to the equation of state of a perfect gas. Consequently, for all other bodies either $(\partial U/\partial V) < 0$ or $(\partial U/\partial V) > 0$ must hold" (Helmholtz, 1903, p. 225; author's translation).

Unfortunately, Helmholtz's conclusion proved to be wrong: The so called *Joule curve,* defined by $(\partial U/\partial V)_T \equiv 0$, exists for all real gases, at least in principle at high temperatures (Schaber, 1965, pp. 25, 58). Starting from any point of its Joule curve the real gas, subject to a free expansion, will undergo the largest possible change in temperature. Schaber (1963) proved the Joule curves of helium, neon, hydrogen, and deuterium. It should be also noted that Van der Waals' thermal equation of state, the well-known prototype for the representation of real fluids, does not inherently contain this curve.

Due to definition (8.51), the left-hand side of (8.52) vanishes identically for all ideal gases. This result seems to be rather banal, but, indeed, it deserves attention in view of the inverse question: How do we design a test that will solve the problem of whether a fluid approximates an ideal gas or not? For reasons noted above, we need only to check the fluid's actual p, ρ, T behavior in comparison with the thermal equation of state. Especially for low gas temperatures, large values of the derivative $(\partial U/\partial V)_T$ should be expected, even though $(8.51)_1$ might be exhaustively fulfilled at least within a certain range of low mass density. From this point of view, the famous experiments concerning the free expansion of a (nearly) ideal gas (e.g., helium at room temperature) occupy the rank of an *experimentum crucis* now as before (cf., e.g., Falk, 1990, p. 154). They were executed by Gay-Lussac in 1807 (cf. Mach, 1896, p. 198) and—in an improved way—by Joule in 1843 to 1845 (cf. Pellat, 1897, p. 196).

In these experiments, the constraints that require the configuration to have a volume V_i are suddenly relaxed, permitting it to expand to an amount V_f. If the configuration contains a gas, the expansion may be adequately accomplished by restraining it in one chamber of a rigid container while the other chamber is evacuated. If the septum separating the two chambers were suddenly fractured, the gas would spontaneously expand to the volume of the whole container. The essential element of the experiments is the fact that the total internal energy of the whole gas configuration remains constant during the free expansion. Thereby, neither heat nor work are transferred to that configuration by any external agency.

The problem is the prediction of the change in temperature and any other parameters of the configuration. To avoid the occurrence of shock waves, in practice the realization of the experiments has to start from an arrangement consisting of two rigid vessels filled with gas of the same temperature each but subject to a considerable pressure ratio. Another point germane to the understanding of non-equilibrium phenomena is that Gay-Lussac first experimented under the condition that after the beginning of the expansion the pressure equalization within the two vessels will adapt itself very rapidly. This extraordinarily fast process certainly implies an adiabatic running of the whole irreversible expansion as a secondary phenomenon. However, this conjecture is only qualitatively justified. It is indeed hard to quantitatively prove Joule's famous conclusion, according to which the internal energy U must be independent of the volume V. For a pure ideal gas such a fact presupposes that the final state of the expansion will have to be settled by a temperature that equals the initial gas temperature. In spite of this apparently simple situation, in theory and experiment (even in principle) it is difficult to determine the precise value of this final temperature.

The measurement procedure to be realized with the help of any precision sensor requires some time interval $\Delta\tau$. This depends not only on the initial gaseous state, but also on some thermal properties of the sensor used. Notwithstanding, $\Delta\tau$ is considered to be consumed to establish the pertaining equilibrium between the local gas state and the sensor. Consequently, we cannot reject the possibility that the adiabatic condition is abolished within $\Delta\tau$. Hence, heat transfer from the gas to the surrounding walls (or vice versa) may occur. Hirn (in 1865) and later Cazin (in 1870) tried to circumvent this difficulty with rather dubious success (cf. Zemansky, 1968, p. 117).

Since the experimental uncertainties prevented them from obtaining reliable results, both Gay-Lussac and Joule put their experimental setup into a heat reservoir. Thereby, two measures were offered for realization:

1. The closure condition prescribed for the internal energy is fulfilled with respect to the composed system consisting of the setup and the rigid heat reservoir.
2. The surface temperature of the wall enclosing the test gas may be kept constant.

Measure (1) without assertion of measure (2) leads to a temperature drop that can be measured in the reservoir, provided that a nonideal gas has been expanded. But precise measurements are generally hard to get because of error arising from an unfavorable proportion between the various heat capacities of the test gas and that of the whole test apparatus. This is particularly true with respect to ideal gases for which that temperature drop should actually vanish. By means of measure (1), the most recent experiments were performed by Baker in 1938 with respect to the equivalent derivative $(\partial U/\partial p)_T$. The scarce measurements substantiated the results of the National Bureau of Standards published in 1932. The authors of those results, Rossini and Frandsen, notably found neither pressure nor temperature ranges in which this derivative was equal to zero (cf. Zemansky, 1968, p. 119).

The more interesting case amalgamates both measures (1) and (2). The mixture guarantees that the gas expansion will always run toward a final state that corresponds precisely to the initial gas state in all its properties of state. Of course, to realize this condition it is necessary to choose a heat reservoir with a very large heat capacity compared with that of the test gas.

Evidently, by this method the behavior of a real fluid can be distinguished from that of an ideal gas by means of the directly observed time. Consequently, we must monitor the time dependency of gas pressure and temperature. This is equivalent to dealing explicitly with the special non-equilibrium phenomena occurring over the whole period of time.

Interestingly, the same is true for Gay-Lussac's original idea, provided that his most important argument—the adiabatic condition—is taken seriously. In other words, the isothermal boundary constraint for the walls of the two gas containers must be replaced by adiabatic walls realized by adequate equipment. Naturally, the time behavior of the expanding gas will be basically different for the two differing wall conditions. By means of heat transfer across the isothermal wall, the expanded gas will return to the initial temperature within a short time compared to the adiabatic case. The latter is distinguished by extremely long time intervals for the expected temperature compensation. This is caused by the fact that such a temperature balance can only be realized by the slowly running nonstationary heat conduction within the gas.

In the following discussion, we will give a detailed account of both these physical processes by way of their mathematically adequate modeling. We will use MNE (7.66) exclusively. However, the equation of motion $(7.66)_1$ will be replaced by its original version (7.61) together with the global continuity equation (5.72). Moreover, constant fluid properties are presumed as well as the thermal equation of state $p = \mathcal{R}T\rho$. The last premise converts the original task into a gauge problem: Introducing a priori an ideal gas as test fluid, attention will now be focused on the theoretical investigation of its time behavior, which may be proven experimentally.

Imagine a thermally insulated vessel with rigid walls, divided into two compartments by a partition. We will then simplify the evaluation of the Joule process by considering only a two-dimensional sectional area of a right parallelepiped taken as the vessel. This vessel is divided into two quadratic compartments, each filled initially with an ideal gas at rest. Both volumes are under different pressures, but are at the same temperature. At the instant $t = 0$, a small aperture will open abruptly and the high pressure gas will flow rapidly to the other chamber until pressure balance is achieved. This process runs in alternative reference to the following boundary conditions to be realized along the walls.

$$\textit{No-slip condition: } \mathbf{v}\big|_{\text{wall}} \equiv \mathbf{0} \tag{8.53}$$

$$\textit{isothermal condition: } \Omega_N\big|_{\text{wall}} \equiv \mathcal{R}T_w$$
$$\textit{adiabatic condition: } \nabla\Omega_N\big|_{\text{wall}} \equiv \mathbf{0} \tag{8.54}$$

The initial pressure ratio can be parametrically varied between $1 < p_h/p_l \leq 5$.

As essential results, the numerical calculations provide the times needed for the compensation of all changes of temperature and pressure resulting from the rapid flow process in both gas chambers. Such time values admit experimental proof; eventually they may serve the purpose of calibrating ideal gases with the intention of allowing falsification for real fluids. The calculations are based on the so-called *explicit MacCormack method,* a second-order predictor-corrector difference scheme, the details of which are well documented in Lauster's book (1995, p. 134). As a result we obtain time-dependent, local values of the gas pressure, temperature, and flow velocity. A first impression of the complex behavior of the gas flow is presented in Figure 8.18.

The following observations are of interest.

- For a certain period of time a characteristic flow pattern arises that demonstrates a stable and, for the moment, nearly symmetric *vortex system* within the right compartment. The local intensity of the flow and vortex phenomena increases with the rising ratio of the values p_1 and p_2 whereby the ideal test

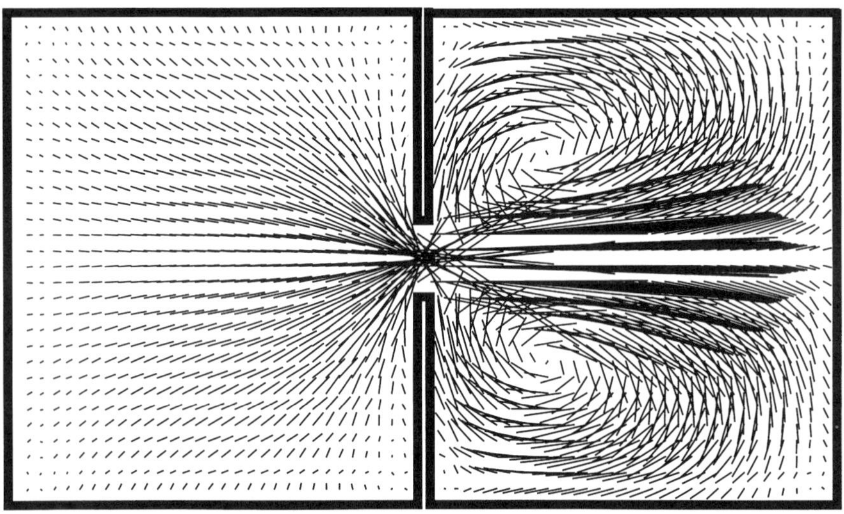

Figure 8.18 Instantaneous velocity vectors for an ideal gas at a pressure ratio $p_1/p_2 = 2$.

gas is initially pressurized within both volumes 1 (left compartment) and 2 (right one).

The corresponding mechanism becomes manifest by the occurrence of two elementary effects. First, one part of the internal energy is stored up for a certain moment inside the vortex system. Then, within a given interval of time the vortex loses an amount of kinetic energy via dissipative transfer processes such as momentum transport, heat conduction, and thermal radiation. It is evident that the gaseous state at rest with its initial temperature in volume 1 cannot actually be reestablished without those irreversible phenomena. This is indeed a crucial point, which is discussed at the end of this subsection.

- There is a profound difference between the details of those transfer processes subject to completely differing wall conditions. Constant wall temperatures intensify the internal changes of state by heat transfer, through which the desired final state will be arrived at within a characteristic time interval. Such a remedial measure is not available if the walls of the whole apparatus are adiabatically insulated. Therefore, the time interval for the temperature compensation is considerably larger than that for the isothermal process realization.

Figure 8.19 is a graphic representation of the numerical results concerning expansion of an ideal gas at isothermal wall conditions. The course of the (nondimensional) temperature refers to average values as a function of time, indicated by a pertaining number of numerical iterations. The initial period of the expansion is distinguished by a definitely asymmetric behavior: The gas within the right compartment is less heated than the gas's lowered temperature in the left one. To reestablish the initial temperature, the evaluation of the flow process requires a number of iterations that is at least about five times larger than the time interval iterated for the minimum. This typical time behavior differs drastically from the one that dominates the pressure equalization inside the vessel. The respective runnings in time happen in a strictly symmetric manner inside both compartments. As a rough approximation, bear in mind that the final pressure state is almost attained along with the two extreme values in the gas temperature.

The same observation concerning the expansion along adiabatic walls, leads to an identical result, apart from a decisive deviation. When the gas temperature within volume 1 takes the expected course up to its minimum value, and the same process, reflectively reproduced, happens within volume 2. Both curves will tend extremely slowly toward the final state, that is, the initial temperature of a gas at rest. The assigned time interval is beyond any reasonable and available computing time. Of course, this extreme time behavior is not surprising. After the resolution of the vortex system within compartment 2, the average flow velocity tends toward zero. Therefore, balance in temperature can only happen by *unsteady heat conduction* along the path from volume 2 to volume 1 and is, above all, gravely restricted by the minuscule sectional area of the overflow valve.

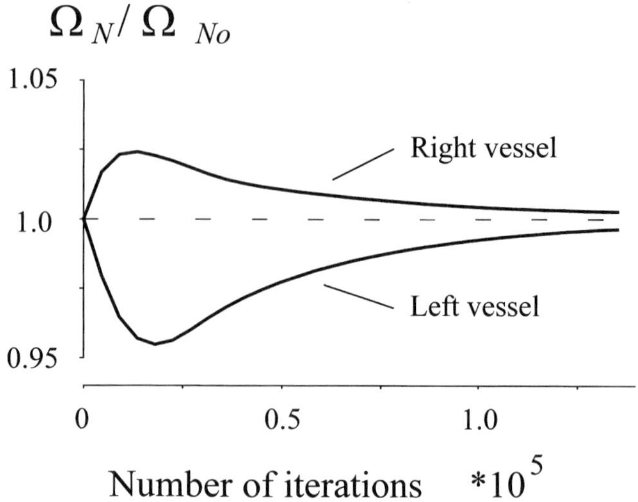

Figure 8.19 Gas temperature inside the vessel for pressure ratio $p_1/p_2 = 2$ and isothermal wall temperature.

Figure 8.20 demonstrates some relevant results of an interesting second case that is representative for pressure ratios $p_1/p_2 > 2.5$. This time, the gas temperature inside compartment 2, starting from the initial state, first drops to a value from which it increases rapidly to its maximum amount. During this initial period, the temperature also decreases within compartment 1. Otherwise, the same time behavior as described above is observed.

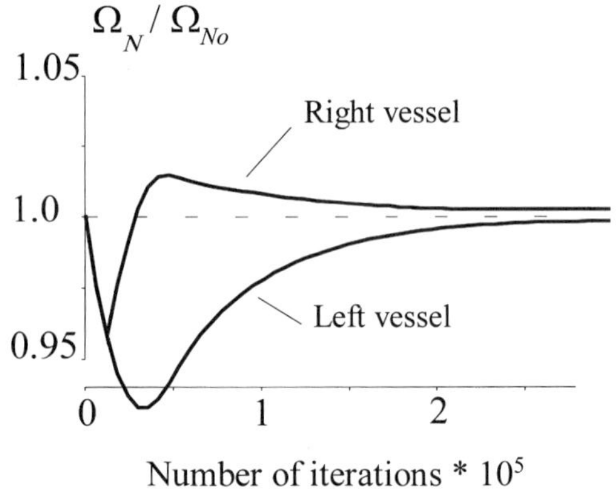

Figure 8.20 Gas temperature inside the vessel for pressure ratio $p_1/p_2 = 2.5$ and isothermal wall temperature.

An essential result of the numerical calculations concerns the determination of some time intervals $\Delta\tau$ characterizing the desired temperature and pressure compensation after the gas expansion under consideration. These time parameters $\Delta\tau$ are amenable to theoretical and experimental tests.

Bühler (1988) dealt with Joule's free expansion experiments with respect to those very time parameters. Gasdynamical aspects such as the free-stream hypothesis, were the basis of his research. To compare $\Delta\tau$ with his results, we first introduce a reference time,

$$t_N := \frac{\ell_f}{u} :\equiv \frac{1}{4}\frac{n}{n-1}\frac{M}{\sigma^2}\frac{\sqrt{\frac{\mathcal{R}}{M}T_1}}{p_1}, \tag{8.55}$$

which follows from elementary gaskinetics, defined as a ratio of Maxwell's mean free path ℓ_f to the mean molecular speed $\bar{u}$. The gas is specified by its mole mass M, polytropic index n, and collision diameter σ presumed to be representative for any intermolecular forces. Gas temperature T and pressure p are related to the initial state of the high-pressure chamber 1.

As an example of the use of Equation (8.55), the mean speed $\bar{u}$ of the argon atoms at 300 K is 275 m/s. The mean free path at this temperature and atmospheric pressure is found to be 7.7×10^{-8} m, if the value 3.3×10^{-10} m for σ is given. Hence, the reference time t_N is 28 nanoseconds.

By means of nondimensional relaxation time intervals defined as

$$\tau_p := \frac{\Delta\tau_p}{t_N}; \qquad \tau_T := \frac{\Delta\tau_T}{t_N}, \tag{8.56}$$

the compensation of temperature as well as the pressure balance may be conveniently represented by the following plots. Consider first the property τ_p as an approximately linear function of the reciprocal pressure ratio p_2/p_1. Related to the adiabatic wall of the whole vessel, it manifests the most rapid process undergone by an expanding ideal gas toward pressure equalization (Figure 8.21).

Of course, it is reasonable to compare these results with experiments designed for the gas in question to be in any state, where it behaves like a real fluid. If this is so, we are able to calculate the final state after the conclusion of the fluid expansion by means of the thermal and caloric equations alone.

The change in fluid temperature due to the change in volume during the expansion, characterized by an unchanged amount of internal energy, i.e., U = constant, may then be determined by the following relationship

$$T_{12}(U; V_{12}) - T_1(U; V_1) = \int_{V_1}^{V_{12}} \frac{p - T\,(\partial p/\partial T)_V}{C_V}\, dV, \tag{8.57}$$

The pressure $p(T, V)$ and the heat capacity $C_V(T, V)$, at constant volume of the real fluid, are assumed to be known. The subscript 12 indicates the final state settled by the aggregate volume $V_{12} = V_1 + V_2$. The final pressure results from the thermal

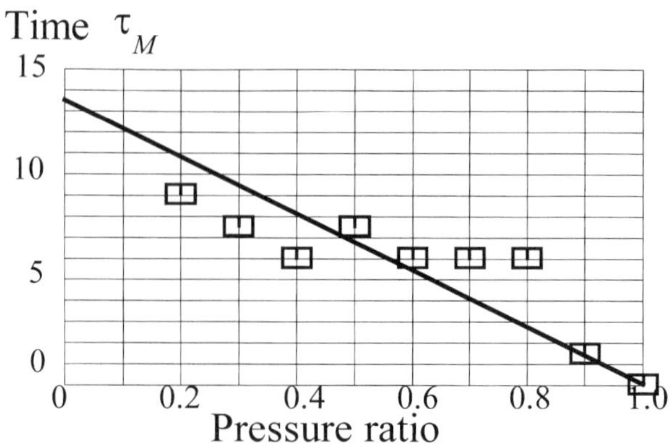

Figure 8.21 Nondimensional time up to pressure balance under adiabatic condition.

equation of state $p_{12}(T_{12}; V_{12})$. Taking into account the well-known thermodynamic relationship

$$S_{12}(U;V_{12}) - S_1(U;V_1) = \int_{V_1}^{V_{12}} \frac{p}{T} dV, \tag{8.58}$$

the corresponding change of entropy can be calculated exactly. Since the integral is always positive, S_{12} is always larger than the initial value S_1; that is, "in the typically irreversible Joule process entropy increases" (Kirillin et al., 1987, p. 267).

However, note that the final state is a state of thermodynamic equilibrium by definition. Therefore, the compensation of temperature, subject to adiabatic wall condition as well, takes a lot of time, quite similar to the case of an ideal gas.

For the function $\tau_p \propto (p_1/p_2)^{-1}$, valid for adiabatic wall insulation, the same linear relationship—but with about 45% larger values of time consumption—is valid for isothermal wall condition. The inverse behavior is observed for the respective temperature compensation. Whereas adiabatic wall conditions entail extremely long relaxation times $\Delta\tau_T$ for all pressure ratios considered, the respective time values τ_T under isothermal wall conditions are the same order of magnitude as all resulting τ_p equivalents (Figure 8.22).

Division of τ_T by τ_p, for the same pressure ratio each, yields a striking result, which has been plotted in Figure 8.23.

The course of these ratio values $\theta := \tau_T/\tau_p$ exhibits a distinctive maximum, thereby offering a new way to solve Gay-Lussac's and Joule's problem experimentally.

Nevertheless, the numerical results presented above deviate considerably from Bühler's calculations. Detailed examination reveals τ_p-values about twelve times larger than those presented in Figure 8.21. If applied to isothermal walls, the respective proportion even works out at about 20 (Bühler, 1988, p. 30). These exorbitant divergences might be due to Bühler's assumptions of the free stream hypothesis and, above all, an isentropic running of the gas expansion.

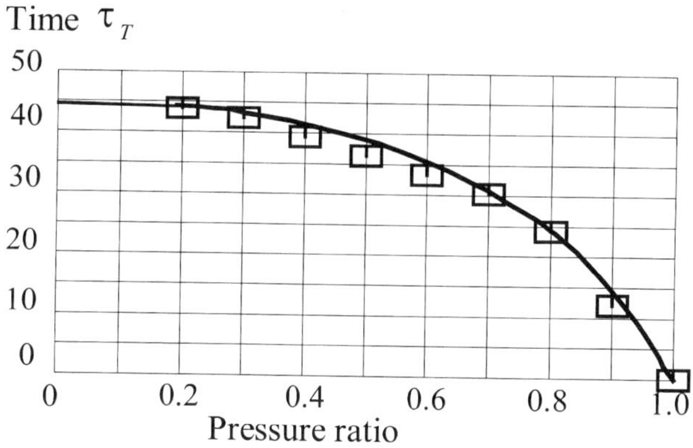

Figure 8.22 Nondimensional time up to temperature compensation under isothermal condition.

We conclude this subsection with some remarks about some physical, historical, and psychological aspects of Gay-Lussac's and Joule's free expansion experiments. Seen from a physical point of view, it is curious that this same basic test was repeated from the time of Gay-Lussac's first realization up to about the outbreak of WW II, even though the test never actually yielded satisfactory results (at least in view of Joule's law of ideal gases). Of course, the experimental difficulties were well-known from the beginning. For this reason, Joule and Thomson undertook a new experiment that bypassed Gay-Lussac's problem in an intelligent and effective way.[16]

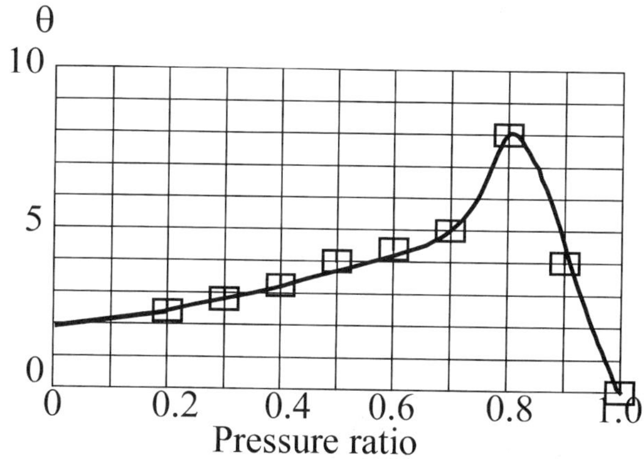

Figure 8.23 Ratios of times τ_T and τ_p under isothermal condition.

[16]This experiment was published in *Proceedings Royal Society London* **143** (1853) p. 357.

The great success of the well-known *Joule–Thomson effect* may explain why a convincing analysis of Gay-Lussac's experiment never took place (Falk, 1990, p. 158). But perhaps it is not only experimental difficulties that prevented deeper insight. There may be other reasons found in the history of science and, moreover, of Europe's early industrialization. That period was dominated by the widely used steam engine in its numerous variants. This basic invention—and its decisive improvements after the expiration of Watt's patent in 1800—is closely tied to all early steps of the evolution of thermodynamics. More precisely, the early steps are equivalent to the first theoretical testing of the primary qualities and functions of any steam engine (Truesdell and Bharatha, 1977). For this very reason that phase of thermodynamics was quite the field of action for both engineers and some amateurs [for example, Carnot, Hirn, and Reech (engineers); Joule and Regnault (factory managing directors); and Mayer and Helmholtz (physicians)].

Truesdell (1980) coined the best term for the state of the art of this new branch of physics in his book with the title: *The Tragicomical History of Thermodynamics 1822–1854*. It was only at the last phase of that history that physicists took over leadership: The mechanical theory of heat evolved in such a way that thermodynamics finally degenerated to thermostatics (Straub, 1990, p. 36). A similar development took place with the revival of the kinetic theory of gases. Thus, for instance, "J. Herapath's paper was rejected by the Royal Society. The story is an ugly one" (Truesdell, 1975, p. 7). Waterston, the second early pioneer, submitted his long memoir to the Royal Society on December 11, 1845. "The paper was rejected. ... burying the genius of Waterston in permanent oblivion" (Truesdell, 1975, p. 11). Even Joule's note on kinetic theory, published in 1851, was similarly buried: "The community of physicists, always given to orthodoxy, was not ready to take its basic idea seriously" (Truesdell, 1975, p. 18).

Several other examples indeed allow the assumption that for many people the history of the kinetic theory began in 1856 with a paper on "Grundzüge einer Theorie der Gase," published by Krönig, the influential editor of *Die Fortschritte der Physik*. Thereby, a way of thinking was initiated that seems to prevail among modern physicists: "regarding papers that do not happen to be read by physicists, like events that do not happen to be observed by physicists, as not existing at all" (Truesdell, 1975, p. 20). From then on the prominent physicists published their basic contributions to today's kinetic theory of gases: Clausius, Maxwell, Stefan, Loschmidt, Boltzmann, Thomson, and Tait in the late 1800s; Jeans in 1904 and 1925; and Chapman and Cowling in 1939.

Nearly all these famous scientists took part in both fields—that is, in the development of the mechanical theory of heat and the kinetic theory of gases. Curiously, no serious efforts have brought the notions of both fields into mutual relations. There is an outstanding example for this incredible fact: Clausius's famous book, published in 1891, identified the kinetic theory of gases (*die kinetische Theorie der*

Gase) with the mechanical theory of heat (*die mechanische Wärmetheorie*), although he never treated any problems involving a trend in time. Here, we find the crux of the conceptual difficulties arising from Gay-Lussac's and Joule's free expansion experiment.

This gas expansion is a dissipative process in all steps of its evolution. There has never been any opportunity to approximate the characteristics of this flow by "quasi-static steps," "infinitely slow changes of state," or other pertinent constructs much-liked by adherents and users of thermostatics applied to actual non-equilibrium phenomena. During all periods of its time behavior that overflow is subject to locally irreversible and dynamic events. For the first moment it runs adiabatically, where this kind of process realization is only guaranteed by the flow rate not by any theoretical concept materialized by adiabatic walls. Furthermore, the subsequent compensation action within the gas can only be realized by local transfer processes triggered by local gradients of the field variables.

It is no wonder that nobody wanted to deal in detail with this basic process other than with thought experiments and remarks for textbooks: Neither time behavior, gas flow rates, transfer processes, and so on nor local gradients of the various field variables are common terms in thermostatics. In contrast, the results of the kinetic theories confined to nonuniform gases presume empirical constitutive laws like those of Newton, Fourier, and Fick. They are applied to flow fields, for which the postulate of local equilibrium is assumed to be valid for each cell of the continuum, even though Gay-Lussac's gas expansion is undoubtedly in a non-equilibrium state at every moment and position.

Obviously, the two sets of notions, one adapted for thermostatics and the other for kinetic gases, are actually incompatible. But in numerous applications of thermostatics this inconsistency does not play a significant role. Unfortunately, this is not true for Gay-Lussac's and Joule's basic discharge experiments, due to the fact that from the very beginning of systematic scientific research concerning thermal processes the wide gateway was closed for many processes like expansion flows. Even now it is common practice to prefer equilibrium states as a basic property of matter. For this reason, non-equilibrium phenomena are either unable to be theoretically described or must be approximated by certain methods of extrapolation sometimes established on rather obscure assumptions. Surely, in all the prevailing theories the most dubious premise concerns the fact that motion as part of every non-equilibrium phenomenon has been separated from the other properties of matter.

Consequently, their changes in space and time are believed to be executed by quasi-static displacements or movements of matter at a locally gliding equilibrium. Attempting to explain such dubious terms, even modern textbooks still visualize them by means of state diagrams plotting "a series of similar elementary processes, i.e., processes during which the working medium fills a great number of successively connected elementary vessels, ... (Figure 7.17d)" (Kirillin et al., 1987, p. 266).

Regrettably, such an explanation prevents thermostatics from advancing to thermodynamics in a rational way. Unlike notions such as the mechanical equivalent of a unit of heat or even the perpetual motion of various kinds, the Gay-Lussac–Joule experiment will obviously be of great importance for our current understanding of nature. Its highly pedagogical relevance is to be gathered from some relatively transparent insights into the immense variety of non-equilibrium phenomena in reality, along with their diverse aspects of irreversibility, instability, and even chaos.

8.4 Complex Flow Phenomena

8.4.1 PROBLEMS OF EXPLORING COMPLEXITY

Crossflows around a cylindrical body, normally arranged to the flow direction, are among the most interesting flow configurations in practice and theory. Today's technology provides many examples for all kinds of fluids and flow regimes. But also theoretical aspects concerning frictionless potential flows also stimulate research work on this subject.

These flow processes become considerably complex, provided that the cylinder is vertically installed inside a straight channel and is able to be rotated as well as heated. Assuming, for example, a rectangular cross section of the channel, an additional parameter may be defined by the ratio of the projection area of the cylinder and that cross section. The resulting flow patterns, strongly influenced by blockage effects due to this aspect ratio, enables us to create optional features of complex matter. Obviously, such opportunities induced the team of experts mentioned in Section 7.3 to turn attention to this special field of research: "Two problems that would be useful to attack are the flow around a backward-facing step and around a rotating cylinder, although the numerical details for the latter are difficult" (Nixon, 1986, p. 25).

But what is complexity in this context? Consider a flow field around any obstacle marked by a large number of vectors that display the local velocities by their directions and lengths even for turbulent flows. Do these large numbers qualify this flow configuration as complex? Intuition tells us no, because we cannot make out any structure or coherence; nor can we perceive any dynamics. We tend to regard such an image as a prototype of disordered, erratic behavior indicating certain short-range interactions. However, the same configuration may appear in different aspects, evoking successive impressions of simplicity and complexity, provided that the velocity vectors are ordered in a definite pattern. Obviously, it is less ambiguous to speak of complex behavior than of complex systems. Notwithstanding, it is not easy to reveal certain representative characteristics proved to be proper for a classification of complex flow phenomena. Some contributions to this basic problem will be concisely dealt with in this section, concerning channel flows of a gas around a rotating vertical cylinder and calculated by the Modified Nehring Equations (MNE).

Details have been published by Lauster (1995, p. 73); thus the following discussion will focus on two topics only, which are relevant for different reasons. First, by variation of the main parameters, MNE are applied to the motion of ideal gases for the quantitative study of the flow configuration designed above and realized by own experiments. Second, the material property $v_\tau := \beta_\tau/\rho$, defined by the characteristic time t_τ [cf. Equation (7.8)] is intended to be determined with the help of special empirical information.

When the flow to be investigated shows complex patterns with respect to space and time, it seems inevitable to characterize such complexity by some primitive indicators. As a rule, these indicators cannot be immediately read off from the common field variables determined either experimentally or numerically. Yet, these variables may sometimes be transformed and summed up so as to result in just this kind of simple characteristic, which is particularly useful for a quantitative indication of that complexity and its theoretical description. If the description is to be done by means of MNE, some special problems must be solved with respect to their parameters. Thus, pertinent nondimensional characteristic numbers like the crossflow Reynolds number

$$Re_D := \frac{U_\infty D}{\nu_\tau} \tag{8.59}$$

must not only be referred to given amounts of a free gas flow velocity U_∞ and a typical length D, but also to a viscosity value n_t that normally differs from the well-known kinematic viscosity of the gas. To dispose of reliable values for n_t, we can make use of some adequate data available for the crossflow around the nonrotating vertical cylinder. This flow configuration is appropriate at best as a reference case for all flows running under the same conditions, but now around rotating cylinders.

Compared with this reference flow, Figure 8.24 sketches three types of crossflows around a rotating cylinder. Note some characteristic features following from Figure 8.24:

- The crossflow around any nonrotating cylinder is represented by a symmetric "dead water region," the length of which depends mainly on U_∞.
- For a rotating cylinder, the shape of the dead water region is decisively influenced by the ratio α of the peripheral speed U_r of the cylinder to the free stream velocity U_∞; that is, $\alpha := U_r/U_\infty$. The plots of Figure 8.24 identify a significant indication by means of the displacements of the upper and lower points of separation.

In the next two subsections, we will analyze the flow behavior around and behind the cylinder as well as near the channel walls.

8.4.2 DEAD WATER REGION BEHIND NONROTATING CYLINDERS

In 1983, Leder extensively measured the dead water regions of several bodies. Test runs in an open wind tunnel covered the range from 10^4 to 10^5 for Reynolds numbers

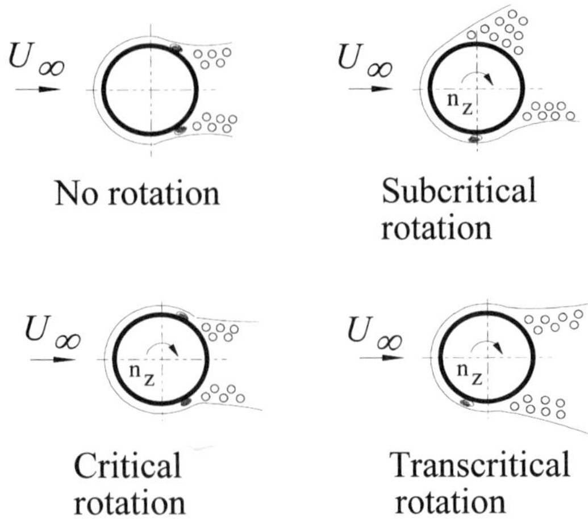

Figure 8.24 Flow patterns of crossflow around a cylinder.

related to dry air and the projection length d of the body normal to the main flow direction. Additionally, Leder considered measurements performed by other authors for Reynolds numbers within the range of roughly 100 to 1000.

Averaging a sequence of snapshots during the initial period of the periodic formation and separation of fluid vortices—known as elements of Kármán's famous vortex street—Leder (1983, p. 45) obtained an interesting characterization of such flows around nonrotating bodies:

1. The steady state of any dead water region seems to be best classified by a pair of counterrotating vortices.
2. For wakes behind bodies arranged symmetrically with respect to the free-stream velocity, the symmetry as well as the two-dimensionality of the flow field has been proved.
3. Such a steady flow may be modeled by the potential flow around a replacement body built up by the original body along with its dead water region (see Figure 8.25).
4. All experiments suggest a definite relationship between the Reynolds number according to definition (8.59) and the nondimensional length x_T/D_{cyl} of the dead water region. It is noteworthy that this nonlinear function can be validated for order of magnitude of about six in Re_D.

Figure 8.26 reproduces Leder's results for crossflow in an unbounded flow field around a vertically fixed circular cylinder.

Gerопp (cf. Leder, 1983, p. 141) offered a satisfactory explanation of this chart, the course of which is supposed to be dependent mainly on the effective viscosity $\mu_{\text{eff}} := \rho\nu_{\text{eff}}$. He identified laminar flows inside the vortex street for Reynolds num-

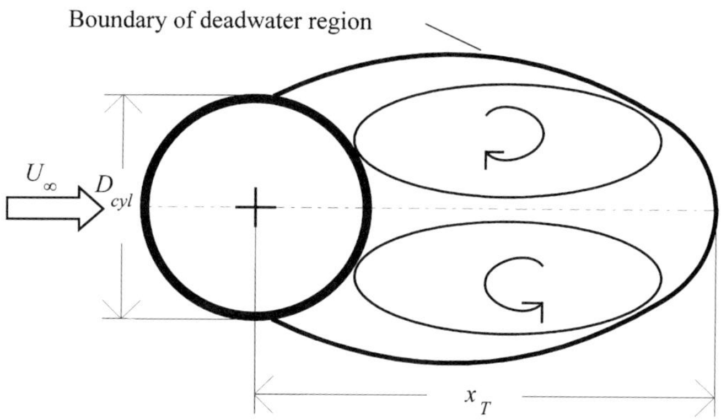

Figure 8.25 Replacement body = cylinder + dead water region.

bers up to about $Re_D \approx 180$, assigned to the maximum of the Leder curve. With increasing values of the Reynolds number, the turbulent character of the dead water region intensifies, accompanied by augmented values of μ_{eff}. Consequently, the local state of the flow is more precisely typified by a Reynolds number $Re_{x_T} := Re_D(n_t/\nu_{eff})$ than by the variable Re_D of the function $x_T/D = f(Re_D)$. Hence, the value of Re_{x_T} is lower than that of Re_D and therefore implies an allocation of x_T/D, which really belongs to a lower Re_D value. Because Leder's curve is based on experimental facts only, it seems appropriate to apply his concept to the issues (1) and (2) posed above.

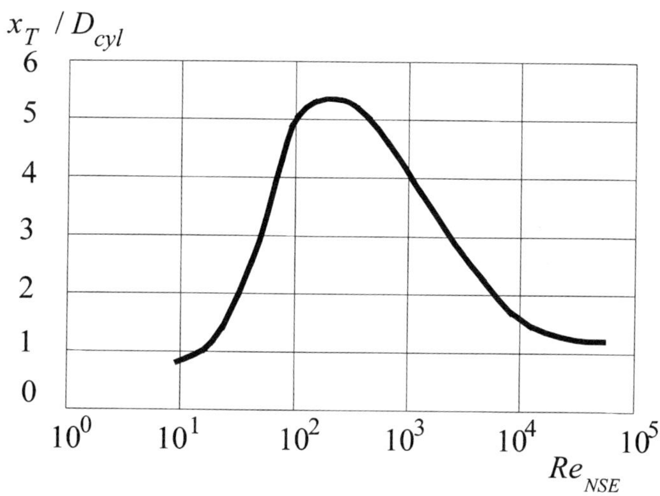

Figure 8.26 Length of dead water region, depending on the Reynolds number for the cross-flow around a cylinder.

Thus, we will aim at the numerical simulation of the function $x_T = F(Re_D)$ concerning the channel flow of an ideal gas around a nonrotating cylinder. In case the curve resulting from MNE turns out to be geometrically similar to Leder's plotted data, both curves will then be related to each other, thus offering the option to convert reciprocally both the different scales defined by Reynolds numbers. Subsequently, the desired material property ν_τ introduced into MNE may be easily calculated by an ensuing regression analysis.

The mathematical modeling of the channel flow around a vertical cylinder is oriented by an experimental setup initiated some time ago and applied to a test program performed for a considerable range of parameter variations (cf. Wurst et al., 1991). Oesterle's experiments concerned extensive measurements using cylinders each with different diameter and even various gases typified by a broad spectrum of mole masses, such as $M = 2$ (helium) or $M = 146$ (sulphurhexafluoride). The gas stream allowed mass flow averaged velocities to be increased up to about 70 m/s; the corresponding initial-valued Reynolds numbers achieved a maximum amount of about 10^7. The influence of free-stream turbulence was considered with regard to an average turbulence intensity of about 1%, determined at the inflow cross section.

The cylinder walls could be heated up to an overall temperature of 100°C at most. Starting from the limiting case of a nonrotating cylinder, the experiments covered a test range from zero rotation up to rotational numbers of about 15,000 rotations per minute. Along the channel, including the regime of interactions between the crossflow and the flow induced by the cylinder rotation, the pressure drop was determined together with heat transfer measurements near the cylinder wall. By means of laser-Doppler-anemometer and hot-wire measurement techniques, extensive experiments were additionally carried out concerning field distributions of local flow velocities and temperature covering the dead water region behind the cylinder (Oesterle, 1996). In particular, boundary layer profiles and locations of separation points were registered along with the occurrence of separation bubbles and turbulent reattachments of local flows. This research work continued in a series of experiments, the results of which were published in the late eighties by Peller and associates (see Peller et al., 1984; Peller, 1986; Peller and Straub, 1988).

Using the so-called MacCormack method, a numerical solution of MNE applied to the flow separation problem of the nonrotating and rotating cylinder succeeded in view of a flow configuration, as sketched in Figure 8.27.

More efficient solvers are now available, but are reserved for future solutions of the present problem (cf. Bruneau and Fabrie, 1994, p. 320). Note that the selection of the geometric data of the flow configuration was motivated by Oesterle's experimental setup and an unproblematic blockage ratio $k_q := H/D$, which amounted to Equation (2.93) in accordance with the values given above. The selected value of length L considered the incorporation of the complete expected dead water region. To solve MNE with respect to the time-space evolution of the flow under consider-

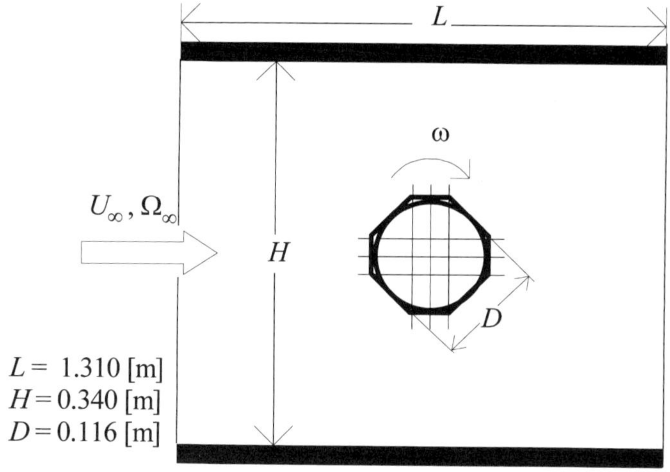

Figure 8.27 Channel and cylinder, together with an approximating octagon (lattice only presented in part; cf. Lauster, 1995).

ation, both initial and boundary conditions must be prescribed. Distributions of the state values at the channel inlet are considered as is the adiabatic constraint along the solid walls of the channel. The walls are expressed by the no-slip condition (7.68) and the mathematical constraint (7.72) of the form $\nabla\Omega \equiv \mathbf{0}$ for an adiabatic wall, provided that the pressure function $\Omega := p/\rho$ is applied to an ideal gas. Such a premise may be reasonable for the experimental conditions noted above. Hence, the condition $\nabla\Omega \equiv \mathbf{0}$ agrees with the local vanishing of the temperature gradient at the wall. At the channel inlet ($z = 0$), the flow velocity is assumed to be constant across the cross section, whereas the gas inflow takes place one-dimensionally with zero gradients of the local temperature. The reverse is true at the exit of the channel section to be regarded: The gas temperature is allowed to be constant over the cross section at $z = L$, where this constant value of the temperature may be identified with that of the atmospheric surroundings.[17] Opposed to this, the vectorial gas velocity is free to find a level consistent with the restrictions of MNE.

The numerical calculations were stopped if steady-state conditions appeared and the number of time steps for all state variables of the whole region seemed to be sufficiently large to ascertain a well-founded time average for each of these field quantities.

The details of the computer calculations are documented in the work of Lauster (1995, p. 134). Their results, presented here, focused on the evaluation of the dead

[17]The temperature of the atmospheric surroundings is assumed to be about 25°C. The corresponding mass density of dry air amounts to $\rho = 1.2$ kg/m^3.

water region. First, the velocity field is transformed by means of Equations (8.34) into the streamline representation. The permanent change of vortex generation at the lee side of the cylinder and the subsequent separation of vortices running toward the exit lead to a stable contour for steady-state flows, if the streamlines are properly averaged with respect to time. Figure 8.28 offers an impressive example, from which the typical length of the dead water region is immediately perceived, along with the two included contrarotating and stable vortices.

The extensive calculations by means of MNE establish the dimensionless length x_T/D as a function of the crossflow Reynolds number Re_{MNE}. According to (8.59), the latter is related as usual to the gas velocity U_∞ for undisturbed flow conditions and to the diameter D of the circular cylinder. However, as mentioned above, it cannot be assumed that the material property ν_τ used in MNE equals the common kinematic viscosity ν_{NSE} known from daily applications, for example, with the NSE. Hence, to set a definite value of Re_{MNE}, an appropriate procedure is needed to obtain in a first approach a constant assumed to be representative for the parameter ν_τ.

As expected, the function $x_T/D = F(Re_{MNE})$ shows a qualitatively similar course with respect to Leder's result. Taking for granted the preliminary assumption that the maximum of this function establishes the correct connection between the reference velocity U_∞ and the respective Reynolds number, we obtain a simple rule to fix ν_τ. Thus, the elementary evaluation

$$\nu_\tau = \frac{U_{\infty\ \mathrm{MNE}} D}{Re_{\mathrm{NSE}}} = \frac{4.4\,[\mathrm{m/s}] \cdot 0.116\,[\mathrm{m}]}{82.5} = 6.187\ 10^{-3}\,[\mathrm{m}^2/\mathrm{s}] \tag{8.60}$$

uses the value $Re_{NSE} = 82.5$ assigned to the maximum of Leder's curve, whereas the other quantities D and U_∞ are related to the values of the actual flow configuration, viz. $D = 0.116$ m and $U_\infty = 4.4$ m/s. Surprisingly, the parameter $\nu_\tau = 6.19\ 10^{-3}$ m^2/s assigned to MNE is about two orders of magnitude larger than the common kinematic viscosity $\nu_{NSE} = 1.56\ 10^{-5}$ m^2/s of dry air at the surrounding conditions. Even

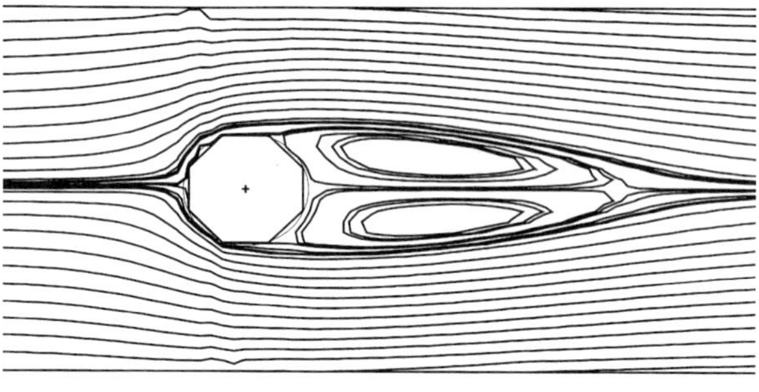

$Re = 82.5$

Figure 8.28 Streamlines of a time-averaged channel flow around a vertical cylinder.

taking into account the numerical simulation as merely a rough approximation, the difference between the frictional parameters is remarkable.

Figure 8.29 repeats Leder's chart, but the simulated results are plotted using adequately changed scales for the coordinate axes.

A comparison with Figure 8.26 leads to the following statements:

- Evidently, the relationship between x_T/D and the corresponding Reynolds number is correctly described by MNE with respect to the shape of the curve. However, the difference in the corresponding ordinate values, plotted in Figure 8.29 the maximum values of about 3.4 compared to 5.4 in Leder's curve, is salient.
- Assigned values of the Reynolds numbers agree by definition in both figures near the maximum, but will diverge more and more with increasing abscissa amounts.

Simple considerations lead to the assumption that the dead water region is subject to some influences by the walls that affect the envelope of von Kármán's vortex street mainly by a supposition of boundary layer and blockage effects. The flow velocities inside this region are on the average lower than those prevailing along the contour of the replacement body. Therefore, the latter is forced to become smaller, as the continuity condition for the mass flow rate cannot be satisfied in another way with respect to the no-slip conditions at the solid walls. As opposed to this, Leder's curve is mainly based on experiments performed by means of an open wind channel of the Göttingen type, for which blockage effects do not play a significant role.

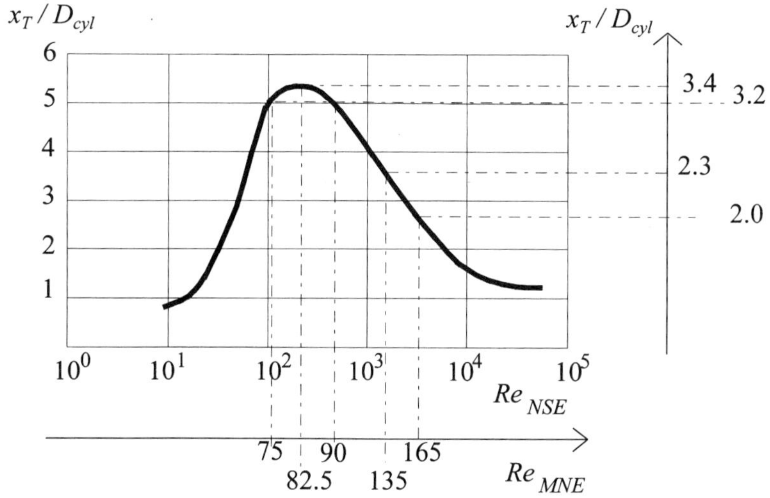

Figure 8.29 Measured and calculated lengths of dead water regions.

Therefore, no displacement effects at the expense of an extension of the dead water region will take place, because the continuity condition can easily be fulfilled by the outer parts of the flow field far from the cylinder.

Let us start from the hypothesis that the curves are similar to each other; that is, they can be developed from one another. Hence, the Reynolds numbers Re_{NSE} and Re_{MNE} may be related by two kinds of regression analysis. The corresponding linear regression curves, each derived from Leder's empirical curve and the MNE simulation, are given as follows:

$$Re_{NSE} = 23.135\, Re_{MNE} - 1642.753; \qquad Re_{MNE} = 0.0426\, Re_{NSE} + 71.716. \qquad (8.61)$$

Both formulas, covering the ranges $10 < Re_{NSE} < 2250$ and $75 < Re_{MNE} < 165$, can be used to check the kinematic viscosity ν_τ calculated by Equation (8.60). The regression between ν_τ and Re_{MNE} plotted in Figure 8.30 succeeds by means of both functions (8.61) and turns out to be a constant.

The value of the kinematic viscosity ν_τ calculated by (8.60) agrees with the mean value $\bar{\nu}_\tau = 6.187\ 10^{-3}\ m^2/s$. The standard deviation is equivalent to ± 2.9%. Although the measure of definiteness lies near 1 (98.55%), the results can only be assessed as a first approximation in reference to the following points:

1. Leder's curve itself is established by means of a compensation procedure based on a set of experimental data.
2. The results are assumed to be valid only under the condition that the introduced similarity hypothesis concerning Figure 8.29 holds.
3. The numerical simulation presented here may only be regarded as a crude image of the actually existing experimental situation.

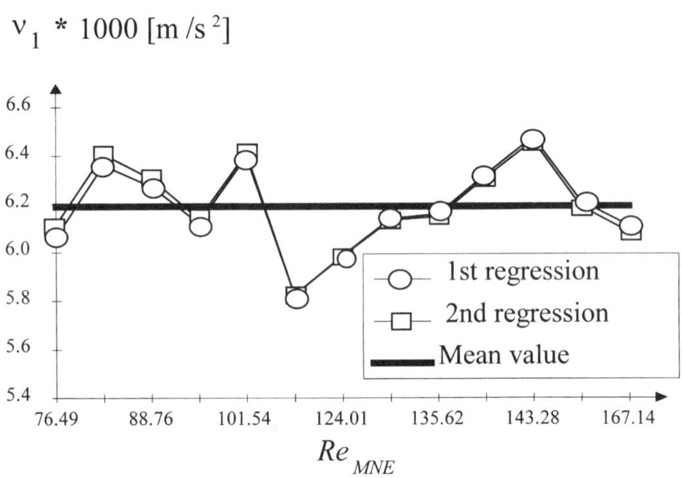

Figure 8.30 Kinematic viscosity as mean value of two regression functions.

Notwithstanding, the results obtained justify the application of the constant value $\bar{\nu}_\tau$ to MNE at least at the present state of the art. The calculated flow fields obviously provide the observer with both qualitatively and quantitatively correct patterns. This is true even if one considers the channel blockage as an additional parameter. For this reason, the model defined in this way by MNE seems suitable for the case of a channel flow around a rotating cylinder.

8.4.3 CROSSFLOW AROUND A ROTATING CYLINDER

Actual experimental data were recently determined and will be published by Oesterle et al. (1997). They were determined by means of a complicated setup, allowing crossflow around three cylinders of different diameter D_z each (50, 85, and 116 mm). High Reynolds numbers of some gases characterized by different Prandtl numbers could be realized, as well as the option to heat the surface of the cylinder equally. At present the closed wind tunnel operates at α values between the bounds $\alpha = 0$ and $\alpha > 3$, defining α as the ratio between the (mass flow rate) velocity of the incoming undisturbed gas stream, corrected for blockage effects by a factor C_b, and the velocity $U_R = n\ D_z/120$ of the cylinder surface. The rotation rate n in rpm can be varied from zero up to 30,000.

Due to these experimental conditions various complex flow fields of viscous fluids may be realized by means of low ratios α. Nevertheless, some characteristic flow patterns will arise, even if the rotation rate increases to large α values. Past experience has shown that some properties of the resulting boundary layers around the cylinder may be appropriately used as a representative measure of complexity. This is especially true with regard to their dependency on α. In a first examination of the flow configuration introduced in this section, we will investigate the displacement of the two separation points as indicated in Figure 8.31.

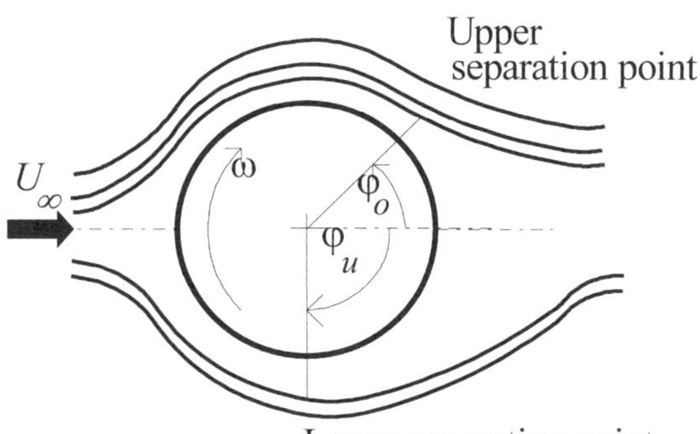

Figure 8.31 Angle position of the separation points of crossflow around a rotating cylinder.

From a physical viewpoint, such a separation point is distinguished by vanishing viscous stress components of the boundary sublayer flow near the cylinder wall. Its geometric position may be determined exactly by identifying the streamline that runs nearest along the cylinder surface toward a certain point where the streamline lifts off from its contour and commences to encompass the dead water region. For the assessment of the numerical results, the angles assigned to the two separation points were experimentally determined by means of a so-called light-cut procedure, assuming a constant initial flow velocity $U_\infty \approx 4.4$ m/s and alternating revolutions n_{cyl} per minute. An upper bound of n_{cyl} was fixed by the value $n_{cyl} \approx 2000$ for the numerical calculations, because earlier experiments of Oesterle (1996) did not indicate any essential changes of those angles due to larger values n_{cyl}.

Figure 8.32 plots our own experimental data along with two values determined for the special case of a cylinder at rest within an infinitely extended flow field. The first value $\varphi_0 = 81.8^\circ$ is evaluated by means of the potential theory, whereas the angle value $\varphi_0 \approx 95^\circ$, recently published by Stücke (cf. Lauster, 1995, p. 87), is recorded as a representative example of some available measurements yielding angle values that commonly become significantly larger than the theoretical result.

To exclude any uncertainty due to the considerable amount of the blockage factor $C_b = 0.34$, the limiting value φ_0 was first evaluated by means of MNE, according to the experimental condition prescribed for Stücke's experiments. The resulting value, $\varphi_0 = 95^\circ$, agrees with the experimental value. For this reason, we can assess our own measurements as strongly influenced by the actual blockage of the crossflow along

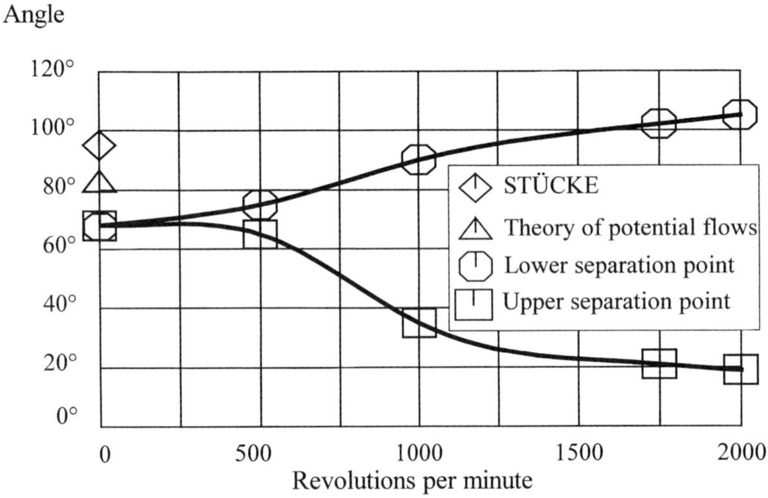

Figure 8.32 Experimental angle positions of separation points.

the channel. Of course, this is particularly true for the value $\varphi_0 = 68°$, measured for the channel flow around the stationary cylinder. The observed values of separation angles indicate a markedly asymmetric course with respect to their changes due to an increasing number of revolutions. Apparently, there is only a rather weak dependency on this rotational number if augmented approximately above 1750. Of course, this conjecture needs to be confirmed by more extended measurements.

By means of the MacCormack procedure, numerical solutions have been obtained for special initial conditions used already in Equation (8.60). The set of MNE, made discrete in an adequate manner, is given by (7.66) along with the boundary constraints (7.68) and (7.71). They concern the two-dimensional no-slip conditions applied to walls moved with regard to the flow velocity components, as well as to the pressure function Ω. Considering in addition an adiabatic cylinder surface, condition (7.71) may then be simplified to

$$\nabla[3\,\Omega + \mathbf{v}^2]_{\text{cyl.-wall}} \equiv \mathbf{0}, \tag{8.62}$$

where $\mathbf{v}$ must be identified with the circumferential velocity of the cylinder surface. All relevant details, concerning particularly the mathematical preparation of the numerical procedure, are available in Lauster's booklet (1995, p. 134). In accordance with the analysis of Leder's curve, the numerical calculations were performed by using the mean value $\bar{\nu}_\tau = 6.187\ 10^{-3}\ \text{m}^2/\text{s}$, which was suggested to settle a realistic image. This expectation was sufficiently confirmed by the streamlines evaluated numerically and related to the experimentally determined data (cf. Oesterle, 1996). Figures 8.33–8.35 show the behavior of the streamlines around the vertical cylinder for three different numbers of revolutions.

For small numbers of revolutions, there obviously exists a distinctly asymmetric course of the streamlines near the cylinder surface. The rotating cylinder

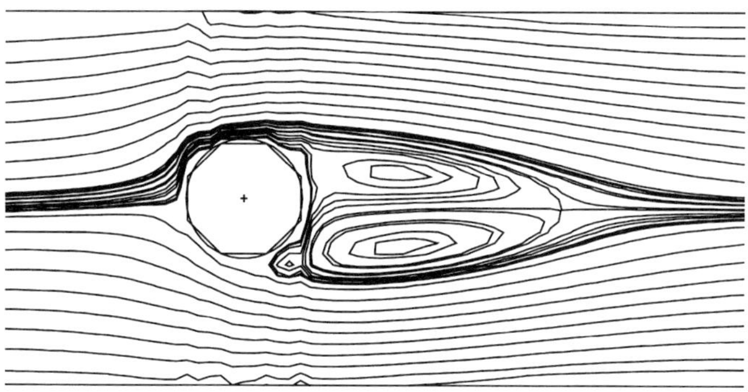

Re = 82.5

Figure 8.33 Streamlines for 500 revolutions per minute.

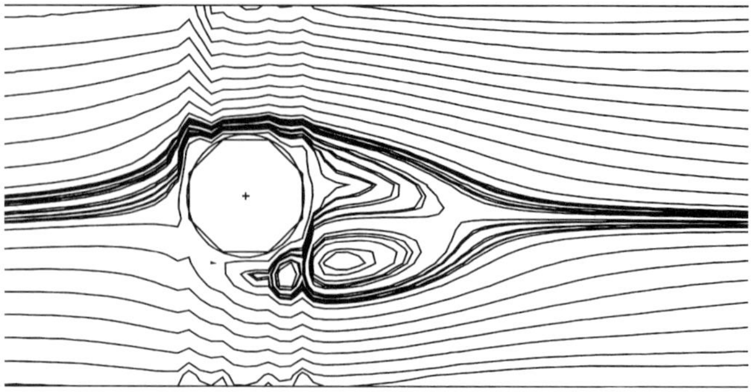

$Re = 82.5$

Figure 8.34 Streamlines for 1000 revolutions per minute.

accelerates the crossflow within the upper domain of the channel, whereas its walls slow the flow velocity down to zero. The latter effect is also true within the lower domain, but the crossflow there is delayed. As a consequence, both stagnation points in front and at the end of the cylinder are pulled together with its motion out of the symmetry plane of the channel. Inside the dead water region, both characteristic vortices are progressively deformed. A notable indication of the complex flow behavior is represented by the typical *separation bubbles* arising at the lower part of the cylinder and increasing with larger numbers of revolutions.

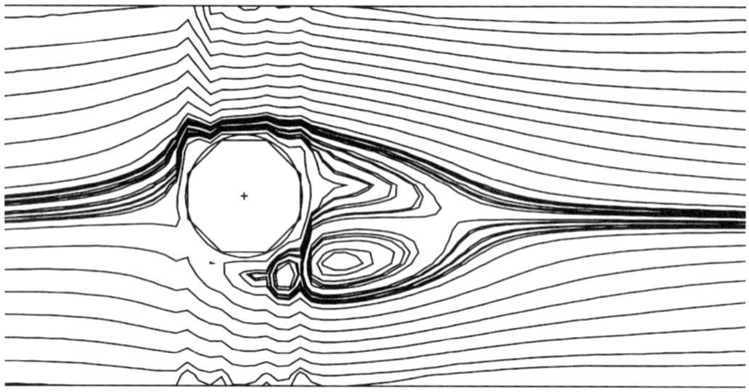

$Re = 82.5$

Figure 8.35 Streamlines for 1500 revolutions per minute.

As plotted in Figure 8.36, the angle positions of the separation points are well simulated up to about 1000 revolutions per minute. For higher numbers of revolutions, however, the values of the upper angle positions deviate slightly from the measurements, whereas for the lower separation points the angle values agree with the experimental data for numbers of revolutions even above 1750 per minute. Maybe those deviations are due to some intrinsic shortcomings in the numerical procedure applied to MNE and the given boundary conditions.

In spite of the good agreement between the calculated and measured values, we recommend that the results be used as no more than a clue for further extended research, especially with respect to improved diagnostic methods and more efficient numerical procedures. Still, these results demonstrate the ability of MNE to realistically describe even complex flow phenomena. Additionally, the analysis presented above indicates the possibility of obtaining adequate information concerning material properties and process parameters, such as kinematic viscosities and Reynolds numbers. In the light of this concrete example, the analysis reveals that, indeed, one of the major discrepancies between conventional fluid dynamics and the Alternative Theory lies in the completely different conception of the actual meaning of dissipation and transport phenomena. Unfortunately, this is not the place to deal more extensively with this basic problem of understanding non-equilibrium phenomena.

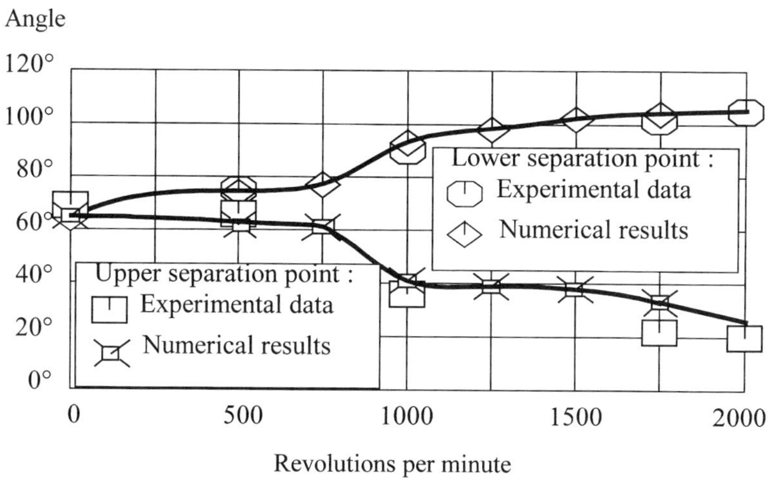

Figure 8.36 Angle positions of upper and lower separation points: a comparison between numerical and experimental data.

Chapter 9 | Gibbs–Falkian Electromagnetism

"Every true experience is finite by nature." —***G. Falk***

9.1 A Quandary Concerning Electromagnetic Field Variables

From the very beginning Falk's thesis quoted above (1990, p. 18, author's translation) indicates a fundamental problem in view of the far-reaching influence of mechanics on electromagnetism in practice. Lorentz's mechanical point of view, especially exposing some considerations on the principles of dynamics, in connection with Hertz's "Prinzipien der Mechanik" (1894, pp. 1–49), still seems to be predominant in the scientific community today. This has been true since 1906, when he published the final version of "The theory of electrons and its applications to the phenomena of light and radiant heat."

In all essential situations, Lorentz operated with the whole set of field quantities introduced by Maxwell's electromagnetic theory. This implies, for instance, that he used the charge density instead of the electrical charges themselves. Consequently, he assigned a spatial extension to each of his electrons, regarding them in principle as very small volumes.[18] This assumption is conceptually fundamental for Lorentz's interpretation of Maxwell's electromagnetic notions and their mutual relations in face of an adequate application to a microscopic view of electromagnetic phenomena. Starting from the idea that in principle the world is empty, aside only from charged particles and the electromagnetic field, it is assumed that outside of the particles neither charge nor electric current will exist. For that reason, it seems rather curious that Lorentz thought his electron was equipped with some definite distributions of charge and current inside its volume. According to this very sophisticated view, both the scalar *charge densities* $C(\mathbf{r}, t)$ and the vectorial *current densities* $\mathcal{I}(\mathbf{r}, t)$, locally depending on time- and space-coordinates, are two of the basic characteristics of the microscopic model under consideration.

[18]For mathematical convenience, however, Lorentz thought of an electron as a particle commonly used in point mechanics. Thus, in addition to its mass, these extensionless particles are loaded with an electrical charge. In this sense, they are subject to the rules of mechanics, if any electric forces attack them.

Taking into account that such a set of infinite distribution functions opens an unpredictable manifold of arbitrary premises, we cannot evade the fact that this model might turn out to be physically ill-founded or even a metaphysical part of the description. There is no doubt that Lorentz felt strongly about this shortcoming of his theory (cf. Falk, 1990, p. 60). However, his theory does provide answers to some questions about the physical idea of electrons and, above all, its mathematical foundation. Hence, it seems reasonable to expect solutions concerning the time and space behavior of both the *electric intensity* $\mathcal{E}$ and the *magnetic induction* $\mathcal{B}$ for prescribed density functions of charge $C(\mathbf{r}, t)$ and electric current $\mathcal{I}(\mathbf{r}, t)$. But an alternative is to assume that Lorentz's set of equations are the characteristic relations of the system called "electromagnetic field." Both options entail mathematically different inferences that are in no way trivial. In the first case, even singular mathematical forms like Dirac's δ-function are admitted for $C(\mathbf{r}, t)$ and $\mathcal{I}(\mathbf{r}, t)$, leading to relationships wherein the common symbols $\mathcal{E}$, $\mathcal{B}$, C, and $\mathcal{I}$ lose their original meaning of field densities in the sense of Faraday and Maxwell. In the second case, the symbols $\mathcal{E}$, $\mathcal{B}$, C, and $\mathcal{I}$ do represent such field densities, the values of which are only determined by regular functions of the space- and time-coordinates, provided that Lorentz's electrons are always extended geometrically.

This is a crucial point for the physical meaning of Maxwell's electromagnetic theory, which is mainly based on the notion of field quantity. Their values are always (at least piecewise) continuous functions of time and space coordinates. Every value of the electric energy *density*, for instance, is a function defined for all points of the space. On the contrary, a value of electric energy is a number assigned to a measurable domain of space.

Another discrepancy exists due to the fact that an essential property of any field quantity is inconsistent with Falk's mapping rules (see Chapter 2). The latter are based on notions like rings or even fields, defined as special mathematical structures (and explained in Section 2.4), whereas *density* quantities will never constitute such terms. Consequently, two density quantities may be added, but multiplications at will are excluded. For this reason only linear functions of this kind of densities are admissible. This means that the set of field quantities does not form a domain that is constitutive for the universally physical quantities introduced in Section 2.4. As a consequence, both modes of description cannot agree, and it is indeed hard to remove this serious contradiction between Maxwell's theory and classical mechanics based on the Newton–Eulerian mass-point definition.

Clearly, the problem exists only if bodies are considered that are assumed to be infiltrated by the electromagnetic field. Such a body is always characterized by its volume V; for the field as such, however, every spatial limitation should be regarded as artificial in order to bound the electrical charge within a finite region even for the case that its spherical extension tends toward an infinite diameter. Yet, conformity with *Falk's finiteness axiom*, introduced in Section 2.1, enables us to elucidate the metaphysical background of statements far beyond finite regions.

Furthermore, there is another constitutive element of Maxwell's and Lorentz's ideas that will even lead to an aporia: The combination of charge, position vector, and local velocity—denoted as "particle" for the sake of brevity—not only establishes the field, but also experiences a force by this field. This means that the particle determines the values of all quantities, thereby constituting exactly that force by which the particle itself is encountered. This quite mechanical image does not agree in any way with Faraday's suggestion to describe electromagnetic phenomena of an unseparated body-field system by mathematical relations. Indeed, the problem outlined above aims directly at the basic reasoning of the natural sciences. Recall that one of the most relevant results of Gibbs–Falkian thermodynamics is that the (total) energy of any physical system is exclusively defined by an assigned set of extensive standard variables. To maintain this universal and efficient method,[19] it seems desirable to transform the Maxwell–Lorentz fundamentals of electromagnetism to the framework of notions delineated in this book. This inference may be justified also in view of the fact that it seems questionable nowadays to sustain Maxwell's imagination concerning point mechanics and ether as the true physical base of electromagnetism.

Furthermore, the history of Maxwell 's theory does not raise any serious objections. In those days the theory was propagated in the UK by an influential group of about 40 Maxwellians, among them such prominent scientists as Lodge, Poynting, Heaviside, Thomson, and Larmor. Moreover, its protagonist in Germany was Helmholtz. This is interesting in that in the 1870s Maxwell could not present any proof for the existence and effect of the field vector $\mathcal{D}$ introduced by him and denoted as *electric displacement*; even the existence of electromagnetic waves was not yet evidenced. Hence, it is rather surprising that Maxwell himself never endeavored to furnish experimental evidence, though the Cavendish Laboratories—founded and headed by him—were well equipped to do so.

The same is true for the Maxwellians in England until the first few years of this century. Perhaps they were overconfident in the inherent truth and simplicity of Maxwell's concept and needed no empirical proof (cf. Meya, 1990, p. 213), particularly since Hertz had allegedly confirmed Maxwell's set of equations by 1894. However, Hertz's famous but extremely arduous and frustrating experiments (Hertz, 1892, pp. 4–21) proved in fact only two items: the finite propagation velocity of electric phenomena in space and time and the correspondence of the properties belonging to light and to electromagnetic waves (Meya, 1990, p. 232). But these central results could also have been explained by the well-known theories of either Helmholtz or the Danish scientist Ludwig V. Lorenz who, in 1867, had incorporated the idea of a retarded remote action into the common electromechanic theory of the Weber–Neumann type. The latter concept was resolutely rejected by Maxwell and may best be demonstrated by his own words:

[19]A certain difficulty is the implicit assertion that the field energy density $\frac{1}{2}(\mathcal{D} \cdot \mathcal{E} + \mathcal{B} \cdot \mathcal{H})$ is the internal energy per unit volume of the electromagnetic field—an assertion which requires a proof and which is not generally true (cf. Chu, 1959, p. 473 and Table 9.1 for the notation used).

> In a philosophical point of view, moreover, it is exceedingly important that two methods should be compared, both of which have succeeded in explaining the principal electromagnetic phenomena, and both of which have attempted to explain the propagation of light as an electromagnetic phenomenon and have actually calculated its velocity, while at the same time the fundamental conceptions of what actually takes place, as well as most of the secondary conceptions of the quantities concerned, are radically different. (Maxwell, 1892, p. X)

But why do all these fervent statements, omissions, and misunderstandings not delay in the least the triumphal procession of Maxwell's theory? Following Weyl, one may suppose that in reality Maxwell's equations only serve to calculate the fields $\mathcal{E}$ and $\mathcal{B}$, provided that any distributions of charge and current densities are prescribed as definite properties of matter. In other words, Maxwell's equations are reduced subsequently to a mathematical scheme defining merely the basic relations between all electromagnetic field quantities assumed to be physically adequate. Of course, the last property is strongly influenced by the knowledge of basic laws founded by men like Ampère, de Coulomb, Gauß, Faraday, Ørsted, and others. And yet, there is no doubt that Maxwell's equations lose their formal character only by supplementing them exactly with such relationships that associate the vectors $\mathcal{D}$, $\mathcal{B}$, and $\mathcal{I}$ with the vectors $\mathcal{E}$ and $\mathcal{H}$, where $\mathcal{H}$ denotes the *magnetic intensity* (cf. Abraham, 1920, p. 217).

9.2 Perspectives and Electromagnetic Units

In the past, natural scientists and all kinds of engineers have paid scant attention to influences of dissipative events on electromagnetic phenomena. With the exception of special publications concerning magnetofluid dynamics, it is remarkable that even modern textbooks written especially for electrical engineers are not concerned with such problems. Earlier literature on the subject was for the most part reserved for scientific journals. For this reason the papers available in the area are incomplete, scattered, and often deficient. However, the advent of new processing concepts, novel techniques, and new materials, as well as a changed sensibility to ecological circumstances have given the topic a greater importance. Thus, present day engineering concepts, for which electromagnetic fields play an important role, include, for example, field energy storage facilities, plants of heat rate, power interconversion, fluid dynamics with ferro- and paramagnetic materials, and processes of mass transfer with polarizable species. Recent advances in superconductivity enhanced the interest in phenomena utilizing magnetic fields. Current research is probing the influence of magnetic fields on biological systems. An interesting aspect may be observed in some kinds of complex systems subject to self-organization, where certain phenomena like the well-known chaotic tree structures evolve from characteristic optimization processes triggered by simultaneous transfer rates with the electromagnetic field. As a whole, new technical perspectives and future innovations, concerning dissipative events along with electromagnetic fields strongly depend on a precise knowledge of the complex mechanisms derived in all details from the basic principles.

In this chapter, we apply Falk's dynamics, combined with the main theorems of the Alternative Theory, to electromagnetic phenomena as an essential part of any process running within a generalized body-field system. The treatment has to begin with some remarks about electromagnetic units. Contrary to the convention of introducing Maxwell's equations axiomatically at the beginning, it is sufficient for the present to presume the mere existence of electric and magnetic phenomena. These occur in form of two separate three-dimensional physical effects—electric and magnetic—mutually linked and each represented by a pair of vectorial field quantities. The latter, denoted by $\mathfrak{B}_1$ and $\mathfrak{B}_2$ enter in such a way that each pair forms a scalar quantity ($\mathfrak{B}_1 \cdot \mathfrak{B}_2$) characterized as an energy density and determined in principle as an available function of time and space.

An abstract formulation like this concisely explains that an arbitrarily fixed value of the position vector **r** will identify not only the respective field quantities and their values, but also the volume V of the body, infiltrated by the fields. In other words, by means of **r** and t each vectorial field quantity $\mathfrak{B}$ is assigned to V in the sense of a phase term, as elaborated in Chapter 4. Forming now the product of these values of V and any field quantity $\mathfrak{B}$, assumed to be homogeneously distributed inside the phaselike volume V of the body, a product variable $[V\,\mathfrak{B}]$ will appear. This is certainly true, because a relationship $\mathfrak{B}_1 \cdot \mathfrak{B}_2 = \mathfrak{B}^a/V$ does not hold, where $\mathfrak{B}^a$ may serve as a hypothetical "absolute" quantity comparable to m^{B+L} in the correct relationship $\rho := m^{B+L}/V$ for the mass density ρ. The reason is that if $\mathfrak{B}^a$ does not exist, the field quantities $\mathfrak{B}_1$ and $\mathfrak{B}_2$ are not related to any particular volume. The same applies to the product $\mathfrak{B}_1 \cdot \mathfrak{B}_2$, although its result—an electric or magnetic energy density—now refers to the volume unit as an abstract dimension. Of course, the "variable" $[V\,\mathfrak{B}]$ now becomes an extensive one via V; its two factors normally vary with time and space.

Before we discuss the consequences, however, note the display of the set of relevant electric and magnetic properties with assignment to some customary unit systems shown in Table 9.1. We use SI units (Système International d'Unitès), the base units of which are taken from the charge-rationalized mksa[20] system of units. This system has a host of virtues, not the least of which includes the practical electrical units of volt, ampere, ohm, and so on, concerning potential differences, current, resistance, and the like. For this reason, the mksa system is now rapidly becoming a standard for the study of electromagnetism. In other areas, notably atomic and nuclear physics, the Gaussian system of units has remained common (cf. Reitz and Milford, 1974, p. 413).

Constitutive relationships connect the fields $\mathcal{D}$ and $\mathcal{H}$ with the pertaining fields $\mathcal{E}$ and $\mathcal{B}$, respectively. Two kinds of equations are used in practice. In any state of a body-field system at rest the linear relations

[20]mksa ≈ meter, kilogram, second, ampère

Table 9.1 Summary of Electromagnetic Field Variables and SI Units

Symbol	*Name*	*SI Unit*	*SI Base Units*
$\mathcal{E}$	electric intensity	volt/meter	m kg s^{-3} A^{-1}
$\mathcal{D}$	electric displacement	coulomb/(meter)^2	m^{-2} s A
$\mathcal{P}$	polarization	coulomb/(meter)^2	m^{-2} s A
$\mathcal{I}$	current density	ampere/(meter)^2	m^{-2} A
$\mathcal{C}$	charge density	coulomb/(meter)^3	m^{-3} s A
$\mathcal{H}$	magnetic intensity	ampere/meter	m^{-1} A
$\mathcal{B}$	magnetic induction	tesla = volt s/(meter)^2	kg s^{-2} A^{-1}
$\mathcal{M}$	magnetization	ampere/meter	m^{-1} A

$$\mathcal{D} := \boldsymbol{\varepsilon} \bullet \mathcal{E} \quad \text{and} \quad \mathcal{H} := \boldsymbol{\mu}^{-1} \bullet \mathcal{B} \tag{9.1}$$

hold, where ε and μ denote the second-order tensors of a dielectric body and its magnetic permeability, respectively. For isotropic matter both tensors reduce to the simple forms $\boldsymbol{\varepsilon} := \varepsilon\mathbf{1}$ and $\boldsymbol{\mu} := \mu\mathbf{1}$, where **1** stands for the unit tensor along with the so-called dielectric constant ε and the permeability (cf. Zahn, 1979, p. 352). Another way to assign the respective fields noted above is by the two defining equations

$$\mathcal{D} := \varepsilon_0\, \mathcal{E} + \mathcal{P} \qquad \mathcal{B} := \mu_0\, (\mathcal{H} + \mathcal{M}), \tag{9.2}$$

where the permittivity and the permeability of free space are denoted by ε_0 and μ_0, respectively. The quantity $(\varepsilon_0\, \mu_0)^{-1/2}$ has the dimensions of a velocity; a solution of Maxwell's equations shows that this quantity equals the propagation speed of light or any other electromagnetic wave in a vacuum. The quantity $\mathcal{M}$ stands for magnetization, a material property defining the state of magnetic polarization of magnetized matter. It follows from $(9.2)_2$ that $\mathcal{B}$ differs from $\mu_0\mathcal{H}$ only in the presence of magnetized matter that behaves like a collection of an equal number of oppositely charged poles—that is, as dipolar matter.

The strength and direction of an electric field are both described at each point by a vector $\mathcal{E}$ in such a way that the force acting on a small stationary test charge ς placed at this point is $\varsigma\,\mathcal{E}$. Polarized dielectrics and ferroelectric substances furnish the electric dipolar analogy. The electrically polarized media are characterized by the polarization vector $\mathcal{P}$ and a vector $\mathcal{D}$, the displacement field defined by $(9.2)_1$. By combining Equations (9.1) and (9.2) we obtain the direct relationships between the vectors $\mathcal{P}$ and $\mathcal{E}$ as well as the vectors $\mathcal{M}$ and $\mathcal{H}$, respectively. These elementary proportionalities were first founded in a simple way by Hertz (1892, pp. 214–224).

9.3 Falk's Dynamics of Electromagnetic Phenomena

Let us start with a compilation of the three basic theorems derived from the principles of the Alternative Theory in Chapter 6:

$$\mathbf{j}_e - \mathbf{JI} \bullet \mathbf{v} - T_* \mathbf{j}_s - \sum_{j=1}^{J} \mu_{*j} \mathbf{j}_j \equiv 0$$

$$\boldsymbol{\tau}_* : \nabla \mathbf{v} + \mathbf{j}_s \bullet \nabla T_* + \sum_{j=1}^{J} \mathbf{j}_j \bullet \nabla \mu_{*j} + T_* \sigma + \sum_{j=1}^{J} \mu_{*j} \Gamma_j \equiv 0 \tag{9.3}$$

$$\rho D\mathbf{v} = \rho \mathbf{f} - \nabla \bullet \Pi + \partial_t \rho \boldsymbol{\varphi}; \qquad \Pi := p_* \mathbf{1} - \boldsymbol{\tau}_* .$$

The presented forms of the divergence theorem $(9.3)_1$, the dissipation theorem $(9.3)_2$, and the equation of motion $(9.3)_3$ are well founded by the Gibbs–Euler function (GEF)—Equation (3.8)—for the class of systems in question. That relation refers to any multicomponent single-phase system classified as a body-field system (BFS) by its Pfaffian

$$dE = \mathbf{v} \bullet d\mathbf{P} - \mathbf{F} \bullet d\mathbf{r} + T_* dS - p_* dV + \sum_{j=1}^{J} \mu'_{*j} dN_j , \tag{6.1}$$

where the field is exclusively confined to the case that is described by the position vector $\mathbf{r}$ only. As before, the asterisk indicates explicitly that (6.1) relates to any changes of non-equilibrium states. The set of equations (9.3) and (6.1) provides the key to extend the system to the class of systems that additionally include electromagnetic phenomena. The incorporation of the adequate electromagnetic variables into the GEF, viz. $E - E_\# = G$(extensive standard variables), will lead to corresponding terms in all three relationships (9.3). The path to this aim was expounded in Chapter 6.

According to the reasoning in the previous section, as well as the basic theorems of Falk's dynamics, a system may be represented by the Gibbs function $\xi_0^{[\mathcal{E}, \mathcal{H}]} = G^{[\mathcal{E}, \mathcal{H}]}(\xi_1, \ldots, \mathcal{E}, \mathcal{H}, \ldots, \xi_r)$. The relationship $\xi_0^{[\mathcal{E}, \mathcal{H}]}$ is derived from the basic relationship (2.72), that is, the GEF $E - E_\# = \xi_0 = G(\xi_1, \ldots, \xi_k, \xi_{k+1}, \ldots, \xi_r)$, by means of the Legendre transformation (2.75), where the two extensive variables ξ_k, ξ_{k+1} are assigned to the two field quantities $\mathcal{E}$ and $\mathcal{H}$. Following, for example, Callen (1966, p. 95), the symmetry between this relationship (2.75) and its inverse is indicated by the following table:

$\xi_0 = G(\xi_k)$	$\xi_0^{[\xi_k]} = G^{[\xi_k]}(\tau_k)$
$\tau_k = \partial\xi_0/\partial\xi_k$	$-\xi_k = \partial\xi_0^{[\xi_k]}/\partial\tau_k$
$\xi_0^{[\xi_k]} = -\tau_k\xi_k + \xi_0$	$\xi_0 = \xi_0\tau_k + \xi_0^{[\xi_k]}$
Elimination of ξ_0 and ξ_k yields	Elimination of $\xi_0^{[\xi_k]}$ and τ_k yields
$\xi_0^{[\xi_k]} = G^{[\xi_k]}(\tau_k)$	$\xi_0 = \mathrm{G}(\xi_k)$

The generalization of the Legendre transformation and its inverse to functions of more than a single independent variable is simple and straightforward.

Using now the usual notation for the quantities ξ_k and τ_k $(k = 0(1)r)$, the Pfaffian of $\xi_0^{[\mathcal{E},\mathcal{H}]}$ becomes

$$dE^{[\mathcal{E},\mathcal{H}]} = \mathbf{v} \bullet d\mathbf{P} - \mathbf{F} \bullet d\mathbf{r} + T_* dS - p_* dV + \xi_k \bullet d\mathcal{E}_* + \xi_{k+1} \bullet d\mathcal{H}_* + \sum_{j=1}^{J} \mu'_{j*}\, dN_j, \quad (9.4)$$

where Equation (5.10) is extended to include the electromagnetic field considered by the two vectorial variables $\mathcal{E}$ and $\mathcal{H}$ and their conjugates ξ_k and ξ_{k+1}. Additional terms may be included to deal, for instance, with conducting magnetic fluids with internal rotation (cf. Shizawa and Tanahashi, 1986). The inverse of the Legendre transformation, leading from the energy-like quantity $E^{[\mathcal{E},\mathcal{H}]}$ to the (total) energy E, yields two product variables, each of the form $[V\mathfrak{B}]$. These variables were introduced above and are identified as being extensive via the body volume V. The corresponding Gibbs main equation of type (3.6)

$$dE = \mathbf{v} \bullet d\mathbf{P} - \mathbf{F} \bullet d\mathbf{r} + T_* dS - p_* + \mathcal{E}_* \bullet d(V\mathcal{D}_*) + \mathcal{H}_* \bullet d(V\mathcal{B}_*) + \sum_{j=1}^{J} \mu'_{j*}\, dN_j \quad (9.5)$$

is the theoretical foundation of the complete description of any multicomponent single-phase system classified as a body-field system, which also includes gravitational and electromagnetic phenomena. Yet, the Pfaffian (9.5) does not allow an unambiguous solution without further explanations concerning the linear momentum conservation. Following Abraham's detailed arguments (1920, p. 26) the total electromagnetic quantity of motion $\mathbf{P}^{\mathrm{elmag}}$ assigned to the whole body-field system is defined by means of an integral over the momentum density $\mathcal{G}$ of volume elements $d\upsilon$ as follows:

$$\mathbf{P}^{\mathrm{elmag}} := \int_{V^{elmag}} \mathcal{G}\, d\upsilon \quad (9.6)$$

The resulting electromagnetic force $\mathbf{F}^{\mathrm{elmag}}$ acting on any body of the system can be expressed by temporal changes of $\mathbf{P}^{\mathrm{elmag}}$ and the sum of *surface force densities* $\mathfrak{T}$ represented by a suitable surface integral

$$\mathbf{F}^{\mathrm{elmag}} := \int_{A^{elmag}} \mathfrak{T}\, dA - \frac{d}{dt}\mathbf{P}^{elmag}, \quad (9.7)$$

where dA denotes a surface element of the area A^{elmag} that encloses the whole system.

Equation (9.7) may be simplified when A^{elmag} is thought to be located far away but enclosing the matter of the body-field system in question. The contribution of the surface forces may be neglected under the additional condition that no propagation processes within the field are able to reach this boundary surface A^{elmag} as long as the body-field interaction takes place. In practice, such an assumption might be justified in the same way as the gravitational influence is justified. Hence, as a first consequence the simple relationship

$$\mathbf{F}^{\text{elmag}} = -\frac{d}{dt}\mathbf{P}^{\text{elmag}} \tag{9.8}$$

evolves, yielding a convenient way to obtain a precise formulation of the principle of momentum conservation.

Subsequent to the pertaining results presented in Section 5.1, the momentum of the body in its time behavior equals the electromagnetic force $\mathbf{F}^{\text{elmag}}$ acting on this body, along with the gravitational forces $\mathbf{F}^{\text{grav}}$. In this case, the conservation equation

$$\mathbf{P}^{\text{elmag}} + \mathbf{P}^{\text{grav}} + \mathbf{P} = \text{constant} \tag{9.9}$$

is assumed to be valid. Then the corresponding differential relation $d\mathbf{P}^{\text{elmag}} + d\mathbf{P}^{\text{grav}} + d\mathbf{P} \equiv 0$ can be used to review the results already obtained in Section 5.1 and mathematically formulated by Equations (5.4)–(5.8). In other words, for polarized body-field systems the respective Gibbs–Euler function $\mathscr{E}(\mathbf{P}; \mathbf{r}; S; V; V\mathcal{D}_*; V\mathcal{B}_*; N_j)$ of the body can thus be reduced to the typical form (5.9). Additionally, the linear momentum $\mathbf{P}$ of the body is subject to (9.9). Furthermore, the displacement vector $\mathbf{r}$ refers to the conjugated force $\mathbf{F}^{\text{body}}$ acting on the body and resulting from the corresponding field forces $\mathbf{F}^{\text{grav}}$ and $-\mathbf{F}^{\text{elmag}}$. This follows from the force balance

$$\frac{d}{dt}\mathbf{P} \equiv \mathbf{F}^{\text{body}} = \mathbf{F}^{\text{elmag}} - \mathbf{F}^{\text{grav}},$$

provided that force laws like (9.8) (i.e., $d\mathbf{P}/dt := \mathbf{F}$) are believed to be correct. As a rule, this is only true for problems concerning mass-point mechanics.

In more general cases, the momentum conservation (9.9) is independent of the determination of the corresponding forces. The latter are related to each other by the definition

$$\mathbf{F}^{\text{body}} := \mathbf{F}^{\text{elmag}} - \mathbf{F}^{\text{grav}} \tag{9.10}$$

and must be evaluated along with the solution of the pertaining set of balances to be solved for given initial and boundary conditions. Thus, forces due to gravitational effects are commonly represented by the earth acceleration $\mathbf{f}$, which is defined to be an exogenous quantity introduced to restrict any motion of matter (cf. Truesdell, 1984, p. 428).

Hence, the energy form of motion—referred to the linear momentum $\mathbf{P}$ of the body—as well as the energy form of displacement—related to its position vector $\mathbf{r}$—may be written as follows:

$$\mathbf{v} \bullet d\mathbf{P} - \mathbf{F}^{\text{body}} \bullet d\mathbf{r} = \mathbf{v} \bullet d\mathbf{P} - \mathbf{F}^{\text{grav}} \bullet d\mathbf{r} + \mathbf{F}^{\text{elmag}} \bullet d\mathbf{r}.$$

According to the electromagnetic variables introduced above, two *force densities* $\mathcal{F}$ and $\rho\mathbf{f}$ will be used in place of the absolute quantities, viz.

$$\mathbf{F}^{\text{elmag}} := V\mathcal{F} \quad \text{and} \quad \mathbf{F}^{\text{grav}} := V\rho\mathbf{f}. \tag{9.11}$$

We are now in a position to write (9.5) in the final form of the respective Gibbs main equation:

$$dE = \mathbf{v}\bullet d\mathbf{P} + V\,[\mathcal{F} - \rho\mathbf{f}]\bullet d\mathbf{r} + T_* dS - p_* dV + \mathcal{E}_*\bullet d\,(V\mathcal{D}_*) + \mathcal{H}_*\bullet d\,(V\mathcal{B}_*) + \sum_{j=1}^{J} \mu^{t}_{j*}\,dN_j . \tag{9.12}$$

Whereas the uniform acceleration **f** in the earth's gravitational field may be treated as a conservative function (i.e., **f** = **f**(**r**)), such an option does not exist for electromagnetic fields. Of course, additional information is always needed to characterize specific problems. Above all, some constitutive equations for the material properties of a body, along with the basic relationships between the electric and magnetic field quantities, are required. The following treatment deals with the incorporation of such general information into (9.12). The systematic representation of this method will lead to an extended set of equations like that of (9.3). A first consequence to be drawn from (9.12) concerns the Gibbs–Duhem–Margules relation (2.76) of the system defined by its GEF, viz. $E - E_{\#} = \mathcal{E}(\mathbf{P}; \mathbf{r}; S; V; V\mathcal{D}_*; V\mathcal{B}_*; N_j)$. The result

$$\mathbf{P}\bullet d\mathbf{v} - \mathbf{r}\bullet d\mathbf{F}^{\text{body}} + S\,dT_* - V\,dp_* + (V\mathcal{D}_*)\bullet d\mathcal{E}_* + (V\mathcal{B}_*)\bullet d\mathcal{H}_* + \sum_{j=1}^{J} N_j\,d\mu^{t}_{j*} \equiv 0 \tag{9.13}$$

proves the existence of an internal fundamental relation $h(\mathbf{v}; \mathbf{F}^{\text{body}}; T_*; p_*; \mathcal{E}_*; \mathcal{H}_*; \mu^{t}_{j*}) \equiv 0$, assumed to be available for the system under consideration. Note, then, that not all of the intensive variables are independent. In practice, however, the relation must be substituted by characteristic relationships determined for the hypothetical state at rest, as expounded in Chapters 6 and 7.

In reference to the baryon–lepton constancy defined by (5.20) and assumed to be valid for the system, the mass m^{B+L} may serve to convert all extensive quantities into specific ones. Consequently, Equation (9.12) takes on the following form

$$\rho de = \mathbf{v}\bullet\rho d\mathbf{i} + [\mathcal{F} - \rho\mathbf{f}]\bullet d\mathbf{r} + T_*\rho\,ds + \sum_{j=1}^{J} \mu^{t}_{j*}\rho\,d\omega_j + \frac{p_*}{\rho}d\rho + \mathcal{E}_*\bullet\rho d\left(\frac{\mathcal{D}_*}{\rho}\right) + \mathcal{H}_*\bullet\rho d\left(\frac{\mathcal{B}_*}{\rho}\right), \tag{9.14}$$

the last two terms of which may be factorized to separate the electric and magnetic variables from the density variable. The resulting Pfaffian

$$\rho de = \mathbf{v}\bullet\rho d\mathbf{i} + [\mathcal{F} - \rho\mathbf{f}]\bullet d\mathbf{r} + T_*\rho\,ds + \sum_{j=1}^{J} \mu^{t}_{j*}\rho\,d\omega_j + \frac{p_*^{bf}}{\rho}d\rho + \mathcal{E}_*\bullet d\mathcal{D}_* + \mathcal{H}_*\bullet d\mathcal{B}_* \tag{9.15}$$

thus seems to contain an extended pressure quantity p_*^{bf} that now includes the contribution of the $\mathcal{E}_* - \mathcal{H}_*$ field:

$$p_*^{bf} := p_* - \mathcal{E}_*\bullet\mathcal{D}_* - \mathcal{H}_*\bullet\mathcal{B}_*. \tag{9.16}$$

However, this conclusion is in error even though it appears (aside from a factor $\frac{1}{2}$) in many textbooks. Particularly with regard to the electromagnetic quantities, any intensive conjugate, according to (9.13), will not only depend on the remaining conjugates of the system, but will also be related primarily to all extensive variables involved. The simplest way to illustrate this fact is by the Pfaffian (9.4): The non-equilibrium pressure p_* immediately follows from the derivative

$$p_* = -\ (\partial \mathbf{E}^{[\mathcal{E},\,\mathcal{H}]}/\partial V)_{\mathbf{P},\,\mathbf{r},\,S,\,\mathcal{E}_*,\,\mathcal{H}_*,\,N_j},$$

where $p_*(\mathbf{P}, \mathbf{r}, S, V, \mathcal{E}_*, \mathcal{H}_*, N_j)$ "is the pressure due to the joint action of the mechanical and electromagnetic changes" (Chu, 1959, p. 474). Let us, moreover, substitute p_* and the two quantities $\mathcal{E}_*$ and $\mathcal{H}_*$ with the pertaining partial differentials following from (9.14).

$$p_* = \rho^2\left(\frac{\partial e}{\partial \rho}\right)_{\mathbf{i},\,\mathbf{r},\,s,\,\omega_j,\,\mathcal{D}_*/\rho,\,\mathcal{B}_*/\rho}; \qquad \mathcal{E}_* = \left(\partial e\Big/\partial\frac{\mathcal{D}_*}{\rho}\right)_{\mathbf{i},\,\mathbf{r},\,s,\,\omega_j,\,\rho,\,\mathcal{B}_*/\rho}; \tag{9.17}$$

$$\mathcal{H}_* = \left(\partial e\Big/\partial\frac{\mathcal{B}_*}{\rho}\right)_{\mathbf{i},\,\mathbf{r},\,s,\,\omega_j,\,\rho,\,\mathcal{D}_*/\rho}.$$

Equation (9.17) implies that the non-equilibrium pressure p_* already contains all respective contributions due to electromagnetic events. In other words, (9.16) or any relationship such as the common expression $p_*^{bf} := p - \frac{1}{2}\ (\mathcal{E}_* \bullet \mathcal{D}_* + \mathcal{H}_* \bullet \mathcal{B}_*)$ (cf. Chu, 1959, p. 475 or Dragoş, 1975, p. 56) is inadequate for any physically consistent theory. There is a high degree of complexity expected if a given number of common standard variables, including electric and magnetic quantities, must be completed by a separate set of equations that remain exclusively confined to the prevalent quantities of the electromagnetic field. Of course, this is the case if one considers the main results of Maxwell's electromagnetic theory. Nevertheless, Equations (9.17) permit the statement that, for example, the distribution $\mathcal{H}_*(\mathbf{r};\, t)$ of any magnetic intensity may be strongly influenced by the actual changes of the system with respect to its motion, entropy, and the other relevant variables. An important example is the climate dynamics covering a large area of the earth and influenced by vehement physicochemical processes on the surface of the sun.

Continuing the analysis, the next step concerns the time t as a curve parameter (cf. Sections 4.3 and 4.4.3). Using the *material derivative* $d/dt := \partial/\partial t + \mathbf{v} \bullet \nabla$ and its abbreviated operator symbol D, Equation (9.15) appears in the form

$$\rho\, De = \mathbf{v} \bullet \rho\, D\mathbf{i} + [\mathcal{F} - \rho\mathbf{f}] \bullet \mathrm{v} + \mathrm{T}_*\rho\, Ds + \sum_{j=1}^{J} \mu_{j*}\rho\, D\omega_j + \frac{p_*}{\rho} D\rho \tag{9.18}$$

$$-\,(\mathcal{E}_* \bullet \mathcal{D}_* + \mathcal{H}_* \bullet \mathcal{B}_*)\frac{D\rho}{\rho} + \mathcal{E}_* \bullet D\mathcal{D}_* + \mathcal{H}_* \bullet D\mathcal{B}_*$$

where the postulate (5.69), which interconnects kinematics and dynamics, is considered with respect to the differential $D\mathbf{r} := d\mathbf{r}/dt$. The left-hand side allows us to open the system toward its surroundings by means of the First Law of Thermodynamics.

For continuum physics, the corresponding relationships have already been given in Section 6.3 and compiled again here.

$$\rho\, De + \nabla \cdot \mathbf{j}_e \equiv 0 \tag{6.32}$$

$$\mathbf{j}_e := \mathring{\mathbf{q}}_* + \mathring{\mathbf{w}}_* \tag{6.33}$$

$$\mathring{\mathbf{w}}_* = \mathbf{v} \cdot \mathbf{\Pi} \tag{6.45}$$

$$\mathbf{\Pi} := p_* \mathbf{1} + \boldsymbol{\tau}_*. \tag{6.42}$$

Compared with the problem treated in Chapter 6, the notation used above differs in meaning if field quantities like the electromagnetic ones are included. Notions like the heat flux vector $\mathring{q}_*$ and the work rate vector $\mathring{w}_*$ lose their original meaning, but this is not true for either the non-equilibrium pressure p_* or the momentum flux density $\mathbf{\Pi}$ and the viscous pressure tensor $\boldsymbol{\tau}_*$. Now these do not refer only to the body, but also to some characteristic contributions due to the electromagnetic field. Hence, a precise analysis must be done with respect to the complicated structure of all the electromagnetic phenomena in connection with the diverse flux densities and production densities of the other variables.

In preparation for this analysis, the problem needs to be based on the Legendre-transformed energy $E_*^{[V]}$ instead of the (total) energy E. Recall Planck's discovery that every identity $f(E; S; V; N) \equiv 0$ between E, S, V, and N would be destroyed by a Lorentz transformation from any reference system at rest to one in motion, since the volume V, unlike E, S, and N, is not a Lorentz invariant (Planck, 1910, p. 125). For this reason, the variable V will be exchanged with the variable p_* for which Lorentz invariance has been proved. Exactly this invariance is required to incorporate any electromagnetic phenomena in a consistent way into Falk's dynamics, as shown in the Gibbs rate equation (6.31). The latter, however, is only valid for a multicomponent single-phase body-field system. Including new terms that refer to the divergence of velocity and a pair of electric and magnetic energy forms, the following Gibbs rate equation results

$$\rho\, De_*^{[\rho]} = \mathbf{v} \cdot \rho\, D\mathbf{i} + [\mathcal{F} - \rho\mathbf{f}] \cdot \mathbf{v} + T_* \rho\, Ds + \sum_{j=1}^{J} \mu_{j*} \rho\, D\omega_j + Dp_* \\ + (\mathcal{E}_* \cdot \mathcal{D}_* + \mathcal{H}_* \cdot \mathcal{B}_*) \nabla \cdot \mathbf{v} + \mathcal{E}_* \cdot D\mathcal{D}_* + \mathcal{H}_* \cdot D\mathcal{B}_*, \tag{9.19}$$

where the divergence $\nabla \cdot \mathbf{v}$ replaces $D\rho/\rho$ by means of the global continuity equation. Furthermore, the Legendre-transformed specific energy $e_*^{[\rho]}$ is related to the (total) specific energy e via $e_*^{[\rho]} = e + (p_*/\rho)$. As a consequence, the balance of $e_*^{[\rho]}$ agrees with the expression already derived in Section 6.3:

$$\rho\, De + \nabla \cdot \mathbf{j}_e \equiv 0 \quad \Rightarrow \quad \rho\, De_*^{[\rho]} + \nabla \cdot [\mathbf{j}_e - p_* \mathbf{v}] = \partial_t p_*. \tag{9.20}$$

The Legendre-transformed energy flux density $\mathbf{j}_{e*}^{[\rho]} := \mathbf{j}_e - p_* \mathbf{v}$ can be related easily to the heat flux vector $\mathring{\mathbf{q}}_*$ and the work rate vector $\mathring{\mathbf{w}}_*$ by means of the set of equations compiled above. Equations (9.19) and (9.20) are the starting points of the following theoretical consideration. If we now substitute the corresponding balance equations for the specific quantities $e_*^{[\rho]}$, **i**, s, and ω_j, according to Equation (6.37), the question is whether there will be balances for the two electromagnetic energy forms $\mathcal{E}_* \cdot D\mathcal{D}_*$ and $\mathcal{H}_* \cdot D\mathcal{B}_*$. Remember the general form of such a balance equation already derived in Section 5.4, viz.

$$\rho\, Dz + \nabla \cdot \mathbf{j}_z = \sigma_z = \partial_t \rho z + \nabla \cdot (\mathbf{j}_z + \rho z \mathbf{v}). \tag{5.82}$$

Both the flux density $\mathbf{j}_z$ and the production density σ_z refer to the specific quantity z of the system.

This basic structure implies the fulfillment of some special mathematical relationships required for the interaction between the four electromagnetic field quantities. Of course, these essential requirements do not refer to mathematics alone, but also to all basic knowledge of electromagnetism assumed to be universally valid. This may clearly be regarded as the basic essence of Maxwell's theory of electricity and magnetism and, particularly, of his celebrated set of equations. Nevertheless, we should note that Maxwell never published a final treatment of his ideas. His "Treatise" presented a rather mixed discourse that allowed different interpretations of his basic ideas of electromagnetism. Most interpretations confirm that Maxwell believed, above all, in a purely mechanical reality behind all electric and magnetic events (cf. Meya, 1990, p. 192). In this context, we quote D. S. Walton's competent statement: "The physical substance is in Maxwell's writings, but the formal expression that we are familiar with is due to Heaviside" (quoted by Catt, 1985, p. 35). This analysis is surely true and, moreover, reduces the complicated history of electromagnetism to its simplest form. As a matter of fact, it also seals the traditional ideas of electromagnetic physics until now, although one should bear in mind that historically the theory of electrodynamics thoroughly evolved from the theory of static fields, both electric and magnetic. Static fields, however, emerged from steady electric currents and steady charge. Thus, both these notions preceded Faraday's concept of a *transverse electromagnetic wave*.

Nevertheless, the evolution that occurred was in no way mandatory, though it is generally agreed upon at present. There may be some reason for the view that the transverse electromagnetic wave is a fundamentally more primitive starting point for an advanced electromagnetic theory than electric charge and current. As early as in 1898 Fleming, the inventor of the diode, emphasized that "although we are accustomed to speak of the current as flowing in the wire, ... [it] is, to a very large extent, a process going on in the space or material outside the wire."[21] This statement reflected the principal conflict between the partisans of two scientific groups. The problem concerned the way to describe the interaction between distant bodies. Historically first and conceptually simplest was what may be termed the mechanical concept of direct action between the bodies across the intervening distance. Faraday

[21]Fleming, J. A. (1898). "Magnets and Electric Currents," p. 80, quoted in *Wireless World*, Dec. 1980, p. 79.

more subtly argued for the idea of a force field produced by one body. From this source, the field extends throughout space-time and acts on other bodies.

Adherents of the traditional line stubbornly stuck to a descriptive view of electromagnetic phenomena by means of mechanical models. Some propagators of Faraday's method favored constructing adequate but abstract relations between completely new terms, for which neither pictorial nor concrete models exist. Although his theoretical work was unscholarly and merely of a qualitative nature, Faraday in a certain sense preceded Mayer's and even Gibbs' basic ideas. With his sometimes obscure notion of force he anticipated the modern version of the energy principle and its realization by means of a finite number of energy forms (cf. Meya, 1990, p. 118; Schirra, 1991, p. 130).

But there was yet another irreconcilable antagonism: Faraday was a brilliant but uneducated technician who threatened the belief of England's upperclass that all scientific progress should of necessity be carried by the rigor and discipline controlled and celebrated by the world of academics in places like Cambridge University. Consequently, the ultimate in scientific rigor was reserved for mathematics. Some authors in the history of science openly claim that, lacking mathematics, Faraday could not and did not really affect his discovery of electromagnetic induction in 1831. Hence, Professor Maxwell, not Faraday the technician, opened the path for a conclusive exploitation of electromagnetism. Thus, Faraday did not have the slightest chance. Not only did the mechanical theory of heat win, but victory was also achieved by the mechanical theory of electromagnetism.

As a consequence, the scientific community generally accepted the view that every kind of electromagnetic field and matter was something that existed as an individual separate thing. Science had to identify the field and the material body under consideration and, furthermore, it had to discover their special properties as well as the force laws of their mutual interaction. This strict mechanical point of view led to some serious problems in connection with extending Lorentz's theory of electrons to Broglie's wave mechanics. It was only Dirac's brilliant theory of light that allowed deeper insight into microphysics of strictly coupled electrical and mechanical phenomena (cf. Broglie, 1939, p. 106).

9.4 Maxwell's Equations

Aside from the requirements noted above in connection with (5.82), two additional propositions must be satisfied with respect to the following issues:

1. Electricity and magnetism are linked together in all states. Thus, it is impossible for any state of an electromagnetic field with the alternative $\boldsymbol{\mathcal{E}} \neq \mathbf{0}$, $\boldsymbol{\mathcal{H}} = \mathbf{0}$ or $\boldsymbol{\mathcal{E}} = \mathbf{0}$, $\boldsymbol{\mathcal{H}} \neq 0$ to be transformed into another one by mere changes of the reference systems.

2. Whereas all equations for the system in question should be subject to the Lorentz invariance, there is no need to require relativity by means of certain modifications of the theory presented here as a first step.

We will see shortly that the first point presupposes the second one so that it concerns at least the physical meaning of the Lorentz invariance. It is true that by this invariance we can replace instantaneous propagation by propagation with a finite flow velocity not greater than speed of light in vacuum. For this reason, the Lorentz invariance holds the rank of an undisputed principle in physics. Additionally, it can ensure exactly the mathematical representation of the simultaneous existence of electric and magnetic phenomena.

The structure of the relationships to be established below is also influenced by the basic laws characterizing all electromagnetic fields; it is additionally affected by some experimental findings found to be universally valid. Hence, four basic laws deserve consideration:

Faraday's law relates the circuit voltage to the flux linkages varying with time. It affects electrical generators or material bodies by an electromagnetic wave traveling through space; neither an electric current flow nor a conductor need to be present.

Ampère's law concerns the magnetic field that curves in a spiral shape around a current flux, corrected for unsteady values of the electric field. Wound magnetic field sources rely on the net current density, while wave propagation is due to the electric field.

Gauss' two laws describe the main asymmetry between electric and magnetic fields with respect to the different charge density. Law I expresses the net charge distribution producing the electric field lines, whereas Law II states that isolated magnetic poles are unknown; as a consequence, each line of magnetic induction forms a loop.

To summarize, all the general requirements concerning physical conditions and universal phenomena discussed above constitute an integrated entity established by a compact set of mathematical relationships. Its central part consists of Maxwell's celebrated equations and governs electromagnetic phenomena (cf. Reitz and Milford, 1974, p. 296).

The electromagnetic theory of a moving medium was first introduced by Cohn subsequent to the version of Lorentz and Abraham. Minkowski was the first who based his theory on the Lorentz transformations. Thus, he proved that Maxwell's equations still hold (cf. Abraham, 1920, pp. 289, 378), regardless of whether the medium is moving and being deformed or not. For the present work, the nonrelativistic version of Minkowski's theory is used.

Consider a fluid particle according to (9.19) that moves with velocity $\mathbf{v}$ much smaller than the velocity of light c in vacuum, so that all terms of order $\mathbf{v}^2/c^2$ can be neglected when compared with unity. Relative to an observer joined to the particle, the five electromagnetic field vectors are $\mathcal{B}_*$, $\mathcal{D}_*$, $\mathcal{H}_*$, $\mathcal{E}_*$, and $\mathcal{I}_*$. Given in differential form, Maxwell's equations appear suitable for the calculation of fields both inside and outside of moving matter. Hence we obtain (cf. Zahn, 1979, p. 489)

Faraday's law: $\nabla \times \mathcal{E}_* = -\partial \mathcal{B}_*/\partial t$ $(9.21)_1$

Ampere's law with Maxwell's displacement current "correction":

$$\nabla \times \mathcal{H}_* = \mathcal{I}_{*f} + C_f \mathbf{v} + \partial \mathcal{D}_*/\partial t \tag{$(9.21)_2$}$$

Gauss' law I: $\nabla \bullet \mathcal{D}_* = C_f$ $(9.21)_3$

Gauss' law II: $\nabla \bullet \mathcal{B}_* = 0.$ $(9.21)_4$

These basic field equations deviate only by the *convective charge density* $C_f\mathbf{v}$ from the classical Maxwell equations governing a system as viewed by an observer from the fixed laboratory reference frame (cf. Thomas and Meadows, 1985, p. 39). The electromagnetic field vectors $\mathcal{B}$, $\mathcal{D}$, $\mathcal{H}$, $\mathcal{E}$, and $\mathcal{I}$, introduced with respect to this *stationary reference coordinate system*, are related to the corresponding vectors of the moving body-field system as follows (cf. Chu, 1959, p. 476):

$$\mathcal{E}_* = \mathcal{E} + \mathbf{v} \times \mathcal{B} \tag{$(9.22)_1$}$$

$$\mathcal{D}_* = \mathcal{D} + c^{-2}\mathbf{v} \times \mathcal{H} \tag{$(9.22)_2$}$$

$$\mathcal{H}_* = \mathcal{H} - \mathbf{v} \times \mathcal{D} \tag{$(9.22)_3$}$$

$$\mathcal{B}_* = \mathcal{B} - c^{-2}\mathbf{v} \times \mathcal{E} \tag{$(9.22)_4$}$$

$$\mathcal{I}_{*f} = \mathcal{I}_f - C_f \bullet \mathbf{v}. \tag{$(9.22)_5$}$$

In Equation $(9.21)_2$, $\mathcal{I}_f$ is the *free current density.* The term *free* distinguishes such a charge from the bound charge associated with the electric dipolar matter. If there is more than one type of charge carrier, the net charge density C_f is equal to the algebraic sum of all the charge densities, whereas the net current density $\mathcal{I}_f$ equals the vector sum of the current densities due to each carrier. Since the net charge is conserved, a conservation equation can be written a priori as

$$\nabla \bullet \mathcal{I}_f + \partial C_f/\partial t \equiv 0. \tag{9.23}$$

Additionally, boundary conditions are imposed on fields at interfaces between any media possessing different properties of $\mathcal{M}$ and $\mathcal{P}$: In a magnetic system these relationships assert the continuity of normal $\mathcal{B}$ and tangential $\mathcal{H}$:

$$[\mathcal{B} \bullet \mathbf{n}] = 0 \quad \text{and} \quad [\mathcal{H} \times \mathbf{n}] = 0, \tag{9.24}$$

where the brackets indicate a difference across the interface.

Note that the set of equations (9.21) represents Minkowski's version of the electromagnetic field equations proven to be valid for any moving system. By means of Lorentz's transformations they agree with the *Maxwell–Heaviside field equations* of systems at rest and even confirm Hertz's relations (9.1) for the basic vectors (cf. Abraham, 1920, p. 379).

The first consequence can be drawn from Equations $(9.22)_1$ and $(9.22)_3$ for the case of pure electromagnetic fields. It is defined by the vector identities $\mathcal{E} \equiv \mathcal{D}$ and $\mathcal{H} \equiv \mathcal{B}$, respectively. Considering the well-known vector rule $\mathfrak{B} \bullet [\mathbf{v} \times \mathfrak{B}] = 0$, applied to $\mathcal{E}$ as well as to $\mathcal{H}$, scalar multiplication of $(9.22)_1$ by $(9.22)_3$ yields the constitutive result

$$\mathcal{E} \bullet \mathcal{H} \equiv \mathcal{E}_* \bullet \mathcal{H}_*, \tag{9.25}$$

whereby point (1) is satisfied. Notice that the electric vectors $\mathcal{E}$ and $\mathcal{D}$ are *polar* vectors, which change sign if the coordinate axes are reflected relative to the origin. Unlike this, the magnetic field quantities $\mathcal{H}$ and $\mathcal{B}$ are *axial* vectors, which are left unchanged by such a transformation. The curl of an axial vector is a polar vector, and vice versa. Because parity is conserved by all electromagnetic phenomena, a polar vector is never equated to an axial vector.

Another inference concerns the mathematical structure of Maxwell's equations: They establish linear relations between field quantities and by means of this linearity allow the decomposition of these very field quantities into such quantities that are only assigned to the field itself and into others directly linked with matter. In case Maxwell's equations provide the common form (5.82) of a balance equation, this notable property exerts an effect that is proven as follows. Multiplying $(9.21)_1$ by the magnetic field intensity $\mathcal{H}_*$ and $(9.21)_2$ again by the electric field strength $\mathcal{E}_*$, respectively, the following relations

$$\mathcal{H}_* \bullet \nabla \times \mathcal{E}_* = -\, \mathcal{H}_* \bullet \partial\, \mathcal{B}_* / \partial t \tag{9.26}$$

and

$$\mathcal{E}_* \bullet \nabla \times \mathcal{H}_* = \mathcal{E}_* \bullet \mathcal{I}_{*f} + C_f\, \mathcal{E}_* \bullet \mathbf{v} + \mathcal{E}_* \bullet \partial \mathcal{D}_* / \partial t \tag{9.27}$$

will arise. Subtracting the last equation from the first one and using the well-known vector identity $\mathcal{H}_* \bullet \nabla \times \mathcal{E}_* - \mathcal{E}_* \bullet \nabla \times \mathcal{H}_* = \nabla \bullet [\mathcal{E}_* \times \mathcal{H}_*]$, the relationship

$$\nabla \bullet [\mathcal{E}_* \times \mathcal{H}_*] + \mathcal{H}_* \bullet \partial \mathcal{B}_*/\partial t + \mathcal{E}_* \bullet \partial \mathcal{D}_*/\partial t = -\, \mathcal{E}_* \bullet \mathcal{I}_{*f} - C_f\, \mathcal{E}_* \bullet \mathbf{v} \tag{9.28}$$

results. This may easily be changed with the aid of the definitions (9.2), where both the polarization $\mathcal{P}$ and the magnetization $\mathcal{M}$ are introduced as electromagnetic properties of the involved body of matter. For a moving body-field system, we obtain the expression

$$\frac{\partial}{\partial t}\left[\frac{1}{2}\varepsilon_0 \mathcal{E}_*^2 + \frac{1}{2}\mu_0 \mathcal{H}_*^2\right] + \nabla \bullet [\mathcal{E}_* \times \mathcal{H}_*] = -\,\mu_0 \mathcal{H}_* \bullet \frac{\partial}{\partial t}\mathcal{M}_* - \mathcal{E}_* \bullet \frac{\partial}{\partial t}\mathcal{P}_* - \mathcal{E}_* \bullet \mathcal{I}_{*f} - C_f \mathcal{E}_* \bullet \mathbf{v}. \tag{9.29}$$

According to Poynting, Equation (9.29) should be read as a local energy balance of an electromagnetic field. Every temporal change of the energy density $\frac{1}{2}\varepsilon_0\, \mathcal{E}_*^2 + \frac{1}{2}\mu_0 \mathcal{H}_*^2$ is due to the assigned energy flux density represented by the so-called *Poynting vector* $[\mathcal{E}_* \times \mathcal{H}_*]$ and caused by the *energy source terms* shown on the right-hand side of (9.29). This equation plays an important part hereafter. Equation

(9.29) proves that Maxwell's equations permit a transformation into a balance equation, as required in view of the last two energy forms of the system representation (9.19). Hence, the latter terms can be expressed by considering Maxwell's equations as follows. The material time derivatives $D\mathcal{D}_*$ and $D\mathcal{B}_*$ are first rewritten with the help of the basic operator (5.68):

$$D\mathcal{B}_* = \partial_t \mathcal{B}_* + \mathbf{v} \bullet \nabla \mathcal{B}_* \qquad \text{and} \qquad D\mathcal{D}_* = \partial_t \mathcal{D}_* + \mathbf{v} \bullet \nabla \mathcal{D}_*. \tag{9.30}$$

Then, scalar multiplication of the material differentials $D\mathcal{D}_*$ and $D\mathcal{B}_*$ by the respective conjugated variables $\mathcal{E}_*$ and $\mathcal{H}_*$ immediately allows us to substitute both the resulting partial time derivatives $(\mathcal{H}_* \bullet \partial\ \mathcal{B}_*/\partial t)$ and $(\mathcal{E}_* \bullet \partial \mathcal{D}_*/\partial t)$ by means of Equations (9.26) and (9.27). If one takes the sum of both energy forms, the expression

$$\begin{aligned} \mathcal{E}_* \bullet D\mathcal{D}_* + \mathcal{H}_* \bullet D\mathcal{B}_* = & -\nabla \bullet [\mathcal{E}_* \times \mathcal{H}_*] - \mathcal{E}_* \bullet \mathcal{I}_{*f} - C_f \mathcal{E}_* \bullet \mathbf{v} \\ & + \mathcal{E}_* \bullet [\mathbf{v} \bullet \nabla\ \mathcal{D}_*] + \mathcal{H}_* \bullet [\mathbf{v} \bullet \nabla\ \mathcal{B}_*] \end{aligned} \tag{9.31}$$

ultimately follows from (9.28). Inserting this equation into (9.19), we obtain the tool for finding the complete set of relations concerning non-equilibrium motion, fluxes, and production densities in accordance with the theorems given in Chapter 6.

The first three terms on the right-hand side of Equation (9.31) may be assigned alternatively to one of those relations. This is not the case, however, for the remaining two terms when being prepared for such an allocation, too. Appropriate transformation rules may be written in a variety of ways. In their well-known textbook, Bird et al. (1960, pp. 730–731) give three adequate formulas that are applied to the expression $\mathcal{E}_* \bullet [\mathbf{v} \bullet \nabla \mathcal{D}_*]$, yielding:

$$\begin{aligned} \mathcal{E}_* \bullet [\mathbf{v} \bullet \nabla\ \mathcal{D}_*] &= \mathcal{E}_* \bullet \nabla \bullet \{\mathbf{v}\ \mathcal{D}_*\} - (\mathcal{E}_* \bullet \mathcal{D}_*)\ \nabla \bullet \mathbf{v} \\ &= \nabla \bullet [\mathbf{v}\mathcal{D}_* \bullet \mathcal{E}_*] - \{\mathbf{v}\ \mathcal{D}_*\} : \nabla\ \mathcal{E}_* - \nabla \bullet [\mathbf{v}(\mathcal{D}_* \bullet \mathcal{E}_*)] + \mathbf{v} \bullet \nabla\ (\mathcal{D}_* \bullet \mathcal{E}_*) \\ &= \mathbf{v} \bullet \nabla(\mathcal{D}_* \bullet \mathcal{E}_*) - \{\mathbf{v}\ \mathcal{D}_*\} : \nabla\ \mathcal{E}_* \end{aligned} \tag{9.32}$$

An analogous relationship holds for the last term of (9.31), $\mathcal{H}_* \bullet [\mathbf{v} \bullet \nabla \mathcal{B}_*]$. Note that this term reveals a dissipative effect due to the electricity in a moving body-field system and is represented by a double point product of two dyads.

9.5 Non-equilibrium Flows in Polarized Fluids

Basic for the understanding of all flows of polarized fluids is the correct formulation of every relevant relationship that determines dynamics. This is obviously equivalent to the knowledge of the electromagnetic pressure tensor from which body and surface forces acting on the fluid can be found. Body and surface couples can also occur due to a certain kind of a rotational motion of the particles relative to the fluid. In this case, a number of other effects need to be considered, including the transfer of angular momentum from particles to fluid. A thoroughly extensive and systematic methodology for treating these problems was published by Shizawa and Tanahashi (1986).

There are considerable difficulties in achieving the desired union of efficient thermodynamic and electromagnetic theories of matter and fields. To dispose of mathematical tools believed to be adequate for moving polarized fluids, basic concepts have often been used that are appropriate only to Maxwell's theory of a medium at rest. As early as 1959, Chu had pointed out another problem: Several authors implicitly assert that the field energy density $\frac{1}{2}(\mathcal{D} \cdot \mathcal{E} + \mathcal{B} \cdot \mathcal{H})$ is nothing but the internal energy per unit volume of the field. This hypothesis is not generally true and needs proof since the expression $\frac{1}{2}(\mathcal{D} \cdot \mathcal{E} + \mathcal{B} \cdot \mathcal{H})$ is well-known in thermodynamics as the Helmholtz energy per unit volume instead of as the internal energy.

Moreover, there is a serious misunderstanding that presumably originated in the early history of thermostatics and electromagnetism. Many authors claimed that the whole set of thermofluid dynamical relations concerning dissipation can only be regarded as an additional postulate, empirically justified at best. Such a statement is sometimes complemented with the advice to bear in mind that it does not have the same universal validity as the well-known postulates in mechanics, electromagnetism, and even in thermostatics. Those thermofluid dynamical relations require the occurrence of irreversible processes in nature, which can be revealed by examining their behavior with respect to time reversal. However, many authors confirm the alleged superiority of the aforementioned scientific branches by concluding that the invariance under time reversal may be easily verified for Maxwell's equations and even for Gibbs' thermostatics. (cf. Green and Leipnik, 1970, p. 229)

All these striking inconsistencies primarily reduce to some serious defects concerning the basic idea of combining a merely formal concept, like Maxwell's equations, with pure thermostatics. To overcome internal friction, along with irreversible work, even misconceptions like the rate of reversible work done per unit mass of the medium in time *dt* are judged to play an essential part of any theory concerning dissipative phenomena (Chu, 1959, p. 479). Rather cumbersome derivations of "laws" are exemplarily given by de Groot and Mazur (1974, p. 205) and are related to irreversible processes in polarized systems. Indeed, it is no wonder that the authors justify their widespread results by reasoning that the theoretical concept of an electromagnetic stress tensor would have to be accepted as an ambiguous property (de Groot and Mazur, 1974, p. 210).

The crucial point is the artificial separation of the matter-field entity in all conventional theories hitherto published. This method is apparently suitable for isolating a mass of fluid from the electromagnetic field that is linked inseparably to the body in reality. Within the conventional theory, the decomposition serves to form separate mathematical expressions for the stress behavior of body and field.

In contrast, such an approach is not needed within the frame of the theory presented here. In other words, the general stress behavior of the matter-field entity has already been treated in Chapters 6 and 7. For this reason, only some specifications of the material stress mechanism under electromagnetic conditions will be described in this section.

The first step toward the desired solution consists of the central relationship (9.19), now supplemented by relation (9.31) and (9.32), and its complementary equation for $\mathcal{H}_* \bullet [\mathbf{v} \bullet \nabla \mathcal{B}_*]$. Substituting these expressions into Equation (9.19) and rearranging yields

$$\begin{aligned}\rho\, De_*^{[\rho]} &= \mathbf{v} \bullet \rho\, D\mathbf{i} + [\mathcal{F} - \rho\mathbf{f}] \bullet \mathbf{v} + T_*\, \rho\, Ds + \sum\nolimits_{j=1}^{J} \mu_{*j} \rho D\omega_j + Dp_* \\ &+ (\mathcal{E}_* \bullet \mathcal{D}_* + \mathcal{H}_* \bullet \mathcal{B}_*)\nabla \bullet \mathbf{v} + \mathbf{v} \bullet \nabla(\mathcal{E}_* \bullet \mathcal{D}_* + \mathcal{H}_* \bullet \mathcal{B}_*) - \nabla \bullet [\mathcal{E}_* \times \mathcal{H}_*] \\ &- \mathbf{v} \bullet \mathcal{E}_* C_f - \mathcal{E}_* \bullet \mathcal{I}_{*f} - \{\mathbf{v}\, \mathcal{D}_*\} : \nabla\mathcal{E}_* - \{\mathbf{v}\, \mathcal{B}_*\} : \nabla\mathcal{H}_*.\end{aligned} \tag{9.33}$$

The first two terms on the second line of (9.33) can be easily combined into a single divergence term $\nabla \bullet [\mathbf{v}\, (\mathcal{D}_* \bullet \mathcal{E}_* + \mathcal{B}_* \bullet \mathcal{H}_*)]$. Equation (9.33) indeed represents the relevant tool for obtaining all information about the generally valid relations between the variables of the body-field system under consideration. Since it seems admissible to incorporate mechanical and electromagnetical effects in the same way into Falk's dynamics, the reliability of (9.33) obviously depends on the accuracy of Maxwell's equations.

Let us now introduce the balance equations, compiled in Table 6.1, into (9.33). Additionally, the material derivative of the non-equilibrium pressure is split into its convective part and partial time derivative to cancel the latter term together with the same quantity arising from the balance Equation (9.20) of the specific energy $e_*^{[\rho]}$.

We must first rearrange all existing terms according to the following categories, each with a physical meaning based on: (1) terms with the gradient $\nabla \bullet$ to the left of them, (2) terms with the velocity $\mathbf{v}$ as prefactor of a scalar product, and (3) any remaining terms. Then, the following expression can be written

$$\begin{aligned}\partial_t p_* &= \mathbf{v} \bullet [\rho\, D\mathbf{i} + \nabla p_* + \nabla \bullet \boldsymbol{\tau}_* - \rho\mathbf{f} - \mathcal{E}_*\, C_f + \mathcal{F}] + \partial_t p_* \\ &+ \nabla \bullet [\mathring{\mathbf{q}}_* - T_* \mathbf{j}_s - \sum\nolimits_{k=1}^{K} \mu_{*k} \mathbf{j}_k + \mathbf{v}\, (\mathcal{E}_* \bullet \mathcal{D}_* + \mathcal{H}_* \bullet \mathcal{B}_*) - [\mathcal{E}_* \times \mathcal{H}_*]] \\ &+ \boldsymbol{\tau}_* : \nabla\mathbf{v} + \mathbf{j}_s \bullet \nabla T_* + \sum\nolimits_{k=1}^{K} \mathbf{j}_k \bullet \nabla\mu_{*k} - \{\mathbf{v}\, \mathcal{D}_*\} : \nabla\mathcal{E}_* - \{\mathbf{v}\, \mathcal{B}_*\} : \nabla\mathcal{H}_* \\ &+ T_*\sigma + \sum\nolimits_{k=1}^{K} \mu_{*k}\, \Gamma_k - \mathcal{E}_* \bullet \mathcal{I}_{*f}.\end{aligned} \tag{9.34}$$

In this equation the Legendre-transformed energy flux density $j_*^{[\rho]}$, as part of the balance equation for $e_*^{[\rho]}$, is replaced by the heat flux vector $\mathring{\mathbf{q}}_*$ and the work rate vector $\mathring{\mathbf{w}}_*$. The latter is determined with the help of the momentum flux density $\mathbf{JI}$ according to (6.45). Substituting the assumed form of the momentum flux density $\mathbf{JI} := p_*\mathbf{1} + \boldsymbol{\tau}_*$ into the energy flux density $\mathbf{j}_{e*}^{[\rho]} = \mathring{\mathbf{q}}_* + \mathbf{v} \bullet \mathbf{JI} - p_*\mathbf{1}$ will then yield the relation $\mathbf{j}_{e*}^{[\rho]} = \mathring{\mathbf{q}}_* + \mathbf{v} \bullet \boldsymbol{\tau}_*$.

To proceed, we recall that the structure of Equation (9.34) confirms the form of relation (6.37), which leads to the divergence theorem (6.44) and the dissipation theorem (6.47). Moreover, the momentum balance of the body

$$\rho\, D\mathbf{i} + \nabla \bullet \mathbf{JI} = \sigma_\mathrm{i} \quad \text{with} \quad \sigma_\mathrm{i} = \rho\mathbf{f} + \mathcal{E}_* C_f - \mathcal{F} \tag{9.35}$$

involves a production density term σ_i—identified by (9.34)—and includes all electromagnetic influences implicitly, along with those of the other variables. Of course, such an implicit dependency refers to the moving body and its dynamics, distinguished by certain values of the local flow velocity **v**, the non-equilibrium pressure p_*, and the viscous pressure tensor $\boldsymbol{\tau}_*$. As a rule, all these quantities are connected with the dissipation velocity $\boldsymbol{\varphi}$, derived as a central property of non-equilibrium in Chapter 6. Since $\boldsymbol{\varphi}$, by definition, is a function of all system variables, with the exception of the linear momentum **i** of the body, the investigated influence of any electromagnetic phenomena becomes apparent by their corresponding influences on p_* and $\boldsymbol{\tau}_*$ via $\boldsymbol{\varphi}(\ldots, \mathcal{D}/\rho, \mathcal{B}/\rho, \ldots)$. At first glance this may be evident from the expression

$$\rho\, D\mathbf{v} - \rho\, D\boldsymbol{\varphi} + \nabla \bullet \{p_*\mathbf{1} + \boldsymbol{\tau}_*\} = \rho\mathbf{f} = \rho\, D\mathbf{v} - \partial_t\rho\boldsymbol{\varphi} + \nabla \bullet \{p_*\mathbf{1} + \boldsymbol{\tau}_* - \rho\,\mathbf{v}\,\boldsymbol{\varphi}\}$$

already given in Section 6.3, where the specific momentum **i** and the momentum flux density **JI** are replaced by the basic relation with respect to the dissipation velocity $\mathbf{v} = \boldsymbol{\varphi} + \mathbf{i}$ and the relationship $\mathbf{JI} = p_*\mathbf{1} + \boldsymbol{\tau}_*$, respectively. It is evident that the non-equilibrium pressure p_* is in principle determined by the internal fundamental relation $h(\mathrm{v};\, \mathbf{F};\, T_*;\, p_*;\, \mathcal{E}_*;\, \mathcal{H}_*;\, \mu_j{}^t{}_*) \equiv 0$ which follows from the Gibbs–Duhem relation (9.13) for the system in question.

Inserting the expression $\mathbf{i} = \mathbf{v} - \boldsymbol{\varphi}$ into Equation (9.35), an analogous relationship will emerge, viz.

$$\rho\, D\mathbf{v} = \partial_t\,\rho\,\boldsymbol{\varphi} + \rho\mathbf{f} + \mathcal{E}_*C_f - \boldsymbol{\mathcal{F}} - \nabla \bullet \Pi, \tag{9.36}$$

along with the definitions

$$\Pi := \mathrm{p}_*\mathbf{1} - \boldsymbol{\tau}_*; \qquad \boldsymbol{\tau}_* := \frac{1}{2}\,\rho\,\mathbf{v}\,\boldsymbol{\varphi}$$

already introduced in (6.57) for the pressure tensor Π and the viscous pressure tensor $\boldsymbol{\tau}_*$ of the moving matter. The latter relationships need further comments, but before we do so, we will note the most important inferences from (9.34) with respect to all occurring flux densities and with regard to the complete set of dissipative production densities. Thus, the divergence theorem can be written as

$$\mathring{\mathbf{q}}_* - T_*\,\mathbf{j}_s - \sum\nolimits_{k=1}^{K} \mu_{*k}\,\mathbf{j}_k + \mathbf{v}\,(\mathcal{E}_* \bullet \mathcal{D}_* + \mathcal{H}_* \bullet \mathcal{B}_*) - [\mathcal{E}_* \times \mathcal{H}_*] \equiv \mathbf{0} \tag{9.37}$$

This identity describes the interdependence of all fluxes that may occur at the surface of any body under consideration, such as the volume element in a fluid flow. For polarized matter, there are two separate contributions: a convective part from both field densities concerning the matter-field unity and an interaction term expressed by the intensities of the fields alone. This last term expresses the influx of field-transmitted energy via the *Poynting vector*. A further result gained from (9.34) refers to the dissipative parts of all energy forms considered for a system. It describes two kinds of effects caused by physically different quantities, such as existing production densities of certain extensive variables as well as the gradients of all intensive variables resulting in:

$$\tau_* : \nabla \mathbf{v} + \mathbf{j}_s \cdot \nabla T_* + \sum_{k=1}^{K} \mathbf{j}_k \cdot \nabla \mu_{*k} - \{\mathbf{v}\mathcal{D}_*\} : \nabla \mathcal{E}_* - \{\mathbf{v}\mathcal{B}_*\} : \nabla \mathcal{H}_* \tag{9.38}$$
$$+ T_* \sigma + \sum_{k=1}^{K} \mu_{*k} \Gamma_k - \mathcal{E}_* \cdot \mathcal{I}_{*f} \equiv 0.$$

Evidently, this zero sum is warranted by the existence of the entropy production density σ, which cannot be dropped by any measures to avoid dissipation. Contrary to this, all other terms may be removed separately by adequate measures, for example, by certain physical constraints to suppress gradients or chemical reactions. In this context, Equation (9.38) reveals that dissipation due to electric and magnetic influences only occurs when local gradients of the field intensities exist along with moving matter or when the premises of an *ohmic resistance* are fulfilled. The latter contribution demonstrates that the invariance of Maxwell's equations for a time reversal does not imply reversibility for all electrodynamic processes. Instead, the reverse is true: At best it is a useful approximation to assume that any non-equilibrium phenomena may be separated into a thermomechanical part including irreversibility and a dissipationless electromagnetic part.

Let us continue the analysis of electromagnetic phenomena within the framework of the Alternative Theory. To repeat the derivation of the field equation (6.83) of specific enthalpy h for polarized matter, recall that such a relation is only defined with respect to the hypothetical state at rest assigned to the system under consideration. Following Subsection 6.4.2, the equation of motion (9.36) must undergo a left-side scalar multiplication with the flow velocity $\mathbf{v}$. Taking into account the specific kinetic energy $e_{\text{kin}} = \frac{1}{2}\,\mathbf{v}^2$ and assuming again a potential approach for the field force density $\mathbf{F}_V = \rho\mathbf{f}$, according to $\rho\mathbf{f} \cdot \mathbf{v} := -\rho\, De_{\text{pot}} + \rho\partial e_{\text{pot}}$, we obtain

$$\rho D(e_{\text{kin}} + e_{\text{pot}}) = \mathbf{v} \cdot [\mathcal{E}_* C_f - \mathcal{F}] - \mathbf{v} \cdot \nabla \cdot \Pi + \mathbf{v} \cdot \partial_t \rho\varphi + \rho\partial_t e_{\text{pot}}. \tag{9.39}$$

To include the special electromagnetic contribution to the total balance of energy some additional preparation is required in view of certain mathematical peculiarities linked with electromagnetic field quantities. Thus, the expression (9.29) describes the change in time of such contributions, which still refer to moving systems. For systems in a state at rest, it is easy to prove that the same Equation (9.29) holds (cf. Abraham, 1920, p. 289), provided that the term $\mathcal{E}_* \cdot \mathbf{v}C_f$ can be applied to gain a mathematical condition for this basic transformation to the level of state at rest. Hence, the relationship

$$\partial_t \left[\frac{1}{2}\varepsilon_0 \mathcal{E}^2 + \frac{1}{2}\mu_0 \mathcal{H}^2\right] = -\nabla \cdot [\mathcal{E} \times \mathcal{H}] - \mu_0 \mathcal{H} \cdot \partial_t \mathcal{M} - \mathcal{E} \cdot \partial_t \mathcal{P} - \mathcal{E} \cdot \mathcal{I}_f \tag{9.40}$$

can be incorporated into the concept of the hypothetical state at rest.

The brackets on the left-hand side of Equation (9.40) encompass the sum of two terms that may be interpreted as the electromagnetic contribution to the total energy of the body-field system. This sum can be added also to the left-hand side of Equation (9.39), under the condition that the latter is converted into the following conservative version

$$\partial_t \rho(e_{\mathrm{kin}} + e_{\mathrm{pot}}) + \nabla \bullet [\rho(e_{\mathrm{kin}} + e_{\mathrm{pot}})\mathbf{v}] = \mathbf{v} \bullet [\mathcal{E}_* C_f - \mathcal{F}] - \mathbf{v} \bullet \nabla \bullet \Pi + \mathbf{v}\, \partial_t \rho \varphi + \rho \partial_t e_{\mathrm{pot}}.$$

After adding the last two equations and extending both sides of the resulting sum by the term $\nabla \bullet [\rho(\frac{1}{2}\rho^{-1}\varepsilon_0 \mathcal{E}^2 + \frac{1}{2}\rho^{-1}\mu_0 \mathcal{H}^2)\mathbf{v}]$, we obtain

$$\begin{aligned}\rho D(e_{\mathrm{kin}} + e_{\mathrm{pot}} + \tfrac{1}{2}\rho^{-1}\varepsilon_0 \mathcal{E}^2 + \tfrac{1}{2}\rho^{-1}\mu_0 \mathcal{H}^2) = {} & -\nabla \bullet [\tfrac{1}{2}\varepsilon_0 \mathcal{E}^2 + \tfrac{1}{2}\mu_0 \mathcal{H}^2)\mathbf{v} + \mathcal{E} \times \mathcal{H}] \\ & - \mathbf{v} \bullet \nabla \bullet \Pi + \mathbf{v} \bullet \partial_t \rho \varphi + \rho \partial_t e_{\mathrm{pot}} - \mu_0 \mathcal{H} \bullet \partial_t \mathcal{M} \\ & - \mathcal{E} \bullet \partial_t \mathcal{P} + \mathbf{v} \bullet [\mathcal{E}_* C_f - \mathcal{F}] - \mathcal{E} \bullet \mathcal{I}_f, \end{aligned} \tag{9.41}$$

where the required application of the term $\mathcal{E}_* \bullet \mathbf{v} C_f$ may be considered with respect to the complete term $\mathbf{v} \bullet \mathcal{F}$ or solely by means of a part of the field force density $\mathcal{F}$.

The crucial point of the analysis is revealed by the full understanding of Equation (9.40). For one thing, it is derived from a representation of Maxwell's equation that is related to the state at rest. But this original form of Maxwell's equations also enables us to apply them to any moving system, provided that the actually occurring *Joule heat* be included in (9.41) by an adequate mathematical description. Chu (1959, p. 477) stressed this decisive difference between the terms $(\mathcal{E} \bullet \mathcal{I}_f)$ and $(\mathcal{E}_* \bullet \mathcal{I}_{*f})$: The latter refers exactly to this real dissipation produced within the moved system, whereas the former does not. Consequently, there is a theoretical need for (9.41) to contain $(\mathcal{E}_* \bullet \mathcal{I}_{*f})$, instead of $(\mathcal{E} \bullet \mathcal{I}_f)$. Fortunately, this required representation can be realized by means of an appropriate definition of the field force density $\mathcal{F}$, the physical meaning of which is still unknown within the framework of the theory presented above. In other words, it is easy to prove the following connections

$$(\mathcal{E}_* \bullet \mathcal{I}_{*f}) = (\mathcal{E} \bullet \mathcal{I}_f) - \mathbf{v} \bullet C_f \mathcal{E} + \mathcal{I}_f \times \mathcal{B}] \quad \text{and} \quad \mathcal{F} := \mathcal{I}_f \times \mathcal{B}, \tag{9.42}$$

which express the physical foundation for the application of (9.40). It is worth mentioning that this change in the levels of description is made possible by a mathematically definite relation of $\mathcal{F}$ to some basic quantities of the electromagnetic field. By the way, for the derivation of (9.42) the identity $\mathbf{v} \bullet \mathcal{E}_* \equiv \mathbf{v} \bullet \mathcal{E}$ is used, which follows immediately from $(9.22)_1$ if the vector property $\mathbf{v} \bullet [\mathbf{v} \times \mathcal{B}] \equiv \mathbf{0}$ is considered.

The next step follows directly from Equation (6.71) together with Equation (6.72), whereby the specific enthalpy h of polarized matter is now introduced by the definition

$$\rho e^{[\rho]} - \rho e_{\#} := \rho h + \rho e_{\mathrm{kin}} + \rho e_{\mathrm{pot}} + \tfrac{1}{2}\varepsilon_0 \mathcal{E}^2 + \tfrac{1}{2}\mu_0 \mathcal{H}^2 + \rho e_{\varphi}. \tag{9.43}$$

According to the method presented in Section 6.1, the state function h is dependent not only on the pertaining thermal variables, but also on polarization and magnetization for the limiting state at rest. The density-transformed energy $e^{[\rho]}$ (without an asterisk) may be formally assigned to an energy balance equation

$$\rho\, De^{[\rho]} + \nabla \bullet \mathbf{j}_e^{[\rho]} = \partial_t \rho, \tag{6.72}$$

where the energy flux density $\mathbf{j}_e^{[\rho]}$ relates to the heat flux vector $\mathring{\mathbf{q}}_*$ and work rate $\mathring{\mathbf{w}}_* := \mathbf{J}\mathbf{I} \bullet \mathbf{v}$ via Equation (6.74).

The last three equations easily turn (9.41) into an expression that is structurally equivalent to equation (6.76),

$$\begin{aligned}\rho\, Dh = &-\nabla \cdot \mathbf{J}_e^{[\rho]} + \mathbf{v} \cdot \nabla p_* + \tau_* : \nabla \mathbf{v} - \nabla \cdot [\mathbf{v} \cdot \tau_* + \rho e_\varphi \mathbf{v}] \\ &- \mathbf{v} \cdot \partial_t \rho \boldsymbol{\varphi} - \partial_t \rho e_\varphi - \rho \partial_t e_{\text{pot}} + \partial_t p \\ &+ \mu_0\, \mathcal{H} \cdot \partial_t \mathcal{M} + \mathcal{E} \cdot \partial_t\, \mathcal{P} + \mathcal{E}_* \cdot \mathcal{I}_{*f},\end{aligned} \tag{9.44}$$

where all dissipative non-equilibrium contributions appear explicitly. The vector $\mathbf{J}_e^{[\rho]}$ comprises the energy flux density $\mathbf{j}_e^{[\rho]}$, along with the corresponding electric and magnetic contributions $[(\frac{1}{2}\,\varepsilon_0 \mathcal{E}^2 + \frac{1}{2}\,\mu_0 \mathcal{H}^2)\mathbf{v} + (\mathcal{E} \times \mathcal{H})]$, according to the definition

$$\mathbf{J}_e^{[\rho]} := \mathbf{j}_e^{[\rho]} - (\tfrac{1}{2}\,\varepsilon_0 \mathcal{E}^2 + \tfrac{1}{2}\,\mu_0 \mathcal{H}^2)\mathbf{v} - (\mathcal{E} \times \mathcal{H}). \tag{9.45}$$

Whereas the convective term is related to the electromagnetic energy density, the Poynting vector $(\mathcal{E} \times \mathcal{H}\,)$ represents the thermal radiation of the body, normally directed to a black surface and referred to time and area units (cf. Born, 1972, p. 12). The numerous terms of (9.44), representing the unsteady changes, attract attention due to the rather problematical mixture of quantities belonging to two different sets of system variables (that is, either indicated by an asterisk or not).

As already demonstrated in Chapter 6, favorable circumstances again allow a remarkable transformation of Equation (9.44) to a special mathematical structure with quantities related to the hypothetical state at rest. Since we have to consider only terms that equal those of Equation (6.76), we can adopt completely the result of the derivations expounded in Section 6.4. As to their mathematical structure, the same relationship reappears, with the additional inclusion of $\mathcal{E} \cdot \mathcal{I}$ and the field force convection $\mathbf{v} \cdot \mathcal{F}$ as well as two terms representing unsteady effects due to magnetization and polarization phenomena.

To summarize these arguments, the following set of relations describes the specific enthalpy of any polarized body-field continuum. It is noteworthy that this level of description takes the place of the reference state at rest corresponding to

$$\begin{aligned}\rho\, Dh = &-\nabla \cdot \mathbf{J}_e^{[\rho]} + Dp + \boldsymbol{\tau} : \nabla \mathbf{v} + \mathcal{E} \cdot \mathcal{I}_f - \mathbf{v} \cdot [\mathcal{E} C_f + \mathcal{I}_f \times \mathcal{B}] \\ &- \mathbf{v} \cdot \partial_t \rho \boldsymbol{\varphi} - \partial_t \rho e_\varphi - \rho \partial_t e_{\text{pot}} \\ &+ \mu_0\, \mathcal{H} \cdot \partial_t \mathcal{M} + \mathcal{E} \cdot \partial_t\, \mathcal{P}\end{aligned} \tag{9.46}$$

where (6.82) is applied to associate both the non-equilibrium values of pressure p_* and the viscous pressure tensor $\boldsymbol{\tau}_*$ with their hypothetical values at rest. The resulting formulas

$$p := p_* - \tfrac{1}{2}\,\rho\,(\boldsymbol{\varphi} \cdot \mathbf{i}); \qquad \boldsymbol{\tau} := \boldsymbol{\tau}_* - \tfrac{1}{2}\,\rho(\boldsymbol{\varphi} \cdot \mathbf{i})\mathbf{1} \tag{9.47}$$

differ from (6.82) in such a way that (1) at least in principle the dissipation velocity $\boldsymbol{\varphi}$ depends on all system variables, except for the linear momentum $\mathbf{P} = m^{B+L}\mathbf{i}$: $\rightarrow \boldsymbol{\varphi} = \boldsymbol{\varphi}(\mathbf{r};\, S,\, V,\, V\mathcal{D},\, V\mathcal{B},\, N_k)$; and (2) the equilibrium pressure p is identical with the pressure assigned to the thermal state at rest and defined by the thermal equation of state: $p = p(T, \rho, \omega_k, \mathcal{M}, \mathcal{P})$. The same conclusion can also be drawn for the caloric equation of state concerning the specific enthalpy $h = h(T, \rho, \omega_k, \mathcal{M}, \mathcal{P})$. Point (1) immediately follows from the concept of dissipation velocity, introduced in Section 6.1, since this idea is not influenced by the type and number of variables. Point (2) also refers to that idea, but the resulting limiting case lim $p_* \to p$ for $\mathbf{i} \to \mathbf{0}$ follows primarily from Equation (6.18) along with the Gibbs–Duhem relation.

Regarding (9.45), the energy flux density $\mathbf{J}_e^{[\rho]}$ is related explicitly to the electromagnetic field. But, in addition, it admits an implicitly functional dependency on polarization and magnetization due to the flux density $\mathbf{j}_e^{[\rho]}$, which is introduced by the relationship

$$\mathbf{j}_e^{[\rho]} = - t_{q*}\mathrm{T}_*\sigma\varphi = - t_q T\varphi[\rho\, \mathbf{v} \cdot \nabla\, s + \rho\partial_t s + \nabla \cdot \mathbf{j}_s] \tag{9.48}$$

already defined by Equation (6.94). Here, in the last part of (9.48), the term $(t_{q*}T_*)$ is replaced by the product $(t_q T)$, indicating that the characteristic time t_q is presumed to be also related to the hypothetical state at rest. Consequently, this transport property t_q now depends on the field variables $\mathcal{E}$, $\mathcal{H}$, too. Note that the explicitly functional dependency of $\mathbf{J}_e^{[\rho]}$ on the electromagnetic variables $\mathcal{E}$ and $\mathcal{H}$ arises because the vector $\mathbf{J}_e^{[\rho]}$ has to be reduced to a finite vectorial difference, $[-(\frac{1}{2}\varepsilon_0\mathcal{E}^2 + \frac{1}{2}\mu_0\mathcal{H}^2)\mathbf{v} - (\mathcal{E} \times \mathcal{H})]$, even for dissipationless flows. This condition is automatically satisfied, since the flux density $\mathbf{j}_e^{[\rho]}$ vanishes for $\boldsymbol{\varphi} \equiv \mathbf{0}$ by virtue of (9.48) and the identity $\mathring{\mathbf{q}}_\star \equiv \mathbf{0}$, which is evident from (6.74):

$$\mathbf{j}_e^{[\rho]} = \mathring{\mathbf{q}}_* + \mathrm{JI} \cdot \mathbf{v} - p_*\mathbf{v} + (p_* - p)\mathbf{v} = \mathring{\mathbf{q}}_* + \boldsymbol{\tau}_* \cdot \mathbf{v} + (p_* - p)\mathbf{v}. \tag{6.74}$$

This result may be compared with the finite limiting value $\Pi = p\mathbf{1}$ for the pressure tensor Π, which is evidently true for reversible flows.

As mentioned in Section 6.5, formula (6.94/9.48) is essential for the theory of constitutive relations, inasmuch as it requires specification of the class of systems to be treated. The last term on the right-hand side contains the entropy flux density connected by the divergence theorem (6.44), but now together with all other relevant current densities of the system, according to (9.37). The latter identity includes the heat flux density $\mathring{\mathbf{q}}_*$ expressed by (6.97) in general and by (6.98) in particular for wall conditions. Both formulas demonstrate again the complex structure of an adequate mathematical theory concerning transport phenomena as the predominant manifestation of non-equilibrium processes, this time accompanied by electromagnetically induced effects.

We will not attempt in this book to treat transport properties that are assumed to be valid only near an adequate equilibrium state. Such problems concern in particular the general subject of some advanced, but conventional, kinetic theories distinguished by the famous *Onsager–Casimir reciprocity relations* (Truesdell, 1984, p. 377) along with the rather dubious *Curie principle*.

The Onsager–Casimir reciprocity relations are relevant for advanced theoretical concepts, such as the extended irreversible thermodynamics (EIT) or modern variational methods for dissipative processes. Their general validity follows from the hypothesis of microscopic reversibility, with the latter being verified by fluctuation theory. Fluctuation theory, however, is based on a hypothesis of mechanical origin alien both to phenomenological thermodynamics and thermodynamic field theory. Some scientists accept Onsager's reciprocity relations in non-equilibrium thermodynamics (regardless of whether they accept his proof or not), others do not accept them and do not use them at all (Gambár and Márkus, 1993, p. 51).

The Curie principle forbids combinations of vectors and scalars, but admits interactions between second-order tensors and scalars. This assertion is part of a generalized statement, for which de Groot and Mazur, for example, found the following formulation: "Thus, in particular for an isotropic system ... it can be shown that fluxes and thermodynamic forces of different tensorial character do not couple" (Truesdell, 1984, p. 388 and Section 7.2).

Within the scope of the Alternative Theory, elaborated to describe non-equilibrium phenomena, there is no reason to consider such a principle in any conceivable way. As will be confirmed below, the viscous pressure tensor $\boldsymbol{\tau}$ of polarized matter with respect to the hypothetical state at rest does not interact with any first-order tensor (i.e., a polar vector) but couples with the divergence of such a vector. By the way, this special kind of interaction supplements the coupling as it appears between $\boldsymbol{\tau}$ and the gradient of the vectorial flow velocity $\mathbf{v}$.

The mathematical representation of $\boldsymbol{\tau}$ was elaborated in full in Section 7.2. It yields equation (7.11),

$$\boldsymbol{\tau} = \beta_\tau \left\langle \{\nabla \mathbf{v}\} - s^{-1}\{\nabla s\mathbf{v}\} \right\rangle + \left(\beta_\tau + \beta_\varphi\right)\left\{\left(s^{-1}(\nabla \bullet s\mathbf{v}) - (\nabla \bullet \mathbf{v})\right)\mathbf{1}\right\} \tag{9.49}$$
$$+ \beta_\varphi \left\{\left((\rho s)^{-1}\nabla \bullet \mathbf{j}_s + \partial_t \ell n\, s\right)\mathbf{1}\right\},$$

which is based primarily on the concept of dissipation velocity $\boldsymbol{\varphi}$. Therefore, there is no explicit dependency of $\boldsymbol{\tau}$ on any electromagnetic field variables. However, an implicitly functional interaction between those variables and $\boldsymbol{\tau}$ does enter twice:

- The Maxwell viscosities β_τ and β_φ, defined by (7.8), now depend not only on the common thermal variables of state, but also on the field intensities $\mathcal{E}$ and $\mathcal{H}$.
- According to the divergence theorem in its version (9.37), a convective current of electromagnetic properties characterizing the polarization and magnetization of matter enters along with the Poynting vector into $\boldsymbol{\tau}$ via the entropy flux density $\mathbf{j}_s$.

Hence, the flow field of polarized matter, represented by the equation of motion (9.36), is in many ways influenced by electromagnetic field quantities.

9.6 Sundry Remarks on an Electromagnetic Dilemma

The conviction that under certain conditions Maxwell's equations provide an undisputed description of any electromagnetic phenomena has required some pretty crude mathematics and, above all, some tremendous prejudices in favor of mechanical ideas. By now we have gained a clear insight into the special nature of those equations, combined with a deeper understanding of their function as a particular, highly simplified, mathematical model for real events in any motions of matter. This puts an end to guessing and deals with unbiased notions and rigorous mathematical proofs. This situation compares with the status Truesdell elucidated for the kinetic theory of nonuniform gases: "The interest of the kinetic theory is purely rational, not physical; while 'intuitive' or 'physical' thought may be helpful in physics, botchwork in a rational theory destroys its rationality and hence its function" (Truesdell, 1984, p. 412). A similar conclusion concerns the fundamentals of electromagnetism. To accept this conclusion, we must be open minded to two statements that can be summarized from the results of this chapter:

1. Electromagnetic phenomena, together with any other thermofluiddynamical processes within a body-field system, can be represented as an integral part of the system.
2. Falk's dynamics, extended to a mathematical representation of continuum physics, leads to results that are not compatible with any theory of polarized matter denoted as non-equilibrium thermodynamics (cf. de Groot and Mazur, 1974) or as thermodynamics of electrically conducting fluids (cf. Chu, 1959).

Point (2) requires some supplementary comments, which can be confined to the problem of how to understand the notions of energy and (linear) momentum if the given body-field system is studied under the most ideal conditions. These conditions are defined in reference to the corresponding Gibbs function of the system in question and are quite equivalent to the conventional case of electromagnetic technology. They are mathematically written as follows

$$Ds \equiv 0 \rightarrow \sigma \equiv 0 \rightarrow \boldsymbol{\varphi} \equiv \mathbf{0};$$
$$D\rho \equiv 0 \rightarrow \nabla \bullet \mathbf{v} \equiv 0; \qquad D\omega_k \equiv 0;\ k = 1(1)K. \tag{9.50}$$

and express processes assumed to be isentropic and incompressible. Furthermore diffusion and any kind of chemical reactions are excluded. On the other side local displacements and frictionless motions of any polarized fluid are admitted.

Considering (9.50), the asterisks with T_*, p_* and the electromagnetic quantities can be dropped. Hence, the respective relationships (9.33) and (9.28) can be simplified considerably to the expression

$$\rho De^{[\rho]} = \rho De_{\mathrm{kin}} + \rho De_{\mathrm{pot}} + \partial_t p + \mathbf{v} \bullet \nabla p + \nabla \mathbf{v}\,(\mathcal{E} \bullet \mathcal{D} + \mathcal{H} \bullet \mathcal{B})$$
$$+ \mu_0 \mathcal{H} \bullet \partial_t \mathcal{M} + \mathcal{E} \bullet \partial_t\, \mathcal{P} + \partial_t E_V^{\mathrm{elmag}},$$

where the first two terms on the right-hand side of Equation (9.33) are replaced by the corresponding kinetic and potential energies. In the last term the abbreviation

$$E_V^{\mathrm{elmag}} := \tfrac{1}{2}\,\varepsilon_0 \mathcal{E}^2 + \tfrac{1}{2}\,\mu_0 \mathcal{H}^2 \tag{9.51}$$

is used for the energy density of the electromagnetic field, which is introduced by the local energy balance (9.40).

By means of the general formula (5.73) for any balance equation of the common type $\rho Dz + \nabla\,\mathbf{j}_z = \sigma_z$, we obtain the expression

$$\partial_t(\rho e^{[\rho]} - \rho e_{\text{kin}} - \rho e_{\text{pot}} - E_V^{\text{elmag}}) + \nabla \bullet [\mathbf{v}(\rho e^{[\rho]} - \rho e_{\text{pot}} - \rho e_{\text{pot}} - E_V^{\text{elmag}})]$$
$$= \mu_0 \mathcal{H} \bullet \partial_t \mathcal{M} + \mathcal{E} \bullet \partial_t \mathcal{P} + \partial_t p + \mathrm{v} \bullet \nabla \left\langle p + \tfrac{1}{2}(\mathcal{E} \bullet \mathcal{D} + \mathcal{H} \bullet \mathcal{B}) \right\rangle$$
$$- \nabla \left\langle E_V^{\text{elmag}} - \mathbf{v}\tfrac{1}{2}(\mathcal{E} \bullet \mathcal{D} + \mathcal{H} \bullet \mathcal{B}) \right\rangle,$$

from which some remarkable conclusions may be drawn with regard to (9.50). At first, for dissipationless motions all instationary effects vanish in agreement with Prigogine's *theorem of minimum entropy production* (cf. Kreuzer, 1983, p. 87) following the *evolution criterion* of Glansdorff and Prigogine (1954). Both terms, representing the local time behavior of magnetization and polarization, can be dropped in all cases only for the identities $\mathcal{M} \equiv \mathbf{0}$ and $\mathcal{P} \equiv \mathbf{0}$. Moreover, the premises (9.50) are consistent with a decomposition of the specific enthalpy h into two shares. The first part, h_1, only depends on the thermal variables, whereas the second, h_2, can be a function of the electromagnetic field. Then, in accordance with the energy relationship (9.43) and along with (9.50), every change of the enthalpy part h_1 in space and time disappears. Consequently, the last equation reduces to the version

$$\rho D h_2 + \nabla \bullet \mathbf{v}\,[E_V^{\text{elmag}} - \tfrac{1}{2}(\mathcal{E} \bullet \mathcal{D} + \mathcal{H} \bullet \mathcal{B})] = \partial_t p + \mathbf{v} \bullet \nabla\,[p + \tfrac{1}{2}(\mathcal{E} \bullet \mathcal{D} + \mathcal{H} \bullet \mathcal{B})],$$

where the divergence term on the left-hand side can be eliminated with $\mathcal{E} = \varepsilon_0 \mathcal{D}$ for $\mathcal{P} \equiv \mathbf{0}$ and $\mathcal{H} = \mu_0 \mathcal{B}$ for $\mathcal{M} \equiv \mathbf{0}$.

Moreover, the specific enthalpy h does not depend on $\mathcal{E}$ and $\mathcal{H}$, that is, $h_2(\mathcal{E}, \mathcal{H}) \equiv 0$, if (1) the steady condition $\partial_t p \equiv 0$ is fulfilled, (2) the pressure p is identified with the exclusive contribution of the electromagnetic field to the total pressure, and (3) the identity

$$p^{\text{elmag}} \equiv -\tfrac{1}{2}(\varepsilon_0 \mathcal{E}^2 + \varepsilon_0 \mathcal{H}^2) \tag{9.52}$$

holds.

The main result of this analysis may be summarized by the statement that the well-known expression (9.52) only appears if many idealized constraints are satisfied. Additionally, polarized matter cannot be dealt with exactly using Maxwell's equations alone. Above all, isentropic processes or reversible changes of state are in principle incompatible with any polarization or magnetization of matter. For this reason, the meaning of Maxwell's equations is questionable, not only in its historical context, but also in its actual significance for science and technology. Bear in mind that the original electromagnetic theory was based on ideas that were decisively influenced by classical mechanics. Boundary conditions like those formulated in (9.50) were completely unknown. Unfortunately, modern textbooks concerned with electromagnetism do not change this tradition in any way. This situation has been stabilized by the ideas developed in Einstein's general theory of relativity and in conventional quantum theory, which both exclude any elementary dissipation a priori.

Of course, approximation is allowed for electromagnetic research and application, provided that its premises are physically well founded. But it seems rather curious that a basic branch in today's physics and engineering rests on incomplete fundamentals. Surely, this fact does not concern those who propagate high-tech. However, their academic education—as a rule—does not prevent such a defect from lasting a long time. Indeed, it is easy to prove this point: There are only very scant sources that seriously study the various problems questioned above. Virtually without exception, the result given in (9.52) is generally applied to body-field systems subject to any electromagnetic forces. It is hard to find profound research where the authors turn their attention to the specialized physical requirements of (9.52). The same seems to be true for the linear momentum and its conservation for the case of moving matter in electromagnetic fields. Therefore, this problem will be briefly discussed here as a concluding contribution to the infinite saga of classical mechanics and its prevalent influence in nearly every branch of science and society.

With regard to Lorentz's theory of electrons and to modern quantum electrodynamics, Maxwell's equations have a curious meaning concerning the level of description (cf. Thomas and Meadows, 1985, p. 22). Whereas the former theories are thought to be appropriate for every level of microscopic views of nature, the latter are supposed to be thoroughly adequate for a macroscopic view of electromagnetic phenomena. However, in sharp contrast with the usual methods applied to macroscopic systems in mechanics, fluid dynamics, and thermodynamics, the common opinion on Maxwell's equations is quite different. Thus, for instance, about hundred years ago even Boltzmann vehemently argued that "Maxwell's equations are time-reversal invariant, implying that a reversal of time, magnetic fields, and polarization at a given time will make the system retrace its previous evolution" (Kreuzer, 1983, p. 326). It is well-known that such an assessment strongly influenced Einstein's theories of relativity. In this context, the tight joint action of energy and momentum characterizes these theories.

Similar to the derivation of the balance (9.43) for the electromagnetic energy density of a body-field system, a relationship for its electromagnetic momentum is obtained that possesses also the formal structure of a balance equation. Its derivation succeeds again by means of Maxwell's equations alone. In other words, the mathematical representation of this new balance is assumed to be completely independent of the concrete physical state of the body-field system, wherein the pertaining electromagnetic events occur.

For this reason, it is not surprising that there are many inconsistencies in view of this curious property of Maxwell's equations. The new balance noted above refers to at least three obviously different quantities, each of them called the electromagnetic momentum density $\mathcal{G}$. In the version first proposed by Abraham (1920, p. 305) in 1909 for nonrelativistic motions, the term $\mathcal{G}$ is directly related to the field according to

$$c^2\mathcal{G} := \mathcal{E} \times \mathcal{H}, \tag{9.53}$$

where the squared speed of light c equals the inverse product $(\varepsilon_0 \mu_0)^{-1}$. Unlike this, Minkowski's suggestion

$$\mathcal{G} := \mathcal{D} \times \mathcal{B} \tag{9.54}$$

is often applied to modern magnetofluid dynamics (cf. Dragoş, 1975, p. 31); it rests on an idea of a "matter-induced" energy–momentum tensor (cf. de Groot and Mazur, 1974, p. 205) as is well-known from Einstein's general theory of relativity. A third version was propagated by Green and Leipnik (1970, p. 296), who preferred the definition $\mathcal{G} := \varepsilon_0 \mathcal{E} \times \mathcal{B}$, especially for applications in plasma physics.

A crucial point is the assertion that there exists a balance equation of the general form

$$\partial_t \mathcal{G} := \nabla \cdot \mathcal{T} - \mathcal{F} \tag{9.55}$$

for $\mathcal{G}$, where the two quantities $\mathcal{T}$ and $\mathcal{F}$ denote the so-called *Maxwellian stress tensor* and the force density exerted on a polarized system. Note that there is a significant difference between the two force densities $\mathcal{F}$ and $\mathcal{F}$.

It is evident that a more detailed formulation of (9.55) will depend on whether definitions (9.53) or (9.54) are applied to derive both quantities $\mathcal{T}$ and $\mathcal{F}$ from Maxwell's equations. Unfortunately, this evaluation cannot be conducted without some unambiguity since only the complete expression on the right-hand side of (9.55), viz. $\nabla \cdot \mathcal{T} - \mathcal{F}$, is rigorously defined (viz. de Groot and Mazur, 1974, p. 209). We will demonstrate this fact for the case of Abraham's definition (9.53). Both corresponding quantities $\mathcal{T}$ and $\mathcal{F}$,

$$\mathcal{T} = \mathcal{D}\mathcal{E} + \mathcal{B}\mathcal{H} - \{\tfrac{1}{2}\varepsilon_0 \mathcal{E}^2 + \tfrac{1}{2}\mu_0^{-1}\mathcal{B}^2 - \mathcal{M} \cdot \mathcal{B}\}\mathbf{1} \tag{9.56}$$

and

$$\mathcal{F} = C_f \mathcal{E} - \mathcal{B} \times \mathcal{I}_f + \{\nabla \mathcal{E}\} \cdot \mathcal{P} + \{\nabla \mathcal{B}\} \cdot \mathcal{M} + \partial_t[\mathcal{P} \times \mathcal{B}] + c^{-2}\partial_t\{\mathcal{E} \times \mathcal{M}\} \tag{9.57}$$

can be derived easily with the well-known tensor rules (cf. de Groot and Mazur, 1974, p. 206). The equations refer to the basic definitions (9.2), introduced for the polarization $\wp$ and the magnetization $\mathcal{M}$. It should be mentioned that for $\mathcal{M} \equiv \mathbf{0}$, Lord Kelvin's suggestion arises, concerning so-called *ponderomotive forces* (cf. de Groot and Mazur, 1974, p. 218 and Hertz, 1892, p. 234).

There is indeed a variety of momentum definitions following from Maxwell's equations and leading to some other relationships for $\mathcal{T}$ and $\mathcal{F}$, even for the same definition of the electromagnetic momentum density $\mathcal{G}$. Still, the pair of equations (9.56) and (9.57) provide a convenient way to recognize that we cannot find an unambiguous solution of the problem raised by the balance equation (9.55) for $\mathcal{G}$. If we form the divergence of the last term of (9.56), the following expression will result:

$$\nabla \cdot \mathcal{T} = \nabla \cdot \{\cdots + (\mathcal{M} \cdot \mathcal{B})\mathbf{1}\} = \nabla \cdot \mathcal{T}^{\text{red}} + \{\nabla \mathcal{M}\} \cdot \mathcal{B} + \{\nabla \mathcal{B}\} \cdot \mathcal{M}.$$

De Groot and Mazur (1974, p. 210) emphasized that adding both resulting terms to the field force $\mathcal{F}$ would alter (9.57) in that the magnetic contribution $\{\nabla \mathcal{B}\} \cdot \mathcal{M}$ of the ponderomotive forces would be simply exchanged by the term $\{\nabla\mathcal{M}\} \bullet \mathcal{B}$. The Galilei-invariance of (9.57) would not be destroyed by such an operation. Yet, due to the relativistic invariance of Maxwell's equations, this does not apply either to the tensor $\mathcal{T}$ or to the reduced tensor $\mathcal{T}^{\text{red}}$. It is common practice to reformulate the obvious arbitrariness of such a questionable idea, as condensed in the formula

$$\partial_t[\rho\mathbf{v} + c^{-2}[\mathcal{E} \times \mathcal{H}]] := -\nabla \bullet \{\rho\mathbf{v}\mathbf{v} + \Pi^m - \mathcal{T}\}, \tag{9.58}$$

where $\mathbf{\Pi}^m$ denotes a mechanical pressure tensor. Apparently, $\mathbf{\Pi}^m$ is immediately conditioned by the choice of the mathematical form of $\mathcal{T}$. However, without further explanation concerning the central point of momentum conservation, it is hard to evolve Equation (9.58) into an order of general principles. Such an effort was made in Section 9.3, where the force balance (9.9) was introduced. Notice that the conservation law of type (9.9) does not need knowledge of the explicit relationship between the electromagnetic momentum $\mathbf{P}^{\text{elmag}}$ and the various leading terms of electromagnetism, like the couples $\mathcal{D}$, $\mathcal{E}$ and $\mathcal{B}$, $\mathcal{H}$, respectively. Hence, for the moment, the force density $\mathcal{F}$ assigned to $\mathbf{P}^{\text{elmag}}$ is an indeterminate function to be definitely established by the theory.

But this is an essential point of the common idea of momentum in the prevailing dogmatism of electromagnetism: The vagueness of the terms "(Maxwellian) stress tensor $\mathcal{T}$" and "electromagnetic force $\mathscr{F}$" discussed above follows from the inherent overdetermination of the classical theory. Nevertheless, the preference of a certain choice of $\mathcal{T}$ or, alternatively, of $\mathscr{F}$ happens either by habit and tradition or by empirical need. The so-called *Lorentz force* (cf. Zahn, 1979, p. 314)

$$\mathscr{F}^{\mathcal{L}\text{orentz}} := \mathcal{C}_f\,\mathcal{E} - \mathcal{B} \times \mathcal{I}_f \tag{9.59}$$

is a well-known example. It is exactly determined by Equation (9.42).

Contrarily, it is common practice to introduce this quantity $\mathscr{F}^{\mathcal{L}\text{orentz}}$ heuristically by means of a definition chosen in addition to Maxwell's equation. In fact, (9.59) represents the macroscopic form of the force produced by any electromagnetic field on an electric charge moving with velocity $\mathbf{v}$ in the magnetic field. Equation (9.59), stimulated by Ørsted's discovery in 1820, is based on Coulomb's law for the electric field and Biot's and Savart's law for the magnetic field. Both are experimentally established laws, which often serve as a start for an inductive development of the electromagnetic field theory. It is evident that the definition (9.59) results from the "normal form" of the Maxwellian stress tensor

$$\mathcal{T} := \mathcal{D}\mathcal{E} + \mathcal{B}\mathcal{H} - \tfrac{1}{2}\{\varepsilon_0\mathcal{E}^2 + \mu_0^{-1}\mathcal{H}^2\}\mathbf{1} \tag{9.60}$$

in agreement with (9.56) for vanishing magnetization $\mathcal{M}$.

Although definition (9.60) will lead directly to the desired result—that is, exclusively to the Lorentz force—there is no correspondence to Minkowski's representation (9.54) of the electromagnetic momentum density $\mathcal{G}$, for which an additional contribution to the force density $\mathscr{F}^{\mathcal{L}\text{orentz}}$ arises (cf. Dragoş, 1975, p. 30; de Groot and Mazur, 1974, p. 206). Some authors identify this contribution with the ponderomotive forces $\mathcal{F}^{\text{pondero}}$ acting on the dielectrics and due to any polarization phenomena. In vacuum or for simple media these forces vanish by definition. However, in moving media their interpretation is more difficult. Despite this fact, such an extended force term $\mathcal{F}^{\mathcal{L}\text{orentz}} + \mathcal{F}^{\text{pondero}}$ can be proved within the scope of Einstein's theory of relativity, which, following Minkowski's reasoning, contains the momentum density term $\mathcal{G} := \mathcal{D} \times \mathcal{B}$ as an element of the electromagnetic energy–momentum quadritensor.

Turning back to the conservation rule (9.58), we will quickly see that this rule can be made formally consistent with the balance equation (9.55). Its mechanical twin is introduced, viz.

$$\partial_t \rho \mathbf{v} := -\nabla \bullet \{\rho \mathbf{v}\mathbf{v} + \mathbf{\Pi}^m\} - [\mathcal{F}^{\mathcal{L}\text{orentz}} + \mathcal{F}^{\text{pondero}}], \tag{9.61}$$

where $\mathcal{F}^{\mathcal{L}\text{orentz}} + \mathcal{F}^{\text{pondero}}$ is the resultant of all electromagnetic forces involved; all other forces, such as gravitational forces, are excluded.

The current state of the electromagnetic theory is obviously characterized by the dilemma that on one hand the definition of $\mathcal{T}$ provides a definite expression for the force $\mathcal{F}$, which is determined independently of the fluid state. On the other hand, this electromagnetic quantity $\mathcal{F}$ may be regarded as part of the definition prescribed for the mechanical pressure tensor $\mathbf{\Pi}^m$ via (9.61). The intricate situation may be outlined as follows. In this differential equation the total electromagnetic force will be expressed only in terms of the electromagnetic field, which are for their part subject to Maxwell's equations. In addition, certain properties of matter, formalized by structural equations of type (9.1) or (9.2), are also involved. These structural equations, however, depend on the state of motion, so that we must consider the complete set of balance equations (continuity, energy, momentum, material functions) together with the electromagnetic laws.

In other words, due to the presence of any flow motion, expressed by $\mathbf{v}$ in the field equations, the electromagnetic force in the momentum equation cannot be determined independently of $\mathbf{v}$. This fact is commonly assumed to be responsible for the interaction between field and motion. For a fluid at rest, this interaction disappears by definition. Strictly speaking, such a theoretical approach presumes that the body-field system can be separated into a "mechanical" and an "electromagnetic" part. This separability condition is frequently satisfied; it fails, however, with regard to a rigorous standard of the theory. Of course, this special kind of separability has its origin in the inability of the traditional theory to explicate a well-founded concept of the immediate dependency of material properties on electromagnetic variables. A unique representation was given in the previous section of this chapter.

The corresponding results, mainly expressed by the various theorems as proved in Section 9.5, lead to a first version of the equation of motion (9.36) along with (9.49). But perhaps the most interesting conclusion can be deduced from Equation (9.42) to complete (9.36). The resulting relationship,

$$\rho D\mathbf{v} = \partial_t \rho \varphi + \rho \mathbf{f} + \mathcal{F}^{\mathcal{L}\text{orentz}} - \nabla \bullet \mathbf{\Pi}, \tag{9.62}$$

clearly substantiates the following four statements.

1. The Lorentz force appears in a theoretically stringent way as a constituent of some transformation rules concerning the hypothetical state at rest.
2. It is true that there is no theoretical need to introduce partial balance equations like (9.55), (9.58), and (9.61). In practice, however, the dogmatism of mechanics prevails in all perspectives.
3. The equation of motion (9.62) represents a descriptive level, where the interactions between field quantities and matter variables are not reduced to two separate subsystems of the whole body-field system.
4. There is no place for a term in (9.62) that can be identified as the ponderomotive force density. Such a property is a stringent consequence of the dissipation theorem in its (9.38) version.

Of course, the results of the analysis presented in this chapter permit some far-reaching inferences for current theories such as conventional electrodynamics or even general relativity.[22] According to the principles advocated here, those theories, affected by these principles may be seen at best as empirically well-proven approximations. However, there is no need to accept such a mechanically dominated view of physical phenomena—at least not with respect to theoretical reasonings.

Concluding the treatment of the subject as a whole, the main result of the analysis presented in this chapter is that a rational theory of electromagnetism may consequently be formulated and carried to completion within the scope of quite a unified theoretical view of macroscopic physics. Its leading idea is not based only on certain philosophical leitmotifs as, for instance, the intention to largely avoid the metaphysical elements of the theory. It is conceived from the very beginning as a thoroughly mathematical theory. The concrete development of this theory is accomplished by rigorous application of the Gibbs–Falkian algebraic formalism with the concept of non-equilibrium established by the Alternative Theory.

It is notable that only a few physical principles are sufficient to present a closed mathematical framework that works well for all applications in mechanics, fluid mechanics, thermodynamics, and electromagnetism. Naturally, this well-founded conceptual claim is confined to the basic terms and formulations that concern the mathematical theories of the respective branches in physics and their independence

[22]Vigier suggested that the measured red shifts of distant galaxies are not caused by the expansion of the universe at all. Instead, they are caused by something quite different—something called a tired-light mechanism. In other words, as light moves through space it becomes red-shifted simply from traveling a certain distance, just like any other physical phenomenon subject to irreversibility. Vigier's assertion that the universe is not expanding is based on the striking conclusion: The red shift is not a Doppler effect, but rather affected by dissipation. In this context, it should be remembered that *all* wavelengths are red-shifted. The prevailing doctrine accounts for this displacement toward the red end of the spectrum as "a consequence of the expansion of the universe." (Harrison, 1986, p.184). However, it should be stressed that according to this doctrine red-shift measures do not tell astronomers how fast the observed galaxies recede, how far away they are, or how long ago they emitted the light now received; all this information must be deduced within the framework of a theoretical model. But, undoubtedly, this model rests on the mechanically dominated concept of electromagnetism criticized above.

from quantum effects. Especially for this occasion this unified theory does not need any approximation in reference to the presupposed principles.

Realize also that every macroscopic theory is limited by peculiar conditions and properties, according to which the system in question may be labeled individually. Initial and boundary conditions for the processes to be described are not only typical, but also furnish the relevant material properties (such as the equations of state or relationships for all pertaining relaxation times). As a rule, such information must be provided with a restricted number of reliable experiments or some adequate auxiliary theories. But the results of these experiments and theories must be consistent with the framework of the unified theory. Therefore, the credo of every true natural science sets high standards for this new concept, too. Conversely, it enables us to disclose grave deficiencies of current theories believed to be valid for a wide range of applications within the scope of the respective scientific discipline. Current Navier–Stokes theories and electromagnetic phenomena of moving matter are prominent examples thereof.

A final remark: There is a prodigious possibility of applying the major part of the basic principles, here assumed to be true for physical phenomena, to some other fields of knowledge. Initial examinations that concern basic *economics* appear to be successful. More work of this nature is underway.

Appendix 1 | Atomism or The Art of Honest Dissimulation[23]

"We need to realize that the cardinals who suppressed Galileo did not need to be competent theologians or scientists; they only needed a much narrower competence, the ability to distinguish between the orthodox and the heretical, in both content and in style."—***I. Catt***

A1.1 Motivation

"Historical investigations of a science's evolution are most necessary unless the accumulated theorems shell degenerate little by little to a collection of half-baked recipes or even to a system of prejudices" (Mach, 1883, p. 237; author's translation). Ernst Mach's notable statement refers to his sketchy outline of the history of classical mechanics. However, it is also quite an appropriate comment concerning the historical development of the original atomism from its beginning to today's particle physics. The particular status of this idea of nature has been, and still is, a great influence on worldwide living conditions of mankind. This is especially true with regard to the notion of matter and its applications in modern science and technology.

We will gain a broader comprehension of the fundamentals of mathematical physics from the changes in the key notions of atomism over time. This is not trivial or superfluous, if we are aware of the fact that the typical singularities in modern particle theories result directly from a metaphysically induced, and hence biased, picture of a particle assumed to be a shapeless mathematical mass-point of zero extension.

That such a unique and unrealistic approach has been so successful may be ascribed to two adequate circumstances: the selection of certain favorable research conditions that admitted some decisive simplifications in the mathematics used and the conscious repression of sophisticated results of the atomistic doctrine that has gone on for centuries. As Heisenberg has commented: "No scientist asks for the essence of the atoms, or why they are assumed to be thus and in no other way" (Heisenberg, 1973, p. 22; author's translation).

This special type of "physical" agnosticism, which seems to suspend rational judgment about some basic notions of physics, indicates a rigorous dogmatism that is in my opinion counterproductive in the long run. Obviously, it is very unwise for the scientific community to disregard the larger part of the pertaining results recently published by historians and experts in Greek natural philosophy. This research reveals that there are some hidden coercive means to abandon Democritus's ideas in

[23] T. Accetto (1641). *Della dissimulazione onesta*. Ed.: B. Croce, Bari 1928.

favor of Euler's rigorous design of mathematical mass-points. Both the historical background and the dogmatic force of such social pressures are similar to the allegedly well-known controversy between the adepts of Copernican astronomy and the earth-centered version. It should be stressed, however, that in those days the threats against the atomists' life and liberty were much more powerful than any of the restrictions imposed on the Copernican view of the world.

Some fundamentals of modern mathematical physics are strongly influenced by this "atom debate," of more than a hundred years (between Bruno's auto-da-fé and Newton's death). I am convinced that the resulting dogmatic constraints are part of a set of special limitations mandatory for any scientific branch. These limitations are assumed to be imposed from outside of the particular branch and its inherent set of axioms. But in reality, they are embedded in the whole of civilization, which is related to its political and cultural history. For any concrete case, it seems hard to identify even the essentials of such external limitations.

For this reason, Gödel's two famous *theorems of incompleteness*, valid originally for problems in number theory only, are of great importance to all sciences. Strictly speaking, at least for one domain of human knowledge these theorems exactly demonstrate that it is principally impossible to axiomatize and formalize this domain completely (Kanitscheider, 1993, p. 183). Such a result surely offers a certain motivation to take Gödel's theorems for a reasonable approach to any system. Thus, they may serve to remove in principle the apparent arbitrariness of the external limitations mentioned above and, vice versa, claim to account for their existence. In this sense, the history of atomism belongs to the science of atomism.

In the following section, we will outline some important items concerning the ideas of atomism in their historical context. Some far-reaching consequences may perhaps be that these ideas, which take into account a key notion of science, were extremely subject to ideological reprisals or mathematical aesthetics. The contemporary concepts of matter applied in daily practice are actually coined just by these long controversies.

By the way, the central conclusions of the presented essay affirm Schrödinger's question: "Is natural science dependent on its social environment?" (author's translation). He addressed this question in his famous lecture to the members of the Prussian Academy of Sciences in 1932. Now, the atomism ideology arises as a further example and correlates with the dogma of the microscopical reversibility.

A1.2 Pre-Socratic Atomism and its Tradition in the Ancient World

The crucial point of the pre-Socratic philosophy of nature lies in the differing ideas of reality held by the Eleatics and the atomists. Parmenides and his followers warned against confusing the images of phenomena with reality. The atomists' message was instead the reduction of reality, as represented by sensational phenomena, to general principles. In this context, we should mention the categories of classification and the categories themselves. In other words, all things that can be conceived and named are included under some particular classifications or genera: *quantity* (e.g., 175 lbs), *quality* (e.g., good, red), *relation* (e.g., dependent, interacting), *modality* (e.g., existent, contingent), and so on.

Certainly, our actual understanding of things is formed by categories. And the things themselves are assumed to be formed by forces acting in nature that operate according to these categories. Regarding this important epistemological notion there were some famous considerations by Aristotle and, later on, systematic approaches by Kant. For the earliest periods of Greek natural philosophy, only rudiments of respective ideas are vaguely identifiable.

Eleaticism is characterized by an extreme monism: Reality is one, eternal, motionless, unchanging, undifferentiable, and, above all, unrecognizable (Capelle, 1968, p. 158). Unlike Parmenides's or Zeno's position, the main representatives of the early Greek atomism, Leucippus, Democritus, Epicurus, and (the Roman) Lucretius, all claimed that *primary qualities* exist in reality independent of an observer, but every person may indeed observe and grasp the reality of the universe. Their reasoning is based on a materialistic, yet purely speculative, worldview distinguished by the greatest influence on natural resources and human society.

Beyond doubt, Anaxagoras, perhaps the first victim of the state authority against free research in Hesperia, held a special position. Sympathizing equally with the Eleatics and Empedocles, he propagated also an idea of an atomlike entity presupposed to be divisible ad infinitum. This basic property was deduced from the belief that "all contains all" however little it may be. Nevertheless, Anaxagoras's brilliant feat was the clear separation of mind and matter for the first time in human history.

The atomists maintained that reality manifests itself by matter composed of atoms. These basic entities were envisioned as minute material particles that were the ultimate constituents of all things. Some primary qualities were assigned to them, such as size (extension), shape (figure), arrangement, and motion. It was taken for granted that an infinity of atoms existed for all eternity; they were held to be separate, impenetrable, weightless, and irreducible. Atoms in themselves, they declared, did not possess qualities such as color, taste, or smell. Such *secondary qualities*, considered to be illusive, were imagined to result from the activity of the atoms upon the senses.

According to Leucippus's hypothesis (Capelle, 1968, p. 297), motion of the single elementary particle and the atomistic collectives moving within an otherwise empty space was considered existent from the very origin of the universe. This means that self-movement was the essential property of every atom. In other words, motion was assumed to be eternal. Consequently, all things in the universe could be accounted for in terms of the distribution and redistribution of these self-moving atoms. Just as atoms were irreducible and unexplainable in terms of anything more fundamental, so motion was claimed to be irreducible and needed no further source of explanation.

Democritus's explication of this elemental impulse was not purely materialistic but rather ethical, with emphasis on human self-responsibility (Capelle, 1968, p.

441). He preferred the strange word *anangké* for the first cause of the permanent motion of all matter understood as the constantly present, irrational, unpurposeful, indirect, and uncontrolled element in the universe. This *primum agens* (that which is acting) of nature induces the mutual affinity between the atoms, which notably does not refer to the size of each atom but to its geometric form. The ancient texts leave no doubt: "The form is in fact of primary importance" (Löbl, 1987, p. 124; author's translation). This is true, particularly, with respect to the formation of material objects, induced by the *concursio fortuita*, (concurrent accident) and due to characteristic interactions between the extended atoms.

The second notion of central relevance in Democritus's reasoning is that of empty space. In the literature, empty space is commonly related to the traditional term of space used in Newtonian physics, for instance. This is, however, questionable because in pre-Socratic texts the discussion of the antagonism of "being versus not-being" prevails. "Being" equals the self-moving, eternal, material atoms; "not-being" (i.e., a nothing) means void.

The void owns no qualities whatever: no powers, no potentiality, no existence in any way. But it was thought to be the place that atoms occupied before they shifted their position to another place. It was also in reference to a void that things could be distinguished, separated, and classified.

Although Epicurus's natural philosophy completely agreed with most of Leucippus's and Democritus's speculations, there was one additional element in his typical sensualism: All sensations were presupposed to be true and could be attributed to atomic motions. From then on, weight belonged to the primary parameters of an atom, together with its size and shape.

For this reason, a divergent concept of motion might be deduced in connection with Epicurus's special term *declinatio*. This term was defined as the downward slope of a trajectory, for which spontaneous and acausal changes were presumed to happen independently of its initial conditions. Thereby, perhaps for the first time, an accidental element arose in natural philosophy, related to the typical rush of an atom. Thus, interactions with other trajectories mainly affected the formation of secondary qualities as real manifestations of matter.

To give a short summary of Democritean worldview, three general axioms have been compiled:

1. All real events happen by necessity and are assumed to be an inherent part of matter, which consists of categorically endowed atoms.
2. Atoms themselves are the basic entities of all, in accordance with Parmenides' famous statement: "That which is, is; and that which is not, is not and can never be." Thus, nonbeing, understood as pure void, is the lowest manifestation in the hierarchy of reality. Conclusively, an atom is indivisible in principle, as indivisibility of matter excludes nonbeing by definition.
3. There exist neither any divine authority nor any teleological causality of the world.

According to Pythagoras's reasoning, early Greek atomism was, above all, quantitatively oriented (Capelle, 1968, pp. 69, 98, 392). With respect to its strictly anti-metaphysical worldview, chaos and what nowadays is usually called *chance* were felt to be much more dominant than any force of religious or even divine origin. Yet, this worldview was vehemently attacked by Plato's highly sophisticated speculations and Aristotle's fundamental critique in the very prime of antique philosophy.

Roman contributions to ancient atomism, eventually handed down to posterity in fragments only, can almost be regarded as mere refinements of the old Democritean ideas. This is true, particularly, for Lucretius's famous didactic poem, but also holds for Cicero's momentous dispute "De natura deorum" about Epicurus's atomistic physics. As a whole, Cicero's critique is rather skeptical if not negative. Nevertheless, his work was believed to be an important bridge between the original atomism and the relevant Roman period of the Early Fathers. He reduced the complete pre-Socratic atomism to bodies, empty spaces and states of motion of the atoms: "... omnium, quae sint, naturam esse corpora et inane quaeque is accidant."[24]

Nevertheless, during the period of Imperial Rome there were two scholars whose publications significantly influenced medieval atomism. The first to be mentioned is Claudius Galenus, a famous physician from Pergamum. He suggested a new definition of an element, distinguished peculiarly by some specific properties and characterized generally by the loss of its individuality if dissolved in mixtures. The second scholar was Servius Honoratus in the fourth century, one of the more influential commentators of all Vergilian works. He obviously derived from one of his poems[25] the four elements water, air, fire, and earth, which some of the great Greek philosophers had believed to be the ultimate and irreducible, primary and simple constituents of the universe. But, in sharp contrast with the whole of ancient natural philosophy, Servius regarded his elements only as *syntheta* composed of *atomi* and *inani*, that is, the atoms and the empty spaces of Democritean doctrine. By the way, in the context of his commentaries on Vergilius's work, Servius declared *ex cathedra* that the word *atomus* be used as feminine. Furthermore, it was in Servius's work that the term *atomus* was extended for the first time to the term *littera*, which itself was assumed to be inseparable.

It is interesting that such extensions are reported by some Roman authors concerning not only matter, but also other phenomena in nature and human life. A striking example can be found in Augustinus's Sermo 362 concerning the apostle Paul's First Epistle to the Corinthians. Augustinus mentioned two kinds of atoms: one referring to matter, the other one to time. The latter term occurs in Tertullian's work for the first time in Latin, but is thought to have come from an early Greek source

[24]In "reality there are only bodies, empty space, and the motive states of atoms." De natura deorum II, § 82

[25]eclogea VI,31-34, see also Pabst, 1994, p. 29.

(Pabst, 1994; p. 40). Augustinus identified this "atom in time" with an instant that could by no means be subdivided any further.

In the late Roman period, the principal attitude toward the core of pre-Socratic atomism was clearly determined by the Early Fathers' response to the strictly antimetaphysical position of the early Greek atomists. Following Pabst (1994, p. 37), we quote the prominent Milan bishop Ambrosius. Based on the famous words from the Bible in the book of Genesis, "In the beginning, God created the heaven and the earth," he declared that the world is the result of God's act of creation alone and, therefore, is in no way caused by accidentally interacting atoms.

In spite of this verdict, a good many Early Fathers examined classical atomism and thus even preserved its tradition because their works were the single source of information for a long time. Due to the decline of the Roman Empire many original papers of ancient authors were lost or unavailable. Except for (the Greek Early Father) Eusebius's studies on Democritus's work, only the two Early Fathers Lactantius and Augustinus were prominent in this context.

Lactantius made an effort to rationally discuss all relevant arguments of the early atomists. In the final analysis all his reasoning amounts to the alternative between the "atomized" chance and the divine necessity. Save for some studies on the substantial part of the pre-Socratic school, Augustinus's contribution is mainly written as a polemic against the epistemology of the atomists. His primary result leads to the conclusion that the basic idea of atoms is in itself contradictory.

A1.3 Atomism in the Dark Middle Ages

After the disaster of the Roman Empire, the representatives of natural philosophy in the European Middle Ages were subject to some considerable restrictions in continuation of the Early Fathers' work. The culmination of the scholastics' atomism was attained about the year 1200. Still, during this period, Plato's and Aristotle's critical comments were virtually unknown in Europe, or were not available in the worst case. It is true that Plato's Timaios was accessible, but only without the relevant section dealing with his geometrized atomism (Pabst, 1994, p. 50).

As for Aristotle, it is well-known that no one in Europe knew of his voluminous work for centuries. Thanks to the efforts of Arabian philosophers, this situation changed drastically at the beginning of the twelfth century as particularly the English philosophers became the leaders of the Aristotelean adoption (Pabst, 1994, p. 237). But before the Aristotelean critique began to undermine and finally destroy the reasonings of the medieval philosophers on atomism, many remarkable studies on this subject had been completed. This research covered a period of about three hundred years, beginning at the end of the eleventh century and reaching far into the fourteenth.

Starting with the leading manual on natural philosophy, some precursors that may be aligned temporally with the early Middle Ages are Isidorus's "Origines,"

Hrabanus Maurus's "De computo," and the Bernese medieval manuscript "Tractatus de atomo," published only very recently by Pabst (1994, p. 327). Regarding these works, two points need to be underlined:

1. All these manuscripts offered a new aspect to the atomic concepts: There exist at least five sorts of so-called canonical atoms. These genera are denoted: in littera, in temporibus, in numeris, in corpore, in minutis radiis solis.
2. During the period from the seventh to the eleventh century, the notion of an atom received a considerable degree of publicity among the scholars' community. Thus, the widespread assertion that the original idea of atomized matter should have lost its relevance is entirely unfounded.

An atom *in littera* is the indivisible letter noted above, an atom *in temporibus* concerns the smallest unit of time, an atom *in numeris* equals the cardinal number **1**, an atom *in corpore* refers to that part of any body that cannot be further subdivided, and an atom *in minutis radiis sole* belongs to a particle collective of radiation (cf. Pabst, 1994, p. 76).

Important contributions to atomistic theories in the high Middle Ages stem from a minority of authors and scholars hardly known by name to the public or even to experts. The following is a survey of those names and titles. Subsequent to these lexicographic data, a summary of the most valuable results will be related to historical and topical events.

The largest propagation of the atomic doctrine was achieved within a period of about fifty years beginning at the end of the eleventh century. It gained a degree of acceptance never found in ancient natural philosophy nor attained again before the seventeenth century (Pabst, 1994, p. 318). This epoch was marked by distinguished thinkers like Odo of Cambrai, perhaps the first medieval atomist ever. Then came Adelard of Bath, presumably the first scholar who acknowledged contemporary Arabian science, and, finally, William of Conches, the author of the "Dragmaticon," the best encyclopedia of natural philosophy available in those years. William appeared as the most important representative of the famous "school of Chartres."

Hugo of St. Victor was certainly a fascinating personality, too. He ranked among the few really influential theologians of the twelfth century; but, strikingly, his atomistic ideas are the most mechanistically adjusted ones published in the Middle Ages. In the case of *Petrus Abaelardus*, one of the very prominent founders of scholasticism, it is hard to grasp how that valuable part of his work concerning mainly mathematical aspects of atomistic traditions could nearly pass unnoticed today.

Thierry of Chartres emerges as the first "modern atomist" to realize that the classical four elements are nothing else but the characteristic state of aggregation resulting from different mechanisms of the atomized matter. His work set the trend for the twelfth-century thinkers' ideas of essential atomic properties. His definition of the term *atom* is quite useful: "Est autem atomus minima pars quantitatis corporis, nec tamen ipsum alicuius quantitatis quia, ob nimiam brevitatem, nullam potest, habere divisionem" (Pabst, 1994, p. 174). In other words, due to its diminutiveness, an atom is indivisible and consequently nondimensional.

Around the turn of the twelfth century, the prominent Salernitan physician and Platonist Urso published some studies on atomism. His work focused on the basic problem of in which way as well as in what ratio the four classical elements amalgamate themselves to a body, provided that an element is typified by special corpuscles as vehicles of its individual properties. This problem was associated with the question of how the properties of the composition could be deduced from the properties of the elements comprised. Compared with Aristotle's arguments, he reduced this problem to a refined theory of mixtures presupposing twelve kinds of corpuscles and their mutual interactions.

Due to the successful progress of Aristotle's philosophy, few relevant contributions to contemporary atomism appeared during the decades after the twelfth century. There was an uncommon exception: Robert Grosseteste initiated the formerly famous Oxford School, emphasized the priority of mathematics with respect to the foundation of natural philosophy, and developed an independent and complex atomistic theory that even included Plato's and Aristotle's suggestions and critiques.

From about the first half of the thirteenth century, Aristotelianism suppressed all approaches to advance the variety of ingenious theories developed by the scholars mentioned above. In this period, a characteristic pattern of behavior evolved to stabilize the delicate relations between some scholars and the high dignitaries of the church. The rule was simple: All theories that presupposed an atomistic world to be the only manifestation of reality were banned. Adherents of pre-Socratic atomism a priori were suspected to range among the persons affected by this anathema (Pabst, 1994, p. 286). This was always insinuated to be true, particularly in view of the metaphysical consequences of Democritean atomism, which were assumed to be totally inconsistent with what scholasticism perceived as divine.

On the other hand, such theories—considered as mere hypotheses and presented mainly by mathematical means—were allowed if the ecclesiastical regulations valid at the time were strictly observed. It is interesting that even within the frame of this rule two notable approaches to a mathematical view of atomism were published. Both authors were fairly unknown within the ecclesiastical and secular hierarchies of former periods.

The first author, Nicholas d'Autrecourt (Nicholai de Ultricuria), a Parisian professor who is considered a medieval David Hume, was a scepticist (like Al-Gazali, the great Arabian critic of his contemporary philosophy and author of *Incoherence of the Philosophers* [*Tahafut*]). In his opus magnum, "Exigit ordo" (completed in 1330), d'Autrecourt discussed the dubious conclusions of the Aristotelian position against ancient atomism in the light of his own atomistic ideas. Remarkably, in the course of his argumentation d'Autrecourt anticipated Leibniz's principle of the best of all possible worlds (Pabst, 1994, p. 288). Regarding the origin of the universe, this entailed an open dissent against Democritus's thoroughly ethical, yet nonreligious epistemology.

d'Autrecourt had a worldview clearly substantiated by two interesting perspectives: Although he agreed with the classical atomic concept, he introduced two kinds of atoms. The first kind formed the substance of a body. The second kind of atom was the vehicle of characteristic properties assigned to the *accidentalia*. This key term of scholasticism meant secondary attributes of any substance, for example, color or taste.

They refer to his concept of a vacuum compared with the original idea of an altogether empty space all over. By his definition "... illud in quo non est corpus, potest tamen esse corpus" ("That where there is no body, can yet be a body"), the conjecture was made for the first time that empty space should exist as a microvacuum only between the atoms of a body, but never outside of an atom complex. As to the "mathematical" view of atomism, d'Autrecourt risked controversy with Aristotle's adherents by his very sophisticated reasoning against their typically geometry-inspired argumentation.

It is sufficient here to make clear that the geometric properties of any elementary particle resulting from every atomistic concept directly affect the notion of a continuum in natural philosophy and mathematics. At that time this affinity was generally accepted to be true, because Aristotle had expounded at great length in book VI of his *Physics* that there could never exist a material continuum, if indivisible atoms are assumed to have neither geometric extension nor any shape.

Although d'Autrecourt accepted this "mathematical" idea of atoms, he disputed the Aristotelean inferences. These, indeed, cogently follow from the geometric image of a body consisting of many material parts bounded inherently. According to this view, Aristotle claimed that an atom formed like a mathematical point could not attribute to the continuous structure of any body because the existence of boundaries of the atom constitutive for any material contact could not be imagined. d'Autrecourt argued—thereby following Henry of Harclay's and Gerardus Odonis's "canonical" way out—that atoms might be additionally endowed with a locality of their own and an individual boundary. For this reason, two atomic points could establish a new mutual relation, by which a new and larger quantity might be constituted and even equipped with new material properties (Pabst, 1994, pp. 294, 298).

This modern point of view was never reached by John Wyclif, the second great representative of atomism in the late fourteenth century. Wyclif, influenced by the heretic William of Occam, was subjected to such threats to life and liberty that he could present only mathematically veiled hypotheses. In his voluminous *Tractatus de logica*, the prominent Oxonian professor as a matter of fact protected an extremely formal idea of atomism defined by the identity of the physical atoms with the mathematically founded non quanta formed like a spatially dimensionless point.

This atomistic concept seems to anticipate the notion of a mass-point, first introduced by Euler little less than four hundred years later. Wyclif established a linear relation between the magnitude of a body and the number of non quanta involved. Here for the first time a measure arose that could lead to the extensive quantities in the sense of homogeneous functions.

In addition to the contributions of the major atomists mentioned above, some relevant texts of unknown authors should be noted. Within the sphere of influence of the Chartres School the tractate titled "De elementis," written probably in the second half of the twelfth century, made an essential contribution to the atomic ideas. Here, for the first time, all properties of an element were attributed to its characteristic intensity of motion.[26]

Experts assure that the treatise "De generibus et speciebus"[27] belongs to the most intricate medieval studies on atomism. Its central passage concerns hypotheses on the quantitative proportion between the atomic constitution of any given body and its essence to be realized by an individual form. With the help of this term *forma*, a new holistic attribute had been suggested that was in sharp contrast to the prevailing mechanistic ideas of matter.

Last but not least let us regard the didactic poem in an Oxonian medieval manuscript,[28] for which the first editor chose the title "Carmen de mundo et partibus." Probably written before 1230, it is a singular document of a transition period. The author resolutely polemized against the Aristotelean anti-atomic position. The central part of his tractate deals with the basic problem of whether atoms exist or not. It is remarkable that atoms are conceived as spatially dimensionless and, for this reason, immaterial: "Ex incorporeis consistunt corpora rebus, ex athomis dico" (Bodies consist of immaterial things, say atoms"), (Pabst, 1994, p. 254).

A concise summary concerning medieval atomism follows.

1. Ab initio early Greek atomism was categorically conceived. *Atomos* meant indivisible, having no parts, uncuttable. Atoms as ultimately irreducible entities were contingently interrelated to form objects. Thus, their primary qualities were presupposed to realize dynamical activity only by motion and interaction in a void. Secondary qualities such as odor or thoughts existed only as conscious content. From the very beginning, the universe would appear chaotic and accidental, devoid of any metaphysical plan.
2. Patristics and scholasticism vehemently waged perpetual war against Democritus's and Epicurus's atomism. The point of attack was the early atomists' cosmology rather than their ideas of properties and functions of atoms. But, considering this ideological boundary, the scholars' chance to make innova-

[26]R.C. Dales, Anonymi De elementis. From a 12th-Century Collection of Scientific Works in British Museum MS Cotton Galba E.IV, in Isis 56 (1965), pp. 174–189. See p. 176

[27]ed. V. Cousin, Ouvrages inédits d'Abélard, Paris 1836, pp. 507–550.

[28]Oxford, Bodleian Library, Digby 41, fol. 93r–100r.

tive contributions to these ideas were seriously impeded by a second restriction: Aristotelean natural philosophy brought every advance in atomistic concepts to a complete standstill in the second half of the thirteenth century at the latest.

3. Beginning with the work of Isidorus of Sevilla, a strange extension of the notion of a material atom was suggested for nonsubstantial items. The introduction of five "canonical" atoms was an anticipated form of a theoretical approach called logical atomism. The major discussions of this discipline centered around concepts such as "facts," "atomic and molecular propositions," "object versus sensory data," and so on. Its philosophic outlook is primarily linked with Bertrand Russell and Ludwig Wittgenstein, and is characterized by such themes as: Language and thought can be analyzed in terms of indivisible and discrete components; there is a close similarity (possibly an isomorphism) between the structure of a formal (ideal) language and the real structure of the world; and the like.
4. Amazingly, a partly new atomistic worldview originated, endangered by the reproach of heresy on the one side and of unscholarly speculation on the other. It was based on several novel aspects and sophisticated elaborations. Some results can, as a matter of fact, be associated with some ideas of modern physics. This seems particularly to be true when compared to most of the ancient atomists' reasonings or even to the rare studies within the tradition of "mathematical" atomism entertained during the pre-Galilean period.
5. As to medieval atomism in the twelfth century, numerous concrete problems in natural philosophy were dealt with, at least qualitatively. It would be entirely wrong to think of these ideas as mere resorts for pertaining antique traditions and teachings. The most important and striking feature of these ideas was the connection between the atomistic concept and the doctrine of the four constituents fire, air, water, and earth. Each of these elements was regarded as a mixture of qualitatively different corpuscles assumed to be the vehicle of the specifically elemental properties. In rare cases, the elements were even directly related to purely atomized matter and considered to be a special state of aggregation. The boldest conjecture was offered by the Anonymus "De Elementis", who suggested explaining all properties of the elements by the dynamics of different self-movements of the atoms.
6. A significant difference between the scholars and the early atomists existed in reference to the old notion of the void. With one exception (d'Autrecourt's microvacuum), all medieval atomists disputed the existence of an empty space. Another key term of Democritean atomism had obviously lost its relevance: the size and shape of an atom assumed to be extended and endowed with a characteristic pattern. For the majority of thinkers who had coined the natural philosophy of the Latin Middle Ages, atoms were the material correlate to mathematical points defined to be geometrically dimensionless. Certainly, as a consequence of profound Aristotelian criticism, this

presupposition of a nonextended atom led to the decay of atomism at the close of the fourteenth century.

A1.4 The Galilei Affair

Regarding the innovative intricacies contributed to atomism, it is evident that the broad spectrum of interests in natural philosophy that existed for about three hundred years defies the often alleged intellectual homogeneity of the Middle Ages. In retrospect, however, this impression of a variety of pertaining results might perhaps be dimmed by the historical fact that, above all, the dispute on the so-called accidental attributes was definitely decided at the Council of Trent (1545–1563). This harsh controversy referring to the grave problems of scholasticism was over the meaning of *accidentia* for the eucharistic doctrine.

After the condemnation of Berengarius's (Lessing, 1773–1781) utterances of doubt against ecclesiastical authorities in favor of reason, the conflict was coupled with an insuperable antagonism to some philosophers. And the Franciscan William Occam's moderate objection to the prevailing doctrine of transsubstantiation, overruled by the authorities, was the latest desperate public attempt to break the concerted front of Catholic dogmaticians and the adherents of the Aristotelean school (Cusanus). The consequences for Europe's intellectual history were bitter. Creative men like Paracelsus complained about the dominant Aristotelism and its theorization that to him seemed so far removed from everyday life.

Aristotelean physics may in toto be characterized as nominalistic. Undoubtedly, its methods offer the advantage that all phenomena and properties of bodies could be explained qualitatively without having to resort to spiritual or invisible structures. Indeed, Peripatetics believed Democritean atoms to be absurd with respect to Aristotle's basic idea that bodies, time, space, and motion were continuities, each defined as potentially divisible ad infinitum.

One should always bear in mind that Bruno's execution in 1600 was the tragic focus for two controversies in natural philosophy around which sturdy politics could be made. Those altercations are rather evenly balanced with regard to cosmology and atomism. In both cases the authorities relentlessly persecuted any dissent. The first concerned the long, public war mainly concerning the Copernican system. It is now obvious that its most prominent victim was Bruno, the Platonist and mystic, not Galileo Galilei—despite the differently voiced tradition. It was only very recently that Pietro Redondi, a renowned expert of the history of sciences, published new information that contributes to a more thorough understanding of what might be called the climatic prelude of the Galilei trial in 1633. At present it is hard to discern, however, whether Redondi's influence will overshoot the mark set by this new information. Nevertheless, his arguments may serve to clarify some intrinsic contradiction with regard to the official accusation whereupon Galilei allegedly took part in the subversion of then current worldview by advocating Copernicus's astronomy.

What about Bruno? What does his most important work, *La cena delle ceneri* (*The Ash Wednesday Supper*), contain to style him a martyr in science? For Bruno, the intrinsic value of the Copernican system was not found in its astronomical relations, but rather in its scope as a metaphoric vehicle for far-reaching philosophical speculations. The latter were focused on the so-called ancient true philosophy, that is, a special kind of Hermitism that was a mystic view based on Neoplatonic ideas published in the second and third centuries.

According to Bruno's Hermitism, a sort of universal divine principle extends to the entities that make up the macrocosmic universe. Not only man, but also planets are believed to be animate and endowed with souls that are able to operate in a magical way to affect social and political life. In our day it is hard to be moved by Bruno's implications concerning the relations between his ancient true philosophy and contemporary politics.

Bruno's enthusiasm for the projects of the French King Henry IV fostered his erroneous belief that he would succeed in converting even the pope to Brunonian Hermitism. His obvious aim was to establish an alliance between the French moderates and the English Protestants against the powerful and religiously orthodox Spaniards (cf. Lerner and Gosselin, 1986, p. 119).

Bruno's initiative almost immediately ended in disaster. He was arrested and imprisoned for seven years in Rome. Because he had no adherents, no influence, and no money, Bruno in prison was no real threat to the papacy. Therefore, his execution might be seen mainly as an act of raison d´état. That was the pope's issue of a minor quid pro quo with the Spanish Habsburgs and their cruel intolerance against any heresy or change of the political status quo.

If one searches for a link between Bruno and Galilei, three aspects will generally be found. First, the prevailing political and religious climate after Bruno's execution—which was, above all, officially justified by his heresy—was allegedly supported by the ideas of Copernicus and his disciples. Furthermore, both men preferred to write their best books in everyday Italian, indicating a thoroughly alarming message:

> Both Bruno's Supper and Galilei's Letter to the Grand Duchess had expressed similar attitudes toward Holy Scripture. Both works maintained that the Bible often speaks according to the common understanding of the people, and in so doing it may actually say things about nature that are not literally true. The view was eccentric enough to have caused Galilei and Bruno to be closely associated. (Lerner and Gosselin, 1986, p. 120)

The third aspect concerns their relations toward Tommaso Campanella, an apostate Dominican monk. In 1600, the year of Bruno's spectacular death, Campanella wrote his *City of the Sun*, a distinctly Hermetic and Copernican work, thoroughly similar to Bruno's metaphorical interpretation. He viewed Galilei as a genius who did science without comprehending the philosophical significance of his discoveries. Consequently, he wrote an Apologia pro Galileo as Galilei was seriously reprimanded in 1616 by cardinal Saint Roberto Bellarmino to comply with the ecclesiastical verdict of the Copernican system. Campanella attempted to prove that Galilei's scientific teachings were in complete accordance with accepted theology. An unprejudiced analysis of Galilei's books plainly shows that he did indeed pay attention to Bellarmino's warning. For this reason, the question arises: What really happened to Galilei?

The close link between politics and religion will be well illuminated by the second kind of deadly quarrel noted above: a guerrilla warfare, the concealed persecution of any suspicious person who propagated ancient atomism. Nowadays it is hard to imagine a situation where Epicurean atomism was forbidden on penalty of death. But this happened in 1624, proclaimed by the city parliament of Paris. Undoubtedly, this cruel threat was the consistent, but excessive, conclusion from the ban canonized by the thirteenth session of the Council of Trent and sanctioned by immediate excommunication:

> "He who maintains that in the Holy Sacrament of the Eucharist the substance of bread and wine remains with the body and the blood of our Lord Jesus Christ, and who further denies this mysterious and unique transformation of the whole of the substance of bread into the body and that of wine into the blood which merely leaves unchanged the outer appearance of bread and wine, a transformation that the Church very suitably denominates as transfiguration, shall duly be damned"[29] (author's translation).

Faced with danger for life, liberty, and automatic excommunication, it would appear incomprehensibly stupid to plead in public for any form of atomism. Yet exactly this was done by Galilei who in 1623 published one of his most important, but nowadays rather unknown, essays. This book, entitled *il Saggiatore* (which means in English, say, "balance of precision" or "the man who weighs gold") envisaged a new style in natural philosophy. New definitions and rules were introduced and exemplified by a kinetic model of matter in which, for the first time and in complete opposition to Aristotelean physics, a decomposition of motion was established. Above all, the author suggested the use of an alternative language in physics, mathematics and, particularly, Euclidean geometry as an appropriate tool for the quantitative description of phenomena. With this approach, he certainly initiated a momentous polemic against the omnipresent Aristotelean doctrine.

The author was deeply convinced of the corpuscular structure of matter. However, he kept to the old idea of the four elements; the notion of an atom was exclusively reserved for light. As a whole, Galilei had never before published such an advanced and generalized theory. In fact, many coeval intellectuals and politicians identified the book with a manifestation of a new philosophy of nature.

[29] Concilium Tridentinum Diariorum actorum, epistularum, Vol. VII, Sect. IV, 2. Book, Freiburg, 1976, p. 216.

It should be stressed that *il Saggiatore* was most successful with all influential circles of church and state. The Barberini family and its leaders, Pope Urbanus VIII in company with three cardinals, supported Galilei in every way. Many people were perplexed that the authorities could express atomism in a specified way in its mathematical form and thus hide away any suspicion of heresy. Yet, the harmony was fallacious: It reflected, above all, the art of honorable disguise—the great invention of the seventeenth century!

The anonymous denunciation against the author of *il Saggiatore* took place early in 1625 before the Holy Office. The document G_3, written in Italian, was detected by Redondi about ten years ago. It complied precisely with all official prescriptions to suffice as an indictment for heresy. The text, by the way, referred to Anaxagoras's and Democritus's atomism and quoted at great length from *il Saggiatore*. Fully intending to commit Galilei for trial, the denouncer asked the Holy Office for an authorized statement. The suspicion related to the worst case of heresy: the infringement of the *dogma concerning the permanence of the eucharistic accidentia*. According to Redondi, the Jesuit Orazio Grassi, Galilei's archenemy concerning cosmology and the Eucharist, seems to have been the denouncer.

The denunciation was well timed in view of a horrible event late in 1624 and unfortunately disappeared into oblivion. It is the story of Marco Antonio *De Dominis*, archbishop of Spalato, accused of heresy concerning the Sacraments of the Eucharist and of matrimony. After his death in custody he was condemned to suffer stake and fire; the auto-da-fé took place on December 21, 1624.

Of course, the denunciation was also a veiled attack against the Barberinis. But in those years, all power was held by Pope Urbanus VIII and his partisans. For this reason, Galilei's friends and allies succeeded in stopping the charge temporarily. Its burden proved to be too heavy, however. At the time of the deadly trial by inquisition against De Dominis it was impossible to suppress a founded suspicion of heresy without provoking reflections on treachery, secret intrigues, and religious fanaticism. Thus, only a spark was needed to ignite the explosive mixture of malicious rumors and incriminating imputations.

It is true that under the reign of Pope Urbanus VIII danger never really threatened those who followed Copernicus, provided they framed his ideas mathematically or hypothetically at least. But the dogma concerning the permanence of the eucharistic accidentia was without doubt the theological spearhead of the Counter-Reformation. To avoid that this miracle would have to be reconciled somewhat with common sense, it was necessary to insist on two rules of faith: (1) There is a permanent distinction of quantity from substance. (2) Phenomena really do exist independently of man and his consciousness.

Here we have another dogma that issued from the late scholasticism after the ultimate victory of Aristotle's doctrines. It was one of the basic commandments of Catholic faith and a deadly weapon against the Lutherans and Calvinists. Moreover, in the seventeenth century it was the key problem of contemporary theology, the central point of true religion and all intellectual controversies. Nobody could expect justice or mercy for the respective heretical offense against the Dogma of Transsubstantiation, which was believed to be inconsistent with the ancient atomism.

In the end, two events led to the disastrous condemnation: The first was Galilei's publication of his famous retribution to the Aristotelean doctrines of cosmology and, particularly, those of motion,

Dialogo sopra i due massimi systemi del mondo tolemaico, e copernicano,

(*Dialogue about the two main world systems, the Ptolemaic and the Copernican*) which appeared in the year 1632. "Galilei's *Dialogue* is a purely philosophical and scientific work, but it too does not really defend the Copernican system in its then-current form" (Lerner and Gosselin, 1986, p. 120).

The second event was an external one. In that period the course of the Thirty Years' War took an unexpected turn for the worse for the Catholic alliance. Thus, the internal opposition against the Barberinis gathered momentum. With a scheming brain and a will to be obeyed, Pope Urbanus VIII hit back with a kind of exorcism. How he got things done in face of dangers and public emotions is best characterized by a strange event that happened in December 1632: In a solemn ceremony the pope praised God for the death of the Swedish king, "the horror of the universe," killed in the battle of Lützen.

On the pope's initiative, a special Court of Inquisition was set up. The indictment exclusively concerned the Dialogo and spoke of a transgression of the agreement back in 1616 between cardinal Bellarmino and Galilei. Galilei's *il Saggiatore* was not mentioned at all. After an extreme break of the common procedural rules by the chief judge himself, Galilei submitted to the charge of Copernican doctrine and renounced it. On June 22, 1633, Galilei was officially convicted of high treason and sentenced to life imprisonment as minimum penalty. The Cardinal Inquisitor Borgia was so indignant at this farce of a trial that he refused to certify the verdict by his signature.

The inquisitorial procedure was not adequately performed in view of a question of faith and conscience, even if it were possible in public interest. Although there is no doubt that Galilei adhered only to pure Copernican astronomy, the papal court could risk prosecuting the most prominent scientist of the day as a victim of a secret affair of state. This meant that while the papal states were laid waste by war, their dictator could at the same time suppress the real scandal and eliminate the internal opposition. Surprisingly, the quadrature of the circle was successful. The Barberinis got down to the root of their grievance by the damnation of the new natural philosophy initiated by Galilei. And this is, in my opinion, the true secret of the "case of Galilei," provided that Redondi's inquests of the affair produced correct results.

While in the Holy Office the juridical ritual was settled, the simultaneously running punitive action alarmed all of Rome. The aim of the purge were the accused persons as well as the denouncers, but also all prominent friends or antagonists of

Galilei's book about nature. One of the principal actors, Pater O. Grassi—a prominent Jesuit, architect, and professor—was one of the first victims. He was dismissed from office and lost all sinecures during Urbanus's pontificate. The Curia prohibited him from publishing any further research for all times, and even his earlier papers were no longer quoted.

Numerous other superior personalities of public life were relegated, deprived of authority, and forced to terminate their otherwise brilliant careers. Among the persons affected by these measures were the incumbent Dominican cardinal, two cardinals *in pectore*, and some bishops, as well as influential patres and monsignores.

However, all these vindictive activities were really only directed at the tip of the iceberg. Its mass of ice beneath the sea of people's daily unbelief and heresy was the new philosophy of nature, manifest in public by Galilei's *il Saggiatore*. This book was an obsession with many persons of the Church and the sciences, but only in view of its atomism. There was no reason for concern: Both mathematicians and sailors were chiefly interested in applied astronomy. Cosmology was virtually irrelevant. Meanwhile, the Jesuits had many more elaborate theories than Copernicus's debatable theses. For example, they had proved long before that it was not impossible to be a Catholic of firm faith and simultaneously a declared adversary of Ptolemaic astronomy. But the Societas Jesu clearly understood that a Catholic could not believe in the dogma concerning the permanence of the eucharistic accidentia and simultaneously wage war against Aristotelism. In addition, the course of the most terrible religious war in European history blocked public debates about irritating and dangerous ideas. After the harmonization at the Council of Trent, a renewed fragmentation of Catholic theology had to be prevented by all means.

Thus, the Barberinis reacted as expected: The Jesuits were ordered to take full responsibility for all the preventive and repressive interventions believed to be mandatory to plant the true faith into the hearts of people. It was imperative to be on one's guard with one's own friends and to act outwardly according to one's principles. No wonder that Urbanus's pontificate was the end of the so-called Catholic reformatory endeavors.

New revisors as members of the permanent supervising agency were appointed and, above all, vested with far-reaching procuratories; the chief revisors of the Societas Jesu constituted a directory board under the immediate command of the Superior General. The apologetic and scientific strategy pursued three paths:

1. the disrepute of Galilei's contributions to astronomy,
2. the refusal of the contemporary geometry of indivisible elements according to Aristotle's verdict of the notion of infinity, and
3. the campaign against the vacuum, the experimental verification of which surely was the most important scientific experience in the seventeenth century.

Point (1) met with public approval after Galilei's condemnation, but actually referred to a secondary theater of war. The other points aimed at the atomic problem, and it must be added that the last point was rather delicate because of the experimental evidence.

At that time the whole of intellectual life in Western Europe was paralyzed in a manner which cannot be reconstructed by today's experiences or considerations. Many people, prominent or not, suffered from private and public persecutions. For this reason, great thinkers like Descartes and Leibniz wrote full-length treatises on the Dogma of Transsubstantiation in connection with their scientific work. Above all, such essays were manifestations of the prevailing social pressure. For fear of being accused of heresy by the Inquisition, Descartes published anonymously. In addition, he refused to elaborate his ideas of atomism, which are documented in his *Traité de la lumière ou le monde*, completed in 1633, but first published postum in 1677. Remarkably, Descartes denied the existence of a vacuum.

The three-path strategy proved to be extremely successful. The main publications of most of the prominent antagonists were put on the notorious Index. Its edition of 1704, for instance, may be regarded as a Who's Who of the intellectual elite within the realm of Catholic doctrine. Descartes' chief works headed this blacklist. Gassendi, whom several contemporaries ranked higher than Galilei, escaped an official verdict by a sophisticated justification of his mathematically refined, but religiously adequate theory of atoms. Yet for the Holy Inquisition Gassendi remained suspect.

During the second half of the seventeenth century, the intellectual controversies in public were ended. The scholarly disputes were changed into chronicles of criminal procedures. Galilei's physics was then officially denounced by the Holy Office. On December 2, 1676, Father A. Pissini (a prominent atomist) had to recant his heresy as an outrage against the Eucharist. Everyone knew the true defendant, the author of *il Saggiatore*, whose fault Pissini had vicariously confessed.

During the 1790s, it became more and more difficult to keep up the pretense under which the Eucharist was believed to be true, due to corpuscle theories that were gaining increasing success. In the Grand Duchy of Florence, even private instructions in Democritean philosophy (i.e., atomism) were therefore strictly interdicted in writing or verbally.

Meanwhile, the Holy Office made an effort to formalize the dangerous incriminations against Galilei that had nothing to do with Copernicanism. As Father A. Baldigiani wrote:

> The cardinals' extraordinary Congregations at the Holy Office, with the pope also being present, were and still are organized. People are talking about the intention to enact a general injunction against all authors in modern physics. Their names are put on long lists, and Galilei is at the top, together with Gassendi and Cartesius, both suspected to be extremely disastrous for the literary republic and the sincerity of religion. Important advisors will be men of faith, who had formerly made an effort to bring about this injunction"[30] (author's translation).

At that time Galilei had been dead for some decades. Therefore, this blacklist could only be meant to give the contemporaries a serious warning about ruining themselves socially by a wrong partisanship. This period of a disastrous alliance between political absolutism and religious suppression required everyone to make use of the art of honest dissimulation, which had been popular since Tasso Accetto had published his book entitled *Della dissimulazione onesta* in 1641. It is noteworthy that in 1928, during the first phase of Mussolini's fascism, the famous Italian poet and writer Benedetto Croce published a new edition of this old pamphlet.[31] In this context Voltaire should certainly be mentioned. He knew precisely the addressee of his notable conjuration: "I shall if need be a confessor, but never will I be a martyr!" (author's translation).

In this sense Galilei himself offered an impressive example of that art, which may sometimes seem the only efficient method of survival, even in the numerous dictatorships of our century (the notorious McCarthy era, for example). In my opinion a sign of this art was practiced by Galilei long after his trial: He allegedly demonstrated some interest without any obligation in his written reply to a letter whose author suggested that he study some relations between atomism and the Eucharist.[32]

Of course, Galilei was the victim of personal rancor and, above all, of political intrigue. Just as with Bruno's affair, Copernican astronomy was to have been the official pretense once more. However, following Redondi's analysis, the true intention was a general settlement aimed at invigorating Catholic orthodoxy. The pope's authority needed a new orientation in view of the oncoming absolutism within all the feudal societies. Hence, it was necessary to first eliminate every kind of internal opposition within the Papal States as well as their own allies in war (the Spanish Habsburgs) and fellows in belief (France).

Whereas Galilei's opponents and all prominent persons charged as adherents to atomism were subject to rigorous if equal prosecution, Galilei himself amazingly enjoyed a privileged treatment by the pope:

> Urban never tried to interfere with the continued lionization of Galileo, even by princes of the church. Nor did Urban make any attempt to interfere in any effective way with Galilei's freedom to publish, which he surely would have done if he had regarded Galileo as dangerous outside the context of the Spanish policy. ...Galileo thus became a symbolic victim in an age that set great store by symbols, just as Bruno had before him. (Lerner and Gosselin, 1986, p. 123)

[30]quot. A. Favaro. Miscellanea galileiana inedita. Studie ricerche. Venice 1887, p. 155 (ms gal., 257, a. 117, Bibl. Nat. Florence).

[31]Very recently, Arcetto's instructions have been edited again (in German: Wagenbach, Berlin 1995).

[32]It seems remarkable that Stengers (1995, p. 406) mentioned this event in reference to her subliminal polemics against Redondi's interpretation. Unfortunately, the offered quotation is incomplete: Neither the authors' names cited by her nor the title of the paper or other respective bibliographic items are correct. The letter in question is nowhere substantiated (cf. Stengers, 1995, p. 1043).

And yet, times meanwhile had changed drastically. People began to visualize the Eucharist and atomism as symbols of the past and the future. We should consider the fact that at the end of Urbanus's pontificate the human population number was supposed to have reached nearly half a billion, which meant a doubling since the origin of the Christian era. Only 150 years later, the population number passed the one billion mark and increased continuously up to more than five billion at present.

Indeed, the crucial point was that Galilei's lionization symbolized the fact that the laws of matter and motion had begun to dominate the lives of people. Notwithstanding, the everlasting war between ideology and atomism continued: Platonism versus Democritus, the Peripatetics versus Epicurus, the Catholic Eucharist versus Galilei, and Descartes, Gassendi, Newton–Eulerian physicists versus Leibniz's monades, and so on. It is not surprising that even orthodox representatives of science attacked their contemporaries when the latter questioned the officially accepted models of matter, especially by means of advanced atomism. The concluding section of this essay will offer some impressive paradigms.

A1.5 On the Early History of the Eulerian Mass-Point

Due to the omnipresent influence of the Aristotelean doctrines at all universities, even in non-Catholic countries (particularly in the Netherlands and in England), scientists of these countries also responded to the reprisals their Catholic colleagues had to undergo. To deal consistently with their corpuscular concepts tolerated officially by the Inquisition, the well-known mathematical theory of particle impact was first to arise in 1668/1669 at the request of the Royal Society to Wallis, Wren, and Huygens.

Both the "Lex Naturae de Collisione Corporum" (Wren) and the "Tractatus de motu corporum ex percussione" (Huygens), still important tools of today's physics, showed many scientists the way out of the dilemma, with all their fear of the Inquisition on one hand and their curiosity about the new philosophy on the other. By this law, the idealized collision dynamics is quantitatively described and related to practice. Its objects become purely mathematical abstractions of genuine and visible bodies.

This direct way out was subsequently pursued by Newton: Resting on the well-tried method of impact mechanisms, he wrote his opus magnum, *Philosophiae Naturalis Principia Mathematica*, in a very paradigmatic manner. The main goal of the work, subdivided into three books, referred to a new concept of motion. Movement was postulated to be caused by forces imposed on bodies. Their velocities were related to space termed *absolute* by Newton himself. He also for the first time used the term *mechanica rationalis*, by which mechanics should be indicated as a rational, that is, mathematical science.

The Newtonian mathematics claimed to be founded on various definitions, primitive properties, and axioms, from which theorems could be derived and applied to problems and their various solutions. The presuppositions noted above resulted most likely from Newton's rigorous discourse with Descartes' *Principia Philosophiae*, presented in his Cambridgean fragment "De Gravitatione et Aequipondio Fluidorum." They appeared in front of the first book of the *Principia Mathematica* in a concise form, and its comprehension by the reader was in no way didactically supported by applicable examples. The third part of the book, entitled "De systemate Mundi," was at first not intended to be published, perhaps for fear the author would get involved in the current controversies on Copernican astronomy.

The second book is very interesting in view of the topics in question. It deals with bodies consisting of an infinite number of points. For this reason, many readers tend to wrongly assume that the *Principia Mathematica* was meant to be a treatise on mathematical mass-points. Recently, Truesdell analyzed its text with special interest in this problem. That Newton did not once use the words *particula* or *corpusculum* in the connotation of the common term *point-like body* or even in the sense of an atom was certainly a surprising result. There was no evidence that he ever envisioned mass-points. Newton's "points" constituting a Newtonian body are weightless. In the seventeenth century, no one would indeed have accepted mechanics confined to bodies that would have to be regarded solely as dimensionless points.

This general reference of mechanics to macroscopic bodies perhaps induced Boyle to offer a new idea of the term *pressure*. He suggested that the pressure of compressed air at the bottom of a mercury column should be illustrated by particles put in layers one upon each other and equipped with a kind of elasticity similar to the same well-known property of large spheres. Hooke and Newton developed different mechanisms of interactions by which Boyle's model could be quantitatively explained. Although they were in error, such concepts concerning extended bodies as objects defined by the laws of mechanics were widely used in those days. Newton applied his own corpuscular mechanics to his theory of light, presented in his second main work, the *Optics*, first published in 1704.[33] It should be stressed that mechanics defined as mathematical theory of the motion of bodies could be handled without interference from the authorities. A mathematical theory of atomistic phenomena was motivated particularly by the old problem concerning the notion of heat. Traditionally assigned to the *accidentia*, it was dangerous to tackle this problem in close proximity to atomism, which had been banned by the Tridentine Dogma. For this reason, this problem was hidden behind a new way of formulating

[33]Perhaps it is noteworthy that in the same year Jonathan Swift published his early pamphlet "A Tale of a Tub." It concerns the contemporary state of the most important European religions and their histories told metaphorically and satirized by actual peculiarities of the period. In this context he attacked the Roman Catholic Church in a rigorous way and, above all, its Dogma of Transsubstantiation. It was none other than Gulliver who, on the occasion of his voyage to the country of the Houyhnhnms, informed his host of the causes of war among the princes of Europe with reference to the Eucharist: "Difference in opinions hath cost many millions of lives: for instance, whether flesh be bread, or bread be flesh; whether the juice of a certain berry be blood or wine" (*The Portable Swift*, 1963, p. 469, Viking Press, New York).

the following question: What are the properties of a gas consisting of a collective of particles embedded in an ether, in accordance with Aristotle's "*quinta essentia*"? In addition, the particles were thought to be in a position to oscillate and rotate and mutually interact by means of the ether.

It is noteworthy that this approach created two essential ramifications concerning the history of mechanics and that of a new discipline called the *kinetic theory of gases*. On the stage of this tragicomedy Leonhard Euler was the principal actor: After his unsuccessful attempt to handle the ether model in 1727, and aside from Daniel Bernoulli's famous gas law of particles like billiard balls, the scientists' indifference delayed the evolution of the kinetic theory for a long time.

Euler took another way: In his *Mechanica*—the first treatise on analytical mechanics—he followed Newton as he did throughout his life. He understood force itself as a primitive quantity, in the sense it had taken on in statics. In this 1736 book, there were three additions to Newton's principles with regard to the notions of "mass-point," of "acceleration," and of the concept of a "vector" used in velocities and other quantities. Furthermore, "Euler realized that the statements of Newton are generally correct only when applied to masses concentrated at isolated points; he introduced the precise concept of *mass-point*, and his is the first treatise devoted expressly and exclusively to it" (Truesdell, 1968, p. 107). By the way, the notion of mass may be applied on a double standard: It was assumed to be a property of inertia as well as the center of attracting forces.

From this time until recently, many scientists considered the mathematical mass-point as a fundamental quantity of nature, or at least of classical and even quantum mechanics. They assume that matter consists of a tremendous number of very small particles obeying the laws of classical or quantum mechanics. Consequently, they believe that the behavior of gross matter can be described, in principle with any desired accuracy, from a knowledge of the laws of microscopic interactions. Thus, continuum mechanics appears as an approximate theory within classical mechanics.

This traditional belief utterly defies reality. To cite an example as to material properties: No corpuscular theory based on Newtonian mechanics yields serviceable data for the specific heats of solids that agree with experimental values. This can only be done by means of quantum concepts. To reduce things to a common denominator, there is "almost the rule that Newtonian mechanics while not appropriate to the corpuscles making up a body agrees with experience when applied to the body as a whole, except for certain phenomena of astronomical scale. Only pedagogical custom has hindered general realization that as a physical theory, continuum mechanics is better than mass-point mechanics" (Truesdell, 1984, pp. 26–27).

But it is not really only pedagogical custom. It is no accident that the basic equation $\mathbf{i} = \mathbf{v}$—that is, the identity between the specific linear momentum $\mathbf{i}$ and the velocity $\mathbf{v}$ of a mass-point—is believed to hold in continuum mechanics, too, although it obviously is not valid for a flow of a fluid at a rough wall. This "law" defines the momentum of a single mass-point for which such dissipative effects are excluded.

Therefore, it is hard to see why such paradoxes should still stick in physics and in teaching the engineering sciences today.

But the branches of mechanics are not compared in practice; they are normally applied by people who do not communicate with each other. For this reason, *Hamiltonian mechanics*, as a special elaboration of mass-point mechanics, nowadays is considered to have scientific priority over all other branches of nonquantum mechanics, undoubtedly because of its formal esthetics. And in this "leading" mechanics the law $\mathbf{i} = \mathbf{v}$ is valid, while wall friction does not exist.

The close of the eighteenth century brought about a new view on atomism. In analogy to celestial mechanics, the atomistic processes were compared with the trajectories of stars assumed to be interacting only with forces that can be expressed by means of power functions of the respective distances. An attractive example is found in the middle of Laplace's voluminous work on celestial mechanics, namely, his fine corpuscular theory of capillarity. By the way, it was none other than Maxwell who also used such a power function to formulate his kinetic theories of gases. As Fourier renounced this convenient supposition, his theory of heat propagation was seen by his colleagues and himself to be imperfect and provisorial.

In England, the first few decades of the nineteenth century did not see much physically original speculation of any kind, which rather met with distrust as did the mathematical theory available. Mechanics had then been established and exhausted by Newton. It seems strange that only the Royal Society, dominated by amateurs of this quite mechanistic tradition, in a certain sense replaced the function of the notorious Holy Office in view of the acceptance of atomism. The scandalous maltreatment of Herapath's and Waterstone's papers concerning tentative approaches to the kinetics of gases made both authors fall into oblivion. But the history of the early kinetic theory of gases also tells of the great difficulties to obtain acceptance of the theoretical concepts. In 1845, for instance, only a few physicists believed that matter was infinitely divisible,[34] while most of those who regarded it as molecular clung to the theory of heat as a kind of vibration. Moreover, any effort of creative imagination, unless put forward by a prominent authority, had become distasteful.

In the second half of the nineteenth century, there was a strong and vehement opposition against all attempts to establish atomism by means of kinetic theories. Particularly the prominent adherents of the so-called energetics such as Thomson (Lord Kelvin) and Ostwald, but also great mechanicians such as Mach or famous mathematicians such as Poincaré, attacked Boltzmann's idea of proving indirectly the existence of atoms by theoretical arguments and, above all, by way of transport phenomena of nonuniform gases. At that time this idea was provocative to a degree that is hard to envisage today. Indeed, the scientific community accepted it only in the traditional fashion expressed in Mach's apodictic: Any theory of atoms

[34]Remarkably, Hegel stated that the question is nonsense, whether there are atoms or not. One can only ask: How is atomism possible and, in addition, how it is impossible? (cf. Bloch, 1979).

is allowed "to be a mathematical model to represent the facts" (Mach, 1883, p. 464; author's translation).

A first reversal of the general mood in the scientific community took place in the period between about 1900 (the year of Planck's introduction of the quantum of action) and 1906 (the year of Boltzmann's tragic death). During this period three principal actors were of great importance to atomism. In 1902, by means of the famous photoelectric effect discovered in 1889 by Hertz, the first quantum phenomenon was experimentally observed by *Lennard*: Incident electromagnetic radiation onto a metal surface caused electrons to escape without conservation of energy. Then, in 1903 the first results of his thoughts on the interpretation of spectral series obviously fitted a rough theoretical approach to the general structure of atoms.

Einstein, the second principal actor, regarded physics atomistically ab ovo. It is well known that his hypothesis of a light quantum, introduced in 1905 as an originally existing atomistic entity and now called a photon, is the extremely atomistic inference from Planck's radiation law. It is less known, however, that Einstein rated the determination of the numerically precise value of the so-called Avogadro number A_N as the key question of atomism to be answered. Consequently, he dealt with this problem concerning the real existence of an atom, not only in his doctoral thesis in 1905, but also in many papers until 1917. (Among them the reader may find three concrete proposals for some precise measurements of the Avogadro number.)

Clearly, the third principal actor is *Perrin*, whose very notable experiments in the final analysis allowed the calculation of the correct value of A_N, which is still assumed to be valid today. In the preface of his distinguished book published in 1905, he recalled the insurmountable contradiction between the idea of homogeneous matter and the hypothesis of the discontinuous world of the atoms. Perrin illustrated his experiences as follows: The density of matter is thought to take on infinite values at a countably infinite number of isolated points and the value zero elsewhere. This picture apparently resembles the ancient speculations. But it is an imagery that distinguishes atomism in the prequantum era. Of course, the final reversal resulted from the quantum paradigm. In 1929 Max Planck gave an exemplary lecture where he described a metaphorical image believed to be true for atomism in the quantum world (Planck, 1929, p. 23).

To summarize, according to quantum mechanics local relations are just as insufficient for an adequate formulation of the laws of motion as the minute scrutiny of all pointlike parts of an oil painting is for its intellectual and artistic understanding. A realistic image is more readily arrived at only if the physical pattern is regarded as a whole. For this reason, quantum mechanics expects each material point of the particle collective to be assigned to the system in question for any time and at all places of the space. This does not only mean the force field around the point, but also its material identity typified by its own mass and charge. It is evident: Nothing more or less is at stake than the notion of the Eulerian mass-point as the most elementary term of classical mechanics. The hitherto central meaning of this mathematical mass-point will have to be sacrificed. Its application is only still justified for special limiting cases.

I think Max Planck is right, now as before!

However, there is no end of the story. Yet it is hard for interested contemporaries to discern ideologic attitudes from scientific inferences. We should remember Werner Heisenberg's urgent warning not to overtax the particle concept in experimental and theoretical physics. Heeding his own warning, he dissociated himself from the obsession to build ever-larger particle accelerators. (The crisis of this giantism has just recently passed on the occasion of the Super Conducting Super Collider project in Waxahachie, Texas.) Moreover, he even defined the early quantum field theory (QFT) by introducing an "exclusion principle." The word *particle* must not apply within the mathematical frame of the QFT concerning the dynamics of matter: Particles are manifestations only of secondary structures of matter. Thus, all qualities of an atom are derived ones, there are no a priori material properties belonging to it—that is, each kind of imagination about atoms should be assumed to be erroneous eo ipso. Corresponding to this, modern QFT "corpuscles" are not bare particles, but result from basic fields, the so-called Heisenberg fields.

I think Werner Heisenberg is right, now as before!

What, however, would he say about the current super string theories, believed to be working well in a ten-dimensionalized space or about the new vacuum concepts with their ideas of the quarks enclosed in a nucleus and moved irresistibly inside the invisible gluon vacuum? Indeed, we find ourselves in the midst of the current story:

Who will find the God particle?

Appendix 2 Mathematical Supplements

"What is matter?
Never mind!
What is mind?
No matter!"—***A. Baez***

A2.1 Supplements to the Noether Theorem

A2.1.1 INVARIANCE IDENTITIES

The various definitions of invariance applied to Noether's problem imply that certain relations are established between the Lagrange function $\mathcal{L}$ (along with the three derivatives $\partial\mathcal{L}/\partial\mathbf{x}^k$, $\partial\mathcal{L}/\partial\mathring{\mathbf{x}}^k$, and $\partial\mathcal{L}/\partial t$, respectively) and the infinitesimal generators τ_ν and $\xi^k_{\ \nu}$. For these expressions a sequence of derivatives concerning the transformation rules is required with respect to the original variables $\mathbf{x}^k$ of $\mathcal{L}$ and to the parameter vector $\varepsilon := [\varepsilon^1 ; \ldots; \varepsilon^\rho]$ around the identity $\boldsymbol{\varepsilon} \equiv \mathbf{0}$. The results are compiled as follows:

$$\left(\frac{\partial t'}{\partial t}\right)_0 = 1; \quad \left(\frac{\partial x'^k}{\partial t}\right)_0 = \left(\frac{\partial x'^k}{\partial x^k}\right)_0 = 0;$$
$$\left(\frac{\partial x'^k}{\partial x^k}\right)_0 = \delta^k_h \text{ with the Kronecker symbol } \delta^k_h = \begin{cases} 0 \text{ for } h \neq k \\ 1 \text{ for } h = k \end{cases};$$
$$\left(\frac{\partial^2 x'^k}{\partial\varepsilon^\nu \partial t}\right)_0 = \frac{\partial\xi^k_\nu}{\partial t}; \quad \left(\frac{\partial^2 x'^k}{\partial\varepsilon^\nu \partial x^h}\right)_0 = \frac{\partial\xi^k_\nu}{\partial x^h};$$
$$\left(\frac{\partial^2 t'}{\partial\varepsilon^\nu \partial t}\right)_0 = \frac{\partial\tau_\nu}{\partial t}; \quad \left(\frac{\partial^2 t'}{\partial\varepsilon^\nu \partial x^h}\right)_0 = \frac{\partial\tau_\nu}{\partial x^h}. \tag{A.1}$$

A2.1.2 THE FUNDAMENTAL INVARIANCE IDENTITY

The following intermediate steps are required to form the total derivative of the invariance definition (4.23) with respect to the parameter set $\varepsilon := \{\varepsilon^1; \ldots; \varepsilon^\rho\}$ around the identity $\boldsymbol{\varepsilon} \equiv \mathbf{0}$:

Step 1: Take Einstein's summation convention with regard to the index k.

$$\left\{\frac{\partial\mathcal{L}}{\partial t}\left(\frac{\partial t'}{\partial t}\right)_o^{-1}\frac{dt'}{d\varepsilon^\nu} + \frac{\partial\mathcal{L}}{\partial x^k}\left(\frac{\partial x'^k}{\partial x^k}\right)_o^{-1}\frac{dx'^k}{d\varepsilon^\nu} + \frac{\partial\mathcal{L}}{\partial\mathring{x}^k}\left(\frac{\partial\mathring{x}'^k}{\partial\mathring{x}^k}\right)_o^{-1}\frac{d\mathring{x}'^k}{d\varepsilon^\nu}\right\}\frac{dt'}{dt} + \left[\frac{\partial}{d\varepsilon^\nu}\left(\frac{dt'}{dt}\right)\right]_0\mathcal{L} = \frac{d\Omega_\nu}{dt}; \tag{A.2}$$

Note that any contribution of (4.23) with respect to the original function $\mathcal{L}(\mathbf{x}, \mathring{\mathbf{x}}, t)$ does not exist, because $\mathcal{L}(\mathbf{x}, \mathring{\mathbf{x}}, t)$ is independent of ε. The abbreviation $\mathring{\mathbf{x}}^k$ stands for the reference velocity dx^k/dt.

Step 2: Evaluate the last term of the left-hand side of (A.2).

$$\frac{dt'}{dt} = \frac{\partial t'}{\partial t} + \frac{\partial t'}{\partial x^k}\mathring{x}^k \quad \text{for} \quad \boldsymbol{\varepsilon} \equiv \mathbf{0}: \frac{dt'}{dt} = 1 \tag{A.3}$$

$$\rightarrow \left[\frac{\partial}{\partial \varepsilon^{\upsilon}}\frac{dt'}{dt}\right]_0 = \frac{\partial}{\partial t}\frac{\partial t'}{\partial \varepsilon^{\upsilon}} + \frac{\partial}{\partial x^k}\frac{\partial t'}{\partial \varepsilon^{\upsilon}}\mathring{x}^k = \frac{\partial \tau_{\upsilon}}{\partial t} + \frac{\partial \tau_{\upsilon}}{\partial x^k}\mathring{x}^k = \frac{d}{dt}\tau_{\upsilon}\,.$$

Step 3: Evaluate the expression

$$\frac{dx'^k}{dt'} \quad \text{in the term} \quad \frac{d\mathring{x}'^k}{d\varepsilon^{\nu}} = \left[\frac{\partial}{d\varepsilon^{\nu}}\frac{dx'^k}{dt'}\right]_0.$$

$$\frac{dx'^k}{dt'} = \frac{dx'^k}{dt}\frac{dt}{dt'} = \frac{dx'^k}{dt}\left[\frac{\partial t'}{\partial t} + \frac{\partial t'}{\partial x^h}\mathring{x}^k\right]^{-1} = \frac{\partial x'^k}{\partial t} + \frac{\partial x'^k}{\partial x^h}\mathring{x}^h = \delta^k_h\,\mathring{x}^h = \mathring{x}^k \tag{A.4}$$

Step 4: Evaluate the expression

$$\left[\frac{\partial}{\partial \varepsilon^{\nu}}\frac{dx'^k}{dt'}\right]_0 = \left[\frac{\partial}{\partial \varepsilon^{\nu}}\frac{\partial x'^k}{\partial t'}\right]_0,$$

assumed to be true because x'^k is independent of $x'^h (h \neq k)$.

$$\begin{aligned}
\left[\frac{\partial}{\partial \varepsilon^{\nu}}\frac{dx'^k}{dt'}\right]_0 &= \left[\frac{\partial}{\partial \varepsilon^{\nu}}\frac{dx'^k}{dt'}\right]_0\left[\frac{dt'}{dt}\right]_0 = \left[\frac{\partial}{\partial \varepsilon^{\nu}}\left(\frac{dx'^k}{dt'}\frac{dt'}{dt}\right)\right]_0 - \frac{dx'^k}{dt'}\left[\frac{\partial}{\partial \varepsilon^{\nu}}\left(\frac{dt'}{dt}\right)\right]_0 \\
&= \frac{\partial}{\partial \varepsilon^{\nu}}\frac{\partial x'^k}{\partial t} + \frac{\partial}{\partial \varepsilon^{\nu}}\frac{\partial x'^k}{\partial x^h}\mathring{x}^h - \frac{dx'^k}{dt'}\left[\frac{\partial^2 t'}{\partial \varepsilon^{\nu}\partial t} + \frac{\partial^2 t'}{\partial \varepsilon^{\nu}\partial x^h}\mathring{x}^h\right] \\
&= \frac{\partial \xi^k_{\nu}}{\partial t} + \frac{\partial \xi^k_{\nu}}{\partial x^h}\mathring{x}^h - \mathring{x}^k\left[\frac{\partial \tau_{\nu}}{\partial t} + \frac{\partial \tau_{\nu}}{\partial x^h}\mathring{x}^h\right].
\end{aligned} \tag{A.5}$$

Inserting Equations (A.3) and (A.5) into relationship (A.2) and applying both definitions (4.26) to the total derivatives $dt'/d\varepsilon^{\nu}$ and $dx'^k/d\varepsilon^{\nu}$, Equation (4.28) will result.

A2.1.3 FURTHER IDENTITIES CONCERNING FUNDAMENTAL INVARIANCE

Further transformation of the fundamental invariance identity (4.28) with respect to the Lagrange function $\mathcal{L}$ will have to make use of the following identities:

$$\begin{aligned}
\tau_{\nu}\frac{\partial \mathcal{L}}{\partial t} &= \tau_{\nu}\left[\frac{d\mathcal{L}}{dt} - \frac{d\mathcal{L}}{\partial x^k}\mathring{x}^k - \frac{\partial \mathcal{L}}{\partial \mathring{x}^k}\mathring{\mathring{x}}^k\right] \\
\frac{\partial \mathcal{L}}{\partial \mathring{x}^k}\frac{d\xi^k_{\nu}}{dt} &= \frac{d}{dt}\left[\frac{\partial \mathcal{L}}{\partial \mathring{x}^k}\xi^k_{\nu}\right] - \xi^k_{\nu}\frac{d}{dt}\frac{\partial \mathcal{L}}{\partial \mathring{x}^k} \\
\frac{d}{dt}\left\langle \frac{\partial \mathcal{L}}{\partial \mathring{x}^k}\mathring{x}^k\tau_{\nu}\right\rangle &= \frac{\partial \mathcal{L}}{\partial \mathring{x}^k}\mathring{x}^k\frac{d\tau_{\nu}}{dt} + \frac{\partial \mathcal{L}}{\partial \mathring{x}^k}\mathring{\mathring{x}}^k\tau_{\nu} + \mathring{x}^k\tau_{\nu}\frac{d}{dt}\left(\frac{\partial \mathcal{L}}{\partial \mathring{x}^k}\right).
\end{aligned} \tag{A.6}$$

The second term of the right-hand side of $(A.6)_3$ contains the acceleration, which is canceled out if the identities (A.6) are used to convert (4.28) into (4.29). It is worth mentioning that this last formulation of Noether's invariance identity is most appropriate for the consideration of the Euler–Lagrange equations of motion.

A2.1.4 APPLICATION OF NOETHER'S THEOREM TO A SYSTEM OF MASS-POINTS

Point mechanics offers a convenient way to study Noether's abstract theory, above all with regard to its relation to canonical transformations.

Let us start from the Lagrange function $\mathcal{L}$ of a mass-point ensemble,

$$\mathcal{L} := \frac{1}{2}\sum_{\omega} m_{\omega} \overset{\circ}{\mathbf{x}}{}_{\omega}^{2} - \phi\left(\left|\mathbf{x}_{\omega} - \mathbf{x}_{\Diamond}\right|\right), \tag{A.7}$$

where m_ω and Φ respectively denote the mass of the ωth mass-point and the function of the potential energy characterizing the whole system with reference to any arbitrary position vector $\mathbf{x}_\Diamond$. By differentiation the relationships,

$$\frac{\partial \mathcal{L}}{\partial \mathbf{x}_{\omega}} = -\frac{\partial \phi}{\partial \mathbf{x}_{\omega}} \quad \text{and} \quad \frac{\partial \mathcal{L}}{\partial \overset{\circ}{\mathbf{x}}{}_{\omega}^{2}} = m_{\omega} \overset{\circ}{\mathbf{x}}_{\omega} := \mathbf{p}_{\omega} \tag{A.8}$$

are obtained, which transform the Lagrangian equation of motion (4.19) into the well-known Newton-Euler equation of motion,

$$\overset{\circ\circ}{\mathbf{x}}_{\omega} = -m_{\omega}^{-1} \frac{\partial \phi}{\partial \mathbf{x}_{\omega}}, \tag{A.9}$$

for the acceleration of the ωth mass-point assumed to be in motion within a conservative force field.

With reference to the rules (4.26) for the infinitesimal transformations

$$t' = t + \tau_{\nu}(t, x)\varepsilon^{\nu} + O(\varepsilon); \qquad \nu := 1(1)\rho$$
$$x'^{k} = x^{k} + \xi_{\nu}{}^{k}(t, x)\varepsilon^{\nu} + O(\varepsilon); \qquad k := 1(1)s, \tag{4.26}$$

the time-coordinate transformation

$$t' = t + \tau \tag{A.10}$$

may be introduced along with the functional transformation

$$\mathbf{x}_{\omega}' = \mathbf{x}_{\omega} + vt + \mathfrak{d} \times \mathbf{x}_{\omega} + \mathfrak{a} \tag{A.11}$$

that has been proved to be consistent with the Lagrangian (A.7) of the system. Equation (A.11) indicates three different contributions, characterizing first a time-dependent displacement settled by a constant translation velocity $\boldsymbol{v}$, then a position-dependent displacement caused by a constant rotation vector $\mathfrak{d}$, and, finally, a translational displacement represented by a permanently existing spatial vector $\mathfrak{a}$.

The invariance of $\mathcal{L}$ may be postulated, considering all variables $\mathbf{x}$, $\mathring{\mathbf{x}}$, and t, but the mathematical form of $\mathcal{L}$ itself may not. This so-called divergence-invariance is assumed to be valid, if there exist ρ so-called gauge functions Ω_ν such that

$$\mathcal{L}(\mathbf{x}'(t')\,;\,\mathring{\mathbf{x}}'(t')\,;t') - \mathcal{L}(\mathbf{x}(t)\,;\,\mathring{\mathbf{x}}_\omega(t)\,;t)\frac{dt'}{dt} = \varepsilon^\nu \frac{d}{dt}\Omega_\nu(\mathbf{x}(t)\,;t) + O(\varepsilon) \quad (4.23)$$

is fulfilled. This means that $\mathcal{L}$ remains unchanged up to exact differentials weighted by parameters ε_ν and up to first-order terms in $\boldsymbol{\varepsilon} := \{\varepsilon^1; \ldots; \varepsilon^\rho\}$ on the right-hand side of (4.23). The summation convention is to apply to index $\nu = 1, \ldots, \rho$.

The Langrangian (A.7) of a mass-point system is distinguished by the index $\nu := \rho \equiv 1$, with the consequence that only one gauge function Ω exists. Now, by consideration of the Lagrangian equation of motion (4.19), the fundamental invariance identity (4.28) in its alternative formulation (4.29) yields the following expression

$$\tau\frac{d}{dt}\Big[\mathcal{L} - \sum_\omega \mathbf{p}_\omega \bullet \mathring{\mathbf{x}}_\omega\Big] + \Big\{\mathfrak{a}\bullet\frac{d}{dt}\Big[\sum_\omega m\omega\mathring{\mathbf{x}}_\omega\Big] + \boldsymbol{v}\bullet\frac{d}{dt}\Big[\sum_\omega m\omega\mathring{\mathbf{x}}_\omega\Big] + \mathfrak{d}\bullet\frac{d}{dt}\Big[\sum_\omega m\omega\mathbf{x}_\omega \times \mathring{\mathbf{x}}_\omega\Big]\Big\}$$
$$= \frac{d}{dt}\Omega(\mathbf{x}_\omega;\,t)\,. \quad (A.12)$$

Since all infinitesimal parameters τ, $\mathfrak{a}$, $\boldsymbol{v}$, and $\mathfrak{d}$ are independent of each other, all bracket terms will vanish separately . Hence, they establish the characteristic set of conservation laws. However, this is only true for energy conservation (first term on the left-hand side), linear momentum conservation (second term), and angular momentum conservation (fourth term). To find the gravity center and its conservation law for the mass-point system, it is certainly allowed to combine the third term with the gauge function $\Omega(\mathbf{x}_\omega;\,t)$ in such a way that the correct law will result. Thus, the gauge function $\Omega(\mathbf{x}_\omega;\,t)$ becomes

$$\Omega(\mathbf{x}_\omega;t) := \sum_\omega m_\omega \mathbf{x}_\omega \bullet \boldsymbol{v} \quad (A.13)$$

and completes the evaluation presented here. The four conservation laws are compiled as follows:

Energy conservation from time translation: $\frac{d}{dt}\Big[\mathcal{L} - \sum_\omega \mathbf{p}_\omega \bullet \mathring{\mathbf{x}}_\omega\Big] = \frac{d}{dt}H \equiv 0$

Momentum conservation from spatial translation: $\frac{d}{dt}\Big[\sum_\omega m_\omega \mathring{\mathbf{x}}_\omega\Big] \equiv 0$ (A.14)

Center of gravity conservation from velocity – based translation: $\frac{d}{dt}\Big[\sum_\omega (m_\omega \mathring{\mathbf{x}}_\omega)\,t - m_\omega x_\omega\Big] \equiv 0$

Angular momentum conservation from rotational displacement: $\frac{d}{dt}\Big[\sum_\omega m_\omega \mathbf{x}_\omega \times \mathring{\mathbf{x}}_\omega\Big] \equiv 0$

It should be mentioned that the bracket of $(A.14)_1$ contains an expression that equals the Hamiltonian H.

A2.2 Some Useful Tensor Rules

The treatise presented contains several sections that can be written in a transparent way by using only symbolic vector analysis and rules for tensors and dyads. Many renowned textbooks are available in this field. Therefore it seems sufficient to compile here only those rules applicable to the subjects in this book.

In reference to the well-known textbook of Bird, Stewart, and Lightfoot (1960, p. 715f) the following notation is used:

- Quantities: of tensor order **o**
  s = scalar (lightface italic; **o** = 0)
  v = vector (boldface; **o** = 1)
  $\boldsymbol{\tau}$ = tensor (boldface Greek; **o** = 2)
  vw = dyad (boldface; **o** = 2)
- Multiplications: with tensor order of result **O**
  scalars: no multiplication sign $\rightarrow a = b\,c; \rightarrow \mathbf{O} = 0$
  vectors and tensors: three kinds of multiplications with special signs:
  scalar: single dot (•) $\rightarrow u = \mathbf{v} \bullet \mathbf{w} \rightarrow \mathbf{O} = \mathbf{o}(\mathbf{v}) + \mathbf{o}(\mathbf{w}) - \mathbf{2}$
  scalar: double dot (:) $\rightarrow s = \boldsymbol{\tau} : \boldsymbol{\pi} \rightarrow \mathbf{O} = \mathbf{o}(\boldsymbol{\tau}) + \mathbf{o}(\boldsymbol{\pi}) - 4$
  vector: cross (×) $\rightarrow \mathbf{u} = \mathbf{v} \times \mathbf{w} \rightarrow \mathbf{O} = \mathbf{o}(\mathbf{v}) + \mathbf{o}(\mathbf{w}) - 1$
- Brackets: indicate the type of quantity produced by the multiplication

  $$(\ldots) = \text{scalar}; \qquad [\ldots] = \text{vector}; \qquad \{\ldots\} = \text{tensor}$$

- Nabla operator : $\nabla := \Sigma_i \mathbf{1}_i\,(\partial/\partial x_i)$ with unit vectors $\mathbf{1}_i$ and coordinate values x_i. The symbol ∇ is a vector operator; it has components like a vector, and it cannot stand alone but must immediately operate on a scalar, vector, or tensor function. The following rules hold by definition:

$$\nabla \bullet s\mathbf{v} = [\nabla s] \bullet \mathbf{v} + (\nabla \bullet \mathbf{v})s; \qquad \{\nabla s\mathbf{v}\} = \{[\nabla s]\mathbf{v}\} + s\{\nabla \mathbf{v}\} \tag{A.15}$$

A first group of equations refers to two basic relations and two ∇ derivatives with respect to a single product of two functions each, where **1** denotes the unit tensor:

$$\begin{aligned} [\nabla \bullet p_* \mathbf{1}] &= \nabla p_*; \qquad (p_* \mathbf{1} : \nabla \mathbf{v}) = p_* (\nabla \bullet \mathbf{v}) \\ \nabla \bullet [p_* \mathbf{v}] &= [\nabla p_*] \bullet \mathbf{v} + p_* \nabla \bullet \mathbf{v}; \qquad (\nabla \bullet [\mathbf{JI} \bullet \mathbf{v}]) = (\mathbf{JI} : \nabla \mathbf{v}) + (\mathbf{v} \bullet [\nabla \bullet \mathbf{JI}]); \end{aligned} \tag{A.16}$$

As arranged above, p_* means a scalar function, **v** a vector function, and **JI** a tensor function.

A second group first concerns some important rules that allow the change of any scalar product of two different vectors into a dyad that is determined by the right-side or left-side multiplication operated by the vectorial multiplier. Then, a rule is offered to reduce any dyad to a vector (Bird, Stewart, and Lightfoot, 1960, p. 730):

$$\mathbf{v}(\varphi \bullet \mathbf{w}) = \{\mathbf{v}\varphi\} \bullet \mathbf{w}; \qquad \mathbf{w} \bullet \{\mathbf{v}\varphi\} = (\mathbf{w} \bullet \mathbf{v})\varphi = \varphi\,(\mathbf{v} \bullet \mathbf{w}) = \{\varphi\mathbf{v}\} \bullet \mathbf{w}. \tag{A.17}$$

An important relationship may be deduced from Equation $(A.15)_2$. The problem is how to convert the dyad $\{[\nabla s]\,\mathbf{v}\}$ to the transposed dyad, viz. $\{\mathbf{v}\,[\nabla s]\}$. There is a general rule (Gyarmati, 1970, p. 178) for the antisymmetric dyad $\mathbf{D}^a$, which forms the total dyad $\mathbf{D}$ along with the symmetric one $\mathbf{D}^s$ according to

$$\mathbf{D} = \mathbf{D}^s + \mathbf{D}^a. \tag{A.18}$$

Assuming that $\mathbf{D} := \mathbf{vw}$, then the transposed dyad $\bar{\mathrm{D}}$ is by definition $\bar{\mathrm{D}} := \mathbf{wv}$. Hence $\mathbf{D}^a$ is given by

$$2\mathbf{D}^a = \{\mathbf{vw} - \mathbf{wv}\} = 2\mathbf{D}^a_{\alpha\beta} = v_\alpha w_\beta - w_\alpha v_\beta. \tag{A.19}$$

In three-dimensional Euclidean space, the antisymmetric part $\mathbf{D}^a$ is intrinsically related to the vector product of two ordinary (polar) vectors. If $\mathbf{v}$ and $\mathbf{w}$ are two polar vectors, their vector product, defined in the usual sense,

$$\mathbf{C}^a := \mathbf{v} \times \mathbf{w} \leftrightarrow C_\gamma^{\,a} = v_\alpha w_\beta - w_\alpha v_\beta \qquad (\alpha, \beta, \gamma = 1, 2, 3;\ \text{cyclic}), \tag{A.20}$$

represents an axial vector $\mathbf{C}^a$, where the subscripts α, β, γ refer to the components of both vectors $\mathbf{v}$ and $\mathbf{w}$. Hence, (A.19) and (A.20) yield $\mathbf{C}^a = 2\mathbf{D}^a$.

Now $\mathbf{D}$ is identified with the dyad $\{[\nabla s]\mathbf{v}\}$, and $\{\mathbf{v}[\nabla s]\}$ is the transposed dyad. Hence, the comparison with (A.19) immediately yields the tensor formula

$$\{\mathbf{v}[\nabla s]\} = -s\{\nabla\mathbf{v}\} + \{\nabla s\mathbf{v}\} - [\nabla s \times \mathbf{v}] \times \mathbf{1}, \tag{A.21}$$

which may be easily transformed by means of a standard formula (Bird, Stewart, and Lightfoot, 1960, p. 726)

$$[\nabla s \times \mathbf{v}] = [\nabla \times s\mathbf{v}] - s[\nabla \times \mathbf{v}] \tag{A.22}$$

into the final relationship

$$\{\mathbf{v}[\nabla s]\} = -\,s\{\nabla\mathbf{v}\} + \{\nabla s\mathbf{v}\} + s[\nabla \times \mathbf{v}] \times \mathbf{1} - [\nabla \times s\mathbf{v}] \times \mathbf{1}. \tag{A.23}$$

The identity (A.23) is most appropriate for the formulation of the theory presented here.

A2.3 On Vorticity in Navier–Stokes Flows

To obtain the mathematical structure of the Navier–Stokes equation of motion from the symmetric viscous pressure tensor of the Navier–St. Venant equation, the identity

$$s\frac{1}{2}\{[\nabla \times \mathbf{v}] \times \mathbf{1}\} - \langle\, \{\nabla s\mathbf{v}\} - \frac{1}{3}(\nabla \bullet s\mathbf{v})\,\mathbf{1}\rangle \equiv \mathbf{0} \tag{A.24}$$

has to be fulfilled a fortiori. The last bracket contains a second-order tensor with zero trace. The same is true by definition for the first bracket term $\{[\nabla \times \mathbf{v}] \times \mathbf{1}\}$. In index notation the last statement may be presented as follows:

$$[\nabla \times \mathbf{v}] = \begin{vmatrix} 1_x & 1_y & 1_z \\ \partial_x & \partial_y & \partial_z \\ u & v & w \end{vmatrix}; \qquad \{[\nabla \times \mathbf{v}] \times \mathbf{1}\} = \begin{vmatrix} 0 & u_y - v_x & u_z - w_x \\ v_x - u_y & 0 & v_z - w_y \\ w_x - u_z & w_y - v_z & 0 \end{vmatrix} \tag{A.25}$$

The vector $\nabla \times \mathbf{v}$ refers to the flow velocity with the components u, v, and w, and the shorthand notation $\partial_x := u_x, \ldots,$ is applied to the zero trace tensor $\{[\nabla \times \mathbf{v}] \times \mathbf{1}\}$.

The second-order tensor $\{\nabla s\mathbf{v}\}$ (given in the same shorthand notation)

$$\{\nabla s\mathbf{v}\} = \begin{vmatrix} (su)_x & (sv)_x & (sw)_x \\ (su)_y & (sv)_y & (sw)_y \\ (su)_z & (sv)_z & (sw)_z \end{vmatrix} \tag{A.26}$$

may be decomposed into its symmetric and antisymmetric parts $\{\nabla s\mathbf{v}\}^s$ and $\{\nabla s\mathbf{v}\}^a$, in accordance with Equation (A.18). The latter tensor corresponds to the vorticity tensor $\frac{1}{2}$ $\{[\nabla \times s\mathbf{v}] \times \mathbf{1}\}$ of the weighted flow velocity $[s\mathbf{v}]$ that is itself related to the relation

$$\{\nabla s\mathbf{v}\}^a = \frac{1}{2}(\{[\nabla \times s\mathbf{v}] \times \mathbf{1}\}). \tag{A.27}$$

whereas the symmetric tensor $\{\nabla s\mathbf{v}\}^s$ is defined by means of $\{\nabla s\mathbf{v}\}$ and its transposed tensor $\{\nabla s\mathbf{v}\}^T \equiv \{s\mathbf{v}\ \nabla\}$ in the following way:

$$\{\nabla s\mathbf{v}\}^s = \frac{1}{2}\langle \{\nabla s\mathbf{v}\} + \{\nabla s\mathbf{v}\}^T\rangle. \tag{A.28}$$

Writing down the six equations derived from (A.24) and using equations $(A.25)_2$ and (A.26) after the elimination of the trace $\frac{1}{3}$ $(\nabla \bullet s\mathbf{v})\mathbf{1}$ from the last matrix, one obtains in detail:

(1) $s(u_y - v_x) - 2(sv)_x \equiv 0$
(2) $s(u_z - w_x) - 2(sw)_x \equiv 0$
(3) $s(v_x - u_y) - 2(su)_y \equiv 0$
(4) $s(v_z - w_y) - 2(sw)_y \equiv 0$
(5) $s(w_x - u_z) - 2(su)_z \equiv 0$
(6) $s(w_y - v_z) - 2(sv)_z \equiv 0$

Evaluation of these six relations yields

$$\begin{aligned} &(sv)_x + (su)_y = 0 \text{ from (1) and (3)} \\ &(sw)_x + (su)_z = 0 \text{ from (2) and (5)} \\ &(sw)_y + (sv)_z = 0 \text{ from (4) and (6).} \end{aligned} \tag{A.29}$$

These three equations are identical with the sums resulting from the evaluation of the right-hand side of (A.28). In other words, they clearly supply the remaining three elements of the principal diagonal of the symmetric tensor $\{\nabla s\mathbf{v}\}^s$ according to

$$\{\nabla s\mathbf{v}\}^{s} = \begin{vmatrix} (su)_x & 0 & 0 \\ 0 & (sv)_y & 0 \\ 0 & 0 & (sw)_z \end{vmatrix}, \tag{A.30}$$

where all elements of the secondary diagonals vanish.

This result, however, is incomplete, because the viscous pressure tensor τ is subject to a symmetry condition that relates to the vorticity tensor $\frac{1}{2}\{[\nabla \times s\mathbf{v}] \times \mathbf{1}\}$ determined by (A.27). According to Equation (7.10) for the antisymmetric viscous pressure tensor $\tau_{\star}{}^{a}$, this symmetry condition equals the relationship

$$\nabla s \times \mathbf{v} \equiv \mathbf{0}, \tag{A.31a}$$

which corresponds with the following set of three equations:

$$\begin{aligned} &(7)\ ws_y = vs_z \\ &(8)\ us_z = ws_x \\ &(9)\ vs_x = us_y\,. \end{aligned} \tag{A.31b}$$

Combining (A.31b) with the identities (1) to (6) will lead us to the following relations:

identity (1) $\rightarrow su_y - sv_x - 2sv_x - 2vs_x \equiv 0$, i.e., $su_y - 3sv_x - 2vs_x \equiv 0$

and, along with (9) $\rightarrow su_y - 3sv_x - 2us_y \equiv 0$;

identity (3) $\rightarrow sv_x - su_y - 2su_y - 2us_y \equiv 0$, i.e., $sv_x - 3su_y - 2us_y \equiv 0$.

Eliminating the term $(2us_y)$ by subtraction the final result

$$u_y - v_x = 0 \tag{A.32a}$$

is obtained.

Similar calculations, using identity (2) along with (8), together with (5) yield a second result

$$u_z - w_x = 0. \tag{A.32b}$$

The last equation results from (4) along with (7) and will turn into

$$v_z - w_y = 0 \tag{A.32c}$$

by using (6).

Evidently, the three equations (A.32a)–(A.32c) form the elements of the axial vector $\nabla \times \mathbf{v}$. Hence, we have proven that the condition

$$\nabla \times \mathbf{v} \equiv \mathbf{0} \tag{A.33}$$

must be satisfied for fluids that obey the Navier–Stokes equations, considering also the conservation law of *angular* momentum. As a result, expression (A.30) for $\{\nabla s\mathbf{v}\}^{s}$ will be identical with the dyad $\{\nabla s\mathbf{v}\}$, following directly from Equation (A.22).

Appendix 3 | Computation Scheme for a One-Component Single-Phase Body-Field System

"We raise a great dust and complain we can't see."—***Bishop Berkeley***

The following table refers to the set of equations compiled below. It should be iteratively solved for consistent initial and boundary conditions prescribed. The equations are valid for a pure substance, subject to certain non-equilibrium processes running in space and time. The specific field force **f** is known and presumed to be conservative. The complete set contains 24 quantities of different tensor order and, consequently, 24 algebraic and differential equations. All relationships are derived or expounded in the text and can be located by referring to the numbers of the respective equations cited in column 3.

Index	*Name of Equation*	*Number of Equation*	*Character of Equation*	*Aimed-at Quantity*	*Information*
(a)	continuity	5.72	field relation	ρ	wanted
(b)	motion	7.12	field relation	$\mathbf{v}$	wanted
(c)	energy	6.85	field relation	T	wanted
(d)	viscous pressure	7.11	constitutive	$\boldsymbol{\tau}$	wanted
(e)	energy flux density	6.94	constitutive	$\mathbf{j}_e^{[\rho]}$	wanted
(f)	entropy flux density	6.46	constitutive	$\mathbf{j}_s$	wanted
(g)	non-equilibrium temperature	6.18	non-equilibrium system	T_*	wanted

(Continued)

Index	*Name of Equation*	*Number of Equation*	*Character of Equation*	*Aimed-at Quantity*	*Information*
(h)	heat flux vector heat-conductivity	6.97	constitutive	$\mathring{q}_*$ λ_*	wanted definition
(i)	heat dissipation	6.89	non-equilibrium system	$e_{\mathbf{q}}$	definition
(j)	specific momentum energy dissipation	6.10 6.15	non-equilibrium system	$\mathbf{i}$ φ, e_φ	wanted definition
(k)	entropy balance	5.81	field relation	σ	wanted
(l)	thermal/caloric state	6.85	state relation	p, s, c_p, h_p	given
(m)	Maxwell viscosities	7.8	non-equilibrium system	$\beta_\tau, \beta_\varphi$	definition
(n)	relaxation times	6.65	constitutive non-equilibrium system	$t_\tau, t_\varphi, t_{\mathbf{q}}, t_{\mathbf{q}*}$	given definition

(a) $$D\rho + \rho\nabla \cdot \mathbf{v} = 0$$

(b) $$\rho D\mathbf{v} = \rho\mathbf{f} - \nabla p + \nabla \cdot \tau + \partial_t \rho\varphi$$

(c) $$\rho c_p DT = -\nabla \cdot \mathbf{j}_e^{[\rho]} + (1 - h_p)\, Dp + \tau : \nabla\mathbf{v} - \mathbf{v} \cdot \partial_t \rho\varphi - \mathbf{v} \cdot \partial_t \rho e_\varphi - \rho\partial_t e_{\text{pot}}$$

(d) $$\tau = \beta_\tau \langle \{\nabla\mathbf{v}\} - s^{-1}\{\nabla s\mathbf{v}\}\rangle + (\beta_\tau + \beta_\varphi)\, \{(s^{-1}(\nabla \cdot s\mathbf{v}) - (\nabla \cdot \mathbf{v}))\,\mathbf{1}\} + \beta_\varphi \{((\rho s)^{-1}\nabla \cdot \mathbf{j}_s + \partial_t \ell n s)\,\mathbf{1}\}$$

(e) $$\mathbf{j}_e^{[\rho]} = -t_{\mathbf{q}*} T_* \sigma\varphi = -t_{\mathbf{q}} T\varphi\, [\rho\mathbf{v} \cdot \nabla s + \rho\partial_t s + \nabla \cdot \mathbf{j}_s]$$

(f) $$\mathbf{j}_s = \mathring{\mathbf{q}}_* / T_*$$

(g) $$T_* := T + \mathbf{i} \cdot \left(\frac{\partial\varphi}{\partial s}\right)_\rho$$

(h) $$\mathring{\mathbf{q}}_* - t_{\mathbf{q}*}\mathbf{i}\; \nabla \cdot \mathring{\mathbf{q}}_* = -\lambda_* \nabla T_* + t_{\mathbf{q}} \rho\, \{\mathbf{iv}\} \cdot T_* \nabla s + t_{\mathbf{q}*} \rho \mathbf{i}\; T_* \partial_t s$$

$$\lambda_* := t_{\mathbf{q}*} T_*^{-1} \rho e_{\mathbf{q}} \mathbf{i}^2$$

(i) $$e_{\mathbf{q}} := -\frac{1}{2}(\mathbf{v}^2 - \mathbf{i}^2)$$

(j) $$\mathbf{i} = \mathbf{v} - \boldsymbol{\varphi}; \qquad e_{\varphi} := \left(-\frac{1}{2}\boldsymbol{\varphi}^2\right) = -\frac{1}{2}(\boldsymbol{\varphi}_1^2 + \boldsymbol{\varphi}_2^2 + \boldsymbol{\varphi}_3^2) = -t_{\varphi}\rho^{-1}T\sigma$$

(k) $$\rho Ds + \nabla \bullet \mathbf{j}_s = \sigma$$

(l) $$p = p(T, \rho); \qquad s = s(T, \rho); \qquad c_p = c_p(T, \rho); \qquad h_p = h_p(T, \rho)$$

(m) $$\beta_{\tau} := t_{\tau}T\rho s; \qquad \beta_{\varphi} := t_{\varphi}T\rho s$$

(n) $$t_{\tau} = t_{\tau}(T, \rho); \; t_{\varphi} = t_{\varphi}(T, \rho); \qquad t_{\mathbf{q}} = t_{\mathbf{q}}(T, \rho) = t_{\mathbf{q}*}\frac{T_*}{T}$$

List of Relevant Symbols

Note: Some symbols, including superscripts and subscripts, that have application only to single problems in Chapter 9 have been omitted. This applies also to local changes in the meaning of some listed symbols.

a	a locally specified constant or parameter
$A, B, C, \ldots$	quantities as elements of a set $\mathcal{G}$
A_r	affinity of the rth reaction
$\mathcal{B}$	domain of space
c_p, c_v	specific heats at constant pressure and volume, respectively
d	operator of the total derivative
D	operator of the material derivative
đ	diminutive
e	specific (total) energy
$e_{\text{kin}}, e_{\text{pot}}$	specific kinetic and, respectively, potential energy
$e_{\tau}, e_{\varphi}, e_{\mathbf{i}}$	specific dissipation energies concerning stress, velocity, and momentum
$e_{\mathbf{q}}$	specific heat energy
E	(total) energy
E_{kin}	kinetic energy
$E_0, E_{\#}$	energy of state at rest; zero-point energy
$\mathbf{f}$	vector of specific force
F	function, Helmholtz function, and zero M–G element in set $\mathcal{W}$, respectively
$\mathbf{F}$	force vector
G	Gibbs free enthalpy, M–G function $G(\xi_1; \ldots; \xi_r)$
$\mathcal{G}$	domain of physical quantities
h	specific enthalpy
H	Hamilton function
H_0	rest enthalpy
I	action integral
$\mathbf{i}$	specific linear momentum

Symbol	Meaning
$\mathbf{j}_e, \mathbf{j}_s, \mathbf{j}_k$	flux density concerning energy, entropy, and diffusion, respectively
k_B	Boltzmann constant, $k_B = \mathcal{R}/N_A = 1.3805 \times 10^{-23}$ J K^{-1}
K_r	equilibrium constant of the rth reaction
$\mathcal{L}$	Lagrange function
M	molecular mass
m^{B+L}	mass with respect to baryon and lepton constancy
n	mole number
N	particle number
N_j	particle number of species j
N_A	Avogadro's number, $N_A = 6.0225 \times 10^{23}$ particle per mole
$\mathbf{n}$	normal vector
p, p_*	pressure with respect to equilibrium, state at rest, and non-equilibrium (*)
p_i, P_i	Hamilton's canonical and generalized variables of momentum ($i = 1(1)s$)
$\mathbf{P}$	(linear) momentum
q_i, Q_i	Hamilton's canonical and generalized variables of position ($i = 1(1)s$)
Q	heat rate
$\mathring{\mathbf{q}}$	heat flux vector
r	number of the degrees of freedom
$\mathbf{r}$	position vector
$\mathscr{R}$	ring $\mathscr{R}[X_0; X_1; \ldots; X_r]$ of real polynomials as the core of $\mathcal{G}$
$\mathcal{R}$	universal gas constant $\mathcal{R} = 8.3143$ J mol^{-1} K^{-1}
s	specific entropy
S	entropy
$\mathcal{S}$	state ensemble of a system
t	time coordinate as a curve parameter
$t_\tau, t_\varphi, t_\mathbf{q}$	relaxation times concerning Maxwell viscosities and heat conductivity
T, T_*	temperature with respect to equilibrium, state at rest, and non-equilibrium
u	specific internal energy
U	internal energy
V	volume
$\mathbf{v}$	flow velocity
W	work rate
$\mathring{\mathbf{w}}$	work rate vector

x, y, z	Cartesian coordinates in the configuration space
X_j	independent generic quantities of a system ($j = 1(1)r$)
z	specific quantity $z := \rho^{-1}Z_V$
Z	state as an element of the set $\mathcal{S}$
Z_V	$Z_V(\mathbf{r}, t)$ denotes some physical quantities of the system per volume unit
α, β, γ	values as elements of the set $\mathcal{W}$
$\beta_\tau, \beta_\varphi$	Maxwell's viscosities
Γ	Gibbs fundamental equation $\Gamma(\xi_0; \xi_1, \ldots, \xi_r) \equiv 0$
Γ_j	chemical production density of the jth species
δ	variation operator
Δ	Laplace operator
ε	set of Noether's parameters
ζ	state variable conjugated to the specific quantity z
κ	isentropic coefficient $\kappa := c_p/c_v$
$\kappa^f{}_r, \kappa^b{}_r$	reaction velocity constant of the forward and backward rth reaction
λ	heat conductivity
λ_r	reaction coordinate of the rth reaction
μ, ν	dynamical and kinematical viscosity, respectively
μ'_j	chemical potential per particle of the jth species
ξ_i	conjugated values ξ_j ($j = 1(1)r$)) of the independent generics $X_0; X_1; \ldots; X_r$
$\xi^k{}_\nu, \tau_\nu$	the infinitesimal generators as to the Noether theorem
$\mathbf{\Pi}$	pressure tensor
ρ	mass density
σ, σ_z	entropy production density, production density assigned to z
$d\tau$	infinitesimal volume element
τ_j	conjugated variable of the variable ξ_j ($j = 1(1)r$))
τ_k	kth member of natural time scale
τ_ℓ, ℓ	mean free time and mean free path, respectively
$\tau_j^{vib}, \tau_j^{tot}$	vibrational and rotational relaxation time
τ, τ_*	viscous pressure tensor with respect to the state at rest and non-equilibrium
φ	dissipation velocity
ω_j	mass fraction of the jth species
Ω_ν	gauge functions in the Noether theorem
$\mathbf{JI}$	(tensorial) momentum flux density
∇	nabla operator
$\mathbf{1}$	unit tensor

References

Abraham, M. (1920). *Theorie der Elektrizität: Elektromagnetische Theorie der Strahlung.* Fourth Ed., Vol. 2. Teubner, Leipzig.

Ahlbom, B., Keviahan, N., and Lefrançois, M. (1991). "Coherence Energies of Turbulent Flow." *Physics Essays* **4**, pp. 406–418.

Ahmadi, G. (1990). "A Rate-Dependent Model for Compressible Turbulent Flows," *J. Non-Equilibrium Thermodynamics* **15**, pp. 87–102.

Allenby, R. B. J. T. (1991). *Rings, Fields and Groups: An Introduction to Abstract Algebra.* Second Ed. Arnold, London.

Alsmeyer, H. (1974). *Messung der Struktur von Verdichtungsstößen in Argon und Stickstoff.* Ph.D. Thesis, Department of Mechanical Engineering, University of Karlsruhe, Germany.

Ames, W. F. (1992). *Numerical Methods for Partial Differential Equations.* Third Ed. Academic Press, New York.

Ames, W. F., Nucci, M. C., Lauster, M. Adams, E., and Straub, D. (1995). "Comparison of Classical and Alternative Fluid Equations Using Symmetry Methods," *ZAMM: Z. angew. Math. Mech.* **75**, pp. 379–388.

Astarita, G. (1990). *Thermodynamics: An Advanced Textbook for Chemical Engineers.* Plenum Press, New York.

Balian, R. (1992). *From Microphysics to Macrophysics: Methods and Applications of Statistical Physics.* Springer, Berlin.

Balmer, R. T. (1983). "Generalized Local Balance Concepts in Non-Equilibrium Thermodynamic Field Theory," *J. Non-Equilibrium Thermodynamics* **8**, pp. 19–28.

Becker, E. and Bürger, W. (1975). *Kontinuumsmechanik: Eine Einführung in die Grundlagen und einfache Anwendungen.* LAMM (H. Görtler, ed.), Vol. 20. Teubner, Stuttgart.

Bethge, K. and Schröder, U. E. (1991). *Elementarteilchen und ihre Wechselwirkungen.* Second Ed. Wissenschaftliche Buchgesellschaft, Darmstadt.

Bird, G. A. (1976). *Molecular Gas Dynamics.* Clarendon Press, Oxford.

Bird, R. B., Stewart, W. E., and Lightfoot, E. N. (1960). *Transport Phenomena.* Wiley International Edition. Wiley, New York.

Birkhoff, G. (1960). *Hydrodynamics: A Study in Logic, Fact, and Similitude.* Princeton Univ. Press, Princeton, N.J.

Blaschke, W. (1923). *Vorlesungen über Differentialgeometrie.* Vol. II. Springer, Berlin.

Bloch, E. (1979) *Die Atomistik bei Hegel und die Atomtheorie der Physik.* Henn, Kastellaun, Germany.

Böhm, W. (1966). *Die metaphysischen Grundlagen der Naturwissenschaft und Mathematik.* Herder, Wien.

Born, M. (1972). *Optik: Ein Lehrbuch der elektromagnetischen Lichttheorie.* Third Ed. Springer, Berlin.

Boukari, M. S. and Lebon, G. (1987). "An Extended Thermodynamic Theory of Fluid Mixtures," *J. Non-Equilibrium Thermodynamics* **12**, pp. 95–121.

Broglie, L.-V. Duc de (1939). *Licht und Materie: Ergebnisse der Neuen Physik.* Goverts, Hamburg.

Bruneau, Ch.-H. and Fabrie, P. (1994). "Computation of Incompressible Flows and Direct Simulation of Transition to Turbulence," in *Fourteenth International Conference on Numerical Methods in Fluid Dynamics,* Proceedings of the Conference held in Bangalore, July 1994, (S.M. Deshpande et al., eds.), pp. 320–324. Springer, Berlin.

Bühler, K. (1988). "Gasdynamische Untersuchung des Jouleschen Überströmversuchs," *Wärme- und Stoffübertragung* **23**, pp. 27–33.

Callen, H. B. (1966). *Thermodynamics: An Introduction to the Physical Theories of Equilibrium Thermostatics and Irreversible Thermodynamics.* Wiley, New York.

Callen, H. B. (1974). "A Symmetry Interpretation of Thermodynamics," in *Foundation of Continuum Thermodynamics* (J. J. Delgado-Domingos, M. N. R. Nina, and J. H. Whitelaw, eds.). pp. 61–79. Macmillan Press, London.

Capelle, W. (1968). *Die Vorsokratiker: Die Fragmente und Quellenberichte.* Kröner, Stuttgart.

Catt, I. (1985). "The Hidden Message in Maxwell's Equations," *Electronics & Wireless World,* pp. 35–36.

Cercignani, C. (1988). *The Boltzmann Equation and Its Applications.* Springer, New York.

Chapman, S. and Cowling, T. G. (1970). *The Mathematical Theory of Nonuniform Gases: An Account of the Kinetic Theory of Viscosity, Thermal Conduction, and Diffusion in Gases.* Third Ed. Prepared in cooperation with D. Burnett. University Press, Cambridge, England.

Chu, B. T. (1959). "Thermodynamics of Electrically Conducting Fluids," *Physics of Fluids* **2**, pp. 473–484.

Clarke, J. F. and McChesney, M. (1964). *The Dynamics of Real Gases.* Butterworths, London.

Clarke, J. F. and McChesney, M. (1976). *Dynamics of Relaxing Gases.* Butterworths, London and Boston.

Constantin, P. and Foias, C. (1988). *Navier–Stokes Equations.* University of Chicago Press, Chicago.

Davies, P. C. W. and Brown, J. R. (1986). *The Ghost in the Atom: A Discussion of the Mysteries of Quantum Physics.* University Press, Cambridge, England.

Davies, P. C. W. and Brown, J. R. (1988). *Superstrings: A Theory of Everything?* University Press, Cambridge, England.

de Groot, S. R. and Mazur, P. (1974). *Anwendung der Thermodynamik irreversibler Prozesse.* Bibliographisches Institut, Mannheim.

Diestelhorst, A. (1993). *Eine axiomatische Behandlung der Quantenmechanik: Der Kommutator als unabhängige Operation.* Ph.D. Thesis, Department of Physics, University of Karlsruhe, Germany.

Domingos, J. J. D. (1984). "Mach Number Effects in Turbulent Flow," in *AGARD-CP-93 on Turbulent Shear Flows,* pp. 30-1–30-10.

Dragoş, L. (1975). *Magnetofluid Dynamics.* Abacus Press, Tunbridge Wells, Kent, England.

Ebeling, W. and Feistel, R. (1992). "Theory of Self-organization and Evolution: The Role of Entropy, Value, and Infomation," *J. Non-equilibrium Thermodynamics* **17**, pp. 303–332.

Eu, B. C. (1986). "Connection between Extended Thermodynamics and Transport Theory," *J. Non-Equilibrium Thermodynamics* **11**, pp. 211–239.

Falk, G. (1990). *Physik: Zahl und Realität—Die begrifflichen und mathematischen Grundlagen einer universellen quantitativen Naturbeschreibung: Mathematische Physik und Thermodynamik.* Birkhäuser, Boston.

Falk, G. and Ruppel, W. (1976). *Energie und Entropie: Eine Einführung in die Thermodynamik.* Springer, Berlin.

Falk, G. and Ruppel, W. (1983). *Mechanik-Relativität-Gravitation: Die Physik des Naturwissenschaftlers.* Third Ed. Springer, Berlin.

Fizsdon, W., Herczynski, R., and Walenta, Z. (1974). "The Structure of a Plane Shock Wave of a Monatomic Gas," in *9th Symposium on Rarefied Gas Dynamics* (M. Becker, M. Fiebig, eds.), p. AxB23. DLR, Köln, Germany.

Friedrich, R. (1993). "Compressible Turbulence," in *2nd Space Course on Low Earth Orbit Transportation* (R. Friedrich, ed.), October 1993, pp. 15-1–15-45. Technical University Munich.

Gambár, K. and Márkus, F. (1993). "On the Global Symmetry of Thermodynamics and Onsager's Reciprocity Relations," *J. Non-Equilibrium Thermodynamics* **18**, pp. 51–57.

Garcia-Colin, L. S. (1984). "Extended Irreversible Thermodynamics and Chemical Kinetics," in *Recent Developments in Non-equilibrium Thermodynamics* (J. Casas-Vázques, D. Jou, and G. Lebon, eds.), Proceedings, pp. 144–165; Barcelona 1983. Springer, Berlin.

Garcia-Colin. L. S. and Uribe, F. J. (1991). "Extended Irreversible Thermodynamics: Beyond the Linear Regime: A Critical Overview," *J. Non-Equilibrium Thermodynamics* **16**, pp. 89–128.

Garfinkle, M. (1992). "Mass Action and the Natural Path," *J. Non-Equilibrium Thermodynamics* **17**, pp. 281–302.

Geropp, D. and Leder, A. (1984). "LDA-Measurements and Some Theoretical Aspects of Separated Flows," in *Turbulent Shear Flows* (L. J. S. Bradbury et al., eds.), pp. 18.7–18.13. Springer, Berlin.

Gibbs, J. W. (1961). *The Scientific Papers.* In Two Volumes: Vol. I, Thermodynamics. Dover Publications, New York.

Glansdorff, P. and Prigogine, I. (1954). "Sur les propriétés différentielles de la production d'entropie." *Physica* **20**, pp. 773–780.

Gordon, S. and McBride, B. J. (1976). *Computer Program for Calculation of Complex Chemical Equilibrium Compositions, Rocket Performance, Incident and Reflected Shocks, and Chapman-Jouguet Detonations.* NASA SP-273, Interim Revision. NASA Lewis Research Center.

Green, H. S. and Leipnik, R. B. (1970). *Sources of Plasma Physics.* Vol. 1, Wolters–Noordhoff, Groningen.

Grossmann, S. (1990). "Turbulenz: Verstehen wir endlich dieses nichtlineare Phänomen?" *Phys. Blätter* **46**, pp. 2–7.

Gumowski. I. (1989). *Oscillatory Evolution Processes: Quantitative Analysis Arising from Applied Sciences.* Manchester University Press, Manchester, England.

Haas, U. and Schönauer, W. (1987). *Numerical Tests with a* k-ε *Model.* Internal Report 28a/87. Forschungsbereich Numerische Strömungsmechanik am Rechenzentrum der Universität Karlsruhe.

Häfele, K. and Schönauer, W. (1985). *The Disappointing Numerical Tests with a Transitional Low Reynolds Number Turbulence Model*. Internal Report 28/85. Forschungsbereich Numerische Strömungsmechanik am Rechenzentrum der Universität Karlsruhe.

Haken, H. (1978). *Synergetics: An Introduction - Non-equilibrium Phase Transitions and Self-Organization in Physics, Chemistry and Biology*. Second Enlarged Edition. Springer, Berlin.

Hao, B.-L. (1989). *Chaos*. World Scientific, Singapore.

Harrison, E. (1986). *Masks of the Universe: A Physicist's Remarkable Portrayal of Mankind's Search for Meaning in the Universe*. Collier Macmillan, London.

Heisenberg, W. (1973). *Schritte über Grenzen*. Piper, München.

Helm, G. (1898). *Die Energetik: Nach Ihrer Geschichtlichen Entwicklung*. Veit, Leipzig.

Helmholtz, H. von. (1903). "Theorie der Wärme," in *Vorlesungen über Theoretische Physik*, Vol. VI (F. Richarz, ed.). Barth, Leipzig.

Hentschel, K. (1990). *Interpretationen und Fehlinterpretationen der speziellen und allgemeinen Relativitätstheorie durch Zeitgenossen Albert Einsteins*. Birkhäuser, Basel.

Hertz, H. (1892). *Untersuchungen über die Ausbreitung der elektrischen Kraft*. Barth, Leipzig.

Hertz, H. (1894). "Die Prinzipien der Mechanik: in neuem Zusammenhange dargestellt," in *Gesammelte Werke*, Vol. III. Barth, Leipzig.

Heß, R. (1981). *Der Einfluß nichtlinearer Transportansätze und temperaturabhängiger Stoffwerte auf die Berechnung von Strömungs- und Ausbreitungsvorgängen in Gasen mit großen Temperatur-, Druck-, und Geschwindigkeitsänderungen*. Ph.D. Thesis, Department of Mechanical Engineering, University (GHS) of Siegen, Germany.

Jordan, P. (1969). *Albert Einstein: Sein Lebenswerk und die Zukunft der Physik*. Huber, Stuttgart.

Kanitscheider, B. (1993). *Von der mechanistischen Welt zum kreativen Universum: Zu einem neuen philosophischen Verständnis der Natur*. Wissenschaftliche Buchgesellschaft, Darmstadt.

Kestin, J. (1978). *A Course in Thermodynamics*. Volume One, Revised Printing. Hemisphere Publishing, New York.

Kestin, J. (1990). "A Note on the Relation Between the Hypothesis of Local Equilibrium and the Clausius-Duhem Inequality," *J. Non-Equilibrium Thermodynamics* **15**, pp. 193–212.

Kirillin, V. A., Sychev, V. V., and Sheindlin, A. E. (1987). *Engineering Thermodynamics*. Translated by S. Semyonov. Mir Publishers, Moscow.

Kremer, G.M. (1989). "Extended Thermodynamics of Molecular Ideal Gases," *Continuum Mechanics and Thermodynamics* **1**, pp. 21–45.

Kreuzer, H. J. (1983). *Nonequilibrium Thermodynamics and its Statistical Foundations*. Clarendon Press, London.

Lakatos, I. (1974). "Falsifikation und die Methodologie wissenschaftlicher Forschungsprogramme," in *Kritik und Erkenntnisfortschritt* (I. Lakatos and A. Musgrave, eds.), pp. 89–189. Proceedings of the International Colloquium on the Philosophy of Science, London 1965. Vol. 4. Vieweg, Braunschweig.

Lamb, H. (1879). *A Treatise on the Mathematical Theory of the Motion of Fluids*. University Press, Cambridge, England.

Landsberg, P. T. (1961). *Thermodynamics: with Quantumstatistical Illustrations*. Interscience Publ., New York.

Lauster, M. (1995). *Zur Theorie der Nichtgleichgewichtsprozesse in kompressiblen Fluiden auf der Grundlage der Gibbs-Falk' schen Thermodynamik.* Verlag Dr. Kovaĉ, Hamburg.

Lauster, M. (1995). "Conservation Laws in Nonlinear Dissipative Systems," *International J. Non-linear Mechanics* **30**, pp. 915–925.

Lauster, M., Höher, K., and Straub, D. (1995). "A New Approach to Mathematical Economics: On its Structure as a Homophormism of Gibbs–Falkian Thermodynamics," *J. of Mathematical Analysis and Applications* **193**, pp. 722–794.

Lavenda, B. H. (1985). *Nonequilibrium Statistical Thermodynamics.* Wiley, Chichester, England.

Lebon, G. (1984). "An Approach to Extended Irreversible Thermodynamics," in *Recent Developments in Non-equilibrium Thermodynamics* (J. Casas-Vázques, D. Jou, and G. Lebon, eds.), Proceedings, pp. 72–104; Barcelona 1983. Springer, Berlin.

Leder, A. (1983). "Laser-Doppler-Untersuchungen und einige theoretische Überlegungen zur Struktur von Totwasser-Strömungen," *Fortschrittsberichte VDI, Series 7, Nr. 28.* VDI-Verlag, Düsseldorf.

Lederman, L. (1994). *The God Particle: If the Universe Is the Answer, What Is the Question?* Dell Publ., New York.

Lerner, L. S. and Gosselin, E. A. (1986). "Galileo and the Specter of Bruno," in *Scientific American* **255**, Vol. 5, pp. 116–123.

Lessing, G. E. (1773–1781). *Berengarius Turonensis.* Zur Geschichte der Literatur, Wolfenbüttel.

Lewis, G. N. and Randall, M. (1923). *Thermodynamics and the Free Energy of Chemical Substances.* McGraw-Hill, New York.

Liu, I-Shih and Müller, I. (1986). "Extended Thermodynamics of Classical and Degenerate Ideal Gases," in *The Breadth and Depth of Continuum Mechanics* (C. M. Dafermos et al., eds.); pp. 69–116. Springer, Berlin.

Löbl, R. (1987). *Demokrits Atomphysik.* Wissenschaftliche Buchgesellschaft, Darmstadt.

Lorentz, H. A. (1907). *Abhandlungen über Theoretische Physik.* Erster Band. Teubner, Leipzig.

Lorenz, E. N. (1963). "Deterministic Nonperiodic Flow," *J. of the Atmospheric Sciences* **20**, pp. 130–141.

Lu, P.-C. (1970). "On the Treatment of Pressures in a Moving Newtonian Fluid." *Bull. Mech. Engng. Educ.* **9**, pp. 307–311.

Lucas, K. (1990). *Applied Statistical Thermodynamics.* Revised Edition. Springer, Berlin.

Mach, E. (1883). *Die Mechanik in ihrer Entwicklung • historisch-kritisch dargestellt.* Brockhaus, Leipzig.

Mach, E. (1896). *Die Principien der Wärmelehre: historisch-kritisch entwickelt.* Barth, Leipzig.

Márkus, F. and Gamár, K. (1989). "Comments On the Relationship Between Thermodynamics and the Wave Approach to Thermodynamics," *J. Non-Equilibrium Thermodynamics* **14**, pp. 355–362.

Maxwell, J. C. (1892). *A Treatise on Electricity and Magnetism.* Vol. I., Third Ed. Clarendon Press, Oxford.

Meya, J. (1990). *Elektrodynamik im 19. Jahrhundert: Rekonstruktion ihrer Entwicklung als Konzept einer redlichen Vermittlung.* DUV: Wissenschafts- und Technikgeschichte. Dt. Univ-Verlag, Wiesbaden.

Mirowski, Ph. (1990). *More Heat than Light: Economics as Social Physics: Physics as Nature's Economics.* University Press, Cambridge, England.

Mittelstaedt, P. (1980). *Der Zeitbegriff in der Physik: Physikalische und philosophische Untersuchungen zum Zeitbegriff in der klassischen und in der relativistischen Physik.* Grundlagen der exakten Naturwissenschaften (P. Mittelstaedt, ed.) Vol. 3., Second Ed. Bibliograph. Institut, Mannheim.

Moore, F. K. (Ed.) (1964). *Theory of Laminar Flows.* High Speed Aerodynamics and Jet Propulsion Vol. IV (Th. von Kármán, H. L. Dryden, H. S. Taylor, eds.). Princeton University Press, Princeton, N.J.

Moulines, C. U. (1987) "The Basic Structure of Neo-Gibbsian Equilibrium Thermodynamics (Some Methodological Problems in the Philosophy of Thermodynamics)," *J. Non-Equilibrium Thermodynamics* **12**, pp. 61–76.

Muschik, W. (1988). *Aspects of Non-Equilibrium Thermodynamics: Six Lectures on Fundamentals and Methods.* Micromechanics Research Laboratory, Department of Mechanical Engineering. McGill University, Montreal.

Muschik, W. (1990). "Internal Variables in Non-Equilibrium Thermodynamics," *J. Non-Equilibrium Thermodynamics* **15**, pp. 127–137.

Mutschler, H.-D. (1990). *Physik • Religion • New Age.* Echter, Würzburg.

Nagel, E. and Newman, J. R. (1987). *Der Gödelsche Beweis.* Fourth Ed., Scientia Nova (H. Lenk et al., eds.). Oldenbourg, München.

Nehring, U. (1983). *Modell-Gleichungen zur Beschreibung instationärer Strömungsvorgänge.* DFVLR-Mitteilungen 88-17. Deutsche Gesellschaft für Luft- und Raumfahrt, Köln.

Nehring, U. (1993). "Eine Impulsdarstellung kompressibler Strömungen," *ZFW,* **17**, pp. 257–264.

Neumaier, M. (1996). *Der Trend zum Gleichgewicht: Experimentelle Untersuchungen mittels einer Versuchsanlage zur Präzisionsmessung des thermischen Zustandsverhaltens von Fluiden.* Dissertation, Department of Aeronautics and Aerospace Technology, Univ. Federal Armed Forces, Munich.

Neumann, C. (1875). *Vorlesungen über die mechanische Theorie der Wärme.* Teubner, Leipzig.

Nicolis, G. and Prigogine, I. (1989). *Exploring Complexity: An Introduction.* Freeman, New York.

Nixon, D. (1986). *A Report on a Workshop on Turbulence Research.* TR-361. Nielsen Engineering & Research, Inc., Mountain View, CA.

Oesterle, M. (1996). "Zur Topologie der Feld- und Grenzschichtstrukturen beim Wärmeübergang am querangeströmten beheizten und rotierenden Zylinder, Experimentelle Untersuchungen für Fluide unterschiedlicher Prandti-Zahl," *Fortschrittsberichte. VDI, Series 7,* Nr. 299. VDI-Verlag, Düsseldorf

Oesterle, M., Lausetr, M., Lippig, V., Waibel, R., and Straub, D. (1997). "Topological Structures Near a Heated Rotating Cylinder in Crossflow. Part I: An Experimental Device for Measurements at a Wide Range of Reynolds Numbers. Part II: Diagnostic Systems and Experimental Results," *Experiments in Fluids* **15**, to be published.

Oran, E. S. and Boris, J. P. (1993). "Numerical Simulation of Turbulent Combustion." in *2nd Space Course on Low Earth Orbit Transportation* (R. Friedrich, ed.), October 1993, pp. 34-1–34-13. Technical University, Munich.

O'Shea, N. and Fletcher, C. A. J. (1994). "Prediction of Turbulent Delta Wing Vortex Flows Using an RNG k-ε Model," in *Fourteenth International Conference on Numerical*

Methods in Fluid Dynamics, Proceedings of the Conference held in Bangalore, July 1994, (S. M. Deshpande et al., eds.), pp. 496–501. Springer, Berlin.

Ottino, J. M. (1990). The Kinematics of Mixing: Stretching, Chaos, and Transport. Cambridge University Press, Cambridge, England.

Pabst, B. (1994). *Atomtheorien des lateinischen Mittelalters.* Wissenschaftliche Buchgesellschaft, Darmstadt.

Pai, S. I. (1977). *Two-Phase Flows. Vieweg Tracts in Pure and Applied Physics.* (K. Oswatitsched.), Vol. 3. Vieweg, Braunschweig.

Peat, F. D. (1989). *Synchronizität: Die verborgene Ordnung.* Scherz, Bern.

Pellat, H. (1897). "Thermodynamique," *Leçons professées à la Sorbonne en 1895–96.* Carré & Naud, Paris.

Peller, H. (1986). "Part 2.1: Boundary Layer Profiles and Location of Separation Points," in *Experiments in Fluids* **4**, pp. 223–231.

Peller, H., Lippig, V., Straub, D. , and Waibel, R. (1984). "Thermofluiddynamic Experiments with a Heated and Rotating Circular Cylinder in Crossflow: Part 1: Subcritical Heat Transfer Measurements," in *Experiments in Fluids* **2**, pp. 113–120.

Peller, H. and Straub, D. (1988). "Part 2.2: Temperature Boundary Layer Profiles," *Experiments in Fluids* **6**, pp. 111–114.

Penrose, R. (1990). *The Emperor's New Mind: Concerning Computers, Minds, and the Laws of Physics.* Oxford University Press, Oxford.

Petrosky, T. Y. and Prigogine, I. (1988). "Poincaré's Theorem and Unitary Transformations for Classical and Quantum Systems," *Physica* **147 A**, pp. 439–460.

Petrosky, T. Y. and Prigogine, I. (1992). "Complex Spectral Representations and Quantum Chaos," in *Research Trends in Chaotic Dynamics and Transport in Fluids and Plasmas* (I. Prigogine, ed.), pp. 80–108. Intern. Conf., La Jolla. Ca. American. Inst. Physics, New York.

Petrosky, T. Y. and Prigogine, I. (1993). "Complex Spectral Representations and Singular Eigenvalue Problem of Liouvillian in Quantum Scattering," in *Proceedings of Quantum Physics and the Universe* (M. Namiki, Y. Aizawa et al., eds.), pp. 231–238, Intern. Sympos. Tokyo 1992. Waseda University, Tokyo.

Pitowsky, I. (1989). *Quantum Probability–Quantum Logic.* Springer, Berlin.

Planck, M. (1907). *Zur Dynamik bewegter Systeme.* Sitzungsbericht der Preussischen Akademie der Wissenschaften, pp. 542–570. Akademie–Verlag, Berlin.

Planck, M. (1910). *Acht Vorlesungen über Theoretische Physik,* gehalten an der Columbia University, New York, 1909. Hirzel, Leipzig.

Planck, M. (1913). *Vorlesungen über Thermodynamik.* Fourth Ed. Veit, Leipzig.

Planck, M. (1922). "Das Prinzip der kleinsten Wirkung," in *Physikalische Rundblicke: Gesammelte Reden und Aufsätze,* pp. 103–119. Hirzel, Leipzig.

Planck, M. (1929). *Das Weltbild der neuen Physik.* Lecture at University Leiden, February 18, 1929. Barth, Leipzig.

Planck, M. (1958). "Das Prinzip der Relativität und die Grundgleichungen der Mechanik," in *Physikalische Abhandlungen und Vorträge,* Vol. II, pp. 115–120, Vieweg, Braunschweig.

Poincaré. H. (1892). *Thermodynamique: Leçons professées pendant le premier semestre 1888–89.* Cours de physique mathématique. Rédigées par J. Blondin. Carré, Paris.

Poincaré, H. (1906). *Wissenschaft und Hypothese.* Second Ed. Teubner, Leipzig.

Prausnitz, J. M. (1969). *Molecular Thermodynamics of Fluid-Phase Equilibria.* Prentice-Hall, Englewood Cliffs, N.J.

Prigogine, I. (1993). "From Classical Chaos to Quantum Chaos," in *Intern. Sympos. on Quantum Physics and Universe* (M. Namiki, Y. Aizawa et al., eds.), pp. 7–26, Tokyo 1992. Waseda University, Tokyo.

Prigogine, I. and Defay, R. (1962). *Chemische Thermodynamik.* VEB Dt. Verlag Grundstoffindustrie, Leipzig.

Prigogine, I. and Stengers, I. (1993). *Das Paradox der Zeit: Zeit, Chaos und Quanten.* Piper, Munich.

Pruβ, A. and Wagner, W. (1996). "New International Formulation for the Thermodynamic Properties of Ordinary Water Substance for General and Scientific Use," *J. Phys. Chem. Ref. Data* **25**, to be published.

Quine, W. V. (1987). *Quiddities: An Intermittently Philosophical Dictionary.* Penguin Books, Harmondsworth, England.

Reichenbach, H. (1931). *Ziele und Wege der heutigen Naturphilosophie.* Meiner, Leipzig.

Reitz, J. R. and Milford, F. J. (1974). *Foundations of Electromagnetic Theory.* Second Ed. World Student Series Edition. Addison-Wesley, Reading, Mass.

Rogers, C. and Ames W. F. (1989). *Nonlinear Boundary Value Problems in Science and Engineering.* Academic Press, New York.

Rotta, J. C. (1972). *Turbulente Strömungen: Eine Einführung in die Theorie und ihre Anwendungen.* Teubner, Stuttgart.

Rotta, J. C. (1984). "On my Experiences with Second Order Closure Models of Turbulent Flow," in *Internal Report IB 222-84 A 04.* DLR-AVA. Göttingen.

Ruggeri, T. (1989). "Galilean Invariance and Entropy Principle for Systems of Balance Laws: The Structure of Extended Thermodynamics," *Continuum Mechanics and Thermodynamics* **1**, pp. 3–20.

Sato, R. (1981). *Theory of Technical Change and Economic Invariance: Application of Lie Groups.* Academic Press, New York.

Sato, R. and Ramachandran, R. V. (1990). *Conservation Laws and Symmetry: Applications to Economics and Finance.* Kluwer, Boston.

Schaber, A. (1963). "Über die Inversion des Joule-Effekts," *Allgemeine Wärmetechnik* **11**, pp. 146–149.

Schaber, A. (1965). *Zum thermischen Verhalten fluider Stoffe: Eine systematische Untersuchung der charakteristischen Kurven des homogenen Zustandsgebiets.* Ph.D. Thesis, Department of Mechanical Engineering, University of Karlsruhe, Germany.

Schirra, N. (1991). *Die Entwicklung des Energiebegriffs und seines Erhaltungskonzepts: Eine historische, wissenschaftstheoretische, didaktische Analyse.* Series Physik, Vol. 8. Deutsch, Frankfurt am Main.

Schlögl, F. (1989). *Probability and Heat: Fundamentals of Thermostatistics.* Vieweg, Braunschweig.

Schmidt, B. (1969). "Electron Beam Density Measurements in Shock Waves in Argon," in *J. Fluid Mech.* **39**, pp. 361–373.

Schmutzer, E. (1972). *Symmetrien und Erhaltungssätze der Physik.* Pergamon Press, Oxford.

Schönauer, W. and Häfele, K. (1987). "Experiences with the Numerical Solution of the 3-d Laminar Boundary Layer Equations in Streamline Coordinates," *Computers & Fluids* **15**, pp. 93–118.

Sieniutycz, S. (1987). "From a Least Action Principle to Mass Action Law and Extended Affinity," *Chemical Engineering Science* **42**, pp. 2697–2711.
Sieniutycz, S. and Berry, R. S. (1993). "Canonical Formalism, Fundamental Equation, and Generalized Thermomechanics for irreversible Fluids with Heat Transfer," *Physical Review* E, **47**, pp. 1765–1783.
Shizawa, K. and Tanahashi, T. (1986). "New Constitutive Equations for Conducting Magnetic Fluids with Internal Rotation (Thermodynamical Discussion)," *Bulletin JSME* **29**, pp. 2878–2884.
Simon, C. E. (1976). *Theory of Shock Structure in a Maxwell Gas Based on the Chapman-Enskog Development through Super-Super-Burnett Order.* Ph.D. Thesis. University of Colorado, Boulder.
Sommerfeld, A. (1964). *Mechanik der Deformierbaren Medien.* Vorlesungen über Theoretische Physik, Vol. 2, Fifth Ed. Akademische Verlagsgesellschaft, Leipzig.
Sonntag, R. E. and Van Wylen, G. J. (1968). *Fundamentals of Statistical Thermodynamics.* Series in Thermal and Transport Sciences. Wiley, New York.
Sparrow, C. (1982). *The Lorenz Equations: Bifurcations, Chaos, and the Strange Attractors.* Springer, New York.
Spurk, J. H. (1989). *Strömungslehre: Einführung in die Theorie der Strömungen.* Second Ed. Springer, Berlin.
Stengers, I. (1995). "Die Galilei-Affären," in *Elemente einer Geschichte der Wissenschaften* (M. Serres, ed.), pp. 394–443. Second Ed. Suhrkamp, Frankfurt/M.
Stonier, T. (1990). *Information and the Internal Structure of the Universe: An Exploration into Information Physics.* Springer, London.
Stratonovich, R. L. (1994). *Nonlinear Nonequilibrium Thermodynamics II: Advanced Theory.* Springer, Berlin.
Straub, D. (1990). *Eine Geschichte des Glasperlenspiels: Irreversibilität in der Physik: Irritationen und Folgen.* Birkhäuser, Basel.
Straub, D. (1992). "On the Foundation of Thermofluiddynamics by Callen's Symmetry Principle and a Realistc Matter Model," in *Turbulente Strömungen in Forschung und Praxis: Festschrift zum 60. Geburtstag Prof. D. Geropp.* IFT-Mitteilungen 10.92 (A. Leder, ed.), pp. 352–376; University-GHS-Siegen.
Straub, D. (1996). "Zeitpfeile in der Natur? Versuch einer Antwort," in *Betrachten-Beobachten-Beschreiben: Beschreibungen in Kultur- und Naturwissenschaften* (R. Inhetveen, R. Kötter eds.), pp. 105–146, Fink, München.
Straub, D., Graue, R., Heitmeir, F., Nebendahl, P., and Wurst, Th. K. (1987). "Fick's First Law Correction by an Exact Solution of the Boltzmann Equation," *Propellants, Explosives, Pyrotechnics* **12**, pp. 105–112.
Straub, D. and Lauster, M. (1994). "Angular Momentum Conservation Law and Navier–Stokes Theory," *International. J. Non-linear Mechanics* **29**, pp. 823–833.
Szabó, I. (1987). *Geschichte der mechanischen Prinzipien: und ihrer wichtigsten Anwendungen.* Dritte, korrigierte und erweitere Auflage. Birkhäuser, Basel.
Tasaki, S., Nardone, P., and Prigogine, I. (1993). "Resonance and Instability in a Cosmological Model," in *International Symposium on Quantum Physics and Universe* (M. Namiki, Y. Aizawa et al., eds.), pp. 645–654, Tokyo 1992. Waseda University, Tokyo.
Temam, R. (1979). *Navier–Stokes Equations: Theory and Numerical Analysis. Revised Ed. North-Holland Publ., Amsterdam.*

Thom, R. (1984). *Paraboles et Catastrophes.* Flammarion, Paris.
Thomas, E. G. and Meadows, A. J. (1985). *Maxwell's Equations and Their Applications.* Hilger, Bristol.
Tipler, F. J. (1980). "General Relativity and the Eternal Return," in *Essays in General Relativity* (F.J. Tipler, ed.), pp. 21–37. Academic Press , New York.
Tisza, L. (1966). *Generalized Thermodynamics.* MIT Press, Cambridge, Mass.
Tokaty, G. A. (1971). *A History and Philosophy of Fluiddynamics.* Foulis, Henley-on-Thames, Oxfordshire, England.
Truesdell, C. (1968). *Essays in the History of Mechanics.* Springer, Berlin.
Truesdell, C. (1974). "The Meaning of Viscometry in Fluid Dynamics," *Ann. Rev. Fluid Mech.* **6**, pp. 111–145.
Truesdell, C. (1975). "Early Kinetic Theories of Gases," *Archive for History of Exact Sciences* **15**, pp. 1–66.
Truesdell, C. (1980). *The Tragicomical History of Thermodynamics 1822–1854.* Springer, New York.
Truesdell, C. (1984b). *An Idiot's Fugitive Essays on Science: Methods, Criticism, Training, Circumstances.* Springer, New York.
Truesdell, C. (1984). *Rational Thermodynamics* (with an appendix by C.-C. Wang). Second Corrected and Enlarged Edition. Springer, New York.
Truesdell, C. and Bharatha, S. (1977). *The Concepts and Logic of Classical Thermodynamics as a Theory of Heat Engines: Rigorously Constructed upon the Foundation Laid by S. Carnot and F. Reech.* Springer, New York.
Truesdell, C. and Toupin, R. (1960). "The Classical Field Theories," in *Flügge's Encyclopedia of Physics,* Vol. III.1. Springer, Berlin.
Vincenti, W. G. and Kruger, C. H. (1967). *Introduction to Physical Gas Dynamics.* Wiley, New York.
Vujanovic, B. D. and Jones, S. E. (1989). *Variational Methods in Nonconservative Phenomena.* Academic Press, New York.
Walz, A. (1966). *Strömungs- und Temperaturgrenzschichten.* Reihe Strömungstechnik (R. Friedrich, ed.). Braun, Karlsruhe.
West, B. J. (1985). *An Essay on the Importance of Being Nonlinear.* Springer, Berlin.
Weyl, H. (1942). *A Course in Hydrodynamics.* Typewritten notes by F. Adler et al. The Institute for Advanced Study, Spring 1942, Princeton University.
Weyl, H. (1977). *Mathematische Analyse des Raumproblems: Vorlesungen; Was ist Materie?: Zwei Aufsätze zur Naturphilosophie.* Wissenschaftliche Buchgesellschaft, Darmstadt.
Whitrow, G. J. (1990). *The Natural Philosophy of Time.* Second Ed. Clarendon, Oxford.
Wilber, K. (1986). *Das holographische Weltbild.* Scherz, Bern.
Woods, L. C. (1975). *The Thermodynamics of Fluid Systems.* Clarendon, Oxford.
Wurst, T. K., Oesterle, M., and Straub, D. (1991). "Windkanal für Wärmeübergangsmessungen," *Zeitschrift für Versuchs- und Forschungsingenieure* **6**, pp. 27–29.
Zahn, M. (1979). *Electromagnetic Field Theory: A Problem-solving Approach.* Wiley, New York.
Zeleznik, F. J. (1981). "Thermodynamics II: The Extended Thermodynamic System," *Journal of Mathematical Physics* **22**, pp. 161–178.
Zemansky, M. W. (1968). *Heat and Thermodynamics: An Intermediate Textbook.* Fifth Ed. International Student Edition. McGraw-Hill, Kogakusha, Tokyo.

Name Index

Subject Index

Mathematics in Science and Engineering

Edited by William F. Ames, Georgia Institute of Technology

Recent titles

T.A. Burton, *Volterra Integral and Differential Equations*

C.J. Harris and J.M.E. Valenca, *The Stability of Input-Output Dynamical Systems*

George Adomian, *Stochastic Systems*

John O'Reilly, *Observers for Linear Systems*

Ram P. Kanwal, *Generalized Functions: Theory and Technique*

Marc Mangel, *Decision and Control in Uncertain Resource Systems*

K.L. Teo and Z.S. Wu, *Computational Methods for Optimizing Distributed Systems*

Yoshimasa Matsuno, *Bilinear Transportation Method*

John L. Casti, *Nonlinear System Theory*

Yoshikazu Sawaragi, Hirotaka Nakayama, and Tetsuzo Tanino, *Theory of Multiobjective Optimization*

Edward J. Haug, Kyung K. Choi, and Vadim Komkov, *Design Sensitivity Analysis of Structural Systems*

T.A. Burton, *Stability and Periodic Solutions of Ordinary and Functional Differential Equations*

Yaakov Bar-Shalom and Thomas E. Fortmann, *Tracking and Data Association*

V.B. Kolmanovskii and V.R. Nosov, *Stability of Functional Differential Equations*

V. Lakshmikantham and D. Trigiante, *Theory of Difference Equations: Applications to Numerical Analysis*

B.D. Vujanovic and S.E. Jones, *Variational Methods in Nonconservative Phenomena*

C. Rogers and W.F. Ames, *Nonlinear Boundary Value Problems in Science and Engineering*

Dragoslav D. Siljak, *Decentralized Control of Complex Systems*

W.F. Ames and C. Rogers, *Nonlinear Equations in the Applied Sciences*

Christer Bennewitz, *Differential Equations and Mathematical Physics*

Josip E. Pecaric, Frank Proschan, and Y.L. Tong, *Convex Functions, Partial Orderings, and Statistical Applications*

E.N. Chukwu, *Stability and Time-Optimal Control of Hereditary Systems*

E. Adams and U. Kulisch, *Scientific Computing with Automatic Result Verification*

Viorel Barbu, *Analysis and Control of Nonlinear Infinite Dimensional Systems*

Yang Kuang, *Delay Differential Equations: With Applications in Population Dynamics*

W.F. Ames, E.M. Harrell II, and J.V. Herod, *Differential Equations with Applications to Mathematical Physics*

V.I. Zhukovskiy and M.E. Salukvadze, *The Vector-Valued Maximin*

D. Straub, *Alternative Mathematical Theory of Non-equilibrium Phenomena*